Dynamics of the Middle Atmosphere

Advances in Earth and Planetary Sciences

Dynamics of the Middle Atmosphere

Proceedings of a U.S.-Japan Seminar
Honolulu, Hawaii, 8-12 November, 1982

Edited by

J. R. Holton

Department of Atmospheric Sciences, University of Washington

and

T. Matsuno

Geophysical Institute, University of Tokyo

Terra Scientific Publishing Company
Tokyo, Japan

D. Reidel Publishing Company

A MEMBER OF THE
KLUWER ACADEMIC PUBLISHERS GROUP

Dordrecht / Boston / Lancaster

Published by Terra Scientific Publishing Company (TERRAPUB),
307 Shibuyadai-haim, 4-17 Sakuragaoka-cho, Shibuya-ku, Tokyo 150, Japan,
in co-publication with D. Reidel Publishing Company, Dordrecht, Holland

Sold and distributed in the U.S.A. and Canada
by Kluwer Boston Inc.,
190 Old Derby Street, Hingham, MA 02043, U.S.A.,
in Japan by Terra Scientific Publishing Company (TERRAPUB),
307 Shibuyadai-haim, 4-17 Sakuragaoka-cho, Shibuya-ku, Tokyo 150, Japan

In all other countries, sold and distributed
by Kluwer Academic Publishers Group,
P. O. Box 322, 3300 AH Dordrecht, Holland

D. Reidel Publishing Company is a member of the Kluwer Academic Publishers Group

ISBN-13: 978-94-009-6392-4   e-ISBN-13: 978-94-009-6390-0
DOI: 10.1007/978-94-009-6390-0

PREFACE

The contents of this volume are based on the U.S.-Japan Seminar on
the Dynamics of the Middle Atmosphere, held at the East-West Center of
the University of Hawaii, Honolulu, Hawaii, 8-12 November 1982. The
seminar was jointly sponsored by the United States National Science
Foundation and the Japan Society for the Promotion of Science. The
participants included eleven scientists from the U.S.A., eleven from
Japan, and one from Australia.

In recent decades the meteorology of the middle atmosphere has
become a subject of increasing research activity. Much of this activity has been stimulated by concern about possible anthropogenic perturbations of the stratospheric ozone layer, and also possible linkages
with climatic fluctuations. Observational and theoretical efforts to
study the dynamics, physics, and chemistry of the middle atmosphere
have recently intensified on an international basis with the beginning
of the Middle Atmosphere Program (MAP). Thus the Seminar was very
timely.

We restricted the topic of the Seminar to dynamics in order to
promote an intensive discussion with a relatively small number of participants. Radiation and chemistry were discussed only in the context
of dynamical problems. The dynamics of the middle atmosphere has a
unique character in that propgating waves of several types are of central importance in controlling the whole circulation. In fact, problems of middle atmosphere dynamics have to a great extent stimulated
the explosive development in wave mean-flow interaction dynamics, which
has occurred in the past decade or so. This development has led to
a better understanding of transport mechanisms and improved models of
the middle atmosphere circulation. Thus, like surfaces on a beach,
the seminar participants were primarily concerned with wave propagation,
wave transport, and wave breaking. The book reflects this emphasis on
wave dynamics.

The volume contains both overview articles and original contributions. Despite the limited number of authors the present volume
covers most of the important problem areas and represents the current
status of our knowledge about middle atmosphere dynamics. We would
like to express our sincere gratitude to the authors for their valuable
contributions.

Finally, we acknowledge the financial support of the seminar provided by the National Science Foundation Office of International Programs and the Japan Society for the Promotion of Science and also would like to thank Ms. Ilze Schubert at the University of Washington and Ms. Kei Kudo at the University of Tokyo for their assistances in the editorial work.

Seattle and Tokyo
July 1983

James R. Holton
Taroh Matsuno

TABLE OF CONTENTS

viii

GRAVITY WAVES

GRAVITY WAVES IN THE MESOSPHERE

R. S. Lindzen[1]

Center for Earth and Planetary Physics
Harvard University, Cambridge, MA   02138

ABSTRACT

The simple theory for the effect of gravity wave breaking on eddy
diffusion and on mean stress is reviewed. Attempts are made to identify
the magnitudes of known mechanisms for generating gravity waves. The
relation between observed periods and unobserved horizontal wavenumbers
is discussed. Finally the role of group velocity, wave packets and
wave inhomogeneity is discussed.

1. INTRODUCTION

The recognition that gravity waves play a major role in mesospheric
dynamics is relatively recent. For many years, the most famous picture
of upper atmosphere wind was the meteor trail observation by Liller and
Whipple (1954). However, as can be seen from reading the Proceedings
of the International Symposium on Fluid Mechanics in the Ionosphere,
Cornell University, July, 1959 (Bolgiano, 1959), the consensus opinion
was that the wind irregularities were due to turbulence. It was Hines
(1960) who forcefully argued that the irregularities were due to internal gravity waves--although the specific origin of the gravity wave was
unclear. Lindzen (1967) showed that the Liller-Whipple observation was
consistent with the first propagating diurnal tidal mode.

Much work was done on internal gravity waves over the next decade.
An excellent summary of this work and thought appears in the proceedings
of a meeting held in Boulder, Colorado in July of 1968 (Georges, 1968).
It is striking to see how much current work was anticipated in these
proceedings.

The potential role of breaking waves in generating turbulence was
recognized (see also Hodges, 1969; Lindzen, 1968, 1971). In the wake of

---

[1]Address as of 1 July 1983: Department of Earth, Atmospheric and Planetary Sciences, Massachusetts Institute of Technology, Cambridge, MA
02139.

the pioneering work of Booker and Bretherton (1967), the fact that gravity waves could modify the mean flow was recognized (Lindzen, 1968a).
Also, the need for friction in producing the reversal of meridional
temperature gradient at the mesopause was noted (Leovy, 1964; Lindzen,
1968b)--as was the possible role of gravity waves in producing this
friction. There was, moreover, by the late 1960's data available which,
although coarse, could have been used to quantitatively assess these
interactions (Theon et al., 1967). Nevertheless, it was not until
Lindzen (1981) that the first crude attempt to parameterize the role of
gravity wave breaking in generating turbulent diffusion and organized
stress was made. What was specific to Lindzen (1981) was the recognition
that the wave stress followed from the turbulent breaking of gravity
waves--but was nonetheless distinct. Also Lindzen (1981) recognized
that the origin of mesospheric gravity waves was primarily in the troposphere. This plus the filtering properties of the intervening stratospheric mean flow permitted fairly tight estimates of gravity wave phase
speeds in the mesosphere. Such estimates are essential components of
quantitative parameterizations.

The results of Lindzen (1981) are reviewed in other papers at this
symposium and are, by now, well known. Thus, in section 2 only the most
perfunctory review of the parameterization will be given--including a
description of some improvement found in Holton (1982) and of some
recent thoughts on turbulence due to sub-breaking waves (Lindzen and
Forbes, 1983).

The remainder of this paper will deal briefly with some shortcomings
in our present descriptions of gravity waves. In section 3 we will consider the explicit generation of internal gravity waves by mountains and
by shear collapse--attempting to set bounds on the possible wave fluxes
associated with various phase speeds. In section 4 we will examine the
relation between wave phase speeds, wave numbers and observed periods
in order to better understand some observations. In section 5 we will
consider the horizontal as well as vertical propagation of gravity wave
packets in order to develop some intuition about the origin of mesospheric gravity waves. In section 5 we will also remark on the horizontal inhomogeneity of gravity waves and its implications.

2. PARAMETERIZATION OF TURBULENCE AND STRESS

The NASA rocket probings of the mesosphere in the 1960's (Theon et
al., 1967) revealed intense wave activity in winter above 50 km. Wave
amplitudes were characteristically such as to marginally produce unstable
lapse rates. In summer such "breaking" amplitudes did not occur until
greater heights ∿ 70 km. Recent radar data (Balsley, 1983) confirms
this seasonal variation.

Lindzen (1981) introduced the simplest model capable of describing
the effects of breaking gravity waves. He considered zonally travelling
gravity waves which were standing waves in the meridional direction,

i.e., waves with the following form

$$e^{ik(x-ct)} \cos (\ell y + \phi) \tag{1}$$

where

x = eastward distance
y = northward distance
t = time
k = eastward wavenumber
c = eastward phase speed
ℓ = northward wavenumber
ϕ = arbitrary phase constant

In a slowly varying medium (only inhomogeneities in the z (vertical) direction were considered), Lindzen (1981) showed, using the WKB approximation,

$$\delta T \approx A\Gamma^{1/2} T^{-1/2} \lambda^{1/2} e^{i\int\lambda dz} e^{z/2H} e^{ik(x-ct)}$$
$$\cdot \cos (\ell y + \phi) \tag{2}$$

where

A = Amplitude factor

Γ = Static stability = $\dfrac{dT}{dz} + \dfrac{g}{c_p}$

T = Basic temperature
δT = Perturbation temperature
λ = Vertical wavenumber

$$\approx |N/(\bar{u} - c)| \ (1 + \ell^2/k^2)^{1/2} \tag{3}$$

$\bar{u}$ = Mean zonal flow

$$N^2 = \frac{g}{T} \left(\frac{dT}{dz} + \frac{g}{c_p}\right)$$

(2) applies beneath the breaking level.
Consistent with the WKB approximation

$$\frac{d\delta T}{dz} \approx iA \ \Gamma^{1/2} T^{-1/2} \lambda^{3/2} e^{i\int\lambda dz} e^{z/2H} e^{ik(x-ct)}$$
$$\cdot \cos (\ell y + \phi) \tag{4}$$

and breaking occurs when

$$\left|\frac{d\delta T}{dz}\right| = \Gamma$$

or, using (4) when

$$A \ \Gamma^{1/2} T^{-1/2} \lambda^{3/2} e^{z/2H} \approx \Gamma \tag{5}$$

If the breaking height, z_{break}, is observed then (5) determines A.

Above z_{break}, it is assumed that sufficient turbulence is generated to prevent $\frac{d\delta T}{dz}$ from growing further.

Lindzen (1981) shows that, in the absence of damping, $\frac{d\delta T}{dz}$ would grow exponentially with a local exponent given by

$$\frac{1}{2H} - \frac{3}{2}\frac{1}{\bar{u}-c}\frac{d\bar{u}}{dz} \tag{6}$$

Now damping gives rise to an imaginary part of c, c_i, which, in turn, produces an imaginary contribution to λ (viz equ. (3)). The damping is then chosen so that λ_i is exactly equal to the growth exponent given by (6); i.e., so that the growth is exactly cancelled. This degree of damping is, in fact, sufficiently small to permit Lindzen (1981) to relate c_i to eddy diffusivity, D_{eddy}, with the following relation

$$kc_i \approx \lambda^2 D_{eddy}, \tag{7}$$

and the requirement that growth be cancelled then leads to

$$D_{eddy} \approx \frac{k\,|u-c|^4}{N^3(1+\ell^2/k^2)^{3/2}}\left|\frac{1}{2H} - \frac{3}{2}\left(\frac{1}{\bar{u}-c}\frac{d\bar{u}}{dz}\right)\right| \tag{8}$$

above z_{break}.

Lindzen (1981) also shows that the condition for breaking (5) implies that at the level of breaking, z_{break}, we have

$$\overline{w'u'} \approx \frac{k}{2}\frac{N^2}{\lambda^3} \quad \text{at } z = z_{break}. \tag{9}$$

where

$$w' = \text{perturbation vertical velocity}$$
$$u' = \text{perturbation zonal velocity}$$

Now, for plane waves of the form (1) in the absence of damping, the Eliassen-Palm theorem requires

$$\frac{d}{dz}\left(\rho_0\,\overline{u'w'}\right) = 0, \tag{10}$$

implying no acceleration of the mean flow. However, in the presence of damping due to wave breaking (10) is replaced by

$$\rho_0\overline{u'w'} = \rho_0(z_{break})\left.\frac{k}{z}\frac{N^2}{\lambda^3}\right|_{z_{break}} e^{-2\int_{z_{break}}^{z}\lambda_i dz} \tag{11}$$

from which it can be shown from (6) that the flow acceleration, F_x, is given by

$$F_x = - \frac{1}{\rho_0} \frac{d}{dz} \left(\rho_0 \; \overline{u'w'} \right)$$

$$= \overline{w'u'} \left(\frac{1}{H} - \frac{3\frac{d\bar{u}}{dz}}{\bar{u} - c} \right) \tag{12}$$

Lindzen (1981) considered the case where the second term in (12) was negligible.

There is little point in reproducing the discussion of relations (8) and (12) given in Lindzen (1981). However, it should be stressed that, in the context of this simplified model, even given observations of z_{break} we must still know c, k and ℓ in order to evaluate D_{eddy} and F_x. We shall discuss these matters further later. For the moment let us note that Lindzen (1981), taking account of the filtering properties of tropospheric and stratospheric winds concluded that for winter $c \approx 0$ while for summer $c \approx 20$ m/s. k was chosen to be the smallest value consistent with

$$k (\bar{u} - c) \overset{\sim}{>} f^1 \tag{13}$$

where

$$f = 2\Omega \sin \phi$$
$$\phi = latitude$$
$$\Omega = 2\pi/day$$

(13) being necessary for vertical propagation in the presence of rotation. It should be noted that the present analysis otherwise ignores rotation. For $(k/\ell)^2$ Lindzen (1981) chose 1.17 as the value which best duplicated observed vertical wavelengths. Certainly the selection of k and ℓ was casual at best. Nevertheless, with these choices, equ (8) predicted the magnitude and the vertical distribution of D_{eddy} (a peak near 70 km and a minimum near 90 km) called for by totally independent chemical measurements and provided an immediate explanation of anomalous D-region absorption during sudden warmings. In addition equ (12) predicted mesopause level accelerations 0 (100 m/s/day). To be consistent with steady flows these accelerations would have to be balanced by meridional circulations 0 (8 m/s) and such meridional circulations are approximately what are needed to produce the observed reversal of pole to pole temperature gradient at the mesopause. These matters are discussed in more detail in Lindzen (1981). In view of these successes one can be reasonably confident that observed gravity waves do play a major role in the general circulation of the middle atmosphere. Indeed it is commonly stated that the choice of k, above, is too small; larger values would lead to larger values of D_{eddy} and F_x.

---

[1]This led to assumed horizontal wavelengths $\sim$ 1600 km.

Nevertheless, it is equally obvious that the above model for gravity waves is naive, inadequate, and uncertain on many counts. Indeed, there are so many potential improvements to be made on the present model (which after all was chosen to be the simplest model capable of describing the relevant physics) that ultimately we will need observational guidance in our choice.

The remainder of this paper will be devoted to discussing a few of the uncertainties and inadequacies. In general, these may be divided into three categories:

   i) Technical: Difficulties in actually using (8) and (12) in numerical models.
  ii) Parameter choice: How to best determine c, k, ℓ, and A (as functions of latitude).
 iii) Basic assumptions: How to best deal with the fact that gravity waves are unlikely to be described by (1), i.e., they are not likely to be steady, homogeneous in longitude, etc.

There is one technical matter (which may have some basic ramification) which we will discuss immediately. Equs (5), (8), and (12) imply a sudden onset of D_{eddy} and F_x at z_{break}. This, in turn, is generally hard on numerical schemes. Lindzen and Forbes (1983) have recently considered the possibility that stable gravity waves (by equ (5)) might cascade energy to smaller scales which were unstable, thus generating eddy diffusion. Lindzen and Forbes (1983) determined an upper bound for such a cascade and found that the resulting diffusion only modestly affected the original wave until the breaking level was reached. Thus, regardless of the validity of the cascade mechanism, Lindzen and Forbes (1983) provide a harmless way to objectively smooth the onset of D_{eddy} and F_x.

3. HOW MUCH GRAVITY WAVE IS THERE?

The parameterization described in section 2 assumes that the breaking observed by Theon et al. (1967) at a single station (Wallops Island) is characteristic of the whole hemisphere (at least around a latitude circle). It is also presumed to be associated with specific phase speeds. Holton (1982) takes even greater freedom with his choice of fluxes and phase speeds. Similar assumptions underlie the work of Matsuno (1982). The question addressed in this section is whether one can plausibly estimate and bound the generation of gravity waves in the troposphere.

All proposed mechanisms for gravity wave generation fall into 3 broad categories:

   i) Mountain waves forced by flow over topography. These waves are associated with phase speeds ~ 0; they need not be exactly zero since the flows are unsteady and can generate phase speeds other than zero.

  ii) Unstable shear zones can radiate internal gravity waves with phase speeds equal to the flow speeds in the shear zones (Lindzen, 1974). We shall refer to this mechanism as shear collapse for reasons that will be made clearer.

iii) Geostrophic adjustment in principle leads to gravity wave generation (Blumen, 1972); however, it is unclear how this mechanism operates on a day-to-day basis or whether it is, in fact, distinguishable from the first two mechanisms. We shall not discuss geostrophic adjustment further.

For the first two mechanisms we will attempt to crudely estimate gravity wave generation in order to see whether it is capable of generating the breaking levels used in section 2.

3a. Mountain waves

Mountain waves are forced by flow over topography. The forcing appears in the lower boundary condition where

$$w'(o) \approx U_o \frac{\partial}{\partial x} h(x) \text{ at } z = 0 \tag{14}$$

where

h = surface elevation
U_o = surface mean wind

We shall assume surface elevation of the form

$$h(x) = h e^{ikx} \cos(\ell y + \phi)$$

Then (14) becomes

$$w'(o) \approx ikU_o h e^{ikx} \cos(\ell y + \phi) \tag{15}$$

From Lindzen (1981) we have

$$w'(z) \approx A e^{z/2H} \lambda^{-1/2} e^{i\int_o^z \lambda dz} e^{ikx} \cos(\ell y + \phi) \tag{16}$$

where, for $c = o$,

$$\lambda^2 \approx \frac{N^2}{U_o^2}(1 + \ell^2/k^2)$$

From (15) we have

$$A = i k U_o h \lambda_o^{1/2} \tag{17}$$

We also have from Lindzen (1981)

$$u' \approx \frac{\lambda}{k\left[1 + \frac{\ell^2}{k^2}\right]} w' \tag{18}$$

From (16) and (18) we get by averaging over x,

$$\overline{w'u'} \approx \frac{k\, U_o\, h^2\, N_o}{2\left[1 + \frac{\ell^2}{k^2}\right]^{1/2}}\, e^{z/H} \tag{19}$$

and from (9) we have

$$\overline{w'u'}\,\Big|_{z_{break}} = \frac{k}{2}\, \frac{U^3\left(z_{break}\right)}{N\left(z_{break}\right)\left[1 + \frac{\ell^2}{k^2}\right]^{3/2}} \tag{20}$$

Equating (20) gives us a simple expression for z_{break}:

$$e^{z_{break}/H} \approx \frac{U^3\left(z_{break}\right)}{N_o\, N\left(z_{break}\right)h^2\, U_o\left[1 + \frac{\ell^2}{k^2}\right]} \tag{21}$$

For simplicity we will take

$$N \approx N_o \approx \frac{2\pi}{300s}$$

We will also take

$$U_o \sim 10 \text{ m/s}$$

which probably is excessive.

From Sankar-Rao (1965) we estimate that for $k \geq \frac{2\pi}{1600\ km}$, $h \sim 20$ m is probably conservative as well (recall that it is a global value). Finally following Lindzen (1981), we take $\left[1 + \frac{\ell^2}{k^2}\right]^{1/2} \approx 1.5$.

Rather than including the functional form of $U(z)$ in evaluating z_{break}, we will anticipate that breaking will occur between 50 km and 70 km (in winter) and evaluate the r.h.s. of (21) for U = 30 m/s, and 50 m/s--values characteristic of this height range. For U = 30 m/s we find $\frac{z_{brk}}{H} \approx 8.83$ ($z_{brk} \approx 62$ km assuming $z \approx 7$ km) and for U = 50 m/s, $\frac{z_{brk}}{H} \approx 10.4$ ($z_{brk} \approx 73$ km). Given that U = 50 m/s is characteristic of z = 50 km while z = 30 km is characteristic of 62 km, we may conclude that mountain wave forcing on a global basis can account for breaking levels no lower than 62 km--especially since we have ignored damping. This suggests that the observed breaking at 50 km over Wallops Island (Theon et al., 1967) is irregular, regional or both. We will return to this later. Note, however, that this result is independent of the choice for k.

3b. Shear collapse

In attempting to estimate gravity wave production due to shear collapse we must deal with a problem for which no solutions exist so far. Of necessity our approach will be even coarser and less direct than our approach to mountain waves. We will begin with the assumption that the general circulation of the troposphere is trying to produce shear layers which we will idealize as Helmholtz discontinuities of magnitude ΔU. Following Lindzen (1974) we will further assume that such unstable shear zones are stabilized (smoothed until the Richardson number equals 1/4) by the emission of gravity waves, half of which are travelling upward. Thus we assume that half the difference in energy between the Helmholtz profile and the neutral profile goes into gravity waves which may (if the intervening profile of U is favorable) penetrate the middle atmosphere. From Lindzen (1974) we have

$$\Delta E \sim \frac{\rho_o \left(z_{shear} \right) \left(\frac{\Delta U}{2} \right)^3}{3N} = \frac{\rho_o \left(z_{shear} \right) (\Delta U)^3}{24N} \tag{22}$$

This energy is produced over a time, τ, over which the general circulation is assumed to be trying to maintain the ΔU jump. We will take

$$\tau \sim 10 \ f^{-1}$$

where

$$f = 2\Omega \sin \phi.$$

Then

$$p'w' = \rho_o \frac{\Delta U}{2} \overline{u'w'} \lesssim \frac{\Delta E/2}{\tau} \sim \frac{\rho_o (\Delta U)^3}{48N\tau} \sim \frac{\rho_o (\Delta U)^3 f}{48N \times 10}$$

by the Eliassen-Palm theorems

and

$$\overline{u'w'} \Big|_{z_{shear}} \gtrsim \frac{(\Delta U)^2 f}{240 \ N} . \tag{23}$$

Above z_{shear}, the constancy of $\rho_o \overline{u'w'}$ implies

$$\overline{u'w'} \gtrsim \frac{(\Delta U)^2 f}{240 \ N} \ e^{\left(\frac{z - z_{shear}}{H} \right)} \tag{24}$$

Again, from section 2 we have that at z_{break}

$$\overline{u'w'} \approx \frac{k}{2} \frac{N^2}{\lambda^3} \approx \frac{k}{2} \frac{(U - c)^3}{N \left(1 + \frac{\ell^2}{k^2} \right)^{3/2}} \tag{25}$$

12

Combining (24) and (25) we then get

$$\frac{(\Delta U)^2 f}{240\ N}\ e\left(\frac{z_{break} - z_{shear}}{H}\right) \gtrsim \frac{k}{2}\ \frac{(U - c)^3}{N\left(1 + \frac{\ell^2}{k^2}\right)^{3/2}} \tag{26}$$

For purposes of estimation we will take

$$\frac{f}{N} \sim 5 \times 10^{-3}$$

$$N \sim \frac{2\pi}{300\ \text{sec}}$$

$$\left(1 + \frac{\ell^2}{k^2}\right)^{1/2} \sim 1.5$$

(26) then becomes

$$e\left(\frac{z_{break} - z_{shear}}{H}\right) \gtrsim 3.4 \times 10^5\ \frac{k(U - c)^3}{(\Delta U)^2} \tag{27}$$

Following Lindzen (1981) the smallest value we may take for k is

$$k \sim 3.7 \times 10^{-6}\ \text{m}^{-1}$$

For ΔU we shall somewhat arbitrarily take

$$\Delta U \sim 5\ \text{m/s},$$

and for $|U - c|$ we will anticipate $|U - c| \sim 50$ m/s. (27) then becomes

$$e\left(\frac{z_{break} - z_{shear}}{H}\right) \gtrsim 6.3 \times 10^3$$

or

$$\frac{z_{break} - z_{shear}}{H} \gtrsim 8.75$$

or

$$z_{break} - z_{shear} \gtrsim 61\ \text{km}.$$

Clear air turbulence is generally associated with the upper troposphere so that $z_{shear} \sim 10$ km and

$$z_{break} \gtrsim 71\ \text{km}$$

This is at least compatible with summer observations, but again un-certainties and the neglect of damping should be kept in mind. It should also be kept in mind that larger values of k will lead to higher breaking levels.

As a final remark we should note that this is probably the most efficient mechanism for generating gravity waves with phase speeds greater than zero. Thus, in invoking phase speeds greater than average tropospheric wind speeds, we must consider what could possibly be trying to maintain an unstable shear layer at that speed.

4. WAVENUMBER AND OBSERVED PERIODS

In the previous sections we noted that there is substantial uncertainty over horizontal wavelengths--largely because our data is restricted to single station records. Coherence studies over short distances are not likely to be convincing since phase fronts in nature are unlikely to be flat. Nevertheless, the value of k plays an important role in determing D_{eddy} and F_x. In this section we will look at the relation between wavelength and period--hopefully to gain some insight into the question of wavelength.

Historically we have considered the generation of waves to be associated with specific periods--the period being thought to be a signature of physical mechanism. This is certainly the case with tides. Unfortunately, all known mechanisms for generating tides are incapable of maintaining breaking in middle and high latitudes on a global scale. The waves discussed in section 3, on the other hand, are characterized by phase speed and not period.[2]

To the extent that periods, τ, are observed, they are related to phase speed by the simple relation

$$\tau = \frac{2\pi}{kc} = \frac{\text{horizontal wavelength}}{\text{phase speed}} \tag{28}$$

Anticipating phase speeds $\sim$ 20 m/s in summer, we see from (28) that a horizontal wavelength of 1000 km will be associated with a ground measured period, τ, of 14 hours. It would be easy to confuse such a period with a tidal period. A clue to the primacy of period or phase speed is given by equ (3) where we see a tendency for all waves with the same phase speed to have the same vertical wavelength independent of k and hence period. This is in line with currently available data.

5. WAVE PACKETS AND GROUP VELOCITY

The preceding sections were concerned with a zonally homogeneous-- or at least, zonally averaged picture of gravity waves. However, there

---

[2]In this connection it is interesting to consider the period, 3 hrs, commonly associated with gravity waves. To the best of my knowledge, this is not based on observation, but was introduced on an ad hoc basis by Hines (1960). Presumably this was a period between the Coriolis period and the Brunt period--at least in the absence of crucial doppler effects.

is no question that in reality gravity waves are excited regionally and even episodically. Ideally, the theorist should be dealing with the statistics of wave packets rather than with plane waves. Indeed a start in this direction was made by Jones (1968). However, the difficulty of the packet approach and the absence of an adequate data base for such an approach make it uninviting for the moment.

Nevertheless some crude analysis is helpful, at least for the interpretation of data. Let us for a moment consider a gravity wave packet travelling in the x-z plane. Locally we have

$$\sigma \approx N \frac{k}{m} \tag{29}$$

where σ = frequency observed in moving frame following mean flow = k (c-U)

$$k = \text{wavenumber in x direction}$$
$$m = \text{wavenumber in z direction}$$
$$N = \text{Brunt-Vaisala frequency}$$

The group velocity in the x-direction is given by

$$C_{gx} = U + \frac{\partial \sigma}{\partial k} = U + \frac{N}{m} = U + \frac{\sigma}{k} = U + (c-U) = c \tag{30}$$

In the z-direction we have

$$C_{gz} = \frac{\partial \sigma}{\partial m} = -N \frac{k}{m^2} = -\frac{\sigma}{m} = -\frac{k (c-U)}{m} \approx -(c-U)^2 \frac{k}{N} \tag{31}$$

As a rule a packet will travel at the group velocity, and from (30) and (31) we have

$$\frac{C_{gx}}{C_{gz}} = -\frac{m}{k} \frac{c}{c-U} \tag{32}$$

Now for vertical wavelengths of 10 km and horizontal wavelengths of 1000 km, m/k = 100. For stationary waves where c ≡ 0, waves will still be travelling vertically; however, when c ≠ 0 packet travel will become almost horizontal, and packets originating in the troposphere will, in travelling vertically 60 km, also travel several thousand km horizontally. Thus, observations at a single location (especially in summer when c ≠ 0) will frequently be unable to trace the origin of upper atmosphere gravity waves.

In addition, as long as the large-scale wind distribution permits meridional propagation, breaking waves in the upper atmosphere at a particular latitude can originate in the troposphere at very different latitudes. The presence of mountains or unstable jets directly below may be neither necessary nor relevant.

We finally note from equ (31) that variations in U, the large-scale wind, markedly affect the vertical group velocity. In general there is a tendency for wave packets to be deflected into regions of large group velocity. It is of some use to consider the fact that the environment into which gravity waves propagate is not zonally homogeneous. Stationary waves (generally of wave numbers 1-3) will produce large regions of enhanced or depressed zonal velocity. This will affect gravity waves and their mean flow interactions in a number of ways:

i) There will be a tendency for waves to be focused into the regions where the stationary wave enhances the magnitude of $(\bar{u}-c)$. This may not seem too important for a very localized wave packet which might presumably only "see" the local phase of the stationary wave. However, from equ (32) we see that the ray paths in regions of low $|\bar{u}-c|$ will become more horizontal, almost certainly allowing focussing to act before the packet reaches the mesosphere. More commonly, gravity waves will be excited over a broad range of longitudes and this question does not arise.

ii) The focussing of waves into regions of larger $|\bar{u}-c|$ should lead to greater amplitudes and lower breaking heights in these regions. However, as can be seen from equ (4), large $|\bar{u}-c|$ leads to smaller λ's and smaller values for $\left|\dfrac{d\delta T}{dz}\right|$ which in turn should elevate the breaking height. Finally, damping is less important when $|\bar{u}-c|$ is larger--again suggesting larger amplitudes.[3] It is not clear, without further calculations, which effect will dominate and under what conditions.

iii) Despite the ambiguity in item (ii) above, it seems clear that gravity wave breaking can occur at different heights at different locations due to inhomogeneities in excitation, focussing by planetary scale stationary waves, etc. This is especially important in view of the suggestion in section 3 that mountain waves are inadequate to maintain the winter breaking levels observed at Wallops Island on a global basis. Such inhomogeneities can give rise to interactions between gravity waves and stationary waves. In regions of large $|\bar{u}-c|$, F_X due to gravity waves will act to attenuate the stationary wave; however, in regions of small $|\bar{u}-c|$, F_X will act to amplify the stationary wave. Again, it is difficult to anticipate the net effect.

iv) Finally, zonal variations in breaking height will act to vertically spread the zonal averaged distributions of D_{eddy} and F_X compared to the distributions derived from section 2.

It is amply clear that the correct calculation of the effects described in this section will be difficult--though important. However, it is almost as clear that the qualitative expectations are general and not too dependent on specific details.

---

[3]Damping (presumably from infrared cooling) complicates matters in other ways. It permits, for example, the deposition of gravity wave momentum fluxes without breaking and without the generation of D_{eddy}.

6. REMARKS

The present paper follows closely the contents of the lecture the author presented at the U.S.-Japan Symposium--except for the omission (for the sake of brevity) of material on tides, which can, however, be found in Lindzen (1981). In this paper I have reviewed the parameterization of Lindzen (1981) wherein the effects of gravity wave breaking on the generation of eddy diffusion and on the deposition of wave momentum flux can be seen in the simplest (and rather idealized) context. This context consists in zonally and vertically propagating waves whose properties are independent of longitude propagating through an axially symmetric zonal flow. Even in this context certain questions arise. In section 3 we inquire whether known sources for gravity waves can account for the breaking levels observed at Wallops Island. We note that known forcing may be inadequate. A possible solution to the resulting problem is that wave amplitudes at Wallops Island may not be characteristic of other longitudes. This matter is discussed further in section 5.

In section 4 we deal with the question of the horizontal wavelength of gravity waves. This is a very important variable in determining D_{eddy} and F_x due to breaking waves. It is also almost impossible to measure directly at isolated stations. However, if one can otherwise determine phase speed, then single station observations of period can yield horizontal wavelength. It is noted that the periods noted by Balsley at this symposium (12-36 hrs) are compatible with horizontal wavelengths ≥ 1000 km.

In section 5 the group velocity of gravity wave packets is considered. We noted that the origin of gravity waves observed in the mesosphere is likely to be far removed horizontally. We also discuss the fact that gravity waves in fact propagate through significant planetary scale stationary waves.

The above hardly begins to deal with the possible extensions of the simple parameterization of gravity wave breaking. In this connection mention should be made of the recent work of Schoeberl and Strobel (1983) on the effects of infrared damping and of Dunkerton (1982) on the possible intermittancy of gravity waves.

ACKNOWLEDGEMENTS

This work has been supported by the National Science Foundation under grant ATM82-05638, and by NASA under grant NGL22-007-228.

REFERENCES

Balsley, B., 1983 (paper in these proceddings).

Blumen, W., 1972: Geostrophic adjustment. Rev. Geophys. and Space Phys.,

10, 485-528.

Bolgiano, R., Jr., 1959: Proceedings of International Symposium of Fluid Mechanics in the Ionosphere, Cornell University. J. Geophys. Res., 64, 2037-2238.

Booker, J. R. and F. P. Bretherton, 1967: The critical layer for internal gravity waves in a shear flow. J. Fluid Mech., 27, 513-519.

Dunkerton, T., 1982: Wave transience in a compressible atmosphere, Part III: The saturation of internal gravity waves in the mesosphere. J. Atmos. Sci., 39, 1042-1051.

Georges, T. M., 1968: Proceedings of Symposium on Acoustic-Gravity Waves in the Atmosphere, Boulder, Colorado, 15-17 July 1968. Available Superintendant of Documents, U. S. Government Printing Office, Washington, D. C., 20402, 437 pp.

Hines, C. O., 1960: Internal gravity waves at ionospheric heights. Can. J. Phys., 38, 1441-1481.

Hodges, R. R., Jr., 1969: Eddy diffusion coefficients due to instabilities in internal gravity waves. J. Geophys. Res., 74, 4087-4090.

Holton, J. R., 1982: The role of gravity wave induced drag and diffusion in the momentum budget of the mesosphere. J. Atmos. Sci., 39, 791-799.

Jones, W. L., 1968: Ray tracing for internal gravity waves. In Georges, T. M., ibid.

Leovy, C., 1964: Simple models of thermally driven mesospheric circulation. J. Atmos. Sci., 21, 327-341.

Liller, W. and F. L. Whipple, 1954: High Altitude winds by meteor-train photography. Rocket Exploration of the Upper Atmosphere, New York, Pergamon Press, 72-130.

Lindzen, R. S., 1967: Thermally driven diurnal tide in the atmosphere. Q. J. Roy. Met. Soc., 93, 18-42.

Lindzen, R. S., 1968: The application of classical atmospheric tidal theory. Proc. Roy. Soc. A., 303, 299-316.

Lindzen, R. S., 1968a: Some speculations on the roles of critical level interactions between internal gravity waves and mean flows. In Georges, T. M., ibid.

Lindzen, R. S., 1968b: Lower atmospheric energy sources for the upper atmosphere. Met. Mono., 9, 37-46.

Lindzen, R. S., 1971: Tides and gravity waves in the upper atmosphere.

18

In Mesospheric Models and Related Experiments, G. Fiocco, ed.
(Dordrecht, Holland: D. Reidel Pub.).

Lindzen, R. S., 1974: Stability of a Helmholtz velocity profile in a
continuously stratified infinite Boussinesq fluid - applications to
a clear air turbulence. J. Atmos. Sci., 31, 1507-1514.

Lindzen, R. S., 1981: Turbulence and stress due to gravity wave and
tidal breakdown. J. Geophys. Res., 86, 9707-9714.

Lindzen, R. S. and J. M. Forbes, 1983: Turbulence originating from
convectively stable internal waves. J. Geophys. Res., in press.

Matsuno, T., 1982: A quasi one-dimensional model of the middle atmosphere circulation interacting with internal gravity waves. J. Met.
Soc. Japan, 60, 215-226.

Sankar-Rao, M., 1965: Continental elevation influence on the stationary
harmonics of the atmospheric motion. Pure and Appl. Geophys., 60,
141-159.

Schoeberl, M. and D. Strobel, 1983: A numerical model of gravity wave
breaking and stress in the mesosphere. J. Geophys. Res., in press.

Theon, J. S., W. Nordberg, L. B. Katchen, and J. J. Horvath, 1967: Some
observations on the thermal behavior of the mesosphere. J. Atmos.
Sci., 24, 428-438.

J. R. Holton and T. Matsuno, Dynamics of the Middle Atmosphere, 19-43.
Copyright © 1984 by Terra Scientific Publishing Company.

GRAVITY WAVE ATTENUATION AND THE EVOLUTION OF
THE MEAN STATE FOLLOWING WAVE BREAKDOWN

R. L. Walterscheid

Space Sciences Laboratory
The Aerospace Corporation

ABSTRACT

We have studied the evolution of the mean state following the breakdown of an upward propagating internal-gravity wave using a simple time-dependent wave-mean-flow model. The simulations were performed for a wintertime case. We have assumed that breakdown occurs when the superposed state (wave plus mean) is statically unstable and have employed the parameterization of Lindzen (1981) to calculate the eddy diffusion coefficient D_{eddy}. Our basic conclusions are as follows: 1) there is a tendency for waves to break initially near their critical levels; 2) large amplitude waves produced a rapid deceleration of the Doppler shifted phase velocity and gave a much smaller value of the eddy diffusion coefficient than would have been obtained from steady-state models; 3) the heat fluxes induced by breakdown destabilize the mean state and may lead eventually to the dynamical instability of the incident wave, the instability occurring sooner (after breakdown) for large amplitude gravity waves than for small amplitude ones; and 4) for small amplitude waves the regions of breakdown resembled apparent regions of turbulence observed by VHF and MF radars. These calculations were performed for horizontal phase velocities c = 0 and wavelengths L_x = 200 km. For larger values of L_x wave-forcing is reduced, and values of D_{eddy} inferred from the large-scale time-independent circulation might be appropriate even for large amplitude waves if values of $L_x \gg$ 200 km prevail.

1. INTRODUCTION

It is now generally accepted that the dynamics and structure of the mesosphere and lower thermosphere are strongly influenced by the propagation of internal gravity waves into this region of the atmosphere from below (Hines, 1960, 1970; Gossard, 1962; Pitteway and Hines, 1963; Hines and Reddy, 1967; Hodges, 1967, 1969; Lindzen, 1967, 1968a, b, 1971, 1981). Discussions have progressed from quantitative recognition of the importance of gravity waves to quantitative evaluations of the turbulent diffusion and momentum deposition (friction) they induce. Gravity waves may generate turbulence and friction because their growth with height is eventually limited by instabilities (breakdown) or nonlinear interactions.

One quantitative theory for the effects of gravity waves in the mesosphere has been developed by Hodges (1967, 1969) and Lindzen (1967, 1968a, 1981). They suggested that the growth in amplitude of upward propagating waves proportionately with the inverse square root of the background pressure would eventually be limited by wave-induced convection or shear instabilities when the amplitude of the waves became sufficiently large that the wave plus the background state had a negative static stability or a Richardson number smaller than 0.25. The lowest altitude at which the instabilities would occur is the wave breaking level Z_{break}. Above this height it is postulated that there is sufficient generation of turbulence by the wave induced instabilities to prevent further amplification of the waves.

The model of turbulence generation by the convective instability and the resultant breakdown of upward propagating gravity waves permits quantitative determination of eddy diffusivity and friction in the mesosphere. Lindzen (1981) (see also Holton (1982)) extended earlier treatments to account for height variations in the background temperature and a vertically varying zonal wind. As described above, it is assumed that a single vertically propagating wave begins to break at the level Z_{break} where the local temperature lapse rate produced by the sum of the wave and background state is adiabatic. This is the wave saturation condition. By using a WKB approximation to describe the vertical structure and propagation of the wave, and making a number of additional simplifications, Lindzen (1981) and Holton (1982) have derived a relatively straightforward expression for calculating Z_{break}. They also developed a formula for D_{eddy} as a function of height Z, for $Z > Z_{break}$, by requiring that turbulent diffusion produced by wave breaking limit the wave amplitude so that the wave remains just saturated above Z_{break}. The resulting expression for D_{eddy} applies above Z_{break} and below any critical level Z_{crit} lying above Z_{break}.

In this study we adopt the wave-breaking wave-saturation point of view of Lindzen (1981) and Holton (1982). However, we apply a time-dependent wave-mean-flow model and consider, self consistently, the modification to the mean state due to the heat and momentum transports induced by the breaking waves. The transports will be due to both wave and turbulent transports. This approach differs from Holton (1982) who used parameterized wave stresses obtained from WKB theory in a time-dependent model of the large-scale circulation. Our model applies to an individual breaking wave, whereas Holton's model properly applies to the collective effects of a large number of breaking waves. Both Lindzen and Holton have inferred large accelerations of the background flow owing to wave breakdown. Lindzen (1981) inferred a wave-induced acceleration of ~ 100 m s^{-1} day^{-1}, and Holton inferred net accelerations (wave-induced plus eddy diffusion) which could be similarly large. Upon the onset of breakdown it is possible that the wave and diffusive fluxes of heat and momentum will rapidly modify the background state and that the diffusion coefficient will likewise be rapidly changing. We also expect strong vertical gradients of background wind and temperature and of D_{eddy} at the vertical boundaries of the breakdown region.

Our approach will be to use the Lindzen (1981) parameterization of D_{eddy} in a time-dependent numerical wave-mean-flow model (Walterscheid, 1981b). We shall also use the breakdown criterion of neutral superposed (wave plus background) static stability used by Lindzen and others. Thus we employ WKB assumptions to the extent that we employ the Lindzen parameterization, which specifies D_{eddy} in terms of instantaneous background and wave quantities, but relax WKB assumptions to the extent that we do not require that the wave be fully adjusted to the instantaneous background state and D_{eddy} profile. Our use of the Lindzen parameterization is heuristic: No other physically reasonable parameterization was available, and we ultimately justify its use according to the extent to which it is able to maintain a state which is not too far from neutral static stability, and in terms of other modeling limitations and simplifications to be discussed below.

2. THEORY

In this section, we discuss the wave-induced modification, the basic state, the criterion for the onset of wave breakdown, and the model for the diffusion coefficient.

When gravity waves break down, the vertical gradients in the induced vertical flux of heat and momentum may cause a rapid modification of the mean state. These effects are illustrated schematically in Figure 1. The horizontal lines delineate the region of breakdown. In the region of breakdown the diffusion coefficient has a

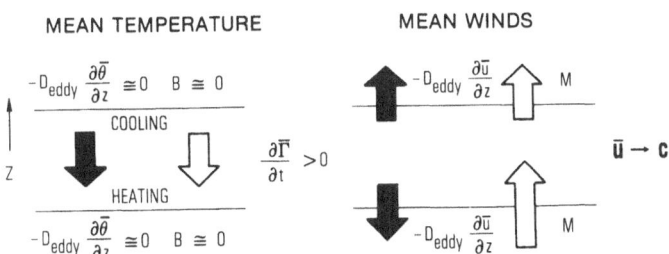

Figure 1. Schematic representation of the modification of the mean state due to wave and diffusive fluxes of heat and momentum in the region of wave breakdown. The region of breakdown is delineated by the horizontal lines; D_{eddy} is the diffusion coefficient; $\bar{\theta}$ and $\bar{u}$ are, respectively, mean-state potential temperature and zonal wind; z the vertical coordinate; and B and M are, respectively, the wave fluxes of heat and momentum.

value D_{eddy} which is assumed to be much larger than the value outside this region. The wave and turbulent momentum fluxes are denoted by M and D_{eddy} $\partial \bar{u}/\partial z$, respectively, and the respective heat fluxes by B and D_{eddy} $\partial \bar{\theta}/\partial z$. (The quantities $\bar{u}$ and $\bar{\theta}$ are, respectively, the mean wind and potential temperature, and z is the log-pressure height coordinate ($z \equiv$ log (1000/p), where p is pressure in mb). In this log-pressure coordinate system D_{eddy} has units of s^{-1}, and $\bar{\theta}$ is related to temperature $\bar{T}$ by $\bar{\theta} = \bar{T}$ exp(κz), where $\kappa = R/c_p$, and R and c_p are, respectively, the gas constant and specific heat at constant pressure for air.) The attenuation of the wave momentum flux causes the mean-flow to be accelerated toward the phase velocity of the wave. The turbulent flux of momentum will, in general, oppose this tendency. The wave and turbulent fluxes will transport heat from the top of the region of breakdown to the bottom (Walterscheid, 1981a). Since the fluxes essentially vanish outside of the breakdown region there will be cooling near the top and a warming near the bottom, thus destabilizing the region.

The diffusion coefficient is given in terms of wave and mean-state quantities as

$$D_{eddy} = \frac{k(\bar{u}-c)^4}{\bar{S}^{3/2}} (\frac{1}{2} - \frac{3}{2}\frac{\partial \bar{u}/\partial z}{(\bar{u} - c)}) \tag{1}$$

(Lindzen, 1981). The symbols used have the following meaning: k and c are, respectively, the wave horizontal wave number and phase speed (c > 0 for eastward propagation), and $\bar{S}$ is a measure of the mean-state static stability ($\bar{S} \equiv R\bar{T}$ ∂log$\bar{\theta}/\partial z$).

We should remark that in applying D_{eddy} we assume that it is constant over a horizontal wavelength. However, only a part of the wave will be unstable. To see this, let Γ' and $\bar{\Gamma}$ be the wave and mean state lapse rates, respectively, and Γ_a be the adiabatic lapse rate, and let $\Gamma_a \gg \bar{\Gamma}$, as in the mesosphere. Then considering the superposed state ($\bar{\Gamma} + \Gamma'$), if Γ' is large-amplitude enough to give a region of low static stability where Γ' is positive, then it is large-amplitude enough to give a region of high static stability where Γ' is negative. In fact, a region of instability where $\Gamma' > 0$ implies a region of temperature inversion where $\Gamma' < 0$. This situation is illustrated in Figure 2. In addition, the $\partial u'/\partial z$ wave is more or less in quadrature with the Γ' wave so that where the superposed state is statically very stable the flow is not apt to be destabilized by large wave shears. Even though the wave is apt to continue growing beyond the height where it has just achieved convective instability (i.e., where $|\Gamma'| = \Gamma_a - \bar{\Gamma}$) at no altitude is the entire wave apt to be involved in turbulence. The stable regions of the wave are apt to have flux Richardson numbers Ri $\gg$ 1, and any residual turbulence generated by the earlier passage of an unstable region will be quickly consumed.

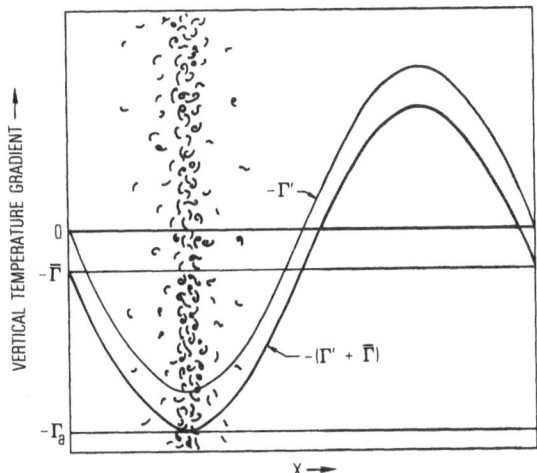

Figure 2. Schematic representation of the variation of the vertical temperature gradient. The quantities Γ', $\bar{\Gamma}$ and Γ_a are, respectively, the perturbation lapse rate, the mean-state lapse rate, and the adiabatic lapse rate. A region of turbulence is indicated near where $\Gamma' + \bar{\Gamma} = \Gamma_a$. The shaded region indicates the region where the temperature of the superposed state $(\bar{T} + T')$ is increasing upward (i.e., where there is a temperature inversion).

This means that D_{eddy} will vary considerably over a horizontal wavelength (or cycle). The dependence of D_{eddy} on perturbation quantities makes the diffusion terms $D_{eddy}\partial u'/\partial z$ and $D_{eddy}\partial\theta'/\partial z$ nonlinear. Thus energy incident at a specific wavenumber and frequency may be scattered into a range of frequencies and wavenumbers. To see this, consider $\delta D_{eddy} \propto \cos(k'x)$ where $D_{eddy} = \bar{D}_{eddy} + \delta D_{eddy}$. Then an incident wave $\propto \cos(kx)$ will force waves $\cos(k \pm k')x$ through the diffusion terms. When $\delta D_{eddy} \gg \bar{D}_{eddy}$ the energy scattered (coupled) into the secondary waves might be quite significant. Insofar as the incident wave is concerned, the scattering process is dissipative, and will play a role in limiting wave growth and maintaining a saturated state. However, unlike real dissipation, the scattering process will not necessarily lead to a local convergence of heat and momentum fluxes when the total wave (incident plus secondary) is considered. This follows because secondary waves may propagate away from the region of generation, and may not contribute locally to the flux convergence of heat and momentum.

Thus treatments based on values of D_{eddy} which are constant in x may substantially overestimate the wave forcing of the mean state when scattering plays an important role in maintaining a saturated state. Since we expect large variations of D_{eddy} over a horizontal wavelength this scattering is apt to be indeed important. However, in this treatment, as in earlier treatments by other authors (e.g., Hodges,

1967; Lindzen, 1981; Holton, 1982; Hines, 1970), we consider D_{eddy} to be constant in x. Henceforward, we consider D_{eddy} to be the effective constant (in x) value required to maintain saturation. It includes the effects of both eddy diffusion and nonlinear scattering. The parameterization (1) obtained by Lindzen (1981) gives an estimate of this effective value. It should at least enable us to estimate an upper limit for the wave-forcing of the mean state.

The interpretation of D_{eddy} when scattering is important also has consequences for the calculation of the diffusive flux of mean-state heat and momentum, especially of time-independent mean-state (henceforward referred to as background-state) heat. Let A_{eddy} (for austausch) be the actual x-averaged eddy diffusion coefficient, rather than D_{eddy}, the effective value which applies only to the wave. According to our previous discussion $D_{eddy} > A_{eddy}$, and quite possibly $D_{eddy} \gg A_{eddy}$. If the stronger inequality holds, then the wave-flux of heat will be much greater than the diffusive flux. To see this we use Eq. (18) from Walterscheid (1981b) and $\overline{u'^2_z} \sim m^2_r \overline{u'^2}$ where m_r is the real part of the refractive index and where $m^2_r \approx \overline{S}(\overline{u-c})^{-2}$ and obtain

$$\overline{w'T'} \sim R^{-1} \overline{S} (u - c)^{-2} D_{eddy} \rho_0 \overline{u'^2}.$$

Assuming $|u'| \sim |\overline{u} - c|$ in the region of breaking (see below) then

$$\overline{w'T'} \sim - R^{-1}\overline{S} D_{eddy} . \qquad (2)$$

For the diffusive flux, on the other hand,

$$\text{heat flux} \sim e^{-\kappa z} A_{eddy} \, \partial\overline{\theta}/\partial z \sim R^{-1}\overline{S} A_{eddy} .$$

If $D_{eddy} \gg A_{eddy}$ then wave flux $\gg$ diffusive flux, and the use of D_{eddy} for A_{eddy} will greatly exaggerate the relative importance of the diffusive flux. In this study, however, we are primarily interested in the forcing of the mean state by the direct effect of the attenuation of the incident wave. Thus in the present study we include the diffusion of background-state sensible heat only to the extent that we allow diffusion to damp the wave-induced temperature change; thus, we do not include the flux of background-state sensible heat. Formally, this is equivalent to assuming that the diffusion of background-state heat is contained in the background-state heat balance. For consistency, the diffusion of mean-state momentum is handled in the same manner. In any case the momentum source due to the diffusion of background-state momentum is comparatively rather small. In future simulations we will consider the limiting case where $D_{eddy} = A_{eddy}$, and include the diffusion of background-state heat and momentum.

A related question is the extent to which the induced wave fluxes act to maintain the background-state balance, rather than contribute to local modifications of the mean state. If gravity waves contribute to the maintenance of the background-state through the stochastic effects of a large number of gravity waves over a large region of the globe, and if gravity waves are locally rather intermittent, then the fluxes induced by an individual gravity wave may act to produce large local changes in the mean state. We adopt this point of view in the present study.

As we mentioned earlier, the wave will continue to grow even after some regions of the wave have become unstable. This means that saturation will not be achieved until sometime after $\epsilon < 0$, where $\epsilon \equiv \Gamma_a - \bar{\Gamma} - |\Gamma'|$. Experiments were performed for various choices of ϵ between 0 and a value for which ~ 30% of the wave satisfied $\Gamma_a - \bar{\Gamma} - \Gamma' < 0$. It was found that the results did not depend sensitively on the choice of ϵ. Experiments were also performed for various choices of critical Richardson number. For values between 1/4 and 1 the results did not differ much from those for $\epsilon = 0$. However, significant changes in the breaking level were found for values ~ 2. Such large values may not be unrealistic. Theoretical and experimental results which may be relevant to the case of propagating atmospheric gravity waves indicate critical values ~ 1 to 3 (Orlanski, 1972; Delisi and Orlanski, 1975). In view of our present lack of a detailed understanding of the instability process and how it proceeds, the criterion $\epsilon = 0$, used by Lindzen and others seems appropriate.

Referring to (1) we see that as $\bar{u} \to c$, $D_{eddy} \to 0$. Thus the wave-induced acceleration of the mean wind may reduce D_{eddy} considerably below its initial values, that is, much below values obtained from a consideration of average mean-wind profiles. Note that the dependence of D_{eddy} on $\bar{u} - c$ is quartic, so that D_{eddy} is very sensitive to $\bar{u} - c$. The stress owing to breakdown is also reduced as $\bar{u} - c$ is reduced. Based on his WKB model, Lindzen (1981) showed that

$$F_x = \frac{\bar{S} \, D_{eddy}}{\bar{u} - c} \tag{3}$$

where F_x is the force per unit mass due to the momentum fluxes. Combining (1) and (2) we see that in regions of moderate shear away from critical levels $F_x \propto (\bar{u}-c)^3$. This dependence means that there is a rather strong negative feedback on the acceleration. This is in strong contrast to the case when D_{eddy} is fixed (Jones and Houghton, 1972). This feedback is counteracted to some extent by the reduced diffusive flux of momentum and by the destabilization of the mean state temperature profile. Since $F_x \propto \bar{S}^{-1/2}$ for fixed $\bar{u} - c$, the tendency for $\bar{S} \to 0$ contributes to an increase in F_x. A similar relationship holds for the induced wave flux of sensible heat (see (2)).

3. NUMERICAL SIMULATION

In this section, we discuss the wave-mean-flow model used to simulate vertical wave propagation and the evolution of the mean state, and describe the specification of the basic state and wave parameters for our numerical experiments.

3.1 Wave-Mean-Flow Model

The wave-mean-flow model is similar to the one described in Walterscheid (1981b). This model describes two-dimensional quasi-static motions of a compressible atmosphere. In the present study we consider motions in a nonrotating atmosphere. We are primarily interested in the short-term transient response of the mean state following breakdown, and indications of rapid accelerations should not be seriously influenced by the neglect of the Coriolis force.

Dependent variables are separated into a mean (x-averaged) part and a perturbation. Thus

$$\Psi(x,z,t) = \overline{\Psi}(z,t) + \Psi'(x,z,t) \tag{4}$$

where Ψ is any dependent variable and $\overline{\Psi}$ is the corresponding mean-state quantity and ψ' is the perturbation. The mean-state quantity is further decomposed as

$$\overline{\Psi}(z,t) = \overline{\Psi}_0(z) + \overline{\Psi}_1(z,t) \tag{5}$$

where $\overline{\Psi}_0$ is the background part and $\overline{\Psi}_1$ is the time dependent wave-induced part. Two systems of equations are developed which describe separately the time tendencies of mean-state and perturbation quantities. The two systems are coupled because predicted mean-state quantities provide the basic state for the perturbation quantities and predicted perturbation quantities provide the wave forcing for the mean state. The perturbations are assumed to have a simple sinusoidal dependence in the x-direction. However, the perturbation equations are fully time dependent, and it has not been assumed that the perturbations have a sinusoidal time dependence.

The model equations are as follows:

(1) Perturbation Equations

$$\frac{Du'}{Dt} + w' \frac{\partial \overline{u}}{\partial z} = ik\phi' - X' \tag{6}$$

$$\frac{D\phi'_z}{Dt} + \overline{S} w' = - Q' \tag{7}$$

$$- iku' + \frac{1}{\rho_0} \frac{\partial}{\partial z} (\rho_0 w') = 0 \tag{8}$$

$$\frac{\partial}{\partial z} \phi' = RT' \tag{9}$$

$$\frac{D}{Dt} \equiv \frac{\partial}{\partial t} - ik\bar{u}$$

$$\bar{S} \equiv R \left(\frac{\partial \bar{T}}{\partial z} + \kappa \bar{T} \right)$$

$$X' \equiv - \nu_m \frac{\partial^2}{\partial z^2} u' - \frac{1}{\rho_0} \frac{\partial}{\partial z} \left(\rho_0 \nu_e \frac{\partial}{\partial z} u' \right)$$

$$Q' \equiv - \kappa_m \frac{\partial^2}{\partial z^2} \phi_z' - e^{-\kappa z} \frac{1}{\rho_0} \frac{\partial}{\partial z} \left(\rho_0 \kappa_e \frac{\partial}{\partial z} R\theta' \right) \quad .$$

The symbols not previously defined have the following meaning: $w(\equiv \dot{z})$ is the vertical velocity, ϕ is the geopotential; ρ_0 is a reference density profile $(\rho_0(0) e^{-z})$; ν_e and ν_m are, respectively, coefficients of eddy and molecular viscosity; and κ_e and κ_m are, respectively, coefficients of eddy and molecular thermal diffusivity. The quantitites κ_e and ν_e are set equal to D_{eddy} plus a background value D_0. It is assumed that $\psi'(x,z,t) = \hat{\psi}(z,t)e^{ikx}$ where ψ' is any perturbation quantity. For computational purposes we replace u', w', etc. by $\hat{u}$, $\hat{w}$, etc.

(2) Mean Equations

$$\frac{\partial \bar{u}_1}{\partial t} = - \frac{1}{\rho_0} \frac{\partial}{\partial z} \left(\rho_0 \overline{w'u'} \right) - \bar{X} \tag{10}$$

$$\frac{\partial \bar{\phi}_{1z}}{\partial t} = - \frac{1}{\rho_0} \frac{\partial}{\partial z} \left(\rho_0 \overline{w'\phi_z'} \right) - \bar{Q} \tag{11}$$

$$\bar{w}_1 = 0 \tag{12}$$

where

$$\bar{X} \equiv - \nu_m \frac{\partial^2}{\partial z^2} \bar{u}_1 - \frac{1}{\rho_0} \frac{\partial}{\partial z} \left(\rho_0 \nu_e \frac{\partial}{\partial z} \bar{u}_1 \right)$$

$$\bar{Q} \equiv - \kappa_m \frac{\partial^2}{\partial z^2} \bar{\phi}_{1z} - e^{-\kappa z} \frac{1}{\rho_0} \frac{\partial}{\partial z} \left(\rho_0 \kappa_e \frac{\partial}{\partial z} R\bar{\theta}_1 \right) - \kappa \nu_d \left[\left(\frac{\partial \bar{u}}{\partial z} \right)^2 + \overline{\left(\frac{\partial u'}{\partial z} \right)^2} \right] \quad .$$

The last term on the right-hand side of the expression for $\bar{Q}$ represents viscous heating $(\nu_d = \nu_e + \nu_m)$. Eq. (10) follows directly from the continuity equation

$$\frac{\partial}{\partial z} \bar{w}_1 - \bar{w}_1 = 0$$

and the assumption that $\rho_0 \overline{w}_1^2$ is bounded as $z \to \infty$.

The perturbation is forced at the lower boundary. After the onset of forcing the perturbation equations, describe the upward propagation of a wave packet (group). The (steady-state) frequency of the wave is prescribed by prescribing the frequency of the forcing. For details see Walterscheid (1981b).

All perturbation and wave-induced quantities are set equal to zero at the initial time. For convenience, the initial temperature ($\overline{T}_0$) is set equal to a constant. This isothermal distribution facilitates the calculation of the parameter Z_{break} used to characterize the forcing of the wave (see below). Since the magnitude of the mean vertical temperature gradient in the mesosphere is only $\approx 2K$ km^{-1}, the assumption of constant $\overline{T}_0$ will not significantly alter the indications of our idealized model. In our simulations $\overline{T}_0 = 240$ K.

The initial wind (background-state) wind profile ($\overline{u}_0$) was given the following form

$$\overline{u}_0(z) = \begin{cases} U & (z < z_1) \\ Uf(z) & (z_1 \leqslant z \leqslant z_2) \\ Uf(z_2) & (z > z_2) \end{cases} \tag{13}$$

where

$$f(z) = [(\Delta z)^2 - (z-z_1)^2]/(\Delta z)^2 . \tag{14}$$

Parameter values were chosen to allow free propagation of slow waves into the mesosphere, and to resemble the wintertime winds used by Lindzen (1981) in the region of wave breakdown. Model parameters are $U = 75$ m s^{-1}, $z_1 = 8.5$ (60 km), $z_2 = 12.0$ (84 km) and $\Delta z = 3$ (21 km). The vertical increment for the integration is $\Delta z = 0.0417$ (0.30 km); the time increment which gives computational stability is $\Delta t = 72$ s. The top of the domain of integration is $z = 20$.

3.2 Numerical Experiments

Lindzen (1981) has argued that owing to the (presumed) meteorological source for the waves, and the phase-speed selectivity imposed by the mean zonal winds, the mean-flow forcing in the mesosphere ought to be dominated by waves with phase speeds near zero. We adopt this point of view and prescribe $c = 0$ for our calculations. For a horizontal wavelength we prescribe $L_x = 200$ km. This value is on the short-scale side of values inferred by Hess and Geller (1976) and Manson et al. (1979), and somewhat on the long-scale side of those measured by Vincent (1983). (However, Vincent could not obtain accurate estimates for wavelengths greater than about 200 km.) It should be remarked that all of these observations refer to propagating waves ($c \neq 0$), and the extrapolation to the case of $c = 0$ is somewhat conjectural. Lindzen (1981) deduced a value of $L_x \sim 1680$ km, while Holton concluded that if turbulence occurs $\sim 10\%$ of the time then a value of $L_x \sim 80$ km seemed appropriate. Since D_{eddy}, and thus F_x, depends linearly on k, the value of k we have chosen may be characteristic of rather strong forcing.

The forcing of the wave is characterized by the parameter Z_{break} (see Section 1). This is the lowest level in the steady-state picture of Lindzen (1981) at which the wave achieves $\varepsilon = 0$. For steady waves in a isothermal atmosphere with shear which is not too strong, the forcing required to specify Z_{break} can easily be calculated.

The actual initial breaking level in the time-dependent (i.e., transient) case will depend, primarily, on how sharply defined the leading edge of the upward propagating wave packet is. Since the forcing will not increase to a steady-state level instantaneously (if at all), and since the vertical phase velocity is frequency-dependent, we do not expect the boundary of the group to be sharp.

Consider an upward-moving wave group forced from below. It may be readily shown that convective breakdown will tend to occur when $|u'| \sim |c-\bar{u}|$ is achieved. To facilitate discussion we assume that breakdown occurs precisely when $|u'| = |c-\bar{u}|$. Let Z_g be the upward-moving level below which the wave has attained steady state and the energy density $\rho_0 u'^2$ is essentially constant with height. (The level $Z_g \approx c_g \tau$, where c_g is the vertical group velocity for $c = 0$ and τ is the time since the onset of forcing.) If the group has a sharp boundary then a small distance above Z_g the energy density will be essentially nil. Thus initial breaking will occur at Z_{break} because before the leading edge of the group reaches Z_{break} the breakdown criterion is not satisfied anywhere, but when the leading edge reaches some altitude just above Z_{break} the wave at Z_{break} reaches its steady-state amplitude, and, by the definition of Z_{break}, breakdown occurs. If, however, the group does not have a sharply defined boundary, the decrease in $\rho_0 u'^2$ above Z_g may be rather slow, and $|u'|$ may even increase for a while. This means that wave breaking may occur in regions well above Z_{break} before Z_g reaches Z_{break} or, equivalently, before breakdown occurs at Z_{break} (see Figure 3). It is clear that the combination of an indistinct group boundary and decreasing westerlies will tend to favor initial breakdown near a critical level, where $|u'| \sim |\bar{u}-c|$ is readily achieved. For $c = 0$ this means initial breakdown will tend to occur in the vicinity of the zero wind level, $\bar{u} = 0$. For the $\bar{u}_0$ profiles we have chosen, the zero wind level occurs near $z = 11.5$ (≈ 80 km). The production of turbulence near a critical level has been studied by Geller et al. (1975) for a Boussinesq fluid.

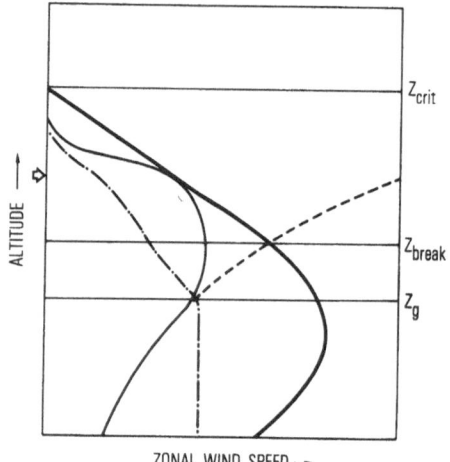

ZONAL WIND SPEED →

Figure. 3. Schematic representation of relative locations of
levels Z_g, Z_{break} and Z_{crit}, and the level of initial breakdown (denoted by the short open arrow). The level $Z_g = c_g \tau$ where c_g is the
vertical group velocity corresponding to horizontal phase velocity
$c = 0$, and τ is the time since the forcing of the wave began. Z_{break}
is the level at which initial breakdown would occur for a monochromatic wave with all energy concentrated at $c = 0$, and Z_{crit} is the
critical level where the mean wind $\bar{u} = c = 0$. The heavy solid curve
is $\bar{u}$, the light solid curve is the amplitude of the perturbation velocity u'. The dashed durve is the constant-energy-density continua-
tion of the $|u'|$ curve below Z_g, and the dot-dashed curve is propor-
tional to the perturbation kinetic energy density.

As $\bar{u} - c$ is reduced after the initial breakdown, the breaking of
waves at lower altitudes is promoted. Combined with the upward movement of Z_g (Figure 3) the tendency for $\bar{u} \to c$ may result in the very
rapid descent of the region of breakdown. In principle, the region of
breakdown may extend well below Z_{break}. In fact, the downward descent
may be limited ultimately by increasing density (decreasing u') rather
than increasing $\bar{u}_0$.

Three experiments were performed corresponding to $Z_{break} = 9$, 10
and 11 (63, 70 and 77 km, respectively). The simulations were run
until the waves became unstable with respect to the basic state (in a
Kelvin-Helmholtz sense) and the solutions began to blow up. The forcing for all of these experiments was brought up to a steady-state
value in a time

$$\tau = L_x / |\bar{u}_0(0) - c|,$$

that is, over the intrinsic period of the wave at $z = 0$. For $L_x = 200$

km, $c = 0$, and $\bar{u}_0(0) = 75$ m s^{-1}, we obtain $\tau \approx 45$ minutes. This interval is certainly smaller than typical time scales for the presumed meteorological sources for fairly large-scale gravity waves, and our group boundary ought to be no less distinct than those occurring in nature most of the time. The forcing was continued at a constant amplitude through the rest of the simulation. Since in all cases the simulation was terminated within a few hours after breakdown, we expect the results to apply to cases where the forcing is of fairly short duration. Some experiments where the duration of the forcing was ~ 3 hours gave results which did not differ significantly from those with constant forcing.

4. RESULTS

In this section we describe the results of the numerical experiments with Z_{break} = 9, 10, 11.

4.1 Z_{break} = 9

Figure 4 shows the evolution of the zonal wind $\bar{u} = \bar{u}_0 + \bar{u}_1$ for the one-hour period following wave breakdown. Shortly after this period the applicability of the D_{eddy} parameterization becomes very questionable (see below).

The short horizontal arrow on the vertical axis indicates the point where initial breakdown occurred. The vertical extent of the region of breakdown grew very rapidly in the first few minutes following initial breakdown. (The initial plotted time on Figure 4 is

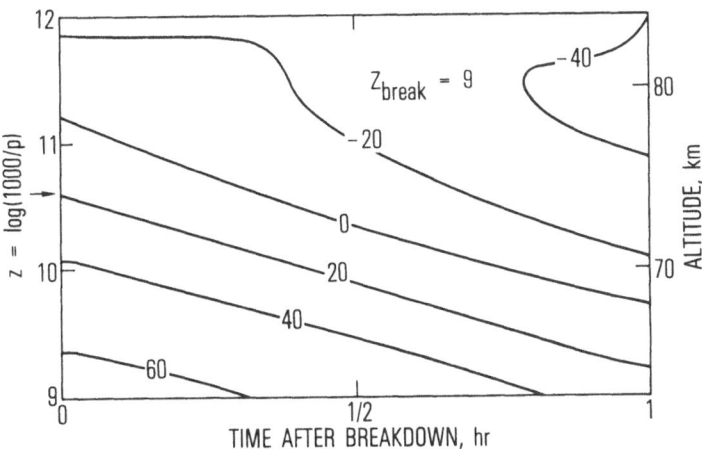

Figure 4. Time-height section of mean zonal wind $\bar{u}$ for Z_{break} = 9. The curves are labeled in ms^{-1}. Positive values are westerly (eastward).

actually the first time step saved after breakdown occurred, and is a few minutes later than the actual time of breakdown.)

The momentum flux convergences forced a rapid easterly accelera- tion of the flow throughout most of the region $\sim$ 60–80 km. The accel- erations were generally on the order of 40 m s^{-1} over the 1-hour peri- od. These accelerations are consistent with the accelerations obtained by Lindzen (1981) if allowance is made for the differences in the horizontal wavelengths chosen, and by Holton if breakdown oc- curs $\sim$ 10% of the time.

The occurrence of easterly acceleration in the easterlies is due to the fact that the upward propagating transient disturbance has easterly (c < 0), as well as westerly (c > 0), wave components. The westerly components encounter critical levels in the westerlies, and are filtered out, while the easterly components are allowed to propa- gate into the easterly winds before they are absorbed. When the easterly wave components are absorbed they provide an easterly accel- eration.

Figure 5 shows the diffusion coefficient D_{eddy} for this case. To convert D_{eddy} to m^2 s^{-1} we multiply by $\bar{H}^2 \approx 5 \times 10^7$ m^2, where $\bar{H} = R\bar{T}/g$ is the mean-state scale height. The diffusion coeffi- cient is the sum of a small constant part $D_0 = 10^{-7}$ s^{-1} (corresponding in geometrical height coordinates to a value of $\sim$ 5 m^2 s^{-1}) and a variable part obtained by use of (1). The small background part was added to avoid excessively strong gradients at the boundaries of the region of breakdown. Physically, it may be taken to be residual turbulence or the diffusive action of small amplitude incoherent grav- ity waves (Weinstock, 1976).

We note the downward progression of the axis of maximum values of D_{eddy} below the critical level (C.L.). This downward progression follows the downward progression of the critical level. We note in addition the increase in the size of the maximum value with time, increasing from a few times 10^{-6} to a few times 10^{-5}. This increase is due to the spreading of the region of breakdown into regions of stronger $\bar{u}$ and also to the destabilization of the region of breakdown due to the heat fluxes.

If the initial wave-breaking level had occurred first near Z_{break}, the values of D_{eddy} near Z_{break} would have been about an order of magnitude greater than those obtained near this level in the present simulation. This is because in our simulations significant deceleration of the westerlies occurred before the region of breakdown descended to Z_{break}. The values of D_{eddy} shown in Figure 5 are in rough agreement with estimates derived from models of eddy diffusion transport (Allen et al., 1980; COESA, 1976). The strong decrease in D_{eddy} over the steady-state picture may not extend to Lindzen's

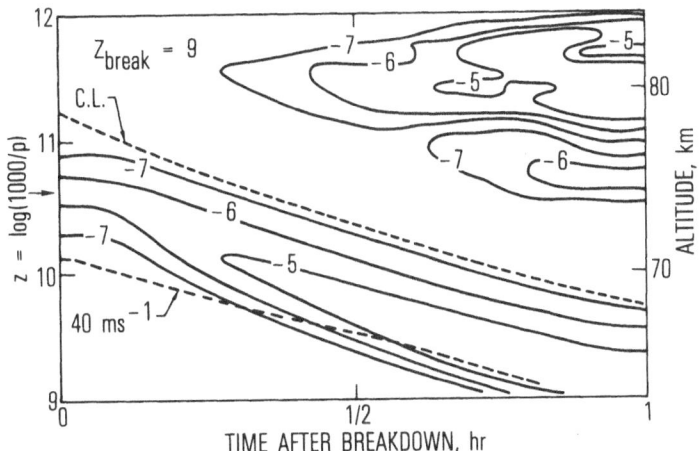

Figure 5. Time-height section of common logarithm of eddy diffusion coefficient D_{eddy} (solid curves) for $Z_{break} = 9$. D_{eddy} is in s^{-1}. The dashed curves show the position of selected $\bar{u} = $ const curves. The $\bar{u} = 0$ level is the critical level for $c = 0$ waves and is denoted C.L.

results, since he assumed a value of L_x an order of magnitude larger than the value we chose. Thus the acceleration of $\bar{u}$ would have been much slower and might be effectively opposed by the processes which we discussed earlier.

The secondary regions of large D_{eddy} above the zero wind level are due to incomplete absorption of wave energy at the zero wind level, owing primarily to the existence of waves with easterly phase speeds. These waves are transients excited by the onset of forcing, and by the rapid variations of D_{eddy} and the mean state. Above the zero wind level u' increases and $\bar{u} - c$ decreases until again $|u'| \sim |\bar{u} - c|$, where c is loosely regarded as an effective phase speed for the group. The values of D_{eddy} in this region should not be taken too seriously, since in calculating D_{eddy} we assumed $c = 0$.

The induced temperature change $\bar{T}_1$ is shown in Figure 6. After about 1/2 hour the wave flux of sensible heat produces definite regions of heating and cooling. A dipolar pair of these regions is clearly associated with the removal of heat near the top of the primary region of breakdown, and its subsequent deposition at the bottom of the region. The larger $|\bar{T}_1|$ for cooling than for warming is due to the greater densities in the lower part of the region. As mentioned, the wave heat fluxes drive the region toward neutral static stability. When the static stability is sufficiently reduced the mean state becomes unstable owing to the sheared mean flow. Sub-critical valves

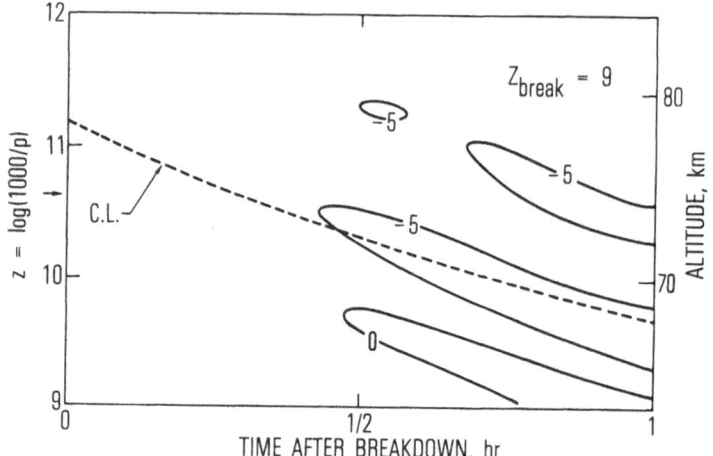

Figure 6. Time-height section of wave induced temperature change ($\bar{T}_1$) for Z_{break} = 9. The dashed curve shows the position of the u = 0 level, which is the critical level (C.L.) for c = 0 waves.

of the Richardson number occur about 90 minutes after breakdown. (In general, we present results only for times earlier than the occurrence of significant mean-state destabilization, since the D_{eddy} parameterization is of very doubtful validity when very large gradients in mean-state temperature have developed.) We remark that since we have neglected the diffusive flux of background-state sensible heat ($c_p\bar{T}_0$) by D_{eddy}, and since, as discussed earlier, this flux contributes to the destabilization of the mean state, we may have significantly underestimated the rate at which the region of breakdown is destabilized.

4.2 Z_{break} = 10

The evolution of $\bar{u}$, D_{eddy} and $\bar{T}_1$ following breakdown is shown for Z_{break} = 10 in Figures 7, 8 and 9, respectively. The basic features of the results are qualitatively the same as for Z_{break} = 9. However, there are significant quantitative differences: 1) the initial breakdown occurs ~ 2 km higher; 2) the easterly accelerations are a factor of ~ 2 slower; 3) the thickness of the primary region of turbulence, and its rate of descent are decreased by nearly a factor of 2; 4) the secondary region of turbulence is greatly reduced in size and intensity; and 5) an unstable mean state is established ~ 50% later owing to the somewhat slower development of large vertical gradients in $\bar{T}_1$. These differences are readily explained in terms of the reduced forcing of the waves, and the consequent reduction in wave amplitude and wave-forcing of the mean state.

Interestingly, the values of D_{eddy} in the primary region of breakdown are not greatly different from the previous case. This is because, on the one hand, the wave still breaks down in a region of fairly large $|\bar{u} - \underline{c}|$, and, on the other hand, the rate of wave forcing is less, so that $|\bar{u} - c|$, hence D_{eddy}, is reduced less rapidly. Since D_{eddy} for this case is comparable to D_{eddy} for the previous case, the diffusive fluxes are also comparable; hence $\bar{T}_1$ in the two cases is roughly comparable - although as we have noted $\bar{T}_1$ is somewhat smaller in this case than in the previous one.

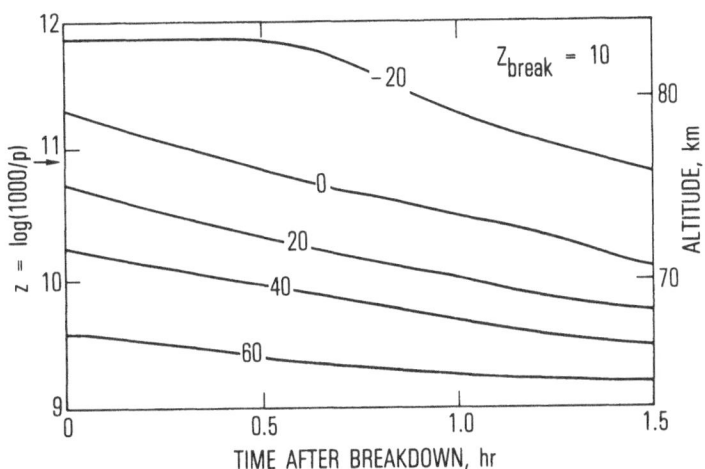

Figure 7. Same as Figure 4 except $Z_{break} = 10$.

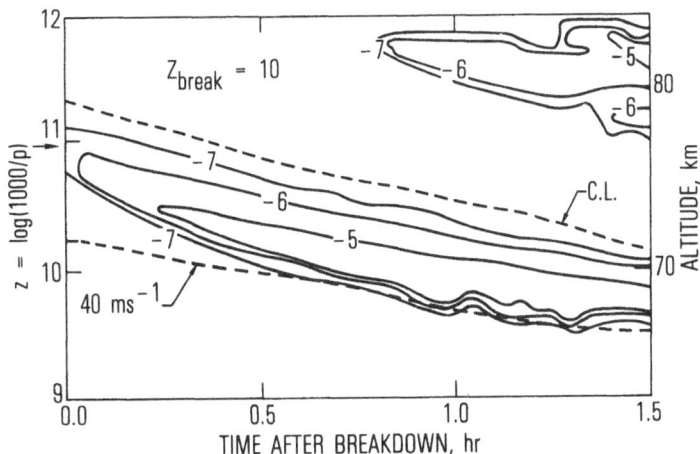

Figure 8. Same as Figure 5 except $Z_{break} = 10$.

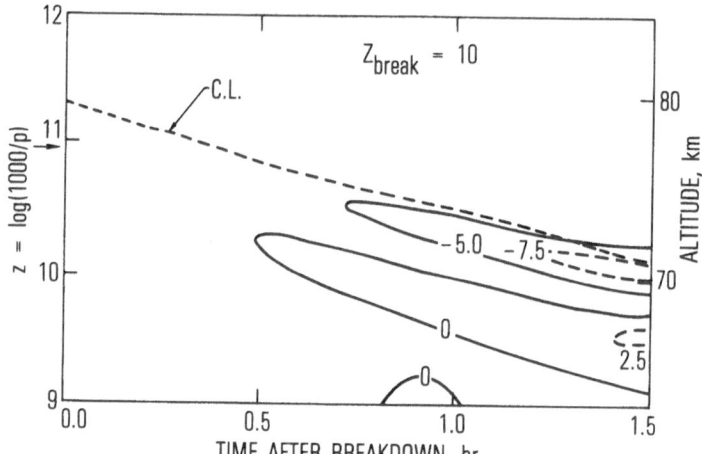

Figure 9. Same as Figure 6 except Z_{break} = 10.

4.3 Z_{break} = 11

The evolution of the mean state for $\bar{u}$, D_{eddy} and $\bar{T}_1$ is shown for Z_{break} = 11 in Figures 10, 11 and 12, respectively. In this case breakdown is limited to a region of small $|u - c|$ near the zero-wind line. Since $u - c$ is small, so too is D_{eddy} and the wave forcing of the mean wind. The region of turbulence is only a few kilometers thick, and exhibits only a slight tendency for downward progression.

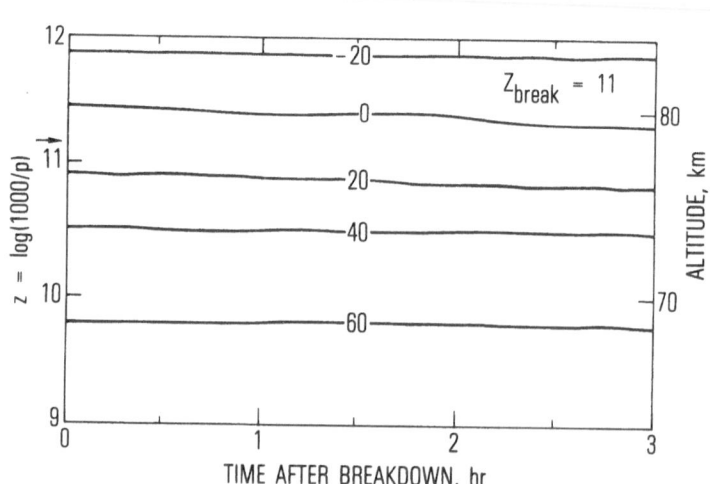

Figure 10. Same as Figure 4 except Z_{break} = 11.

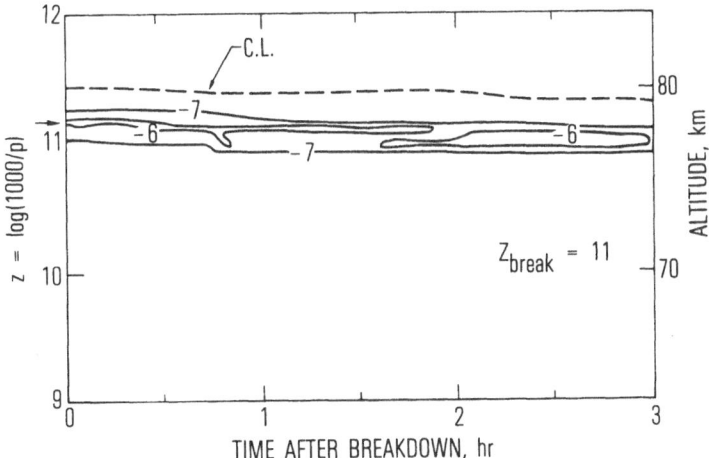

Figure 11. Same as Figure 5 except $Z_{break} = 11$.

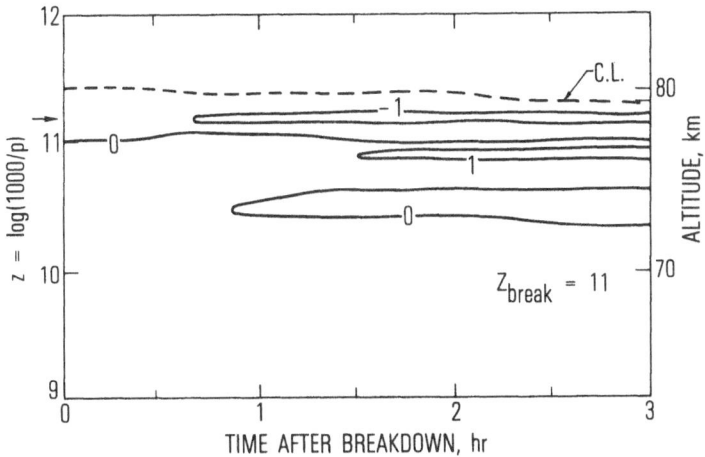

Figure 12. Same as Figure 6 except $Z_{break} = 11$.

The forcing of $\bar{T}_1$ is also weak, but there is still a notable tendency to destabilize the layer of turbulence. This is because 1) the region of turbulence is stable in height and thus the time integrated effects at a given altitude are greater for a given heat-flux convergence, and 2) the region is thinner so that a larger vertical gradient is associated with a given temperature difference between the heated and cooled region. At 3 hours after breakdown the temperature difference is ~ 7 K over a distance of ~ 2 km giving a vertical temperature gradient of ~ - 4 K km^{-1}. About two hours later the basic

38

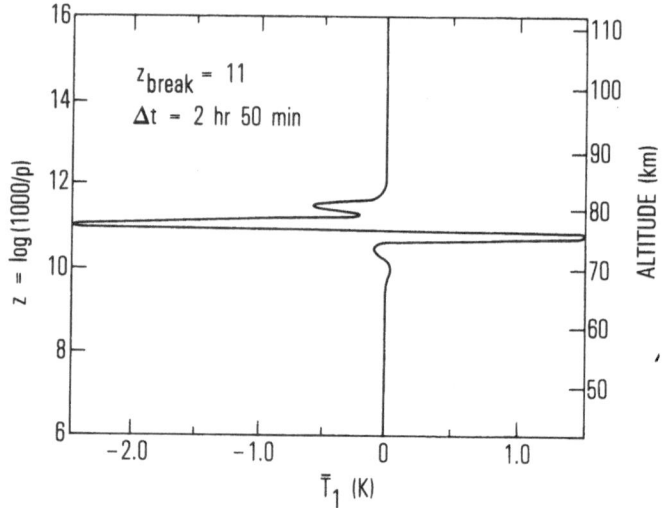

Figure 13.   Vertical   variation   of   wave-induced   temperature
change $\bar{T}_1$ at 2 hours and 50 minutes after initial breakdown.

state lapse rate becomes nearly neutral and the basic state becomes
dynamically unstable owing to the sheared mean flow. Figure 13 shows
the vertical profile of $\bar{T}_1$ near the end of the 3-hour interval shown
in Figures 10, 11 and 12.

4.4 Maintenance of Neutral Stability

Below Z_{crit} ($\bar{u} = 0$) the Lindzen parameterization was able to
limit wave growth and confine unstable superposed lapse rates ($\varepsilon < 0$)
to $\lesssim 25\%$ of the volume of the wave, until after the mean state itself
became significantly destabilized (roughly halfway in time between the
onset of breakdown and the onset of the dynamical instability of the
mean state). Values as large as $\sim 25\%$ seem physical even though $|\varepsilon|$
in this case is rather large. We do not expect growth of the wave to
be prevented until an appreciable fraction of the wave is involved in
turbulence. The fact that large negative values of ε indicate large
superadiabatic lapse rates in our model is an artifact of the requirement in a single-wave model that the disturbance remain sinusoidal in
x even after breakdown. More generally, the form of the total disturbance (primary plus secondary) should adjust to keep the superposed
lapse rate close to adiabatic in the region of breakdown, even though
the amplitude of the primary (incident) wave may indicate large negative values of ε. It should be mentioned that for the cases $Z_{break} =$
9 and 10 the region of $\varepsilon < 0$ extended well above Z_{crit} (a few scale
heights for $Z_{break} = 9$). Above Z_{crit} the Lindzen parameterization
gave $D_{eddy} < 0$; thus we set $D_{eddy} = 0$, and since the wave was effec-
tively undamped $|\varepsilon|$ became substantially larger above Z_{crit} than

below. The negative values of D_{eddy} were due to the combination of strong easterly shears and the use of $c = 0$ in the parameterization of D_{eddy}, whereas the disturbance above Z_{crit}, as we have mentioned, is apt to be dominated by waves with $c < 0$. It seems likely that a more complete formulation of D_{eddy} would have been able to prevent the region of $\varepsilon < 0$ from extending much above the $\bar{u} = 0$ level.

Although an improved parameterization is clearly needed, the modeling limitations and the uncertainties over what the stability criterion should be make the use of the Lindzen parameterization appear warranted at this time. In any case, the inability to maintain a neutral superposed state did not appear so egregious as to invalidate the general indications of our mechanistic wave-mean-flow model.

5. COMPARISON WITH OBSERVATIONS

Enhanced regions of radar echo power from mesospheric heights might be attributable to regions of turbulence. Observations with the Mesosphere-Stratosphere-Troposphere (MST) radar at Poker Flat, Alaska, show wintertime echoes with vertical scales of ~ 5-15 km. These structures exhibit only slight vertical propagation, and the details of the structure appear stable over periods of a few hours (Balsley et al., 1983). Observations with the SOUSY-VHF radar at Lindau, West Germany, show somewhat similar behavior, except that the structures appear to be thinner with vertical dimensions of ~ 1-3 km. Typical lifetimes are ~ 1-3 hours, and the structures are rather fixed in height with perhaps some slight tendency for downward progression (Czechowsky et al., 1979). Hocking (1979) has observed similar structures with the MF radar at Adelaide, Australia. At times he observed quasi-periodic oscillations with periods 10 to 120 minutes in the heights of turbulent layers.

The lack of a pronounced downward progression of turbulent layers seems to indicate a relatively weak forcing of the mean state. Forcing much weaker than that obtained for $Z_{break} = 10$ seems to be indicated. Weaker forcing might be obtained from smaller amplitude waves, larger wavelengths, or nonlinear scattering. When the last two circumstances limit wave-forcing of the mean state, the deeper regions of turbulence might be due to fairly large amplitude waves. Otherwise, the thin regions of turbulence might be due to fairly weak gravity waves which break near favored regions for breakdown (i.e., near a critical level).

In fact, the simulations for $Z_{break} = 11$ resemble the observations fairly well, especially those of Czechowsky et al. (1979). The duration of the echoes might be due to the slow vertical progression of the wave near a critical level, or it might be due to a constant flux of weak gravity waves into a region where the mean state has been destabilized by preceding waves. The latter process may be viewed as a form of preconditioning. Holton (1982) has suggested that present models of wave breaking could be improved by considering the effects of a spectrum of waves, extending the approach of Matusuno (1982), who considered a spectrum of waves, but prescribed D_{eddy}.

6. SUMMARY AND CONCLUSIONS

We have studied the evolution of the mean state following gravity wave breakdown using a time-dependent wave-mean-flow model. Following Hodges (1967) and Lindzen (1981) we have assumed that breakdown occurs when the superposed state is statically unstable. Furthermore, we have employed the parameterization of Lindzen (1981) to calculate the diffusion coefficient D_{eddy}. In the present simulations we did not include the diffusion of background-state (i.e., $\overline{\Psi}_o(z)$ state) heat and momentum (see discussion in Section 2). Our choices of mean-state initial conditions and wave parameters were appropriate for mid-latitude winter conditions, and were essentially the same as those chosen by Lindzen (1981) (including c = 0) except for horizontal wave-length. Our value for the horizontal wavelength was 200 km and is between that suggested by Lindzen (~ 1700 km) and by Holton (~ 80 km), and may be somewhat on the low side of wavelengths inferred from observations. Since wave-forcing increases as k, our simulations may refer to cases of fairly strong forcing.

Our main findings are as follows. (The parameter Z_{break} refers to the lowest level at which breaking would occur in a steady-state model.)

(1) Initial breaking occurs above Z_{break}. Because waves break more readily where the Doppler shifted phase velocity is small, there is a tendency for waves to break initially near their critical levels, which in our simulations is near where $\overline{u}_0 = 0$.

(2) For Z_{break} = 9 and 10 (63 and 70 km, respectively) the forcing of the mean state was quite rapid, resulting in a rapid easterly acceleration. The tendency to reduce $|\overline{u} - c|$ resulted in a large decrease in D_{eddy} over the values that would have been inferred from a steady-state model.

(3) The heat flux induced by breakdown destabilizes the mean state (in the sense of small Richardson number) and may lead eventual-ly to the dynamical instability of the incident wave. The destabili-zation of the mean state proceeds more rapidly for forcing by large amplitude waves. The instability process needs to be more fully ex-plored with models which include wave-wave interactions (Fritts, 1978, 1979).

(4) The simulation with rather weak forcing (Z_{break} = 11, i.e., 77 km) gives a region of turbulence which resembles apparent regions of turbulence observed by radar. The similarities include regions a few kilometers thick, lasting a few hours, and exhibiting no pro-nounced tendency for downward propagation.

ACKNOWLEDGMENTS

This work was supported by the Aerospace Sponsored Research Program. I wish to thank Prof. G. Schubert for a number of valuable discussions.

REFERENCES

Allen, M., Y. L. Yung, and J. Waters, 1981: Vertical transport and photochemistry in the terrestrial mesosphere and lower thermosphere (50-120 km), J. Geophys. Res., 86, 3617-3627.

Balsley, B. B., W. L. Eklund, and D. C. Fritts, 1983: VHF echoes from the high-latitude mesosphere and lower thermosphere: Observations and interpretations, submitted to J. Atmos. Sci.

COESA, 1976: U. S. Standard Atmosphere, 1976, Govt. Printing Office, Washington. 227 pp.

Czechowsky, P., Rüster and G. Schmidt, 1979: Variations of mesopheric structures in various seasons, Geophys. Res. Letts., 6, 459-462.

Delisi D. P and Orlanski, I., 1975: On the role of density jumps in the reflexion and breaking of internal gravity waves, J. Fluid Mech., 69, 445-464.

Fritts, D. C., 1978: The nonlinear gravity wave-critical level interaction, J. Atmos. Sci., 35, 397-413.

Fritts, D. C., 1979: The excitation of radiating waves and Kelvin-Helmholtz instabilities by the gravity wave-critical level interaction, J. Atmos. Sci., 36, 12-23.

Geller, M. A., H. Tanaka and D. C. Fritts, 1975: Production of turbulence in the vicinity of critical levels for internal gravity waves. J. Atmos. Sci., 32, 2125-2135.

Gossard, E. E., 1962: Vertical flux of energy into the lower ionosphere from internal gravity waves generated in the troposphere, J. Geophys. Res., 67, 745-757.

Hess, G. C., and M. A. Geller, 1976: The Urbana meteor-radar system: Design, development and first observations, Aeronomy Rep. 74, Univ. of Ill., Urbana.

Hines, C. O., 1960: Internal atmospheric gravity waves at ionospheric heights, Can. J. Phys., 38, 1441-1481.

Hines, C. O., 1970: Eddy diffusion coefficients due to instabilities in internal gravity waves, J. Geophys. Res., 75, 3937.

Hines, C. O., and C. A. Reddy, 1967: On the propagation of atmospheric gravity waves through regions of wind shear, J. Geophys. Res., 72, 1015-1034.

Hocking, W. K., 1979: Angular and temporal characteristics of partial reflections from the D-region of the ionosphere, J. Geophys. Res., 84, 845-851.

Hodges, R. R., Jr., 1967: Generation of turbulence in the upper atmosphere by internal gravity waves, J. Geophys. Res., 72, 3455.

Hodges, R. R., Jr., 1967: Eddy diffusion coefficient due to instabilities in internal gravity waves, J. Geophys. Res., 74, 4087-4090.

Holton, J. R., 1982: The role of gravity wave induced drag and diffusion in the momentum budget of the mesosphere, J. Atmos. Sci., 39, 791-799.

Jones, W. L., and D. D. Houghton, 1972: The self-destructing internal gravity waves, J. Atmos. Sci., 29, 844-849.

Lindzen, R. S., 1967: Thermally driven diurnal tide in the atmosphere, Quat. J. Roy. Meteor. Soc., 93, 1842.

Lindzen, R. S., 1968a: The application of classical atmospheric tidal theory, Proc. R. Soc. Lond. Ser. A., 303, 229-316.

Lindzen, R. S., 1968b: Lower atmospheric energy sources for the upper atmosphere, Meteor. Mon., 9, 37-46.

Lindzen, R. S., 1971: Tides and gravity waves in the upper atmosphere, in Mesospheric Models and Related Experiments, edited by G. Fiocco, D. Reidel, Hingham, Mass.

Lindzen, R. S., 1973: Wave-mean flow interaction in the upper atmosphere, Boundary Layer Meteorol., 4, 327-343.

Lindzen, R. S., 1981: Turbulence and stress owing to gravity wave and tidal breakdown, J. Geophys. Res., 86, 9707-9714.

Manson, A. H., C. E. Meek, and R. J. Stening, 1979: The role of atmospheric waves (1.5 h - 10 days) in the dynamics of the mesosphere and lower thermosphere at Saskatoon (51°N, 10°W) during four seasons of 1976, J. Atmos. Terr. Phys.. 41, 325-335.

Matsuno, T., 1981: A quasi-one-dimensional model of the middle atmosphere circulation interacting with internal gravity waves, J. Meteor. Soc. Japan, 60, 215-226.

Orlanski, I., On the breaking of standing internal gravity waves, J. Fluid Mech., 54, 577-98.

Pitteway M. L. V., and C. O. Hines, 1963: The viscous damping of atmospheric gravity waves, Can. J. Phys., 41, 1935-1948.

Vincent, R. A. and I. M. Reid, 1983: HF Doppler measurements of mesospheric gravity wave momentum fluxes, J. Atmos. Sci., in press.

Walterscheid, R. L., 1981b: Dynamical cooling induced by dissipating internal gravity waves, Geophys. Res. Lett., 8, 1235-1238.

Walterscheid, R. L., 1981b: Inertio-gravity wave induced accelerations of mean flow having an imposed periodic component: Implications for tidal observations in the meteor region, J. Geophsy. Res., 86, 9698-9706.

Weinstock, J., 1976: Nonlinear theory of acoustic-gravity waves 1. Saturation and enhanced diffusion, J. Geophys. Res., 81, 633-652.

J. R. Holton and T. Matsuno, Dynamics of the Middle Atmosphere, 45-64.

NONZONAL GRAVITY WAVE BREAKING IN THE WINTER MESOSPHERE*

Mark R. Schoeberl

Laboratory for Planetary Atmospheres
Goddard Space Flight Center,
NASA/Greenbelt, Maryland 20771

Darrell F. Strobel

Naval Research Laboratory
Washington, D.C. 20375

ABSTRACT

The steady state gravity wave model of Schoeberl et al. (1983) is extended to compute wave breaking by disturbances originating at the earth's surface. For winter and summer mean zonal wind profiles, no waves reach the mesosphere unless $|\bar{u}-c| \geq 20$ m s^{-1}. Gravity waves with $c = 0$ can only reach the winter mesosphere if planetary scale waves are present in the troposphere and the lower stratosphere to provide strong zonal wind channels for upward wave propagation. This results in non-zonal wave breaking in the mesosphere which could provide in situ forcing of planetary waves. Dissipation of gravity waves by molecular viscosity and conduction can provide significant deceleration and heating/cooling in the 85-105 km region.

1. INTRODUCTION

    Lindzen (1981) and Matsuno (1982) have suggested that breaking gravity waves in the mesosphere could be the physical mechanism for deceleration of the zonal wind above the stratopause which is required in even the simplest zonal mean models of the middle atmosphere (Leovy, 1964). Since Lindzen's publication, a great deal of additional theoretical research has been done. For example, Holton (1982), using essentially Lindzen's parameterization, has investigated the range of gravity waves phase speeds required to decelerate the zonal wind in a β-channel model. Dunkerton (1982) looked at the

---

*Contribution Number 5 of the Stratospheric General Circulation with Chemistry Modeling Project at NASA/GSFC.

stochastic process of simplified gravity wave breaking at critical lines and verified that the steady state approximation used by Lindzen (1981) and Holton (1982) had statistical justification. Dunkerton also noted that because of the small horizontal scale of the gravity waves, the location of a critical line could vary in the zonal direction. Thus, the zonal mean critical line may not approximate the locus of wave breaking. This idea is pursued further in this paper where we discuss the zonal variation of the transmissivity of gravity waves through the stratosphere.

Recently Schoeberl et al. (1983) (here in after, S1) showed that breaking gravity waves and the turbulence that it generates could produce significant departures from radiative equilibrium in the mesosphere. In this paper we report additional investigations using the model of S1. The model has been expanded to incorporate the effects of molecular viscosity and conduction, variable lapse rate and topographic forcing of gravity waves. As mentioned above, we also consider local changes in the zonal wind which may produce selective transmission of gravity wave fluxes into the mesosphere.

2. MODEL

The gravity wave breaking model of S1 assumes, as in Lindzen (1981), a steady state solution to the dynamic equations with turbulent diffusion of heat and momentum. The diffusion coefficient is operationally determined to prevent local (mean plus gravity wave amplitude) lapse rates from exceeding the dry adiabatic lapse rate. The assumption is that the turbulence generated by the convective instability which occurs when the wave breaks acts to redistribute the potential temperature and dissipate the wave.

The dissipation of the wave produces a momentum flux convergence so that the flow tends to be accelerated or decelerated to the phase speed of the wave. Thermal transport occurs through two processes: first, downward heat transport by the gravity wave as it is dissipated by diffusion and radiation. Second, turbulent diffusion acts directly on the stable zonal mean state to produce a downward heat flux. This occurs as turbulent diffusion removes gradients in the mean potential temperature. Finally, there is some additional heating as the wave energy is deposited in the breaking zone.

Recently, Chao and Schoeberl (1983) have shown that convective wave breaking is inconsistent with the eddy diffusion formulation used by Lindzen (1981). They suggest that for breaking gravity waves a turbulent Prandtl number of infinity be used. This eliminates the downward eddy transport of heat; but since the eddy diffusion in the momentum equation must now be twice as large to halt the wave growth with altitude, the effect of turbulent transport of heat on the mean flow will still be important.

Both eddy transport of heat and the turbulent transport of the zonal mean potential temperature produce similar effects (Schoeberl et al., 1983, Eqs. 24,25): the upper regions are cooled as heat is transported downward and the lower regions are heated. Since the basic state flow is fixed in this study we shall use a turbulent Prandtl number of 0.71 to illustrate the effect of zonal mean thermal transport by turbulence even though it appears in the system of equations below as eddy heat transport.

a. Model equations

The relevant equations for linear gravity wave propagation from S1 and used in this study are the momentum equations:

$$im (\bar{u} - c) u' + w' \frac{\partial \bar{u}}{\partial z} + im\phi' = \frac{1}{\rho} \frac{\partial}{\partial z} \rho D \frac{\partial u'}{\partial z} \tag{1}$$

$$im (\bar{u} - c) v' + i\ell\phi' = \frac{1}{\rho} \frac{\partial}{\partial z} \rho D \frac{\partial v'}{\partial z} \tag{2}$$

the continuity equation:

$$imu' + i\ell v' + \frac{1}{\rho} \frac{\partial}{\partial z} \rho w' = 0 \tag{3}$$

the thermodynamic equation:

$$im (\bar{u} - c) \phi'_z + w' N^2 = - \alpha\phi'_z + \frac{P^{-1}}{\rho} \frac{\partial}{\partial z} \rho D \frac{\partial}{\partial z} \phi'_z \tag{4}$$

and the hydrostatic equation

$$\frac{RT'}{H} = \phi'_z \tag{5}$$

where

$$N^2 = R/H (\frac{\partial \bar{T}}{\partial z} + \kappa\bar{T}/H), \quad \kappa = R/c_p$$

u', v', w' are the perturbed zonal, meridional, and upward (x, y, z) velocities; and ϕ is the gravity wave geopotential. ($\bar{}$) designates fields which are zonal averaged. $\bar{v}$ and $\bar{w}$ are assumed to be small (see S1) and have been neglected. T is temperature and N is the bouyancy frequency. D is the diffusion coefficient; m is the zonal wave number of the disturbance which is taken to have the form $e^{im (x-ct)}$; ℓ is the meridional wave number. P is the Prandtl number which is 0.71 for molecular processes, close to 0.74 adopted in S1. The value 0.71 is used here and thus incorporates both molecular and turbulent effects. The rest of the notation is the same as Holton (1975). For the molecular calculations we take m=ℓ, but for the gravity wave transmission studies we take ℓ=0 for reasons discussed later.

The coefficient of molecular thermal conductivity , $\lambda*$, is given by

$$\lambda* = 48.72 \ T^{0.69}$$

and the coefficient of molecular viscosity, μ , is related to the conductivity by

$$\lambda^* = 2c_v\mu$$

where c_v is the specific heat at constant volume (Banks and Kockarts, 1973). Molecular diffusion can be incorporated into the diffusion coefficient by adding $2.78[T]^{0.69}e^{z/H}$ to D. The brackets [] represents global averaging and indicates that molecular diffusion is a linear process in the model.

Equations (1-5) may be combined to form a single equation in ϕ' and its vertical derivatives. We impose a radiation condition at the upper boundary and specify a temperature along the lower boundary. The turbulent diffusion coefficent is adjusted until the maximum lapse rate is the dry adiabatic lapse rate.

b. Surface forcing

We consider the simplest possible surface forcing, orography. If the height of the Fourier analyzed topography is h_o then the forcing of gravity waves with c = 0 at the surface by a steady wind is

$$w_o' \cong im\bar{u}\,h_o \tag{6a}$$

Waves with c $\neq$ 0 can be produced by variable mean winds interacting with orography and/or variations in the mean wind interacting with a gravity wave already present. We have estimated both processes and find that the latter is negligible in comparison with the former.

If an ensemble of measurements of the zonal mean wind were made near the earth's surface we would be able to compute a spectrum of fluctuations such as Vinnichenko (1970). The forcing for a gravity wave with phase speed c would be $w_{+c} = im\,\bar{u}_c h$ where u_c is amplitude of the zonal wind with frequency $\pm$ mc. To our knowledge, detailed measurements of $\bar{u}_c$ have not been made. Therefore, for comparison purposes, we take $\bar{u} = \bar{u}_c$ and h_o = 300 m which is an order of magnitude larger than that obtained by Fourier decomposition of the topography but is probably more typical of gravity waves excited by local topographic variations.

With Eq. (6a) the upward momentum flux at the surface is approximately given by

$$(\overline{u'w'})_{surf} = \frac{N|h_o|^2 m\bar{u}}{2} \tag{6b}$$

or

$$(\overline{u'w'}) \approx 5 \times 10^{-2}\ m^2\ s^{-2}$$

for $\bar{u} = 10$ m s^{-1}, $\lambda_x = 1000$ km, $h_o = 300$ m, and $N = 2 \times 10^{-2}$ s^{-1}. The momentum flux convergence above the breaking zone is approximately given by Lindzen (1981) as

$$M = H^{-1} (\overline{u'v'})$$

If the breaking zone is at z_b we may use the Eliassen-Palm relation $(\rho \, \overline{u'v'})_z = 0$ to obtain

$$M = H^{-1} \exp(z_b/H) \; (\overline{u'w'})_{surf} \tag{6c}$$

If $z_b = 70$ km, $M \approx 1360$ m s^{-1}day^{-1}, in the absence of wave dissipation below the breaking zone.

c. Local effects

Observations (see S1) indicate that the predominate zonal wavelength for breaking gravity waves is $<$ 1000 km. At 60°N this is wave number $>$ 20; thus it is likely that vertically propagating gravity waves could "see" local variations in the flow due to planetary scale waves. Wave breaking would thus occur in a nonzonal distribution since the vertical propagation characteristics of the wave would vary with longitude.

If we generalize the zonal average in Eqs. (1-5) to include planetary scale winds, $\bar{u}$ now represents planetary wave and zonal mean fields combined. For the sake of simplicity we consider waves with zero meridional wave number. We may thus neglect meridional advection of latitudinal gradients of the gravity wave wind field ($\bar{v} \frac{\partial}{\partial y} u'$, $\bar{v} \frac{\partial}{\partial y} v'$) which could be important since $\bar{v}$ (which now includes planetary wave fields) may be as large as $\bar{u}$. Meridional variation of $\bar{v}$ will also be neglected. For the calculations of gravity wave dissipation by molecular viscosity and conduction $m = \ell$.

d. Model parameters

The temperature profile from the US Standard Atmosphere (1962) is used as the basic state. The lapse rate for this profile is shown in Figure 1a. The zonal wind profile for winter and summer calculations are shown in Figure 1b; the zonal wind at 1 km is used to force the gravity waves in (6). We have also performed computations using CIRA (1972) basic temperature state at 60°N, the results are essentially the same as those obtained from the standard atmosphere.

The radiative damping, α, used is the wavelength dependent scheme described in the Appendix B of S1 which was computed with the radiation model of Apruzese et al. (1982). Below 30 km the Newtonian cooling coefficient is given by $a = \alpha(z) + b(z)/\lambda^2$ (days^{-1}) where $a(z) = 2.2 \times 10^{-2} + 1.18 \times 10^{-3}(z(km) -16.5)^2$, $b(z) = 1.21 + 2.08 \times 10^{-2}(z(km) -16.5)^2$ and λ is the vertical wavelength of the

LAPSE RATE

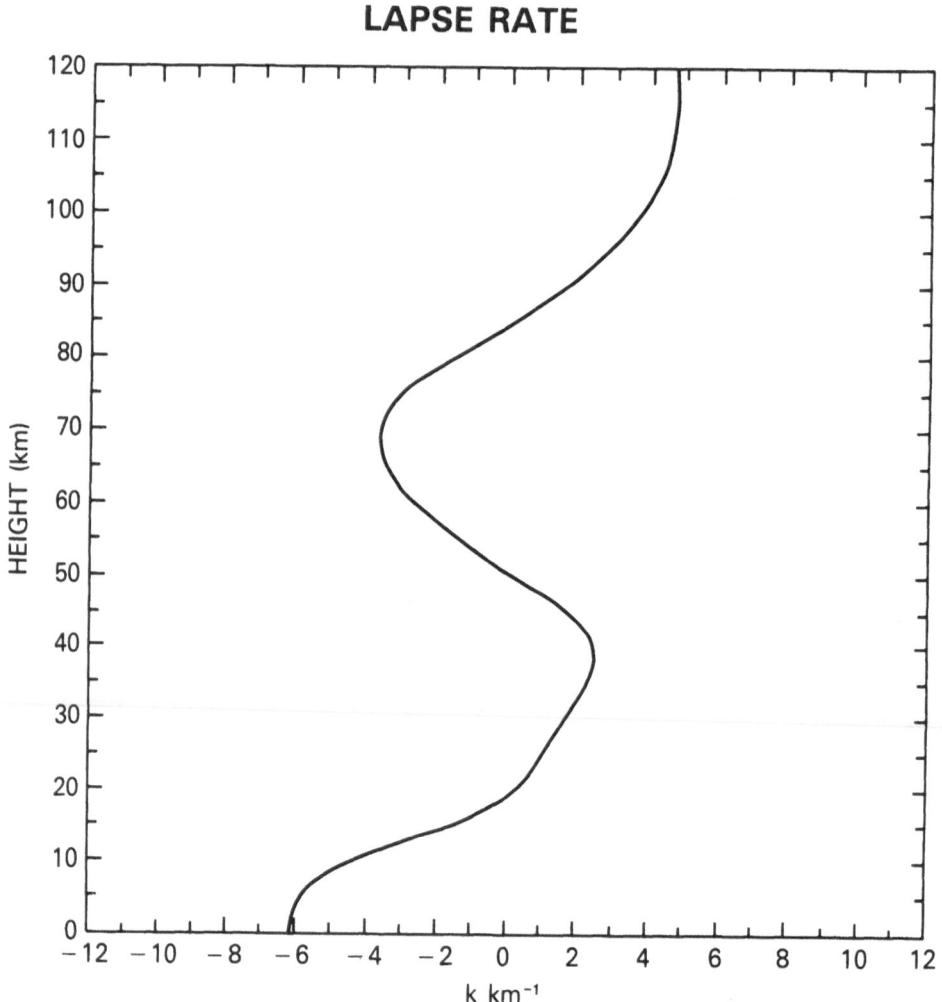

Figure 1a Lapse rate as a function of altitude for the U.S. Standard Atmosphere (1962).

ZONAL MEAN WIND

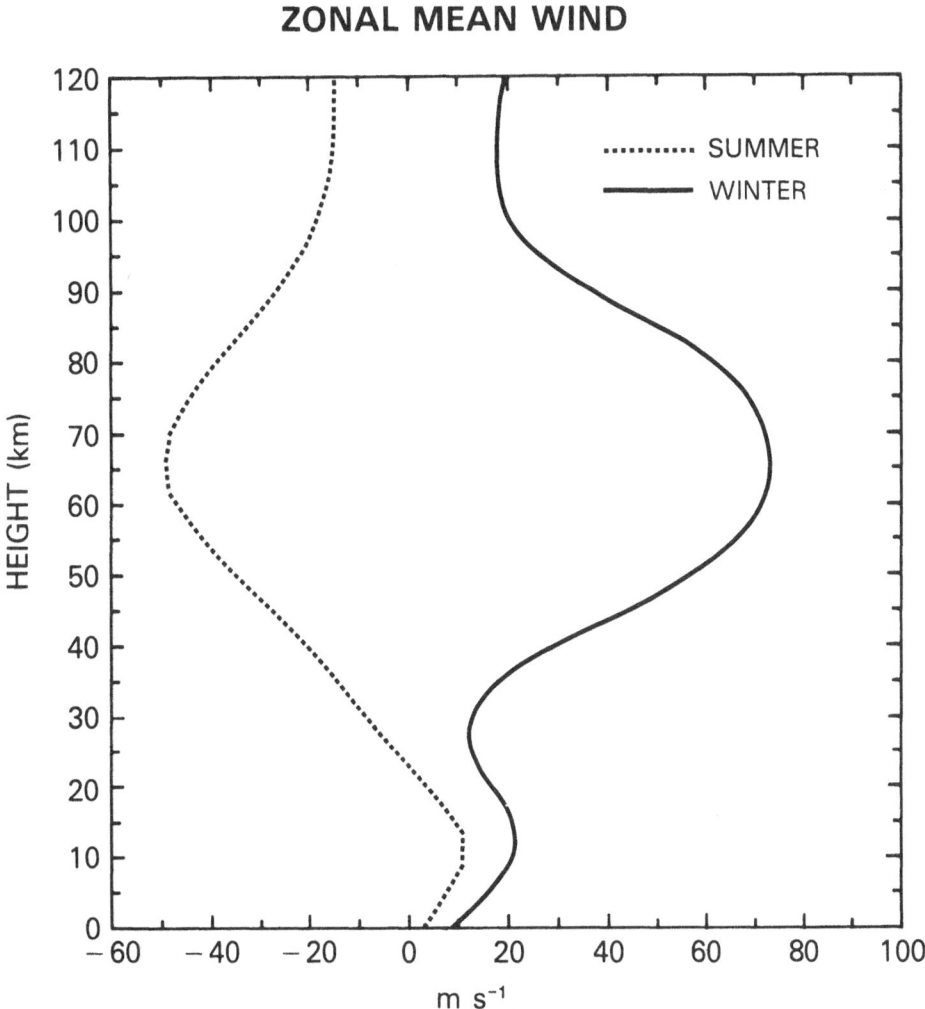

<u>Figure 1b</u>   Winter and summer zonal mean wind profiles used in this study.

thermal disturbance in kilometers. In addition to the radiative damping, a background diffusion of 10^2 cm^2s^{-1} is present at all levels.

In order to simulate the effect of gravity wave transmission through wind fields distorted by planetary waves we use an ad hoc model of the planetary wave field. The planetary wave zonal wind is given by

$$u' = u(z) = \left[3 + 25(1-e^{-z/30})\right] \cos (\alpha'z+s)m \ s^{-1} \quad z < 50 \text{ km}$$

$$u' = u(50)e^{-(z-50)}m \ s^{-1} \qquad\qquad z > 50 \text{ km}$$

$$\alpha' = 2\pi/(10\sqrt{u} + 20) \tag{7}$$

where s is longitude and z is,in km. $\bar{u}$ is the zonal mean wind in $m \ s^{-1}$. The expression for α is based on the dispersion relation for planetary waves. Equation (7) allows the disturbance to grow with height up to the stratopause. The wave is then damped out so that we may selectively test for transmission of gravity waves through the lower and middle atmosphere. Figure 1c shows the total wind, zonal mean plus planetary wave. The variation in the zonal wind produced by (7) is well within the range of observations for winter conditions (M. Wu, private communication, 1983).

3. RESULTS

a. Molecular effects

Only the thermal aspects of the interaction of a vertically propagating gravity wave with the increasingly viscous zone above the mesosphere were described qualitatively by Walterscheid (1981). To compute the maximum cooling and deceleration of the mean flow we have used a gravity wave amplitude at the threshold for breaking near the base of the molecular viscosity and heat conduction region. This gives the upper limit for gravity wave effects on the mean flow due to molecular diffusion. The magnitude of the surface forcing is irrelevant.

Figures 2a and 2b shows the results of the dissipation of a gravity wave with 1000 km zonal and meridional wavelengths. The deceleration is substantial ~ 45 $m \ s^{-1}$ day^{-1} at 97 km (Fig. 2a); a peak cooling rate of ~ 3.5K day^{-1} occurs at 98 km (Fig. 2b) in qualitative agreement with Waltercheid (1981). Note that heating occurs below the cooling zone due to the convergence of the downward heat flux with a peak value of about 1K day^{-1}. Both eddy flux convergence of heat and conversion of wave energy to heat by viscous dissipation are shown in the figure. For other horizontal wavelengths, the cooling rates, deceleration, and heating rates scale as the horizontal wave number (Schoeberl et al., 1983).

ZONAL WIND

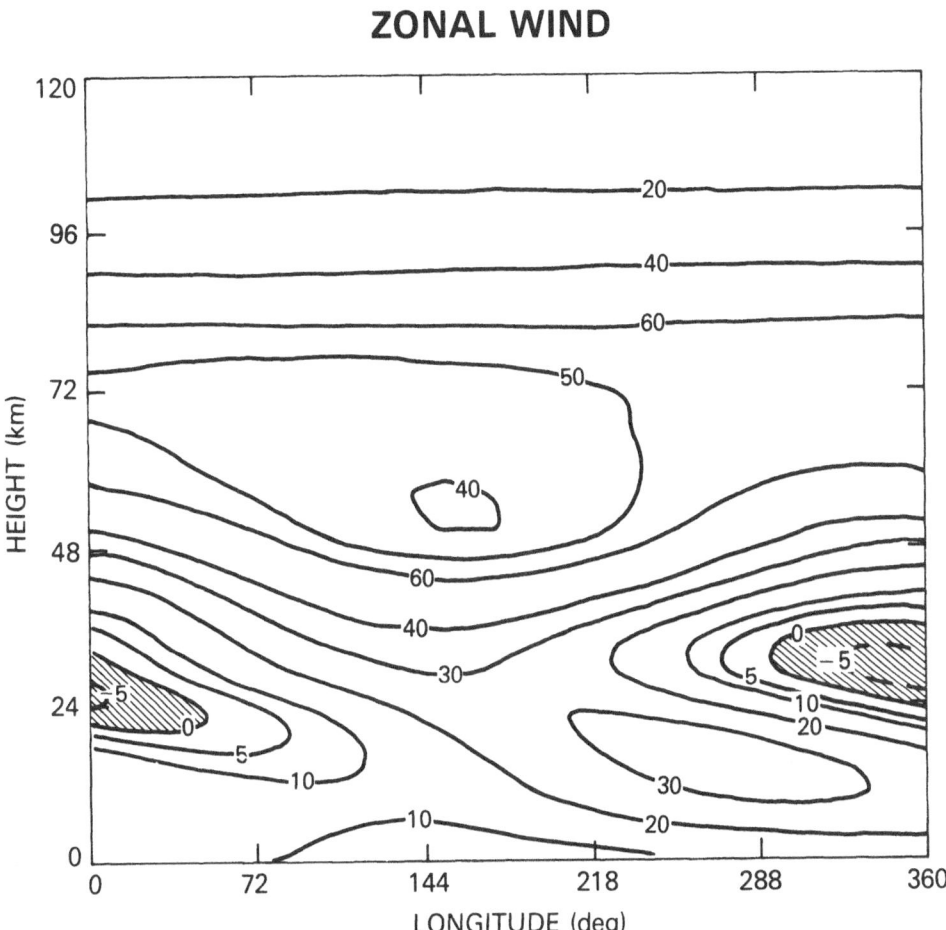

<u>Figure 1c</u>   Zonal wind model for winter including planetary wave wind
field as a function of longitude and height.

ZONAL DECELERATION

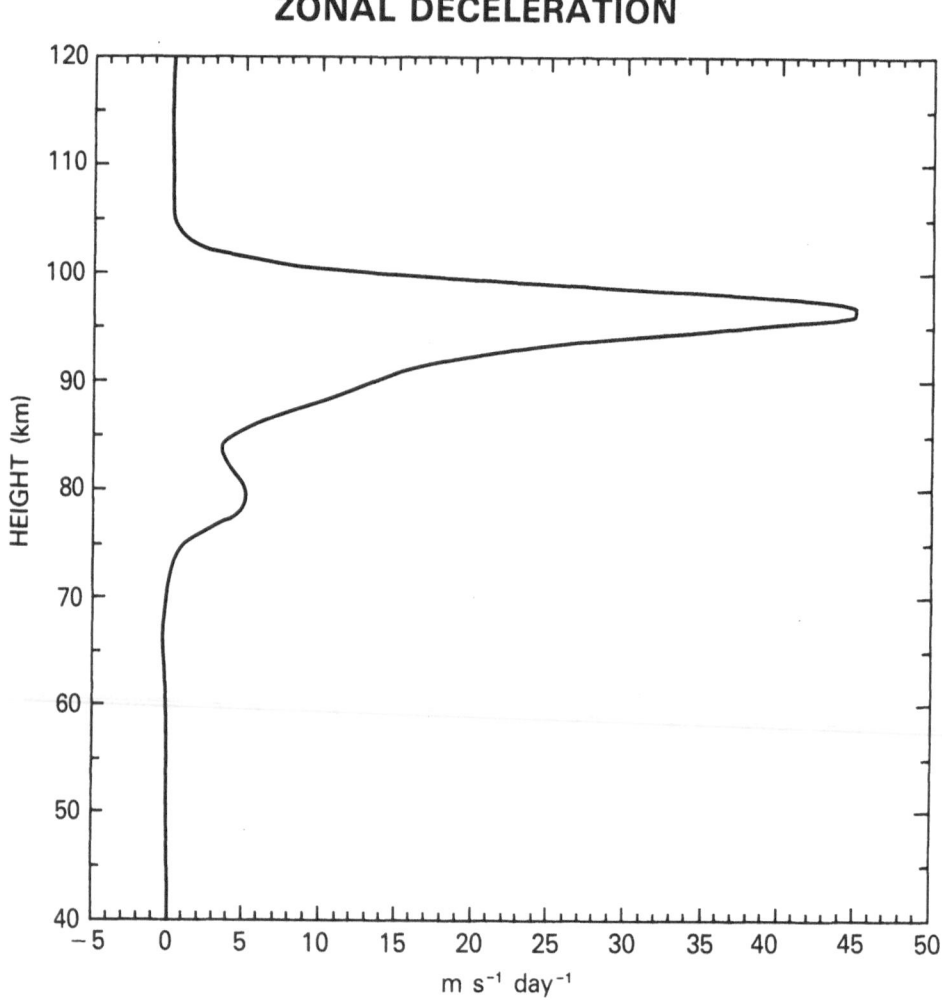

Figure 2a   The zonal deceleration of a gravity wave with 1000 km zonal and meridional wavelengths with $c = 0$ due to molecular viscosity and conduction in m s^{-1} day^{-1}.

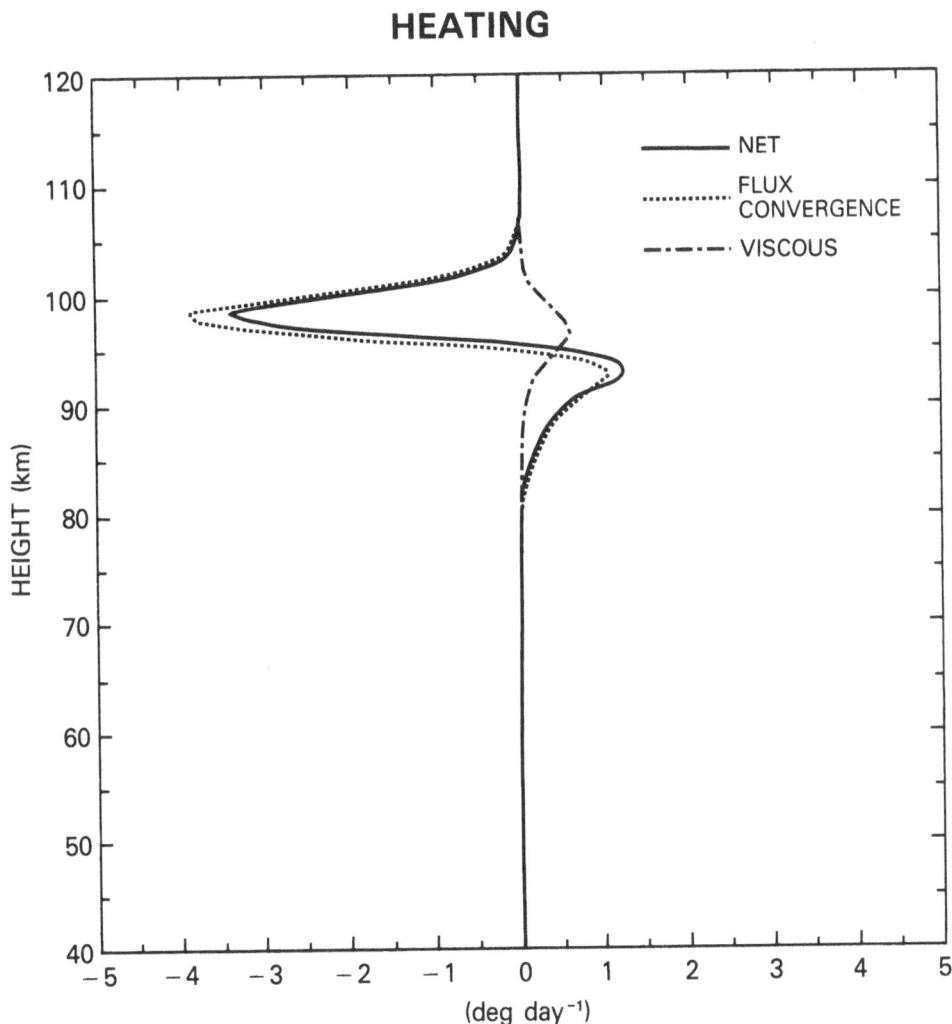

Figure 2b Gravity wave heating due to molecular viscosity and conduction. See text for discussion.

b. Nonzonal wavebreaking

Lindzen (1981) noted that gravity waves could only propagate vertically to the mesosphere if they did not encounter a critical line where the wave energy would be absorbed. Thus he suggested that breaking gravity waves in the summer mesosphere must have phase speeds greater than the maximum $\bar{u}$ in the troposphere since a critical line would exist in the stratosphere for slower disturbances. S1 showed that the limit suggested by Lindzen (1981) is more stringent as a consequence of vertical wavelength dependent Newtonian cooling. They found that $(\bar{u} - c) > 22$ m s^{-1} for 1000 km waves (more precisely the condition is $|\bar{u} - c| > 4\lambda_x^{1/4}$ with the horizontal wavelength λ_x in km and $m = \ell$) or the wave would be radiatively damped before it could attain the breaking amplitude.

Vertically propagating gravity waves can only reach the mesosphere without severe attenuation in duct regions where $\bar{u} - c$ is large. For the zonal mean winter and summer profile (Fig. 1b) no waves with $\lambda_x = 1000$ km and for which $-10 < c < 30$ m s^{-1} reach the mesosphere. Using the wind profile shown in Fig. 1c, wave energy reaching the mesosphere is restricted to specific longitudinal regions. Fig. 3a shows the deceleration associated with selective transmission of $c = 0$ waves as a function of longitude and height. The values shown in the figure are smaller than the approximate result (6c) since dissipation is included. Note that wave breaking occurs only above a transmission channel at 220^0 longitude where $|\bar{u} - c|$ is not less than 20 m s^{-1}. Figure 3b and 3c show the computed turbulent diffusion coefficient and the net heating, respectively. As expected, these quantities are very nonzonal as well.

The large diffusion shown in the figure results from the numerical model's attempt to compute simultaneously both the height of the diffusion zone and the magnitude of the diffusion coefficient. The iterative process used in the model converges slowly and is quite costly since a complete recomputation of the wave structure is involved (a $[120]^2$ matrix inversion). As a result, only a few iterations were used to estimate the magnitude of diffusion and heating which are larger than those computed by Schoeberl et al. (1983). The computations shown here should therefore be regarded as illustrative.

Figure 4 shows the deceleration obtained when the Newtonian cooling is set to zero. Only waves which encounter critical lines ($\bar{u} = 0$) in the stratosphere (see Figure 1c) are excluded from the mesosphere. Furthermore the deceleration is larger since the waves are not radiatively damped below the breaking zone, and the values shown are closer to those given by (6c). Wave breaking is confined mostly to the mesosphere because of the negative lapse rate.

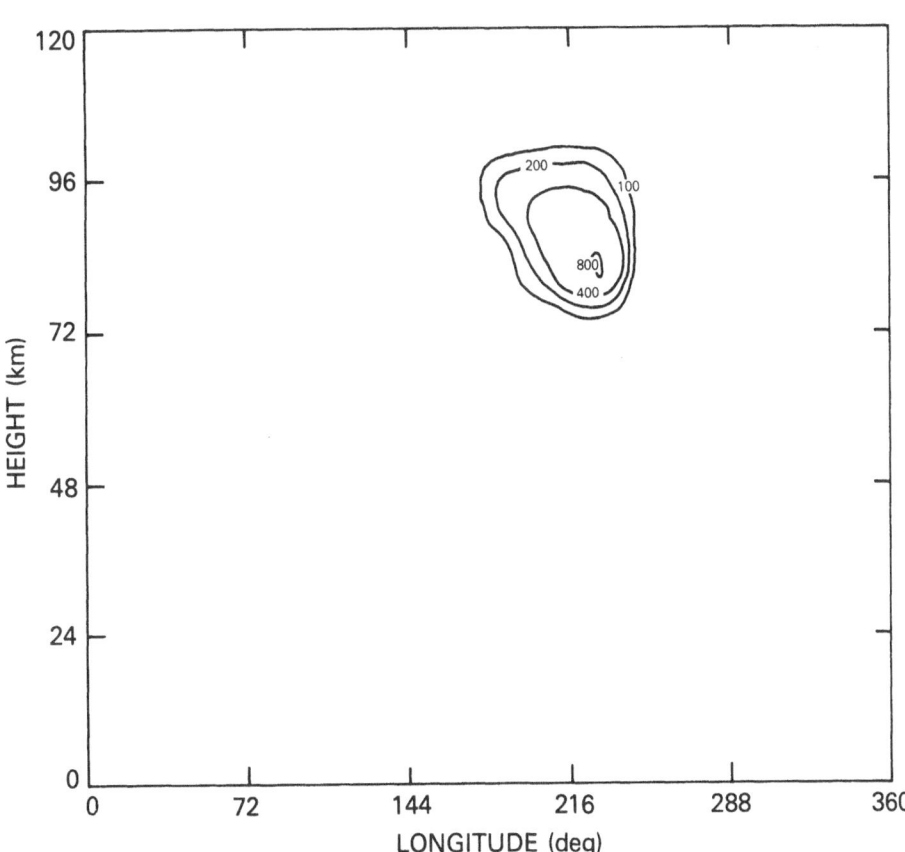

ZONAL DECELERATION (m sec⁻¹ day⁻¹)

Figure 3a Nonzonal wave breaking for $\ell=0$ gravity wave with orographically calculated amplitude. The zonal deceleration in units of (m s⁻¹ day⁻¹) as a function of longitude and altitude.

58

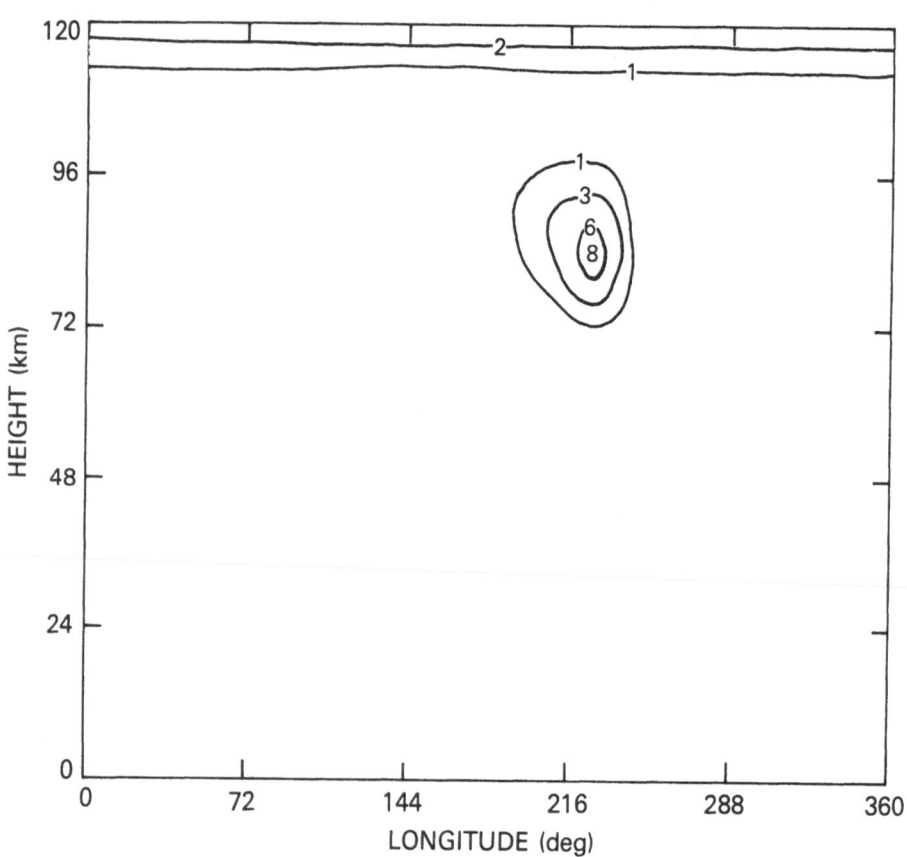

DIFFUSION COEFFICENT (x10⁸ cm² sec⁻¹)

Figure 3b   Same as (a) but for the turbulent diffusion coefficient (cm² s⁻¹).

HEATING (deg day⁻¹)

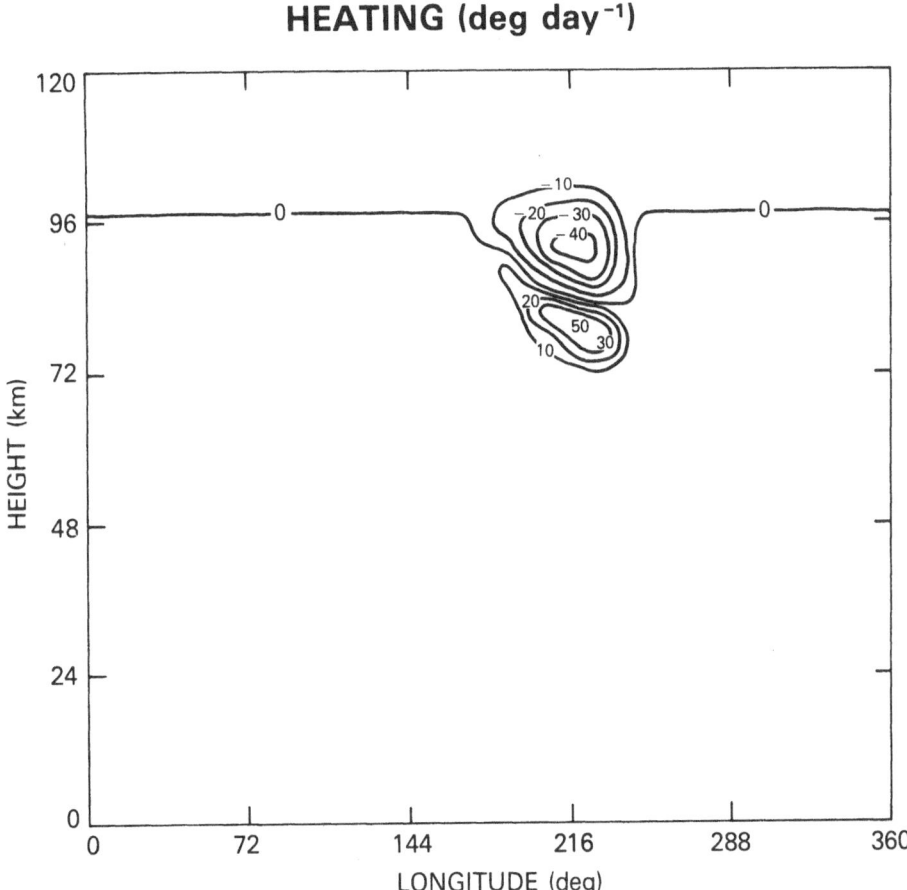

Figure 3c   The net heating (K day⁻¹).

60

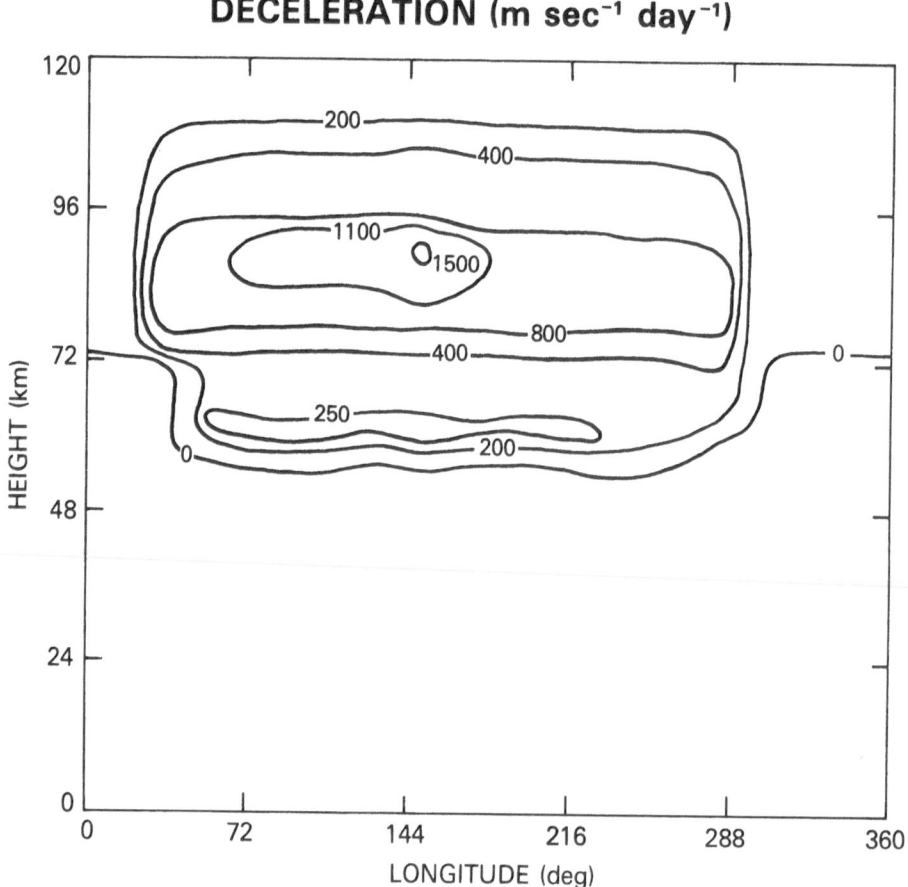

Figure 4   As in Fig. (3b) but with no Newtonian cooling.

Westward propagating gravity waves have a_{-1} better chance of penetrating the mesosphere. For $c < -25$ m s^{-1} the wavebreaking becomes essential zonal below the stratopause, since $|\bar{u}-c| > 20$ m s^{-1} everywhere.

The computations above only consider waves propagating in the zonal direction. Matsuno (1982) also examined waves which propagate in the meridional direction. These waves effectively see the zonal wind as $\bar{u}\cos\theta$ where θ is the angle of propagation measured from the zonal direction. For nonzonal wave breaking the effective background wind is $(\bar{u}\cos\theta + \bar{v}\sin\theta)$. If planetary wave perturbations of wind field are included (as above) then the magnitudes of $\bar{u}$ and $\bar{v}$ are nearly the same so nonzonal attenuation will also occur for waves propagating at an angle θ to the longitudinal direction.

Sl and Dunkerton (1982) suggested that intermittency was required to reduce the very large deceleration of the mean flow predicted by Lindzen's steady state model. Indeed, the spatial intermittency implied by nonzonal wave breaking is entirely consistent with this suggestion. For example, the longitude average of the zonal deceleration at 82 km in Fig. 3a is ~ 50 m s^{-1} day^{-1}, which is comparable to the value required by Apruzese et al. (1982).

c. Effects on the mean flow

The nonzonal wavebreaking, if confined to strong wind channels set up by the planetary waves, will decelerate both planetary waves and the mean flow in the mesosphere. Lindzen's (1981) parameterization gives the zonal deceleration as

$$M = -\frac{m(\bar{u}-c)^3}{2N}\left(\frac{1}{H} - \frac{3\partial\bar{u}/\partial z}{(\bar{u}-c)}\right)$$

evaluated at the breaking level for breaking gravity waves with zero meridional wave number. The shear correction term pointed out by Dunkerton was added in Holton (1982). M is held constant above the breaking level. If a planetary wave were present in the mesosphere then $\bar{u} \rightarrow \bar{u} + u'$ where u' is the planetary wave wind speed in the wave breaking zone. Expanding M and neglecting the shear term gives

$$M = -\frac{m(\bar{u}-c)^3}{2NH} - \frac{3mu'(\bar{u}-c)^2}{2NH} + \ldots\ldots$$

thus the local drag on the planetary wave field is simply proportional to u' (i.e. like Rayleigh friction). The damping of the planetary wave will produce additional deceleration of the zonal flow through planetary wave-mean flow interaction. If there is a zonal variation in the wave breaking level, then the expression above can only be accurate above the maximum breaking height. Furthermore, if there is longitudinal variation in the flux of gravity waves the expression has dubious utility.

The deposition of gravity wave energy as heat will also affect the planetary scale wave. The heating associated with the gravity wave dissipation is approximately given by

$$C \tilde{=} \frac{e}{c_p} \left(\frac{m(\overline{u}-c)^4}{2HN} \right)$$

(Schoeberl et al., 1983, Eq. 20), where P^{-1}, $\alpha = 0$ and e is the efficiency for the conversion of gravity wave energy to heat). Including the planetary wave zonal wind fields as above gives

$$C = \frac{1}{2} \frac{me}{c_p NH} \left((\overline{u}-c)^4 + 4 (\overline{u}-c)^3 u' + \ldots \right)$$

Thus the heating of the planetary wave is proportional to the planetary wave zonal wind. This is very unlike the usual thermal processes which are proportional to the planetary wave perturbation temperature.

The results of the previous section show that it is not necessary for a planetary wave to be present at wave breaking altitudes to produce a non-zonal drag; only planetary wave ducting of gravity waves below the stratopause is required. Gravity wave breaking can also dissipate planetary waves in the mesosphere and could account for the reduced amplitude observed above the winter stratopause. Planetary waves could also be forced in the mesosphere by the strong nonzonal deceleration shown in Fig. 3a. For stationary waves the mesospheric forced planetary wave would be approximately 180° out of phase from its stratospheric component. The overall effect on the zonally averaged westerlies in the mesosphere would, of course, be deceleration since the secondary circulation produced by the dissipation of the gravity wave-forced planetary wave would also decelerate the westerlies. Finally, acceleration or deceleration of the mean flow could result from the interference of the mesospheric forced wave with planetary waves propagating upward from below.

4. CONCLUDING REMARKS

Lindzen's (1981) gravity wave breaking theory and its parameterization provides a mechanism through which the troposphere and mesosphere are dynamically coupled. The requirement that the zonal mean wind in the mesosphere be decelerated toward tropospheric velocities is implicit in the Rayleigh friction parameterization of Leovy (1964) and Schoeberl and Strobel (1978). The turbulent diffusion associated with the breaking wave which dissipates gravity wave energy, also produces significant heat and constituent transport. Molecular diffusion and conduction in the thermosphere produce the same effect on gravity waves as turbulent diffusion generated by breaking gravity waves. The maximum deceleration for a molecularly dissipated gravity waves with $\lambda_x = 1000$ km in the lower thermosphere is 45 m s^{-1} day^{-1} at 98 km. The maximum cooling of ~ 3.5 K day^{-1} is produced in the 95–100km zone with heating below

95 km of ~ 1.2 K day^{-1}. Thus the dynamical coupling of the troposphere is extended to the lower thermosphere.

Sl noted that gravity wave breaking can be significantly reduced by radiative damping. In this study the wavelength dependent radiative damping is shown to be strong enough to prevent any gravity waves from reaching the mesosphere for winter and summer zonal mean wind profiles unless $|\bar{u}-c| \geqslant 20$ m s$^{-1}(4\lambda_x)^{1/4}$ when $m=\ell$) from the troposphere to the mesosphere. This requirement would restrict all but the fastest gravity waves from the summer mesosphere. Indeed preliminary MST measurements (Balsley, Ecklund, and Fritts, this volume) do indicate a lack of breaking gravity waves in the summer mesosphere. Additional support for this requirement comes from Southern Hemisphere late autumn observations of gravity wave momentum fluxes in a westerly zonal flow by Vincent and Reid (1983), who find that the dominant gravity waves in the mesosphere have phase speeds of $\sim - 50$ m s^{-1}.

In the winter middle atmosphere, gravity waves would penetrate to the mesosphere along strong wind channels set up by stationary planetary waves even during sudden warmings. This leads to nonzonal wave breaking which produces nonzonal deceleration and heating as illustrated in Fig. 3. The nonzonal wave transmission in addition to nonzonal wave forcing may provide a mechanism for in situ forcing planetary scale waves in the mesosphere as well as a means to dissipate planetary waves propagating vertically from below. It also complicates the response of the mean flow to gravity wave breaking. Thus a straightforward application of Lindzen's parameterization as in Holton (1982) may not be appropriate for zonal mean models of the middle atmosphere in winter.

Finally, we note that the computations described above could be improved by ray tracing of gravity wave packets through the stratosphere. The model used here assumes that such ray paths are principally vertical which is exact for stationary waves since the horizontal group velocity is just c. For waves with non zero phase speeds, the ray packets may move horizontally several thousand kilometers before breaking.

ACKNOWLEDGEMENTS

We wish to thank J. Apruzese for computing the wavelength dependent Newtonian cooling coefficients below 30km, and J. R. Holton and T. Matsuno for excellent comments on the preliminary version of this paper. D.F.S. was supported by the Upper Atmosphere Research Office of the National Aeronautics and Space Administration and the Office of Naval Research. Computation for this study were, in part, performed in the Space Plasma Computer Analysis Network.

REFERENCES

Apruzese, J.P., M.R. Schoeberl, and D.F. Strobel, 1982: Parameterization of IR cooling in a middle atmosphere dynamics model. I. The Effects on the zonally averaged circulation. J. Geophys. Res., 87, 8951-8966.

Banks, P.M. and G. Kockarts, 1973: Aeronomy Part B, Academic Press, New York.

Chao, W. and M. R. Schoeberl, 1983: A Note on the linear approximation of gravity wave saturation in the mesosphere, (Submitted to J. Atmos. Sci.).

Dunkerton, T.J., 1982: Stochastic parameterization of gravity wave stresses, J. Atmos. Sci., 39, 2490-2506.

Holton, J.R., 1982: The role of gravity wave induced drag and diffusion in the momentum budget of the mesosphere, J. Atmos. Sci., 39, 791-799.

Holton, J.R., 1975: The dynamic meterology of the stratosphere and mesosphere, Meteor. Monog. No. 37, American Meteorol. Soc., Boston Mass.

Lindzen, R.S., 1981: Turbulence and stress owing to gravity wave and tidal breakdown, J. Geophys. Res., 86, 9707-9714.

Matsuno, T., 1982: A quasi-one dimensional model of the middle atmosphere circulation interacting with gravity waves, J. Met. Soc. Japan, 60, 215-226.

Schoeberl, M.R., and D.F. Strobel, 1978: The zonally averaged circulation of the middle atmosphere, J. Atmos. Sci., 35, 577-591.

Schoeberl, M.R., D.F. Strobel, and J. P. Apruzese, 1983: A numerical model of gravity wave breaking and stress in the mesosphere, J. Geophys. Res., 88, 5249-5259.

Vincent, R.A., and I.M. Reid, 1983: HF doppler measurements of mesospheric gravity wave momentum fluxes, J. Atmos. Sci., 40, 1321-1333.

Vinnichenko, N.K., 1970: The kinetic spectrum in the free atmosphere - 1 second to 5 years, Tellus, 22, 158-166.

Walterscheid, R.L., 1981: Dynamical cooling induced by dissipating internal gravity wave, Geophys. Res. Lett., 8, 1235-1238.

J. R. Holton and T. Matsuno, Dynamics of the Middle Atmosphere, 65-75.

CLIMATOLOGY OF GRAVITY WAVES IN THE MIDDLE ATMOSPHERE

Isamu Hirota

Geophysical Institute

Kyoto University

ABSTRACT

An attempt is made at the statistical analysis of small-scale disturbances in the stratosphere and mesosphere with the aid of meteorological rocket observations at many stations from 77°N to 8°S for several years. By applying a high-pass filter to daily rocket data in the height range of 20-65km, wind and temperature fluctuations with characteristic vertical scales close to or less than 10km are obtained, which are considered to be due to internal gravity waves. Their magnitudes are presented as a function of latitude and season in a statistical manner. It is found that the gravity wave activity shows a notable annual cycle in higher latitudes with the maximum in wintertime, while it shows a semiannual cycle in lower latitudes with the maxima around equinoxes. It is also found from the standard deviation around the monthly mean that the temporal variability of gravity waves is large.

1. INTRODUCTION

Recent progress in the theory of vertically propagating, internal gravity waves indicates that they play an important role in the dynamics of the middle atmosphere in partly determining the large-scale temperature and wind structure through their energy and momentum transport.

One of the most interesting problems of the wind structure is the cause of very weak zonal wind velocities near the mesopause level in both the winter and summer hemisphere. As a plausible mechanism for the classical "Rayleigh friction" theory (Leovy, 1964) applied to this problem, Houghton (1978) suggested qualitatively that internal gravity waves might be responsible for producing such a zero-wind level. Lindzen (1981) made an initial attempt to develop a theoretical framework to parameterize the effect of the breaking of gravity waves on the mean zonal flow. Matsuno (1982) and Holton (1982) expanded this idea, by using numerical models, to show the nature of the interaction of the middle atmosphere circulation with internal gravity waves. Dunkerton (1982a,b) also showed that the equatorial mesopause semiannual oscillation could be produced by the selective transmission of gravity waves

65

66

and their interaction.

However, in these numerical models, it is assumed a priori that physical parameters such as the wave amplitudes and phase velocities are known, without having substantial evidence for the gravity wave properties in the actual atmosphere.

As regards the observation of small-scale disturbances in the stratosphere and mesosphere, there were some attempts to make temperature soundings in the 1960's, with the use of the acoustic grenade technique by rocket (e.g., Theon et al., 1967; Smith et al., 1968). From these observations it was found that wave-like structures were prominent in the mesospheric temperature profiles in wintertime, the vertical wavelength being typically about 10km. The evidence was, however, only fragmentary because the rocket sites were limited to a few stations and the total number of soundings was very small.

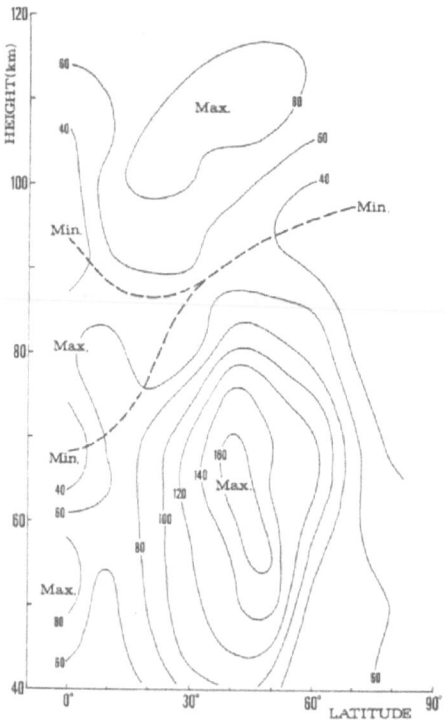

Figure 1. Annual range of the monthly mean zonal wind based on CIRA 1972. Units are m/sec.

On the other hand, as is seen from the result of statistics (e.g., CIRA, 1972), the mean zonal wind system in the upper part of the middle atmosphere varies substantially with latitude and season (Figure 1). The minimum wind line near 90km in middle and high latitudes appears associated with the annual cycle of the mesospheric and thermospheric zonal wind reversal, whereas those in tropical latitudes near 90 and 70km are nodal levels of the stratopause and mesopause semiannual oscillations.

In view of these facts, if we wish to determine the gravity wave-mean flow interaction in the real atmosphere, it is necessary to know the global distribution of gravity waves in a climatological sense.

As a preliminary attempt, in the present paper we show the result of the statistics for gravity wave activity in the stratosphere and mesosphere as a function of latitude and month, with the aid of meteorological rocket network observations for several years. Since the space-time distribution of rocket soundings is not systematic, it is almost impossible to detect the horizontal wavelength, period and phase velocity of gravity waves from the conventional meteorological rocket data only. Therefore our attention is paid mainly to the wave amplitude at individual stations.

2. DATA AND METHOD OF ANALYSIS

High altitude meteorological data supplied by the World Data Center A are used in this study for the 4 year period from 1977 to 1980. Thirteen stations covering a wide range of latitudes are selected for which a large number of rocket data are available. The average number of observation days at each station is about 360 for the 4 years. Roughly speaking, the data density in time is once per 4 days on the average.

Rocket observations of zonal wind component U, meridional wind component V and temperature T are used at an interval of 1 km. The height range is between 20 and 65km in most cases.

The procedure used in the present analysis is as follows:

(1) First, in order to remove the contribution of large-scale components i.e., the mean field, planetary waves and tides, a high-pass filter is applied to the daily data with respect to height. By this filtering, the fluctuations with characteristic vertical scales less than about 10km are separated. Since the mean field, planetary waves (except for equatorial waves) and tides have vertical scales of a few ten kilometers or more, they can be safely filtered out. Figure 2 shows an example of the wind and temperature fluctuations at Thule (77°N). In the filtered data wave-like disturbances with a characteristic vertical scales close to or less than 10km are seen, and these are considered to be due to

68

internal gravity waves.

(2) U, V and T obtained by the above procedure are expressed as
$X(z, day)$ for the height range $z_2 < z < z_1$. Then as a measure of
the wave intensity we define

$$X_0^2 = \frac{1}{z_1 - z_2} \int_{z_2}^{z_1} [\, d^2X / dz^2 \,]^2 \, dz$$

for each day and each component. The quantity X_0 itself has no physical meaning. Rather we are interested in the relative magnitude of
X_0 which depends on space and time. The second order derivatives are
calculated in practice by the three-point method with resolution of 1km.
The reason why we use the second order derivatives as a measure of the
wave intensity is that the small-scale fluctuations are thus emphasized.
However, even if we use X or X_z in stead of X_{zz}, there is no difference
in the results at least qualitatively. This is because the dominant
gravity waves have no significant variation in their vertical scales
from season to season and from station to station.

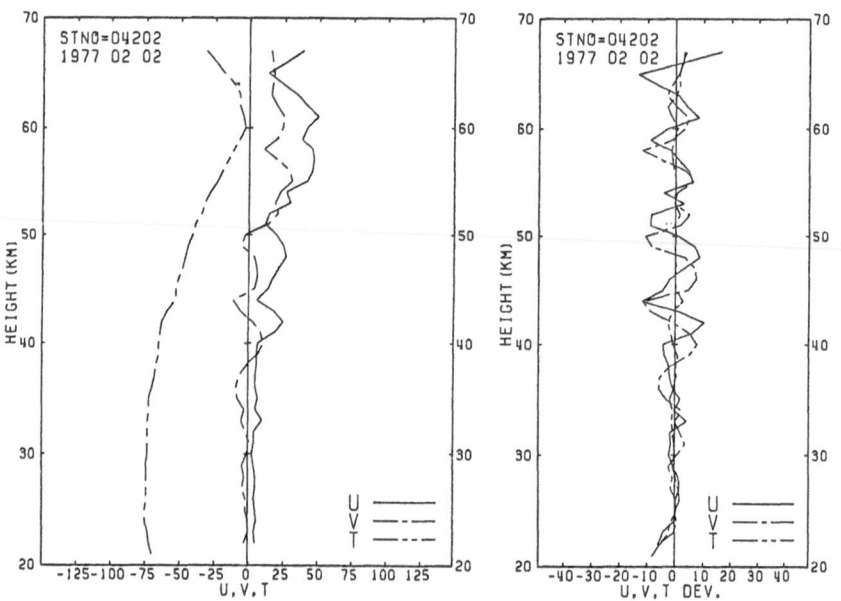

Figure 2. An example of the wind and temperature fluctuation
at Thule (77°N) for February 2, 1977. Raw data
(left) and filtered data (right). Units are
m/sec for U and V, and °K for T.

(3) For each station and for each month (of the 4 years), the root mean square and the standard deviation are calculated for X_0 of U, V and T, respectively. They are therefore given as functions of latitude and season.

Hereafter, we consider these values to be representative of the gravity wave activity in the stratosphere and mesosphere.

3. RESULTS

3-1. Orientation of the wave

Although the wave frequency cannot be determined from our rocket observations, both the earlier result of special soundings (Smith et al. 1968) and the recent observation by radar (Vincent and Reid, 1983) indicate that the dominant period of internal gravity waves in the middle atmosphere is of the order of a few hours or less. Therefore, when we observe the wind fluctuations associated with the passage of a gravity wave train at a station, the relative magnitude of the east-west component U_0 and north-south component V_0 must be determined by the orientation of the wave train in a horizontal plane. The orientation should be related to the excitation mechanism in the lower atmosphere, i.e., surface topography and/or synoptic-scale disturbances, as well as the direction of the prevailing background flow.

The scatter diagram of U_0 vs. V_0 (Figure 3) for two typical stations shows that both component are randomly distributed even for the same mean flow condition. This may provide supporting evidence for

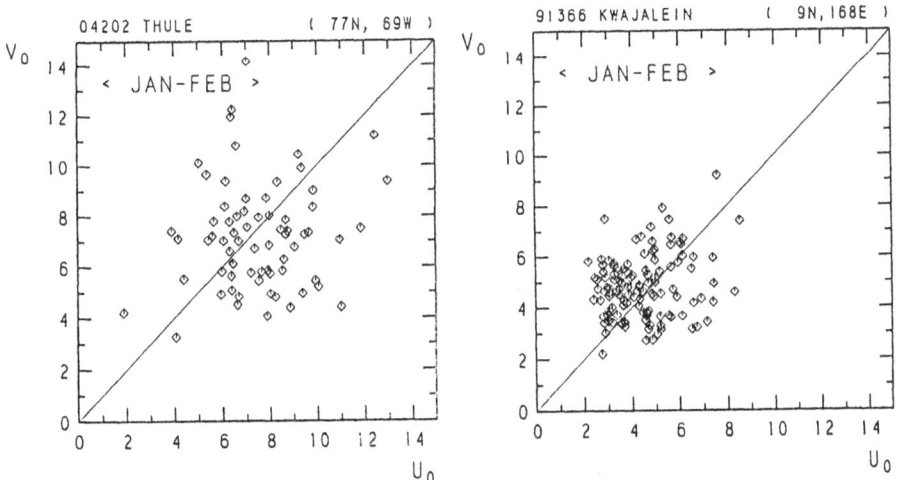

Figure 3. Scatter diagram of U_0 vs. V_0 for Thule (left) and Kwajalein (right) for January-February.

the isotropy of gravity waves assumed in Matsuno's model (1982).

Moreover, the r.m.s. values of U_0 and V_0 are almost equal in magnitude. Hence, in the following, we use the averaged value of the two components, i.e., $[(U_0^2 + V_0^2)/2]^{1/2}$ as a measure of the intensity of wind fluctuations.

3-2. Seasonal variation

Figure 4 shows the seasonal variation of r.m.s. amplitudes of wind and temperature for four typical stations. The standard deviation around each monthly mean is denoted by vertical bars.

In high latitudes (Thule, 77°N and Primrose Lake, 54°N), it is easily found that both the wind and temperature r.m.s. values show a notable annual cycle with the maximum in winter. This is the statistical reconfirmation of the result of earlier studies based on a small number of soundings (Theon et al., 1967; Smith et al., 1968). Roughly speaking, the maximum value in winter is twice as large as the minimum in summer for wind, while the annual variation in percent for temperature seems to be larger than that for wind. This may be because amplitude of observed temperature fluctuations varies with Brunt-Väisälä frequency, assuming wave with fixed energy density.

On the other hand, in lower latitudes (White Sands, 32°N and Ascension Island, 8°S), such a notable annual variation cannot be observed. Results of Fourier harmonic analysis indicate that the semiannual component is rather larger than the annual component, especially in the temperature variation. The maximum value of the semiannual variation appears near the equinox. This fact may be related to the seasonal variation of the mean zonal wind in the middle atmosphere as was shown in Figure 1.

The seasonal and latitudinal dependency of the gravity wave activity defined by the r.m.s. values is summarized in Figure 5. The values at two stations adjacent in latitude are not always close to each other, so that in this figure the latitudinal variation is subjectively smoothed to some extent. Moreover, there is no data in the latitude zone between 38°N and 54°N. Nevertheless, the contrast in the seasonal change is clearly seen between high and low latitudes.

3-3. Annual mean

As was shown in the previous subsection, the magnitudes of wind and temperature fluctuations significantly depend on latitude as well as season. In order to see the latitudinal dependency more clearly, the annual mean of the gravity wave activity is presented in Figure 6, where the two curves are also smoothed.

It is interesting to note that there appears to be a strong cont-

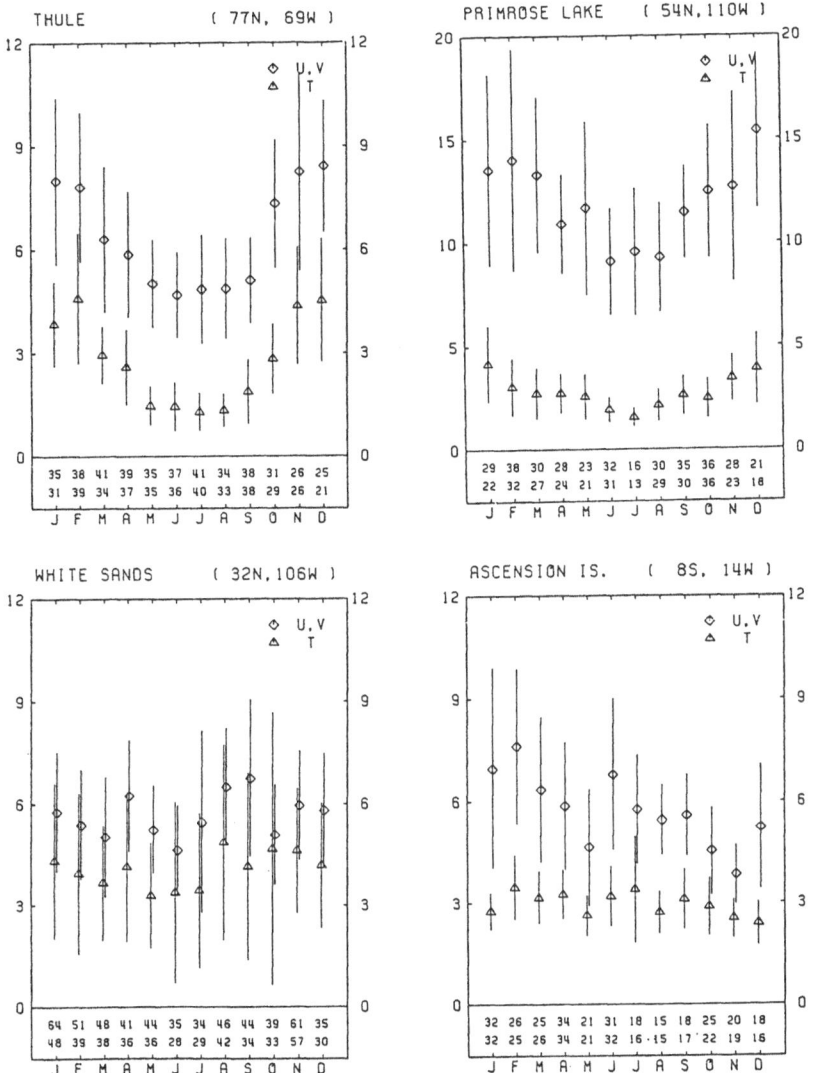

Figure 4. Seasonal variation of r.m.s. values of gravity wave activity at four stations. Vertical bars denote the standard deviation. Units are $m \cdot sec^{-1} \cdot km^{-2}$ for wind and $^{\circ}K \cdot km^{-2}$ for temperature. Figures above the abscissa are total number of rocket observations used in this statistics for wind (above) and temperature (below).

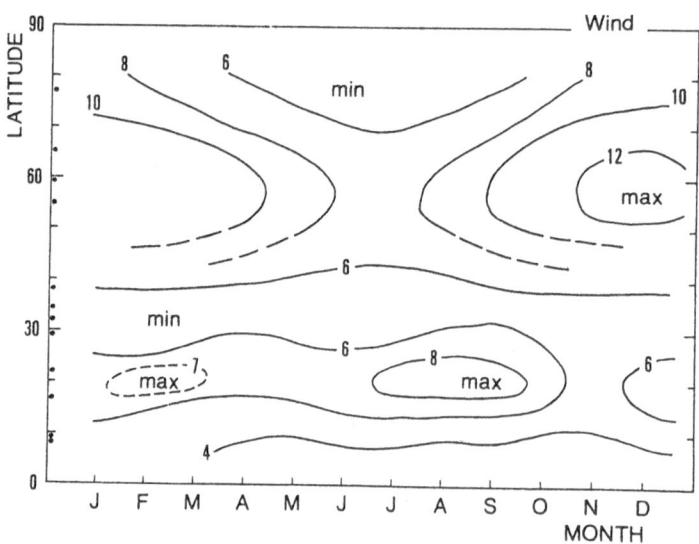

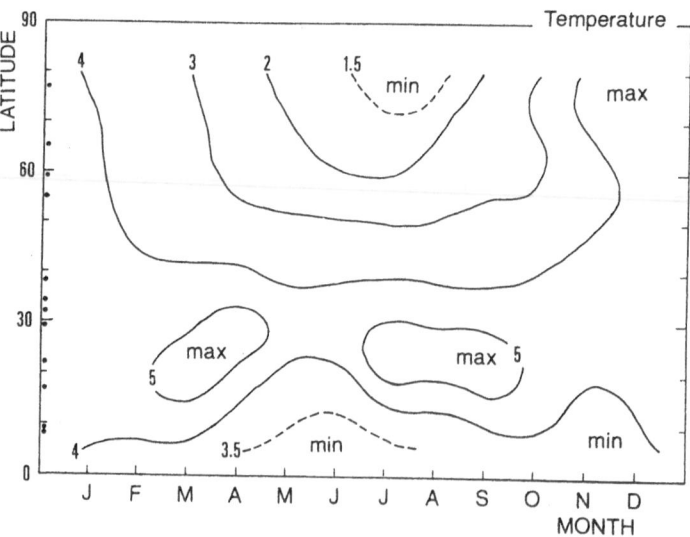

Figure 5. Latitude-time section of r.m.s. values for wind
(above) and temperature (below) in the same units
as those of Figure 4.   Circles on the ordinate
denote the rocket stations.

rast in the relative magnitudes of wind and temperature between higher
and lower latitudes: the maximum value for wind is observed around 50-
60°, whereas the maximum for temperature appears around 20-30°.

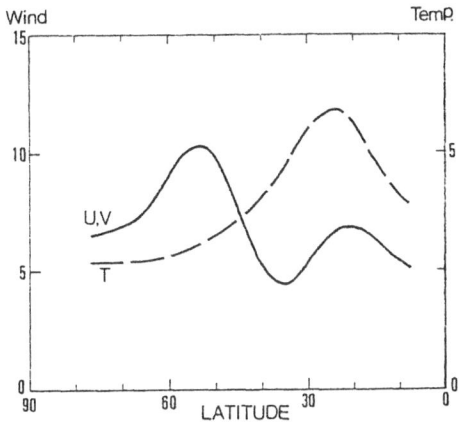

Figure 6. Latitudinal distribution of the annual mean value
of wind and temperature fluctuations. Units are
same as Figure 4.

This fact suggests that the structure of gravity waves in the
middle atmosphere is different between higher and lower latitudes.
Presumably this difference is attributed mainly to the latitudinal variation in temperature stratification in the stratosphere and also to the
excitation mechanisms of gravity waves in the troposphere.

It is also noteworthy that the maxima around 25°N are located to
the south of the axis of the mean zonal wind maximum in the middle atmosphere. This may be due to the selective transmission of gravity waves
with large Doppler-shifted phase velocities (Plumb and McEwan, 1978;
Matsuno, 1982).

3-4. Temporal variability

Another interesting aspect of gravity wave activity is transiency.
In their numerical models, Matsuno (1982) and Holton (1982) assumed
steady wave forcing. However, in the real atmosphere, gravity waves
are likely to be as variable as their generation processes in the lower
atmosphere which have a wide range of time scales from hours to days.
Dunkerton (1982a) made an attempt to parameterize the effect of transient and stochastic gravity waves in his simplified wave-mean flow interaction model.

Such transient variability can indeed be observed in Figure 4: as regards the wind fluctuations, the standard deviation is about one third as large as the monthly mean value throughout a year.

As was mentioned earlier, the data density in time is about once per 4 days on the average, and therefore we cannot infer exactly the typical time scale of the wave transiency from the present data set. Nevertheless, it can be said that the day-to-day variability of gravity waves in the middle atmosphere is very large and the stochastic model of Dunkerton (1982a) should be more realistic than a steady wave model.

4. CONCLUDING REMARKS

Although the present study is preliminary, statistical results of meteorological rocket data analysis indicate some interesting features of internal gravity waves in the stratosphere and mesosphere such as the isotropy, seasonal variation, latitudinal dependence and transiency.

Therefore, it is suggested that in order to improve theoretical models of gravity wave-mean flow interaction we need much more observational effort to determine the characteristics of gravity waves.

Recent progress in MST radar techniques (e.g., Vincent and Reid, 1983) makes it possible to estimate the vertical transport of momentum due to internal gravity waves by measuring all three components of the middle atmospheric motions. However, only isolated observation at a very few number of MST radar stations will be available.

Thus, more detailed statistics of many rocket observations on a global basis will also be important to determine the morphology of gravity waves in the near future.

ACKNOWLEDGEMENTS

The present author wishes to thank Drs. J.R.Holton and D.C.Fritts for their helpful comments on the original manuscript. Thanks are also due to Mr. M.Shiotani for his assistance in the data analysis.

REFERENCES

Dunkerton, T. J., 1982a: Stochastic parameterization of gravity wave stresses. J. Atmos. Sci., 39, 1711-1725.

-------, 1982b: Theory of the mesopause semiannual oscillation. J. Atmos. Sci., 39, 2681-2690.

Holton, J. R., 1982: The role of gravity wave induced drag and diffusion in the momentum budget of the mesosphere. J.Atmos.Sci., 39, 791-799.

Houghton, J. T., 1978: The stratosphere and mesosphere. Quart. J. Roy. Meteor. Soc., 104, 1-29.

Leovy, C. B., 1964: Simple models of thermally driven mesospheric circulation. J. Atmos. Sci., 21, 327-341.

Lindzen, R. S., 1981: Turbulence and stress owing to gravity wave and tidal breakdown. J. Geophys. Res., 86, 9707-9714.

Matsuno, T., 1982: A quasi one-dimensional model of the middle atmosphere circulation interacting with internal gravity waves. J. Meteor. Soc. Japan, 60, 215-226.

Plumb, R. A. and A. D. McEwan, 1978: The instability of a forced steady wave in a viscous stratified fluid; A laboratory analogue of the quasi-biennial oscillation. J. Atmos. Sci., 35, 1827-1839.

Smith, W. S., L. B. Katchen and J. S. Theon, 1968: Grenade experiments in a program of synoptic meteorological measurements. Meteor. Monogr., Am. Meteor. Soc., 9, 170-175.

Theon, J. S., W. Nordberg and L. B. Katchen, 1967: Some observations of the thermal behavior of the mesosphere. J. Atmos. Sci., 24, 428-438.

Vincent, R. A. and I. M. Reid, 1983: HF Doppler measurements of mesospheric gravity wave momentum fluxes. J. Atmos. Sci., 40, (in press).

J. R. Holton and T. Matsuno, Dynamics of the Middle Atmosphere, 77-96.
Copyright © 1984 by Terra Scientific Publishing Company.

VHF ECHOES FROM THE ARCTIC MESOSPHERE AND LOWER THERMOSPHERE,
PART I: OBSERVATIONS

B. B. Balsley and W. L. Ecklund

Aeronomy Laboratory
National Oceanic and Atmospheric Administration

D. C. Fritts

Geophysical Institute
University of Alaska

ABSTRACT

VHF echoes obtained from the mesosphere and lower thermosphere with
the MST radar at Poker Flat, Alaska during the past 3 years exhibit
pronounced seasonal variations. The winter echoes are relatively weak
and occur in the height range 55-80 km whereas the summer echoes are much
stronger and appear in a narrow height range near the mesopause. These
data suggest that the winter echoes are a result of the saturation of
upward propagating gravity waves while the summer echoes arise as a
consequence of a shear instability.

1. INTRODUCTION

An earlier study of VHF radar echoes from the high-latitude mesosphere and lower thermosphere using the Poker Flat MST Radar in Alaska
(Ecklund and Balsley, 1981) revealed the following details: during
"winter" periods (defined here as the period between roughly early
September and early May) the echoes are relatively weak and occur primarily during daylight hours over a height interval 55 km-80 km. During
"summer" periods (i.e., early May to to early September), however, the
echoes are much stronger, with the maximum echo strength centered at
about 86 km. The summer echoes are more-or-less continuous throughout
the day, presumably because the summer mesosphere is almost always sunlit
in summer at these latitudes. General features of the echo seasonal
variability are shown in Fig. 1 (the seasonal variation of the echo
height range) and in Fig. 2 (average echo intensity profiles for winter
and summer), both reproduced from Ecklund and Balsley, 1981).

Subsequent studies at Poker Flat have uncovered a number of additional characteristics of the mesospheric echoes and their seasonal variability. These features are outlined in the following sections. A more

77

complete treatment of the specific features outlined in this paper along with their interpretation as outlined in the following paper (Fritts et al., 1983) can be found in Balsley et al. (1983).

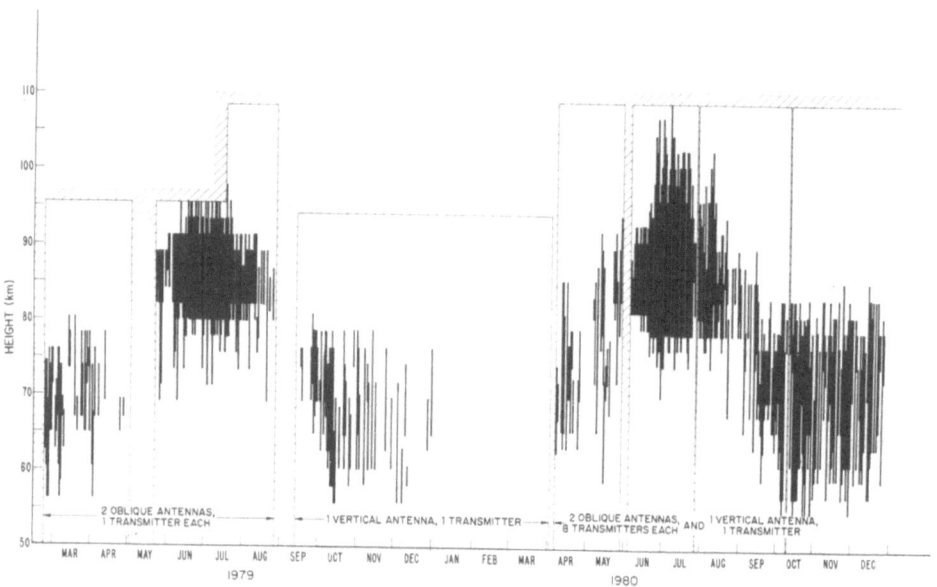

Figure 1. Showing the height distribution of echoes as a function of time of year (from Ecklund and Balsley, 1981). Hatched areas indicate times or height ranges of no data. Note the relatively sharp transition between the summer and winter echo height distributions.

2. WINTER ECHO CHARACTERISTICS

An overlay of hourly-averaged profiles of both echo power and horizontal wind (both obtained on antennas directed roughly northward and eastward), as shown in Fig. 3, suggests that the winter echoing region is dynamic and complex. The echo power profiles (left panels) exhibit a time variable structure having a vertical scale of a few km. Velocity profiles (right panels) also exhibit a marked variability from hour-to-hour, although the vertical spatial scales appear to be somewhat larger. The time sequence of these velocity profiles (Fig. 4) shows that their hour-to-hour variations exhibit a high degree of order: the profiles appear to manifest a wave-like structure with a vertical scale of about 15 km, a period of about 10 hours, and a downward phase velocity of approximately 20 cm/sec. On the other hand, the concurrent sequence of power profiles (i.e., backscattered echo strength vs height) shown in Fig. 5 exhibits no clear evidence for vertical motions or wavelike structure. Apparent vertical motions in power profile peaks

79

could occur, for example, as a result of the breaking action in the steep shear regions of upward propagating waves.

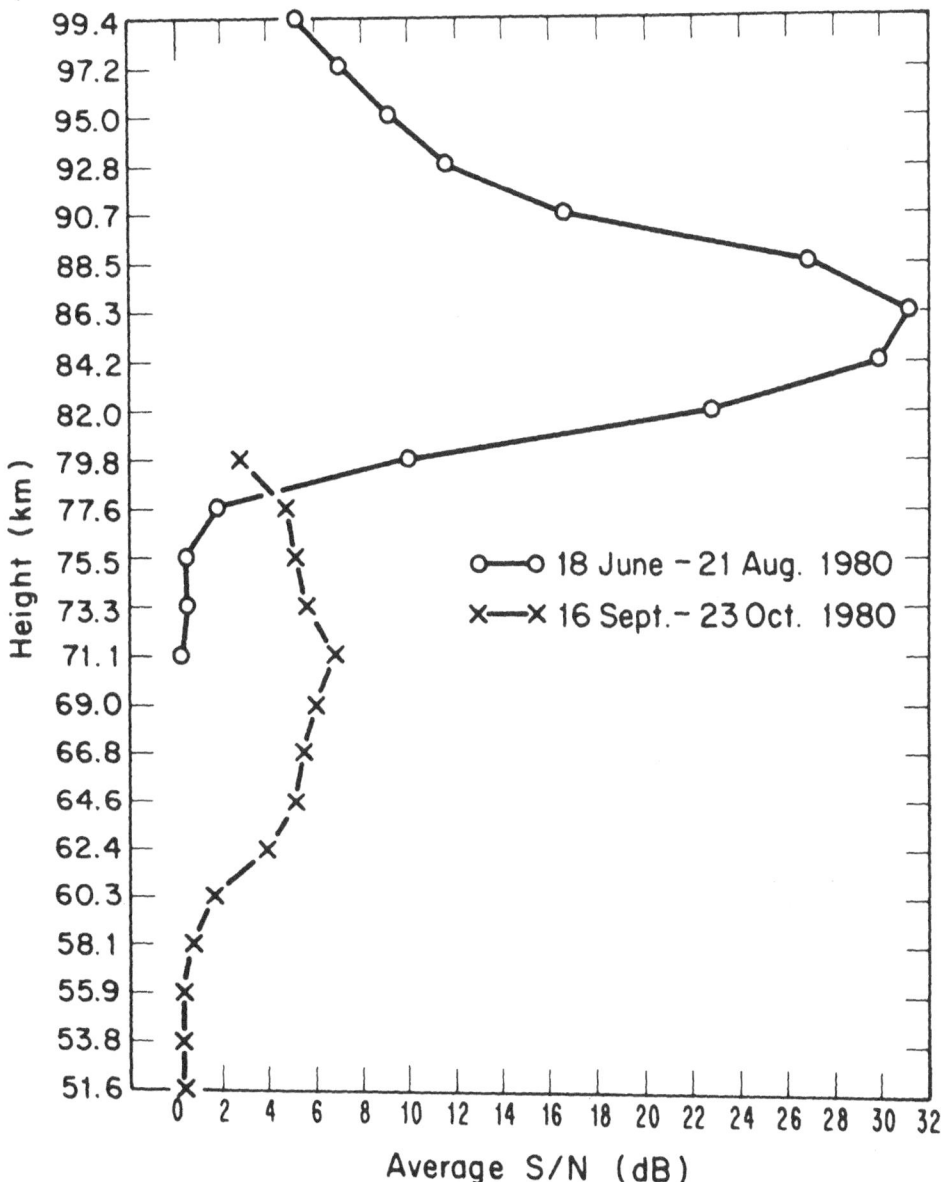

Figure 2. Time-averaged height profile of a signal-to-noise for typical summer (18 June–21 Aug 1980 curve) and winter (16 Sept–23 Oct 1980 curve) conditions (from Ecklund and Balsley, 1981.

80

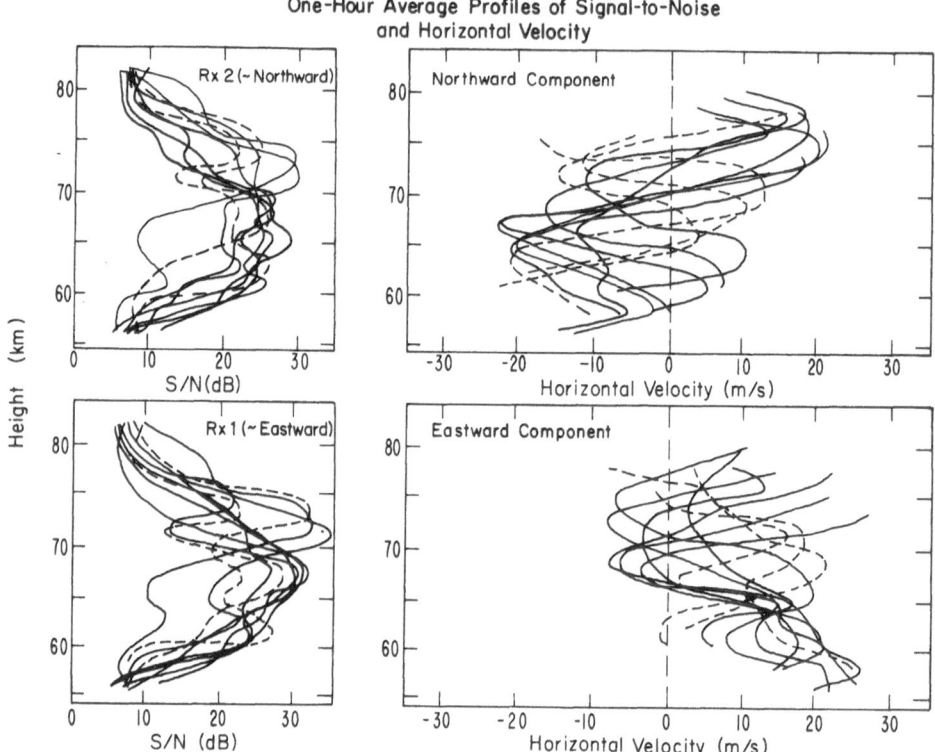

Figure 3. Superposition of one-hour averaged signal-to-noise and horizontal wind profiles on orthogonal antenna systems for the winter echoes
(11 Oct 1981).

In Fig. 6 the average height profiles of both spectral width (here
presumably proportional to vertical shear in the wind profile over the
sampled height range) and wind variability (on time scales of less than
ten hours to reduce effects of tidal variations) show an approximate
constancy with height. Assuming that both of these parameters arise
from velocity variations associated with upward propagating gravity wave
(c.f., the following paragraphs), their height constancy suggests that
the waves are breaking throughout the region (Hodges, 1967; Lindzen,
1967), and are not experiencing the normal exponential growth expected
for upward-propagating conservative gravity waves.

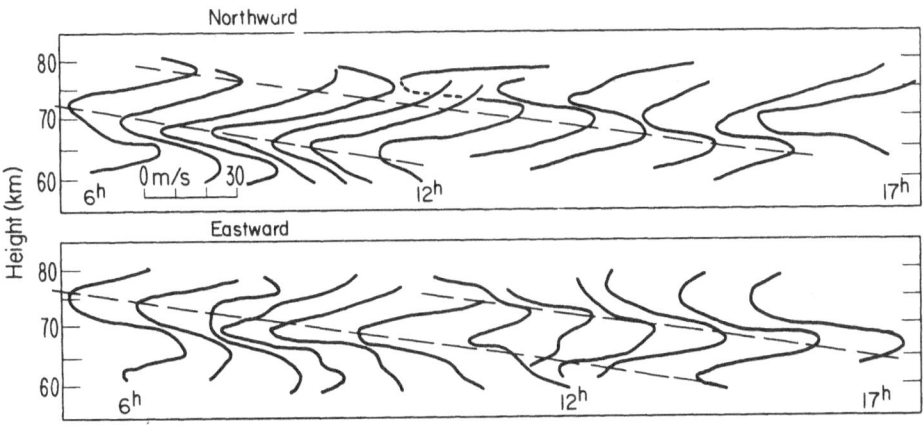

Figure 4. Time sequence of the horizontal velocity profiles shown in Fig. 3. Velocity scale shown in lower left corner of upper panel. Dashed lines indicate approximate height of velocity extreme of the quasi-sinusoidal contours. (Note that left-to-right profile placements are not precisely uniform in time.)

3. SUMMER ECHO CHARACTERISTICS

Hourly-averaged height profiles of the echo power and horizontal wind for the summer echoes appear in Fig. 7. Note that the power profiles are stronger and much less variable in summer than in winter, and exhibit the sharp maximum in signal-to-noise (S/N) near 86 km, as discussed earlier. The wind profiles (right panels) appear to have smaller-scale vertical structure than corresponding winter profiles, and the envelope of the fluctuations appears to grow with increasing height. The time sequence of wind profiles (Fig. 8) does not exhibit the sinusoidal character of the winter wind profiles. The general character of these profiles seems much more chaotic. On the other hand, the summer echo power profile time sequence (Fig. 9) shows a great deal of similarity from profile to profile, with only slight evidence of a height variability of the echo peak.

In contrast to the winter time, the average profiles of spectral width and wind variability shown in Fig. 10 exhibit a smooth and unmistakable increase with height between approximately 84 km and 92 km. Values below and above this range appear to be roughly constant.

82

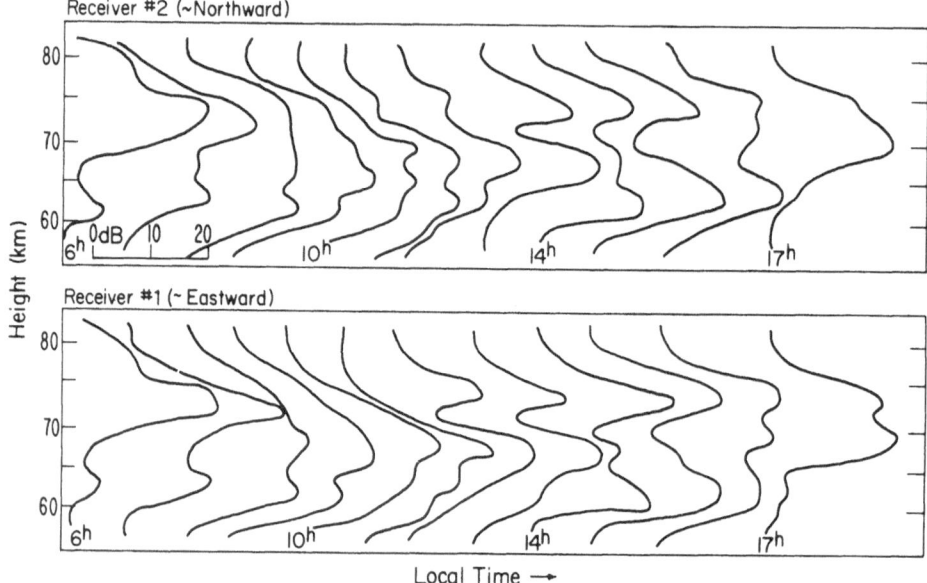

Figure 5. Time sequence of the signal-to-noise profiles shown in
Fig. 3. Scale of S/N values shown in lower left corner of the upper
panel.

Figure 11 shows that the average signal-to-noise in summer varies
diurnally, with a distinct minimum occurring between 21[h] and 24[h] below
about 90 km. As indicated in the following figure (12), which depicts
echo time-height contours for a forty-eight hour period, this premidnight
echo minimum may not occur every day. Further examination of the summer
echo contours in Fig. 12 shows that 1) there appears to be a quasi-
periodicity of the echo strength of a few-hours at any given height, and
2) the echo contours exhibit apparent downward motions with increasing
time with a velocity of about 1 m/s.

4. ADDITIONAL CHARACTERISTICS OF THE SUMMER/WINTER MESOPAUSE REGION

Traditionally, the primary source of the small-scale turbulence
required to produce VHF radar echoes in the mesosphere is thought to
arise from tropospheric gravity waves which grow exponentially as they
propagate upward, until they begin to breakup in the upper mesosphere.
This process is thought to be seasonably variable (c.f., Lindzen, 1981),
primarily because of variatons in the intervening wind field. Basic
concepts are illustrated in Fig. 13, which shows examples of zonal wind

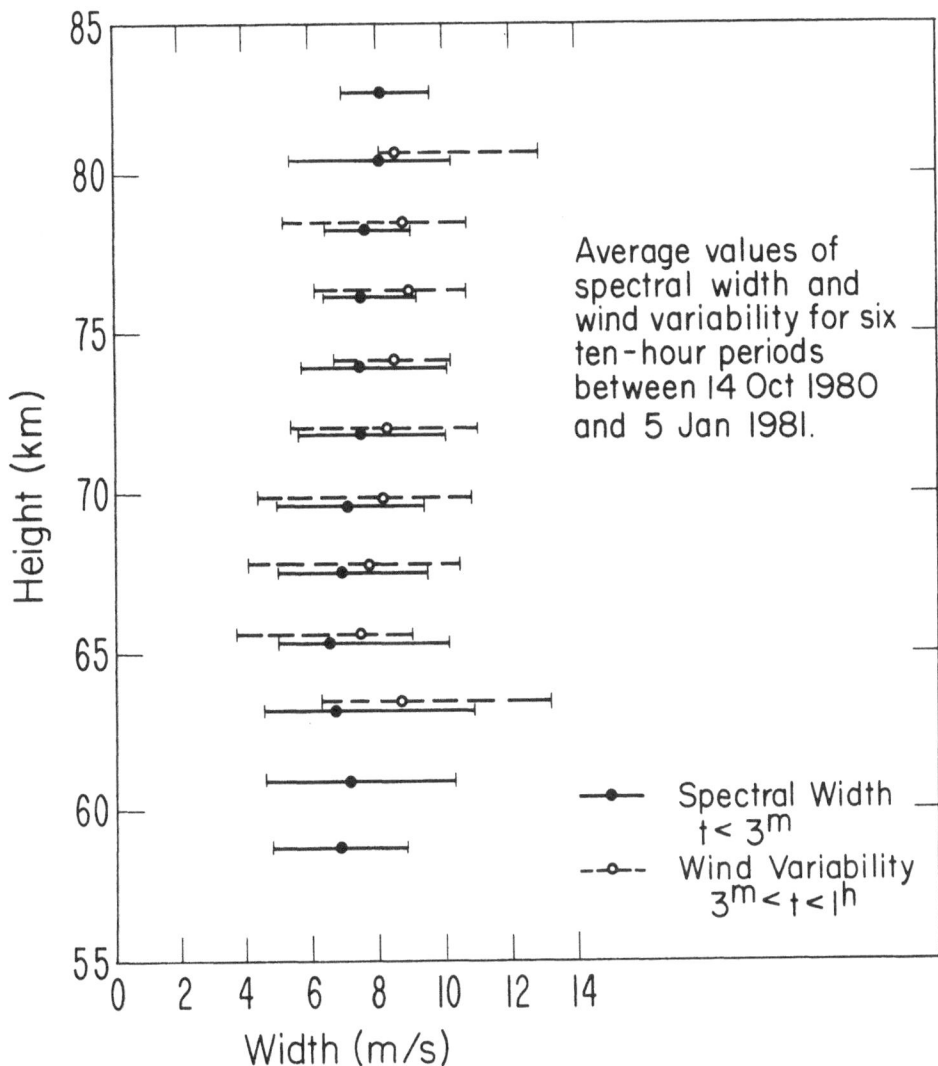

Figure 6. Height profiles of average spectral width (solid circles) and wind variability (open circles) for the winter echoes. Spectral width is full half-power width. Variability was obtained between 3^m and 1^h and represents the variability of the radial wind component (i.e., along the antenna beam). Horizontal wind variability can be estimated by multiplying the abscissal values by a factor of four.

84

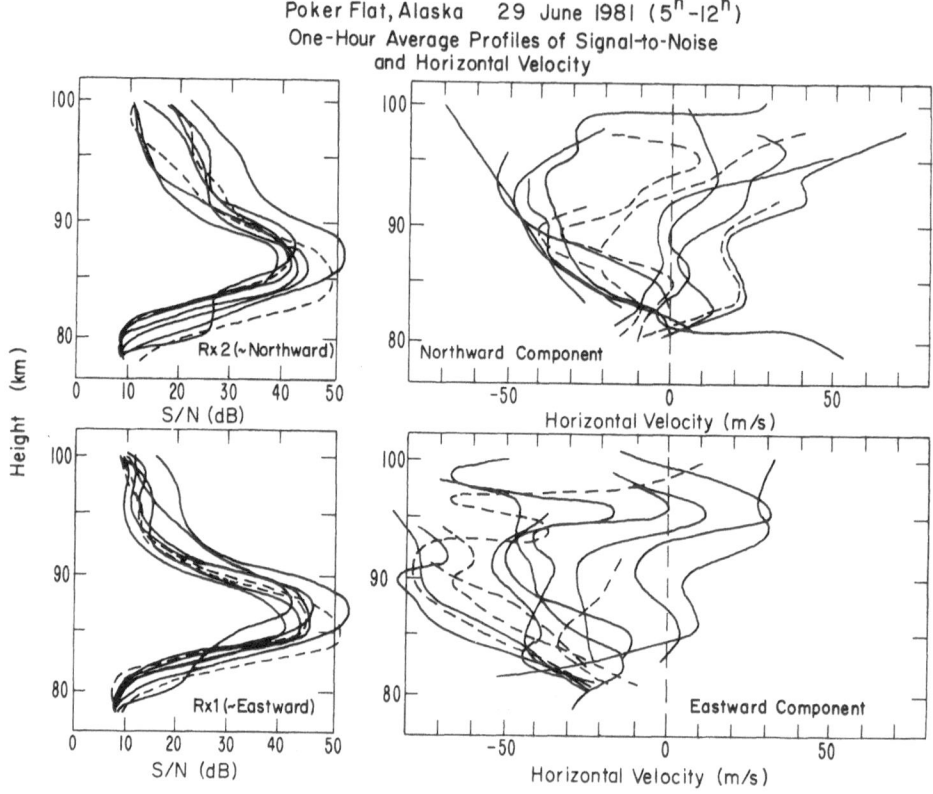

Figure 7.   Superposition of one-hour averaged signal-to-noise and horizontal wind profiles on orthogonal antenna systems for the summer echoes
(29 June 1981).

profiles for both summer and winter solstices, along with comparable
average profiles of Fairbanks rawinsonde data below about 30 km and
examples of Poker Flat MST radar wind data at mesospheric heights. Also
indicated are 1) a Gaussian-shaped estimate of the horizontal velocity
distribution of tropospheric gravity waves and 2) the minimum height of
gravity wave breaking predicted by Lindzen (1981). "Allowed" velocity
distributions of upward propagating gravity waves are indicated by the
width of the vertical arrows. Waves with phase velocities beyond these
ranges will be inhibited at their respective critical levels in the
atmosphere (i.e., that height where the horizontal phase speed of the
wave equals the horizontal mean wind velocity). For example, upward
propagation of gravity waves having a zonal velocity of 20 m/s in winter
will encounter a critical level at about 20 km.

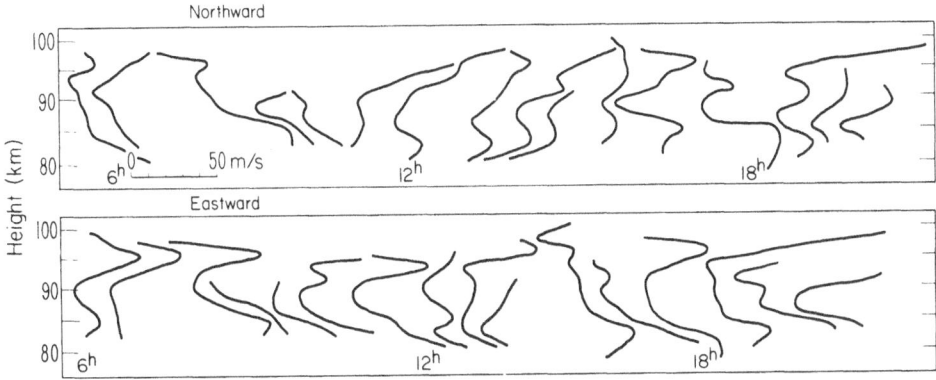

Figure 8. Time sequence of horizontal velocity profiles shown in Fig. 7. Velocity scale shown in lower left corner of upper panel.

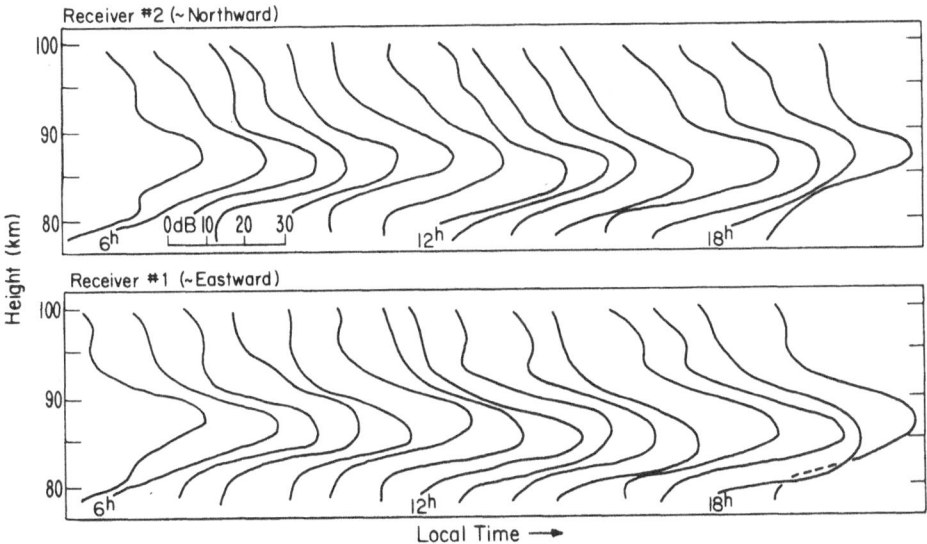

Figure 9. Time sequence of the signal-to-noise profiles shown in Fig. 7. Scale of S/N values shown in lower left corner of the upper panel.

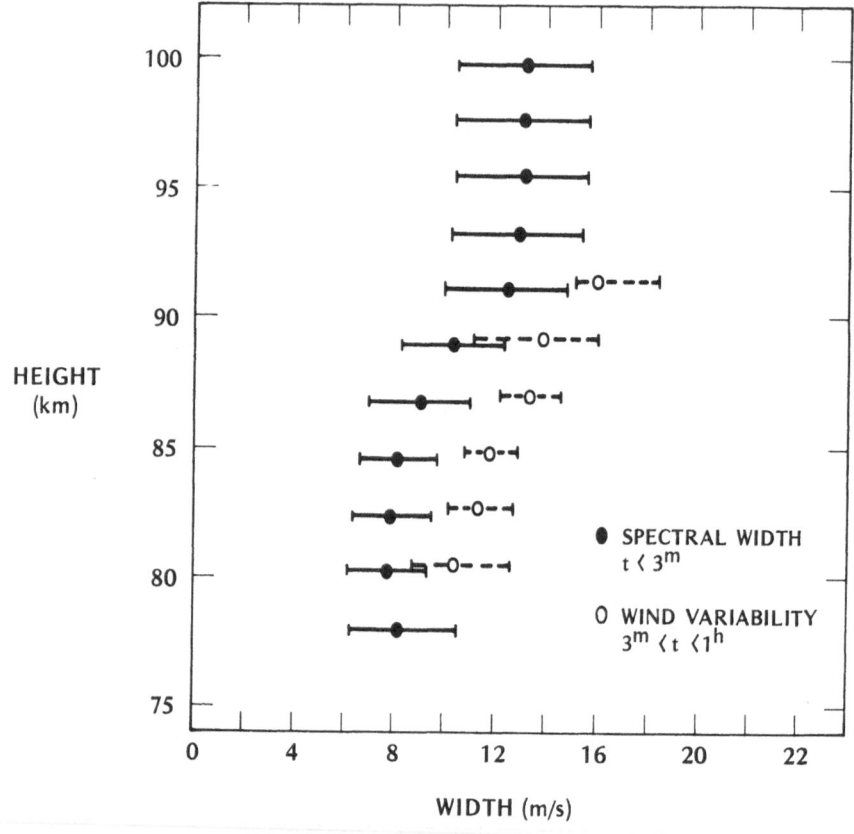

Average Values of Spectral Width and Wind Variability
for Eight 24-Hour Periods between 24 June and 8 July
1981.

Figure 10.   Similar to Fig. 6 except for the summer echoes.

Evidence that the seasonal variations of the mesospheric echo
heights are indeed controlled by the mean zonal wind appears in Fig. 14,
where the echo height transitions (e.g., early September and early May)
correlate well with the reversal times of the intervening zonal wind
field between 20 km and 60 km.

Further evidence that breaking gravity waves are responsible for the
winter radar echoes is given in Fig. 15. The point to be made in this
figure is that the height range of echoes is also a region of linearly

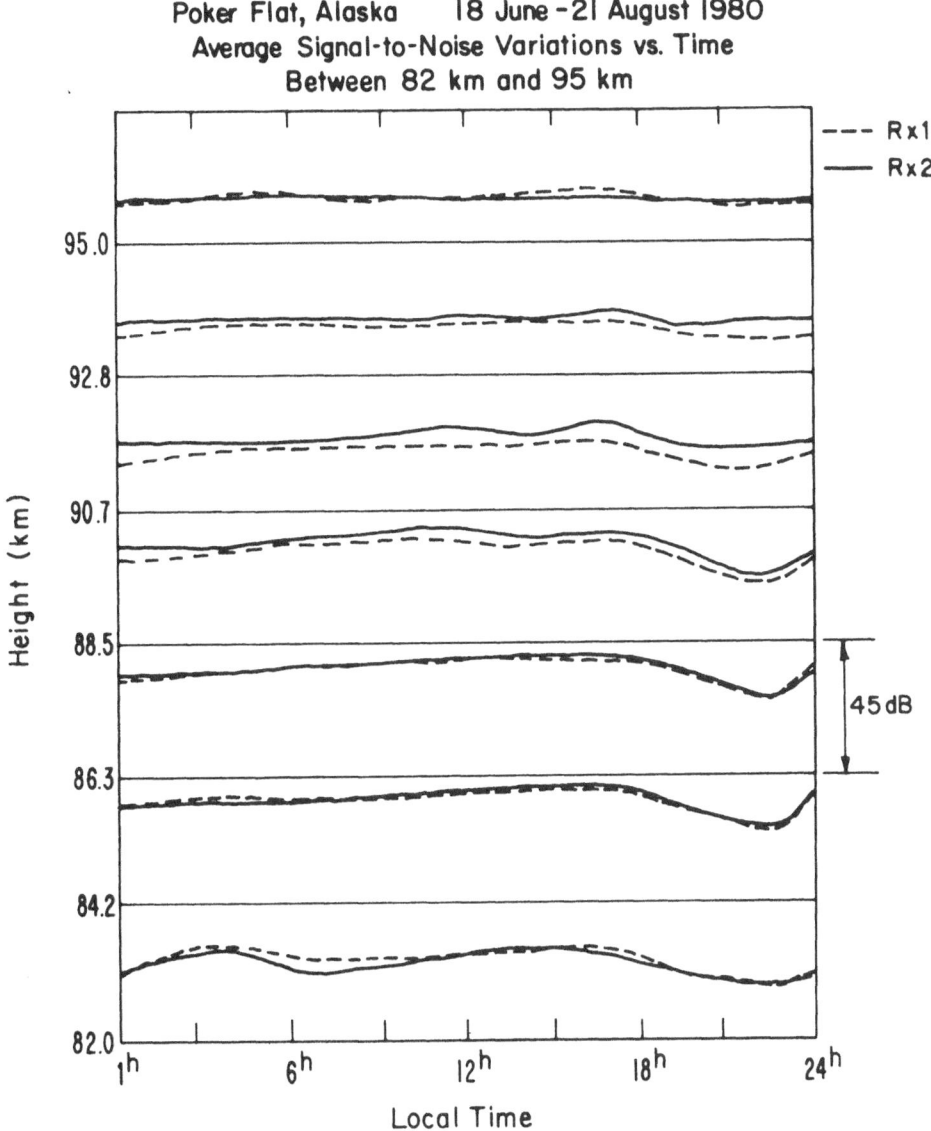

Figure 11. Diurnal variation of average signal-to-noise between 82 km and 95 km for both receivers 1 and 2. Note the pronounced minimum S/N visible in the lower four heights between 21^h and 24^h.

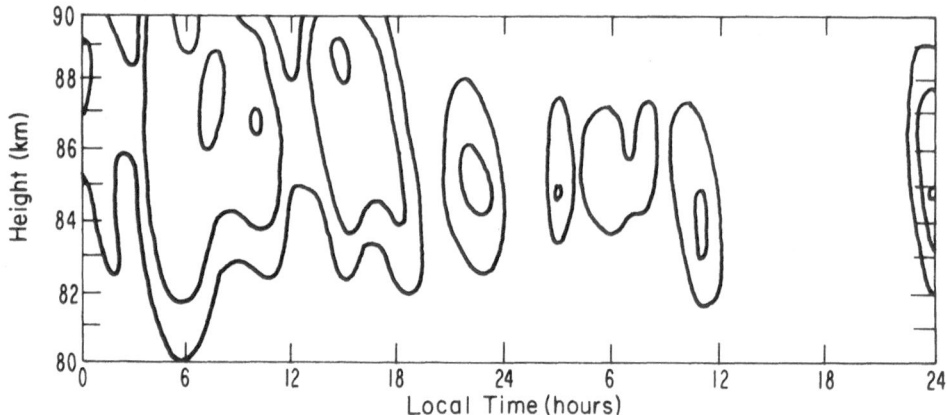

Time-Height Cross-Section of Signal-to-Noise at Poker Flat, Alaska for
26-27 July 1981. (Contours of S/N increase inward and are 30, 40, and 50 dB)

Figure 12. Time-height cross-section of signal-to-noise at Poker Flat
for 26-27 July 1981. Contours of S/N increase inward and are 20, 40,
and 50 dB.

decreasing zonal wind velocity. This is in agreement with the idea
(Holton, 1982; Matsuno, 1982; Dunkerton, 1982) that the gravity wave
energy injected at each height as a result of wave breaking will accelerate the flow toward the gravity wave phase velocity. Note that this
feature is not as apparent in the case of the summer echoes, however, as
indicated in Fig. 16, where a region of increasing velocity can be observed at about 83 km.

An additional indication of gravity wave activity in winter is seen
in the set of wintertime temperature profiles (Fig. 17) obtained at
Barrow, Alaska (71°N) by Theon et al. (1967). Here the temperature
structure is highly variable, with vertical scales of $\sim$ 10 km, and superadiabatic lapse rates occurring at some locations. No such evidence is
seen during comparable summer periods, where the profiles exhibit a
reasonably constant decrease with height to the mesopause minimum. Also
note the VHF echo profiles for summer and winter taken from Fig. 2 and
depicted in the right-hand panels of Fig. 17. A close correspondence
between the summer echo maximum and the minimum mesopause temperature is
evident. A close relationship between the summer echo and mesopause
temperature and vertical temperature gradient is supported by data shown
in Fig. 18, which compares seasonal variations of both the mesopause
temperature and temperature gradient with the occurrence of the summer
echoes. Mesopause temperatures less than approximately 170°K, and upper

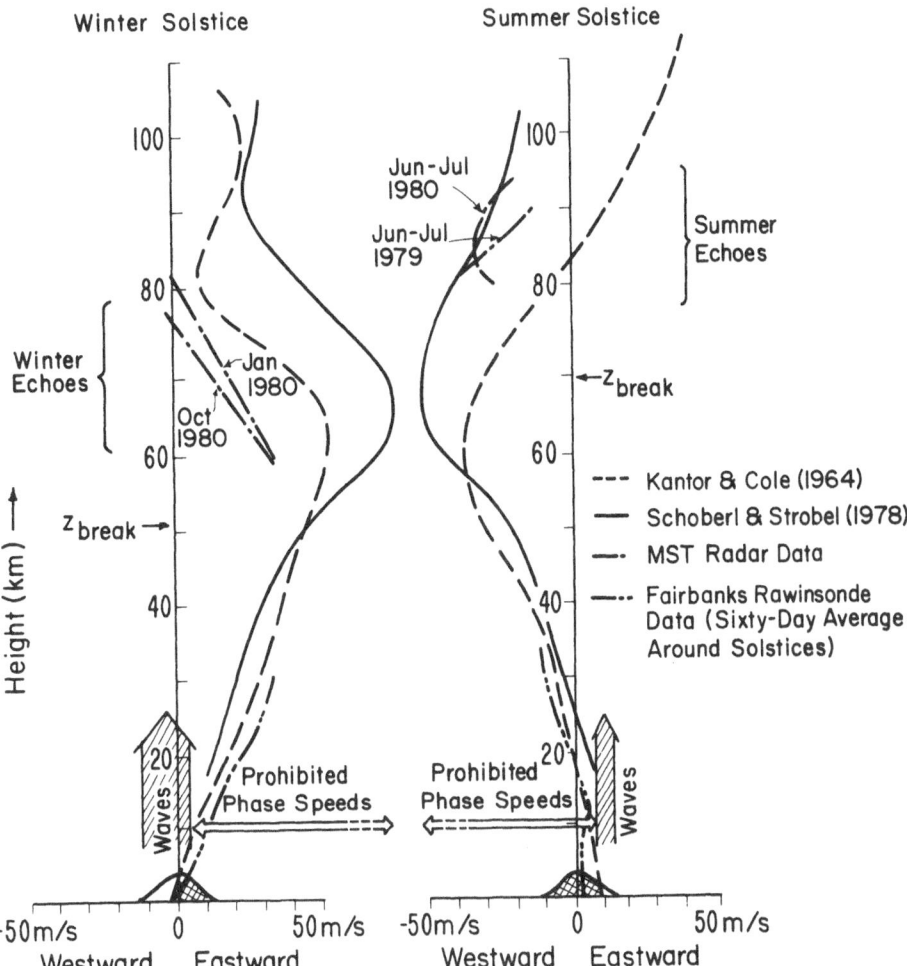

Figure 13. Zonal wind profiles for both summer and winter solsitial periods of 65°N from various sources. Height ranges of both the summer and winter echoes are shown, along with a schematic representation of the range of allowable phase velocities for tropospherically-generated gravity waves that can propagate into the stratosphere and above.

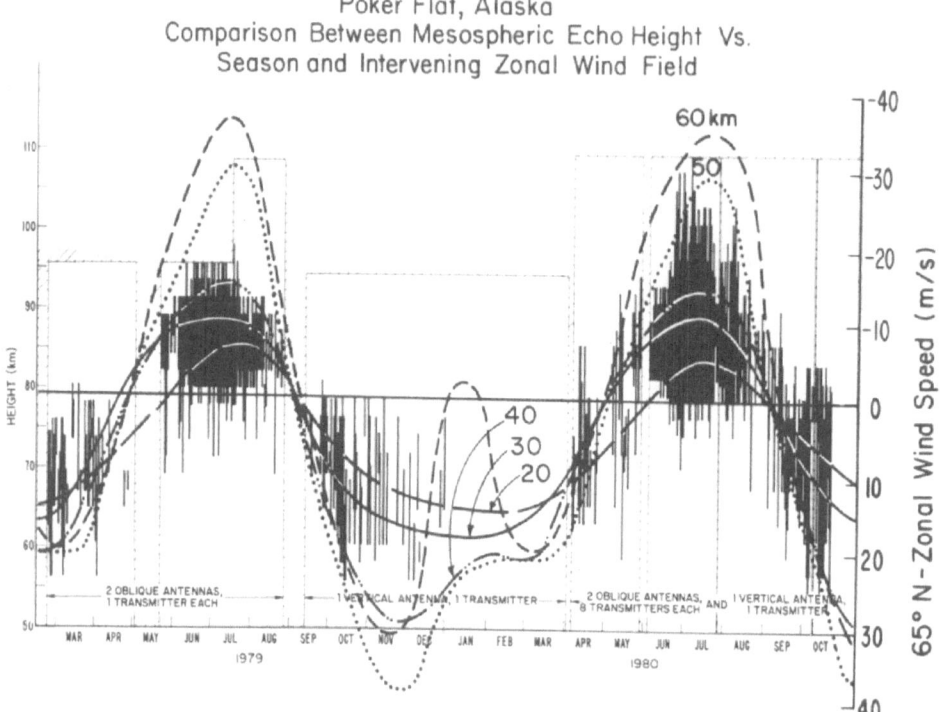

Poker Flat, Alaska
Comparison Between Mesospheric Echo Height Vs.
Season and Intervening Zonal Wind Field

Figure 14. Overlay of Fig. 1 and the mean zonal wind (65°N) at ten km intervals between 20 km and 60 km. Zonal wind values are scaled on right ordinate (positive values are eastward (westerly)). Note the correspondence between the mean zonal wind direction and the height of the echoes.

mesospheric temperature gradients greater than about 2.6°K/km appear to correspond to times of occurrence of summer echoes.

The final piece of evidence connecting the summer echoes and the mesopause temperature profile is shown in Fig. 19, which shows vertical profiles of the Brunt-Väsäïlä frequency, N. Profiles obtained from both climatological tables and the actual summer Barrow profiles (Fig. 17) are shown. Low values of N^2 indicate regions that are potentially unstable to convective activity. A region of sharp increase (e.g., above 83 km for the Barrow profile) has implications in regard to vertical wavelength variations of upward propagating waves. The importance of this feature will be discussed in a subsequent paper (Fritts et al., 1983).

In summary, the results presented here and in Ecklund and Balsley (1981) argue strongly that the wintertime mesospheric echoes observed by

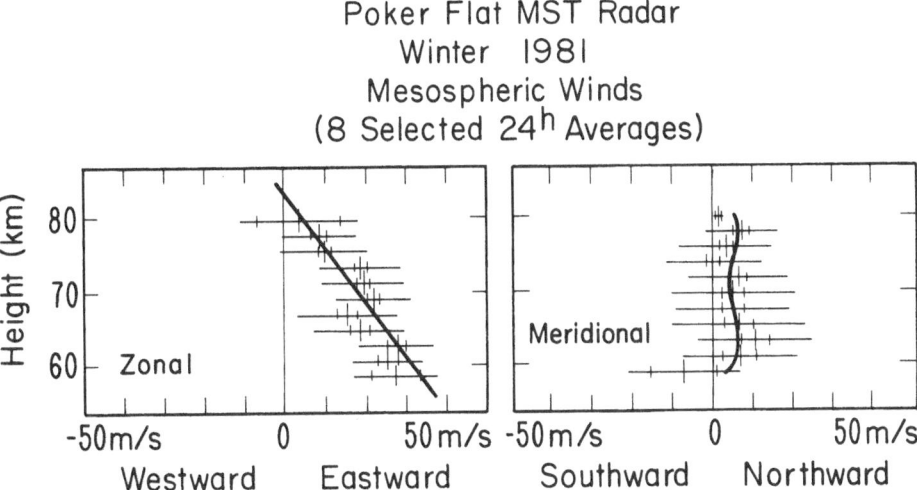

Poker Flat MST Radar
Winter 1981
Mesospheric Winds
(8 Selected 24^h Averages)

Figure 15. Eight selected 24^h averaged mesospheric wind profiles during winter 1981. Note the smoothly decreasing zonal profile with height.

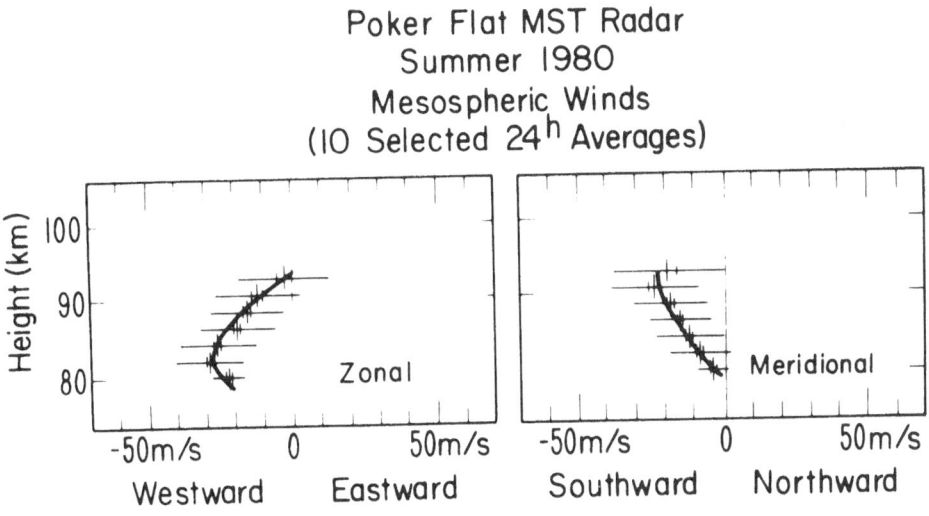

Poker Flat MST Radar
Summer 1980
Mesospheric Winds
(10 Selected 24^h Averages)

Figure 16. Ten selected 24^h averaged mesospheric wind profiles during summer 1980. Note the pronounced cusp in the zonal profile at ∿ 82 km and the strongly increasing meridional profile.

92

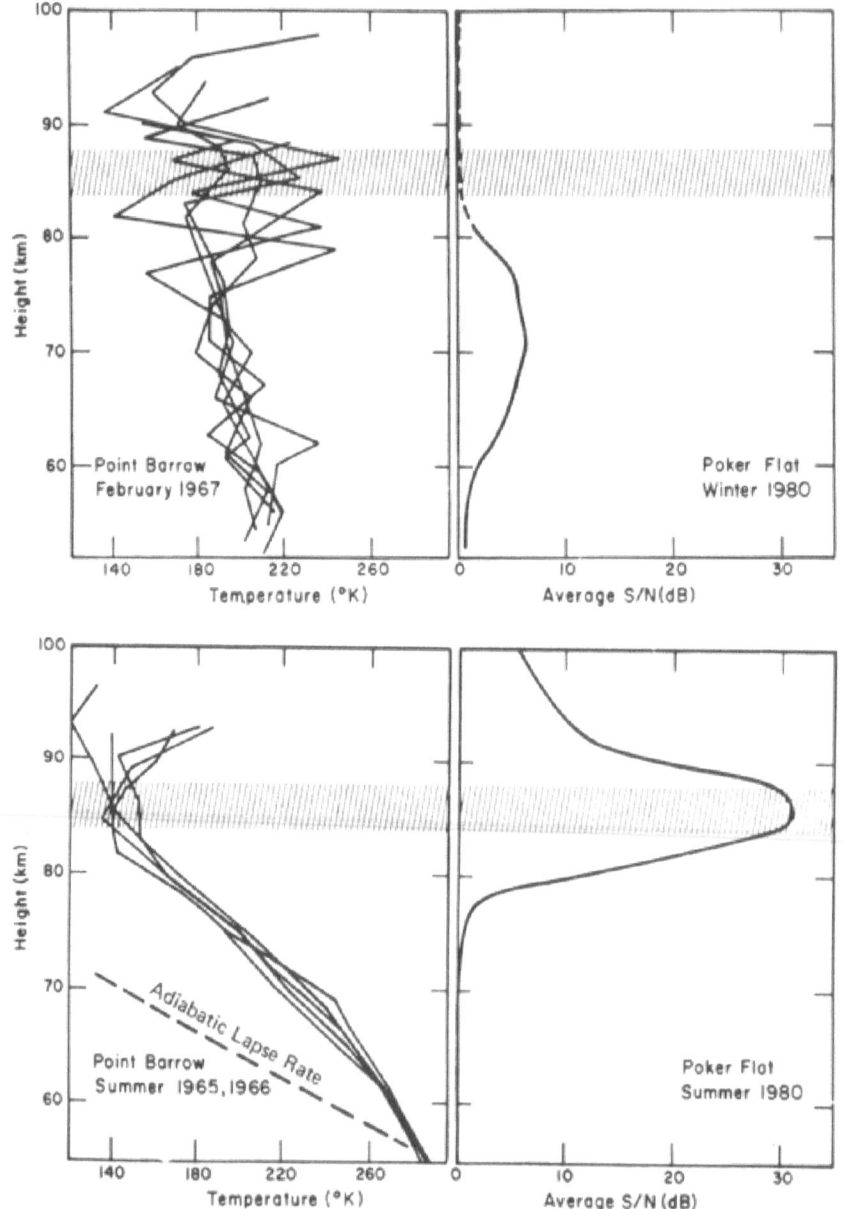

Figure 17. Left-hand panels depict a series of wintertime (upper panel) and summertime (lower panel) temperature profiles at Pt. Barrow, AK ($\simeq 70°N$) obtained from Theon et al. (1967). Right-hand panels show the winter (upper panel) and summer (lower panel) echo profile as in Fig. 2.

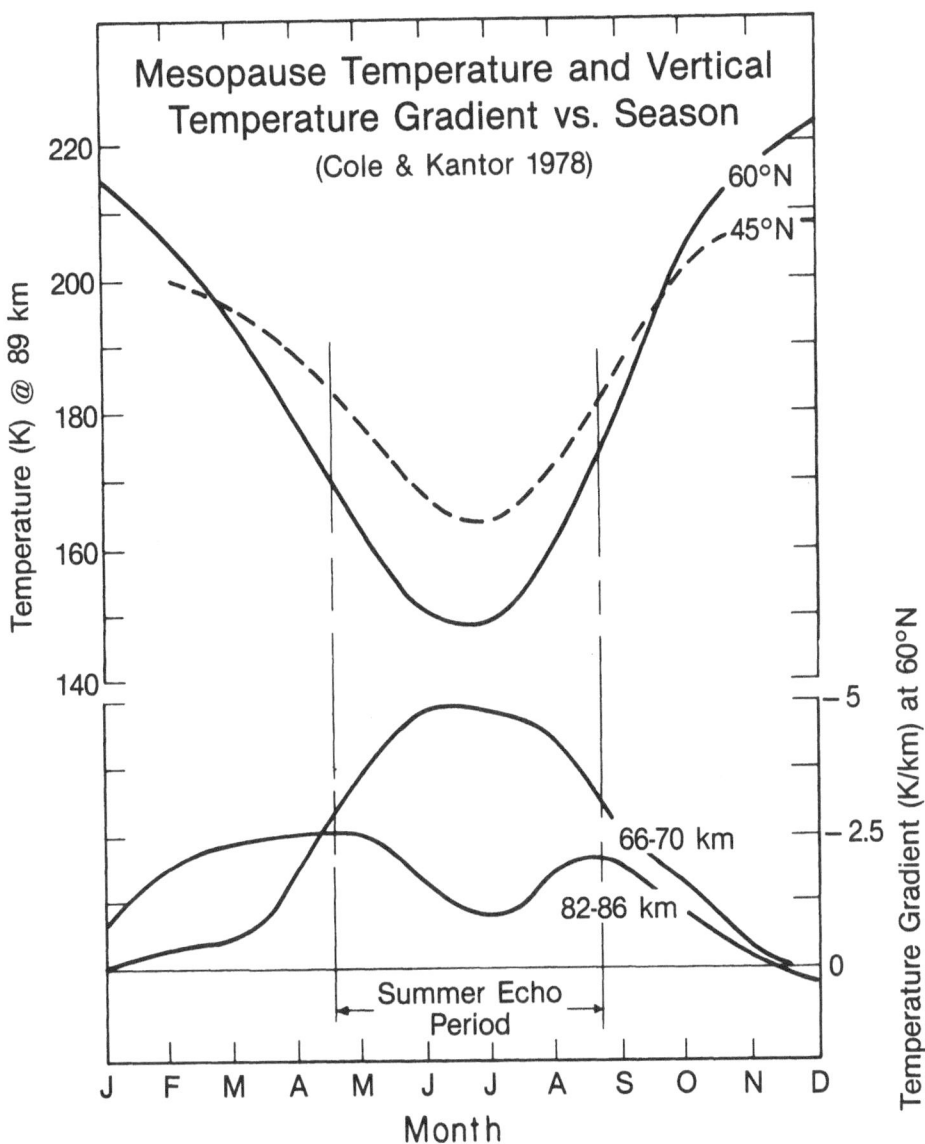

Figure 18. Plots of the seasonal variation of the mesopause temperature at 45°N and 60°N (upper panel) and temperature gradient between 66 km and 70 km and between 82 km and 85 km (lower panel).

94

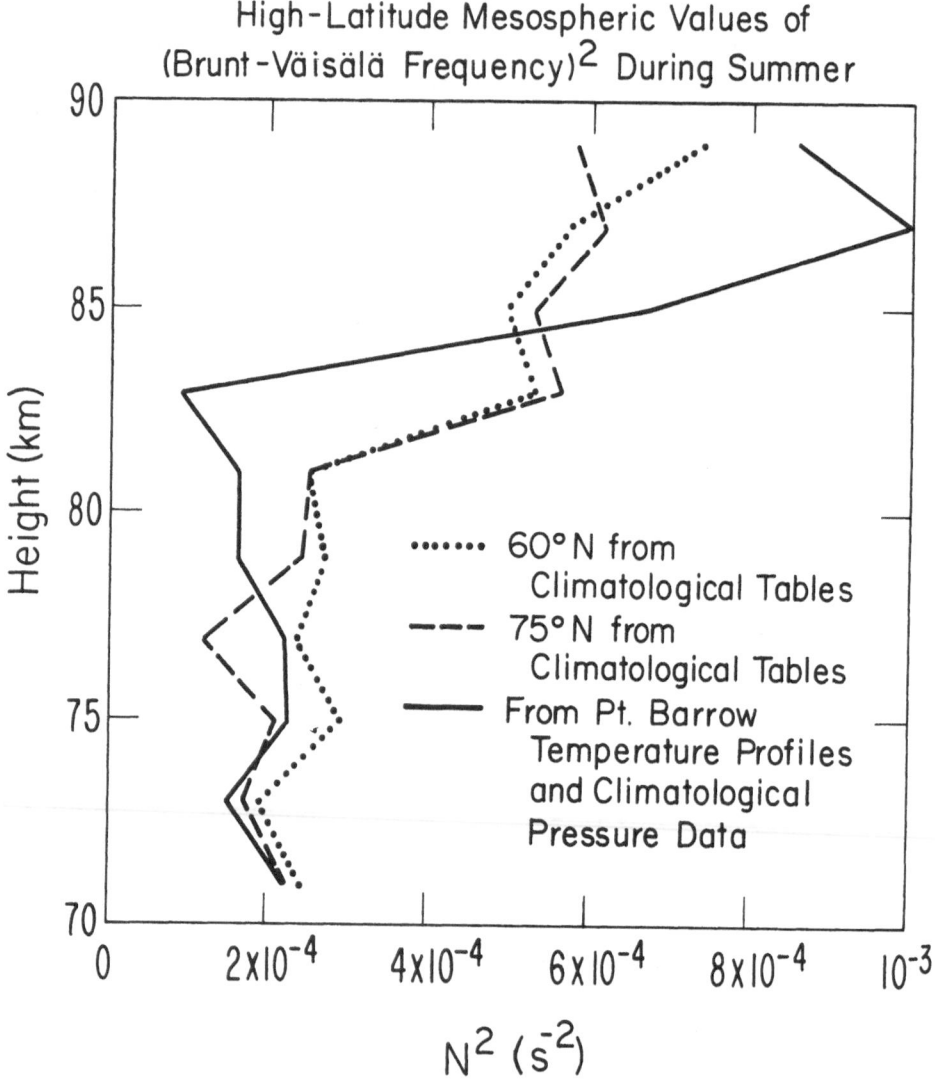

Figure 19. Height profile of the Brunt-Väisälä frequency)² in the high
latitude mesosphere and lower thermosphere during summer, obtained from
climatological tables (Cole and Kantor, 1978) and from the Pt. Barrow
temperature profiles of Theon et al. (1967).

the Poker Flat MST Radar arise from upward-propagating, tropospherically-generated gravity waves. These waves propagate into the region and produce the small-scale "turbulent" structures seen by the radar as a result of wave-breaking. Our results also suggest strongly that the summertime echoes arise from a somewhat different process. Any satisfactory explanation for the summer mechanism must account for the correspondence between the echo maximum and the mesopause temperature minimum as well as the quasi periodicity and apparent downward motion of the echo contours. A plausible explanation for the summer echoes in terms of a shear instability operating on long period waves and tides in the region of large temperature stratification is given in a companion paper (Fritts et al., 1983).

ACKNOWLEDGEMENTS

We are happy to acknowledge valuable conversations with R. S. Lindzen (M.I.T.), T. J. Dunkerton (Physical Dynamics), J. R. Holton (Univ. of Washington) and with our Aeronomy Laboratory colleagues, D. A. Carter, K. S. Gage, T. E. VanZandt, and J. Weinstock. This material is based upon activities supported by the National Science Foundation under agreement No. ATM-8023126. Any opinion, findings and conclusions or recommendations expressed in this publication are those of the authors and do not necessarily reflect the view of the National Science Foundation.

REFERENCES

Balsley, B. B., W. L. Ecklund, and D. C. Fritts, 1983: VHF echoes from the high-latitude mesosphere and lower thermosphere: Observations and interpretations. To appear in J. Atmos. Sci.

Cole, A. E., and A. J. Kantor, 1977: Arctic and Subarctic Atmospheres, 0 to 90 km. Report No. AFGLTR770046, 48 pages. Available NTIS, Springield, VA 22161.

Cole, A. E., and A. J. Kantor, 1978: Air Force Reference Atmospheres. Report No. AFGL-TR-78-0051, 78 pages. Available NTIS, Sprinfield, VA 22161.

Dunkerton, T. J., 1982: Stochastic parameterization of gravity wave stresses. J. Atmos. Sci., 39, 1711-1725.

Ecklund, W. L., and B. B. Balsley, 1981: Long term observations of the Arctic mesosphere with the MST radar at Poker Flat, Alaska. J. Geophys. Res., 86, 7775-7780.

Fritts, D. C., B. B. Balsley, and W. L. Ecklund, 1983: VHF echoes from the arctic mesosphere and lower thermosphere, Part II: Interpretation. This issue.

Hodges, R. R., 1967: Generation of turbulence in the upper atmosphere by internal gravity waves. J. Geophys. Res., 72, 3455-3458.

Holton, J. R., 1982: The role of gravity wave induced drag and diffusion in the momentum budget of the mesosphere. J. Atmos. Sci., 39, 791-799.

Lindzen, R. S., 1967: Thermally driven diurnal tide in the atmosphere. Quart. J. Roy. Met. Soc., 93, 18-42.

Lindzen, R. S., 1981: Turbulence and stress due to gravity wave and tidal breakdown. J. Geophys. Res., 86C, 9707-9714.

Matsuno, T., 1982: A quasi one-dimensional model of the middle atmosphere circulation interacting with internal gravity waves. J. Meteorol. Soc. Japan, 60, 215-226.

Theon, J. S., W. Nordberg, L. B. Katchen, and J. J. Horvath, 1967: Some observations on the thermal behavior of the mesosphere. J. Atmos. Sci., 24, 428-438.

J. R. Holton and T. Matsuno, Dynamics of the Middle Atmosphere, 97-115.
Copyright © 1984 by Terra Scientific Publishing Company.

VHF Echoes from the Arctic Mesosphere and Lower Thermosphere,
Part II: Interpretations

D. C. Fritts
Geophysical Institute
University of Alaska
Fairbanks, Alaska

and

B. B. Balsley and W. L. Ecklund
Aeronomy Laboratory
NOAA/ERL
Boulder, Colorado

ABSTRACT

Data obtained with the MST radar at Poker Flat, Alaska (65°N)
during the past three years reveal a marked seasonal variation of
echo characteristics in the mesosphere and lower thermosphere.
Winter echoes tend to be weak and occur in the height range ~6 0-80
km while summer echoes are very intense and are confined to the vicinity of the mesopause. It is shown in this paper that the available
winter data are consistent with the saturation of upward-propagating
gravity waves whereas the summer data suggest a shear instability of
the low-frequency motions at the summer mesopause.

1. INTRODUCTION

In recent years, it has been recognized that gravity wave and
tidal motions play a fundamental role in the dynamics of the middle
atmosphere. The radiative heating calculations of Wehrbein and
Leovy (1982) and Apruzese et al. (1982) and mean meridional wind
observations of Nastrom et al. (1982) imply large zonal wind accelerations due to Coriolis torques. However, wind and temperature observations (Theon et al., 1967) suggest that these accelerations are
counteracted in the mesosphere and thermosphere by the zonal drag
associated with dissipating wave motions (Lindzen, 1981). Thus,
gravity wave and, to a lesser degree, tidal motions appear to be
essential in the maintenance of the observed mean temperature and
zonal and meridional wind distributions in the middle atmosphere.

One observational technique that has considerable potential in
the study of both large- and small-scale motions in the middle atmos-

phere is the MST (Mesospheric, Stratospheric, and Tropospheric) radar (Balsley and Gage, 1980). Such facilities are now able to provide reasonably high spatial and temporal resolution data on atmospheric motions and turbulent intensities at several locations. Some of the data collected in the mesosphere and lower thermosphere with the MST radar at Poker Flat, Alaska (65°N) during the first few years of operation are described in a companion paper by Balsley et al. (1983), hereafter denoted I.

The purpose of the present paper is to provide a preliminary interpretation of the winter and summer MST data based upon the analysis that has been performed to date. The most significant characteristics of the winter temperature and velocity data described in I are reviewed and shown to be generally consistent with current models of gravity wave saturation in Section 2. One reasonably monochromatic event, in particular, is analyzed to determine gravity wave parameters, momentum flux, and the induced mean flow acceleration. The significant features of the summer temperature, velocity, and noctilucent cloud observations are reviewed in Section 3 and suggested to be indicative of a shear instability of the various tidal components and low-frequency gravity waves present near the summer mesopause. The apparent shear instability of low-frequency motions at the high-latitude summer mesopause is distinct from the saturation, or convective instability, of gravity waves observed in the winter mesosphere and at the summer mesopause at lower latitudes. The conclusions of this study are presented in Section 4.

2.   WINTER DATA

a.   Review of Data Characteristics

Perhaps the first indication of gravity wave saturation in the mesosphere was the winter rocket grenade temperature data collected by Theon et al. (1967). This data provide evidence of strong, nearly continuous wave activity and superadiabatic (convectively unstable) lapse rates above ~ 50-60 km at all locations.

The winter MST data presented in I provide additional evidence of gravity wave saturation above ~ 60 km at 65°N. The broad, nearly uniform average signal-to-noise (S/N) distribution for winter shown in Figure 2 of I indicates the presence of ~ 3 m turbulent irregularities at levels where saturation (and gravity wave dissipation) is expected. Spectral width and wind variability profiles shown in Figure 6 of I are seen to have almost constant amplitudes across the mesospheric echo region, suggestive of a wave amplitude-limiting mechanism. Finally, the seasonal mean zonal velocities are seen in Figure 15 of I to decrease monotonically to ~ 0 across the echo region, providing strong evidence of a zonal wave drag mechanism. These features of the mesospheric echo will be discussed in more detail later in this section.

There is also evidence in the winter data presented in I of reasonably monochromatic wave saturation events. The most obvious example of this is the low-frequency wave motion observed in the horizontal velocity components in Figure 4 of I. This data suggest a wave motion with $\lambda_z \sim 15$ km, $T \sim 10$ hr, and an amplitude that does not appear to increase with height. Because of its monochromatic nature, this event is analyzed in detail in the last part of this section.

b. Gravity Wave Saturation Theory and Interpretations

The breaking or dissipation of atmospheric tides and gravity waves due to their exponential growth with height was first suggested by Lindzen (1967) and Hodges (1967), respectively. More recently, it has been recognized that a strong zonal drag, presumably due to dissipating wave motions, is required to explain the observed mean temperature and velocity structure in the mesosphere. This recognition has stimulated new interest in the saturation of gravity waves (Lindzen, 1981; Weinstock, 1982), and produced new model results (Holton, 1982; Matsuno, 1982; Dunkerton 1982a; Geller, 1983; Schoeberl et al., 1983) that are in better agreement with the mean meridional wind observations of Nastrom et al. (1982) and the radiative heating calculations of Wehrbein and Leovy (1982) and Apruzese et al. (1982).

The basic idea underlying the theory presented by Lindzen (1981) is that gravity waves grow exponentially with height until the perturbation lapse rate becomes sufficiently large to cancel the mean (stable) lapse rate, at which time convective instabilities occur that act to dissipate the wave motion and cancel further wave growth. The induced wave dissipation results in a vertical divergence of the gravity wave momentum flux, causing an acceleration of the medium toward the phase velocity of the wave. The gravity wave saturation theory presented by Weinstock (1982), while based upon the very different notion of strong nonlinear (wave-wave) interactions among the various components of a broad spectrum of waves, results in very similar predictions. In general, then, saturation refers to any mechanisms that act to limit or reduce wave amplitudes due to their superposition or growth with height. For simplicity, however, we will assume that mesospheric saturation is a convective instability and narrow spectrum in nature.

The requirement by Lindzen (1981) that the perturbation lapse rate exceed that of the mean state for wave dissipation to occur can be shown to be identical to the phase speed condition for convective instability, $u' > c - \bar{u}$, derived by Orlanski and Bryan (1969), where u' is the horizontal perturbation velocity, $\bar{u}$ is the mean flow velocity, and c is the horizontal phase speed of the wave. This condition is also obtained in the "quasi-linear" model of Dunkerton (1982b).

In a uniformly stratified environment, two effects cause u' to vary, both the density decrease with height and the variation of $\bar{u}$

(and $c-\bar{u}$) with height. Below the level of wave saturation then, we see that, analogous to Lindzen's (1981) treatment,

$$u' \sim m^{1/2} e^{z/2H} \tag{1}$$

where m is the vertical wavenumber of the perturbation which varies as $m \sim (c-\bar{u})^{-1}$ and H is the local atmospheric scale height. Above the level of saturation, u' is limited by the phase speed conditon of Orlanski and Bryan (1969),

$$u' = c - \bar{u} \tag{2}$$

This implies wave dissipation with an eddy diffusion coefficient and an induced mean flow acceleration of the form (Lindzen, 1981; Weinstock, 1982),

$$D_{eddy} \sim \frac{k(\bar{u}-c)^4}{N^3}\left[\frac{1}{2H} - \frac{3}{2}\frac{1}{(\bar{u}-c)}\frac{d\bar{u}}{dz}\right] \tag{3}$$

and

$$\frac{\partial\bar{u}}{\partial t} = \bar{u}_t \sim -N^2\frac{D_{eddy}}{(\bar{u}-c)} \sim \frac{-k(\bar{u}-c)^3}{N}\left[\frac{1}{2H} - \frac{3}{2}\frac{1}{(\bar{u}-c)}\frac{d\bar{u}}{dz}\right] \tag{4}$$

Far from critical levels (where $\bar{u}=c$), the first term is dominant and D_{eddy} and $\bar{u}_t$ are nearly constant with height. Near a critical level, however, $(\bar{u}-c)$ dominates and D_{eddy} and $\bar{u}_t$ decay rapidly. There is some evidence of each effect in the MST data presented in I.

The occurrence of superadiabatic lapse rates in the temperature data of Theon et al. (1967) and of MST mesospheric echoes (indicative of $\sim$ 3m turbulent structures) beginning $\sim$ 60 km provide evidence of a causal relationship between the convective instabilities produced by saturating gravity waves and the small-scale turbulent structures associated with enhanced wave dissipation. Such convective instabilities of internal gravity wave fields have also been observed in the laboratory shear flow studies of Koop (1981) and the numerical studies of Fritts (1982). These observations serve to confirm the underlying wave dissipation mechanism of Orlanski and Bryan (1969) and Lindzen (1981).

The uniform character of the average winter S/N profile between 60 and 80 km (Figure 2 of I) in the presence of a mean westerly velocity decreasing $\sim$ uniformly with height (Figure 15 of I) suggests a fairly broad or highly variable spectrum of saturating gravity waves, in line with the ideas of Weinstock (1982). The disappearance of the winter MST echo near the level at which the mean zonal wind passes through zero (z ~ 80km) appears coincidental given that super-adiabatic lapse rates are seen in the rocket grenade temperature data of Theon et al. (1967) to much higher levels. Thus, MST echoes probably cease as a result of the disappearance of ~ 3m turbulent structures (due to the increase of the Kolmogorov inner scale, L_0, with height) rather than the approach of a narrow gravity wave spectrum to critical levels near 80 km.

Additional evidence of a broad spectrum of gravity waves is provided by the nearly uniform distributions of spectral width and wind variability with height (Figure 6 of I). If the echo disappearance was due to the critical level approach of a narrow gravity wave spectrum, we should see these profiles decreasing in amplitude near 80 km. These profiles also suggest that gravity wave amplitudes fail to grow with height above the level of breaking, in accordance with the saturation theories of Lindzen (1981) and Weinstock (1982).

Finally the decrease of the mean zonal wind with height across the winter MST echo region provides strong evidence of the gravity wave induced drag examined by Holton (1982), Matsuno (1982) and Dunkerton (1982a). Indeed, such a drag can only result from the dissipation of vertically propagating waves and the resulting vertical divergence of wave momentum flux.

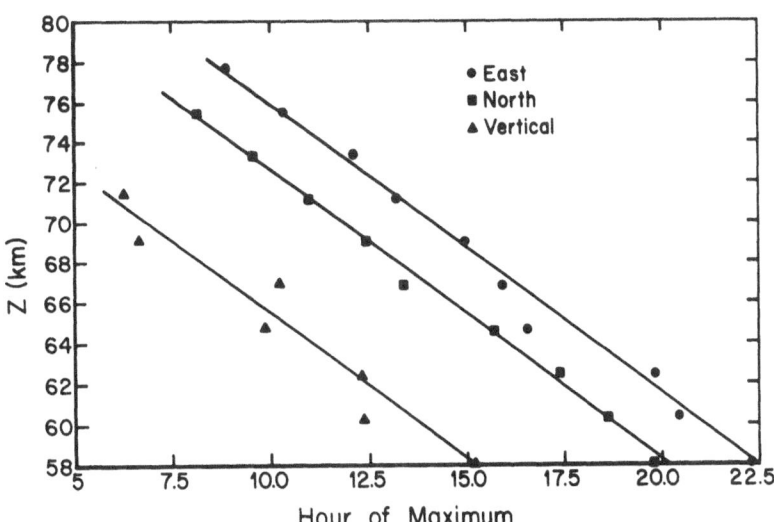

Fig. 1. Time of maximum amplitude for the east, north, and vertical components of the 10-hour wave obtained with the least-squares fits. The downward phase progression is clearly visible.

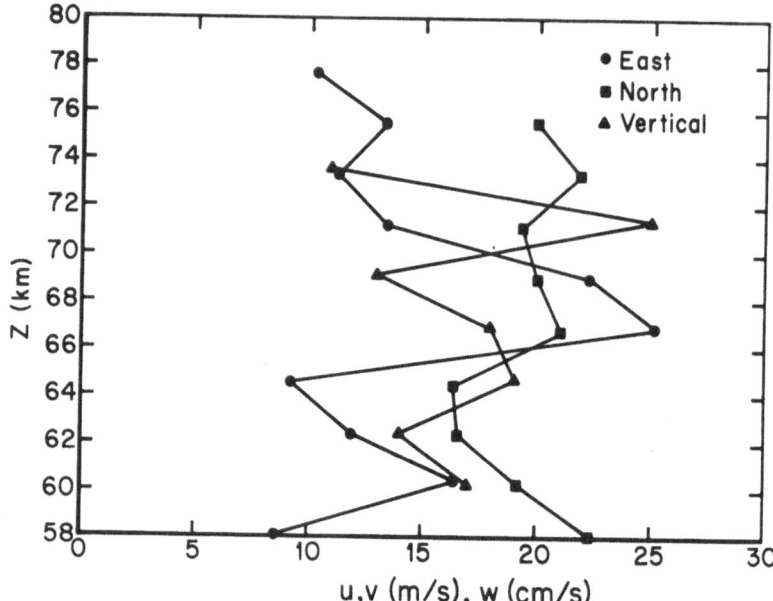

Fig. 2. Amplitudes of the east, north, and vertical components of the
10-hour wave obtained with the least-squares fits at each
range gate. Note the lack of growth with height.

c. Estimation of Gravity Wave Parameters, Momentum Flux, and Mean
Flow Acceleration

As mentioned previously, there is evidence in the winter MST data
of occasional, apparently monochromatic wave motions with near-inertial
frequencies. One such wave is evident in the horizontal velocity data
presented in Figure 4 of I. This wave has a clearly defined, downward
phase progression, a period $T \sim 10$ hr, and zonal and meridional velocity components that are in approximate quadrature. In order to
determine the nature of the wave motion more accurately, the eastward,
northward, and vertical velocity components in each range gate were
fit with a sinusoid with a 10-hr period. The variations of the time
of maximum amplitude and the amplitude with height are shown in
Figures 1 and 2.

The zonal and meridional velocity components exhibit a very uniform phase progression with height because the 10-hour wave clearly
dominates the horizontal velocity field. Vertical phase variations
are less uniform due to the presence in the velocity data of considerable activity at higher frequencies. Nevertheless, the phase progression of the vertical velocity with height agrees with that obtained
for the two horizontal components. The zonal velocity is found to
lag the meridional velocity by $\sim 85°$, with the vertical velocity
leading the meridional by $\sim 175°$.

The amplitude data shown in Figure 2 suggest nearly constant amplitudes with height with average zonal, meridional, and vertical amplitudes of 14.2, 19.6, and 0.18 m/s, respectively. Together with the phase data discussed previously, this data suggest a maximum horizontal perturbation velocity of 19.7 m/s at ~ 5° E of N. The maximum (upward) vertical velocity is then in phase with the horizontal velocity maximum at ~ 5° W of S, implying that this is the direction of wave propagation.

Provided that the horizontal phase velocity of the wave exceeds the mean motion in the direction of wave propagation, the downward progression of phase corresponds to an upward transport of energy and momentum. We proceed now to estimate the various wave parameters and verify that this is the case. To a good approximation, the perturbation velocities and wavelengths in the direction of propagation are related by

$$\frac{\lambda_x}{\lambda_z} \approx \frac{u'}{w'} \ . \tag{5}$$

With

$$\lambda_z \approx 14.5 \text{ km} \tag{6}$$

from Figure 1, we estimate that

$$\lambda_x \approx \frac{u'}{w'}\lambda_z \approx 1590 \text{ km.} \tag{7}$$

The horizontal phase velocity is then

$$c = \frac{\lambda_x}{T} \approx 44 \text{ m/s}, \tag{8}$$

which is substantially larger than the observed mean motions in the direction of wave propagation of

$$\bar{u} \sim 3 \text{ m/s.} \tag{9}$$

The inferred $c-\bar{u}$ of ~ 41 m/s agrees reasonably well with that estimated from the gravity wave dispersion relation,

$$c - \bar{u} \approx \frac{N}{m} \left[1 + \frac{m^2 f^2}{k^2 N^2} \right]^{1/2} \sim 50 \text{m/s}, \tag{10}$$

assuming the mean state in the upper mesosphere at 65°N to have $N \sim 0.02 \text{ s}^{-1}$.

The mean flow acceleration due to wave dissipation in the direction of wave propagation is given by

$$\bar{u}_t = - \frac{1}{\rho} \frac{\partial}{\partial z} (\rho \overline{u'w'}) \tag{11}$$

The mean flow itself is assumed to be in geostrophic balance. For the 10-hour wave,

$$\overline{u'w'} \approx 1.77 \text{ m}^2/\text{s}^2, \tag{12}$$

being approximately constant with height. In that case, equation (11) gives

$$\bar{u}_t \approx \frac{\overline{u'w'}}{H} \approx 25 \text{ m/s/day}. \tag{13}$$

Clearly, then, this 10-hour wave was not acting to appreciably decelerate the mean (westerly) zonal flow. Instead, the major effect of wave dissipation was the acceleration of the mean (northerly) meridional flow.

Given that this wave motion dominated the horizontal velocity field for ~ 3 days, the required large zonal flow decelerations during that time must have been accomplished by higher-frequency wave motions with larger vertical velocities (evident in the data) and appreciably smaller horizontal velocities (implying smaller horizontal wavelengths) than were associated with the 10-hour wave. Although this one gravity wave saturation event may not be representative, the implication that gravity waves with higher frequencies and substantially smaller horizontal scales must play a major role in the zonal flow deceleration in the winter mesosphere is supported by the observation that for the 10-hour wave,

$$u' \underset{\sim}{<} 0.5 (c - \bar{u}) \tag{14}$$

a value considerably smaller than required for the convective instability of a monochromatic wave. Significant zonal flow decelerations due to relatively small-scale gravity waves have also been inferred by Vincent and Reid (1983).

tidal component	ω_n(/24hr)	zonal		meridional	
		u_n(m/s)	ϕ_n^z(rad)	v_n(m/s)	ϕ_n^m(rad)
diurnal	-2π	16	$\dfrac{-19\pi}{12}$	18	$\dfrac{-7\pi}{6}$
semidiurnal	4π	15	$\dfrac{4\pi}{3}$	15	$\dfrac{5\pi}{6}$
terdiurnal	6π	10	$\dfrac{7\pi}{8}$	10	$\dfrac{\pi}{2}$
16-hour	3π	16	$\dfrac{9\pi}{8}$	14	$\dfrac{5\pi}{8}$

Table 1. Observed amplitudes and phases for the diurnal, semi-diurnal and terdiurnal tides and the 16-hour wave at 86 km (from Carter and Balsley, 1982).

3. SUMMER DATA

a. Review of Data Characteristics

The summer velocity and temperature data presently available at high latitudes appear to contrast with the winter data and a strictly gravity wave saturation interpretation. Unlike the broad (weak) average S/N distribution obtained in the winter mesosphere, the average summer S/N distribution has a pronounced peak near the summer mesopause and decays rapidly above and below (Figure 2 of I). Regions of strong echo intensity are observed to occur ~ 2-4 times per day, appearing first several km above the mesopause and descending to the vicinity of the mesopause before disappearing (Figure 12 of I). The average variation of S/N with time-of-day shown in Figure 11 of I exhibits a significant minimum in the four range gates nearest the mesopause between 18 and 24 hr local time, suggestive of a strong daily periodicity in the echo-producing mechanism. Horizontal velocity data reveal the presence of large-amplitude motions, substantial velocity shears, and a range of vertical wavelengths with 12-15 km scales dominant. Finally, the power spectral density obtained by Carter and Balsley (1982) indicates the relative importance of the various tidal components in the wind field at the summer mesopause. The observed mean amplitudes and phases of the various tidal components and the 16-hour wave are listed in Table 1. These observations suggest that a dominant source of the mesopause echo is the instability of the large-scale, low-frequency motions present at the summer mesopause.

Like the MST data, the rocket grenade temperature data obtained in the summer mesosphere and lower thermosphere by Theon et al. (1967) at Barrow (71°N) are in sharp contrast to the winter data. Whereas the winter data show the presence of large-amplitude wave motions and superadiabatic lapse rates, the summer data show a remarkably smooth, near adiabatic lapse rate in the upper mesosphere with a fairly well-defined mesopause. The different summer and winter temperature structure which is most pronounced at high latitudes, suggests strong filtering of the upward propagating gravity waves in summer, as noted by Lindzen (1981). Neither below nor above the summer mesopause, where large vertical displacements would manifest themselves in the temperature data, is there any indication of convective instability, however. This suggests that wave dissipation is driven by a dynamical or shear instability rather than a convective instability of the low-frequency motions at the summer mesopause. The frequent observation of Kelvin-Helmholtz (KH) billows in noctilucent cloud (NLC) formations lends further support to this idea (Haurwitz and Fogle, 1969). KH billows are known to evolve as a consequence of a large velocity shear in a statistically stable environment.

The evidence of a shear instability near the summer mesopause, of course, does not preclude intermittent gravity wave saturation as well. In fact, the necessary conditions for both are expected to occur almost simultaneously in many instances (Fritts, 1982). Thus it is assumed that the instability is of a dynamical nature in light of the observations cited above, but it is recognized that convective instabilities would result in very similar wave dissipation and turbulence production.

It will be shown in the remainder of this section that the assumption of a dynamical instability of the observed tidal and low-frequency gravity wave motions can account for several significant features of the summer MST data, including the average S/N distribution, its diurnal variation, and the occurrence, frequency, and downward motion of regions of enhanced S/N.

b. Mesopause Gravity Wave and Tidal Structure

In this section we approximate the structure of large-scale gravity waves and the observed tidal components in the vicinity of the summer mesopause in order to estimate the occurrence of dynamically unstable velocity shears. It was suggested earlier that shear instability may provide the dominant wave dissipation mechanism at the arctic summer mesopause and explain the occurrence of the summer mesopause echo.

Because of the apparent dominance of the power spectral density at 85 km by the various tidal components, we estimate an "average" wave structure using the results of classical tidal theory and the observed mean tidal amplitudes and phases near the summer mesopause (Carter and Balsley, 1982). The WKB solution of the vertical structure equation for atmospheric tides assuming no local heating (Chapman and Lindzen, 1970) is

$$k^2 = \frac{1}{Hh_n} \left[\frac{\gamma - 1}{\gamma} + \frac{dH}{dz} \right] - \frac{1}{4H^2} \tag{15}$$

where k, h_n, γ , and H are the vertical wavenumber, the tidal equivalent depth, the ratio of specific heats, c_p/c_v, and the local scale height, RT/g, respectively. Assuming the mesopause temperature distribution is sufficiently global, the vertical structure depends on h_n and on the value and vertical gradient of the temperature near the mesopause. A similar vertical structure equation is satisfied by low-frequency gravity waves which may, in many cases, closely resemble various tidal components.

The shear instability model assumes a temperature distribution approximating that observed by Theon et al. (1967) at Barrow (71°N) with a temperature minimum of ~ 133 K at 84 km and temperature gradients of ± 6 K/km above and below the mesopause. This distribution is shown with a solid line in Figure 3.

Based largely upon the appearance of persistent 12-15 km wavelengths in the horizontal velocity data in Figure 8 of I and the minimum vertical wavelength estimates of ~ 12-16 km for the diurnal, semidiurnal, the terdiurnal tides and the 16-hour wave, we assume a minimum vertical wavelength of 12 km for all motions. This corresponds to a selection of an equivalent depth in equation (15) of $h_n \sim 0.7$ km. The resulting vertical wavenumber distribution is shown in Figure 3 with a dashed line. The maximum just above 85 km occurs where the temperature is small and its gradient is large. This implies a region in which the various wave motions have small vertical wavelengths and large shears and provides an explanation for the preferential occurrence of instability (due to the superposition of wave and mean shears) at and above the mesopause. Evidence of this type of wave structure near the mesopause was obtained at Poker Flat on 13 July 1982 during a polar cap absorption event. The observed velocity profiles, shown in Figure 4, suggest strong vertical scale compression and shear enhancement.

The stability of a shear flow, of course, depends on both the stratification and the shear as the Richardson number,

$$Ri = \frac{N^2}{(u_z^2 + v_z^2)} \tag{16}$$

where N is the Brunt-Vaisala frequency and u_z and v_z are the total velo-

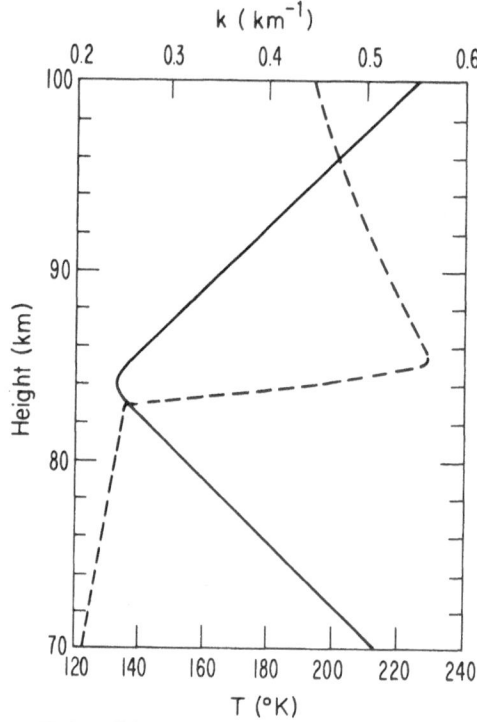

Assumed Temperature Profile (Solid Line)
and Resulting Vertical Tidal Wave
Number (Dashed Line) vs Height.

Fig. 3.   Vertical profiles of the model temperature (———) and the tidal
vertical wavenumber (- - -) obtained with equation (15).   The
maximum vertical wavenumber corresponds to a vertical wavelength ~ 12 km.

city shears in the zonal and meridional directions. Generally a
shear flow is dynamically unstable when Ri < 1/4. The zonal and
meridional shears resulting from the various tidal components are
given by

$$(u_z, v_z) = (u_{z_0}, v_{z_0}) + \text{Re}\left[\sum_n (u_n, v_n) ike^{ikz} e^{i(\omega_n t + \phi_n z,m)}\right] (17)$$

Here u_{z_0} and v_{z_0} are the observed mean shears, u_n and v_n are the mean
zonal and meridional tidal amplitudes at the mesopause, ω_n are
the tidal frequencies, and $\phi_n^{z,m}$ are the mean zonal and meridional

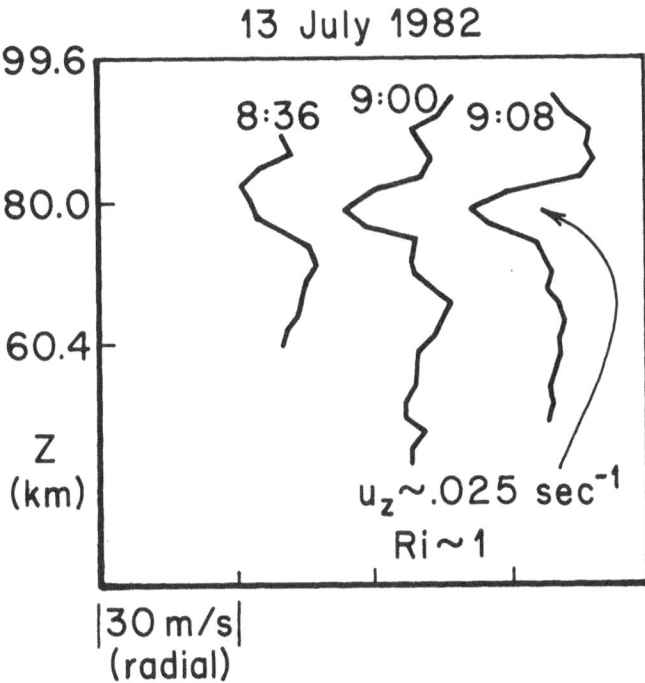

Polar Cap Absorption
Radial Velocity Profiles

13 July 1982

Fig. 4. Radial (NE) velocity profiles at 15° off vertical during the
polar cap absorption event on 13 July 1982. Note the large
amplitudes and shears (Ri ~ 1) occurring near the mesopause.

phases at 86 km. The observed values of these parameters for each
of the tidal components and the 16-hour wave are listed in Table 1.
The vertical structure must, in general, also reflect the decrease
in density with height. However, this effect is assumed to be balanced
by dissipation above the mesopause.

The Richardson number is then calculated as a function of time
and height using the velocity shears given by equation (17) for the
diurnal, semidiurnal, and terdiurnal tidal components and the assumed
vertical structure derived earlier. The inverse of this quantity is
contoured in a time-height cross section in Figure 5a with contour
levels of $Ri^{-1} = 0.25$ and 0.5. The time-height cross section

shows several regions of large inverse Richardson number (where the flow is most likely to be dynamically unstable) appearing first above the mesopause and moving downward with time, in qualitative agreement with observations (Figure 12 of I).

Shown in Figure 5b is the maximum inverse Richardson number for a single tidal component with an amplitude of 20 m/s. The decrease below the mesopause implies that instabilities should become less likely there providing a possible explanation for the decay of average S/N with height below the mesopause.

The tidal components above predict a Richardson number distribution with a period of 24 hrs. Observations, however, show that there is considerable daily variability in the occurrence of large S/N. Much of this daily variability may probably be attributed to the modulation of tidal shears and other tidal parameters by low-frequency gravity waves (Walterscheid, 1981). This daily variability is illustrated with the inclusion of the observed 16-hour component in the time-height cross section of the inverse Richardson number in Figure 6. The contour levels here are Ri^{-1} = 0.25, 0.50, 0.75, and 1.0. Compared wtih Figure 5a, we see that the 16-hour wave

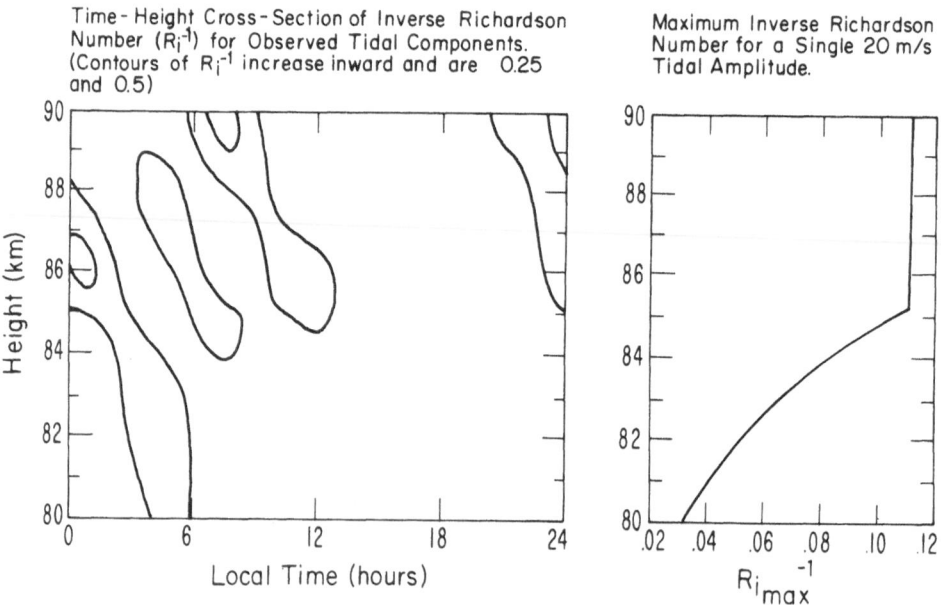

Fig. 5. Time-height cross section of the inverse Richardson number resulting from the mean tidal components (a) and the maximum inverse Richardson number produced by a single tidal mode with an amplitude of 20 m/sec (b), assuming the vertical wavenumber profile shown in Figure 1. The contour values in (a) are Ri^{-1} = 0.25, and 0.5, increasing inward.

results in both a pronounced asymmetry in the daily variation of
large shears as well as more intense shears than predicted for the
tidal components alone. There are now ~ 4 occurrences of large
seasonal mean shears (with $Ri^{-1} > 0.5$) near the mesopause in each
two-day period, in addition to several regions with somewhat less intense shears.

The failure of unstable shears (with $Ri^{-1} > 4$) to occur is likely
a consequence of using seasonal mean tidal and 16-hour wave amplitudes
and phases and neglecting all other low-frequency (and high-frequency)
motions. Such motions would surely act to increase the predicted
inverse Richardson numbers and contribute to the daily variability
of the large velocity shears.

Assuming that we can associate large but stable seasonal mean
shears with the generation of turbulence, however, the model does a
very good job of anticipating the occurrence, approximation frequency,
and motion of the observed regions of enhanced S/N. The time-height
cross section in Figure 6 also indicates that the model provides an
explanation for the observed decrease in average S/N near the mesopause
between 18 and 24 hours each day. This is one feature that is impossible to explain in terms of random, low-frequency gravity waves.

The shear instability model described here is itself unable to
account for the observed decrease in S/N above the mesopause. Such a
decrease, however, is a natural consequence of the increase of kine-

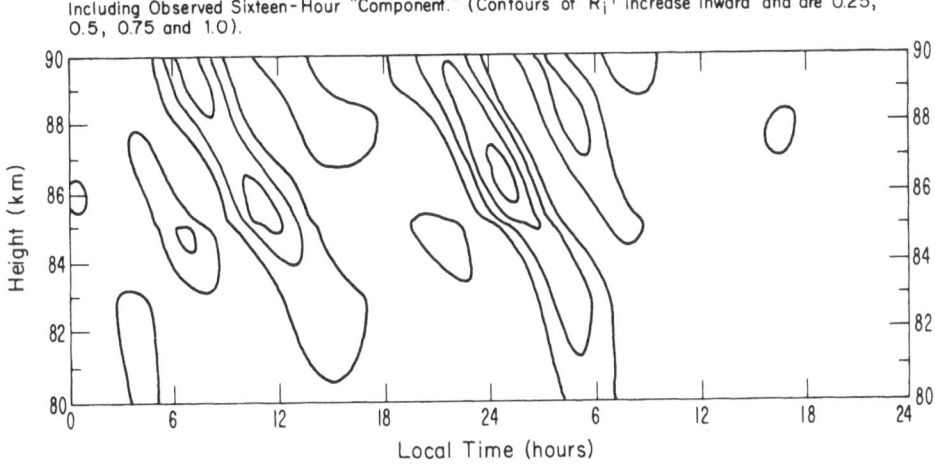

Time-Height Cross-Section of Inverse Richardson Number (R_i^{-1}) for Observed Tidal Components, Including Observed Sixteen-Hour "Component." (Contours of R_i^{-1} increase inward and are 0.25, 0.5, 0.75 and 1.0).

Fig. 6. As in Figure 5a, but including the observed mean 16-hour wave.
Contour values are $Ri^{-1} = 0.25$, 0.5, 0.75, and 1.0, increasing
inward.

matic viscosity, and the resulting increase of the Kolmogorov inner scale of turbulence, with height. Indeed, the required turbulent dissipation rate for the occurrence of ~ 3 m structure at the summer mesopause

$$\varepsilon = \nu^3/L_o^4 \approx 0.02 \ m^2/sec^3 \tag{18}$$

is in reasonable agreement with that expected from tidal and gravity wave dissipation.

4. CONCLUSIONS

We have presented a preliminary interpretation of the early data collected with the MST radar at Poker Flat, Alaska in the mesosphere and lower thermosphere. Together with winter and summer temperature profiles obtained with the rocket grenade technique and observations of NLC at the summer mesopause, these data suggest distinct wave dissipation mechanisms at arctic latitudes during summer and winter.

The winter data appear to be in general agreement with the saturation theories of Lindzen (1981) and Weinstock (1982); enhanced radar echoes (indicative of increased wave dissipation) are observed above ~ 60 km, saturated wave amplitudes appear to be nearly constant with height, and the zonal wind is seen to decrease across the region of enhanced dissipation, consistent with the model results of Holton (1982), Matsuno (1982), and Dunkerton (1982a). Data collected at Poker Flat, Alaska suggest that saturation is nearly continuous but may vary considerably in intensity and character.

An analysis of one saturation event involving a gravity wave with a period of 10 hr found the wave to be nearly circularly polarized with vertical and horizontal wavelengths ~ 14.5 and ~ 1590 km, a phase velocity ~ 44 m/s to the south, and a Reynolds stress ~ 1.77 m^2/s^2, resulting in a mean flow deceleration of $\bar{u}_t$ ~ 25 m/s/day. Based both on the observed amplitude and direction of propagation of the 10-hour wave and on the simultaneous observation of large-amplitude, higher-frequency vertical motions, it was concluded that gravity waves with substantially smaller horizontal wavelengths must be contributing in a significant way to both gravity wave saturation and zonal flow decelerations.

The summer temperature, MST, and NLC data were found to suggest a dynamical instability of the low-frequency motions present near the summer mesopause at high latitudes. This interpretation was based on the absence of superadiabatic lapse rates in the summer temperature data, the frequent occurrence of KH billows in NLC formations at the mesopause, and a strong daily periodicity in the average S/N near the mesopause. A very simple model of the observed gravity wave and tidal structure near the mesopause was found to reproduce many features of the data. These included the observed wave structure near the mesopause, the decrease in average S/N below the mesopause, and the occurrence, approximate frequency, and motion of the observed regions of enhanced S/N.

As mentioned earlier, the apparent dominance of shear instability at the high-latitude summer mesopause does not preclude gravity wave saturation (convective instability) altogether. In fact, convective instabilities must accompany the saturation of KH instabilities themselves and may appear at smaller vertical scales if higher-resolution temperature data were available. But the frequent occurrence of KH billows at the summer mesopause guarantees a major role for dynamical instabilities.

ACKNOWLEDGEMENTS

This research was sponsored by the Air Force Office of Scientific Research (AFSC) under Grant AFOSR-82-0125 and the Atmospheric Section of the National Science Foundation.

REFERENCES

Apruzese, J. P., M. R. Schoeberl, and D. F. Strobel, 1982: Parameterization of IR cooling in a middle atmosphere dynamics model 1. Effects on the zonally averaged circulation, J. Geophys. Res., 87, 8951-8966.

Balsley, B. B., and K. S. Gage, 1980: The MST radar technique: potential for middle atmosphere studies, Pure and Applied Geophys, 118, 452-493.

Balsley, B. B., W. L. Ecklund, and D. C. Fritts, 1983: VHF Echoes from the Arctic Mesosphere and Lower Thermosphere, Part I: Observations, Adv. Earth and Planet Sci. (this issue).

Carter, D. A. and B. B. Balsley, 1982: The summer wind field between 80km - 93km observed by the MST radar at Poker Flat, Alaska (65° N), J. Atmos. Sci., 2905-2915.

Chapman, S., and R. S. Lindzen, 1970: Atmospheric Tides, Gordon and Breach Science Publishers, New York.

Dunkerton, T. J., 1982a: Stochastic parameterization of gravity wave stresses, J. Atmos. Sci., 39, 1711-1725.

Dunkerton, T. J., 1982b: Wave transience in a compressible atmosphere, part III: the saturation of internal gravity waves in the mesosphere, J. Atmos. Sci., 39, 1042-1051.

Fritts, D. C., 1982: The transient critical-level interaction in a Boussinesq fluid, J. Geophys. Res. 87C, 7997-8016.

Geller, M.A., 1983: Dynamics of the middle atmosphere, Space Science Reviews, to appear.

Haurwitz, B. and B. Fogle, 1969: Wave forms in noctilucent clouds, Deep Sea Res. 16, 85-95.

Hodges, R. R., Jr., 1967: Generation of turbulence in the upper atmosphere by internal gravity waves, J. Geophys. Res., 72, 3455-3458.

Holton, J. R., 1982: The role of gravity wave induced drag and diffusion in the momentum budget of the mesosphere, J. Atmos. Sci.,

39, 791-799.

Koop, C. G., 1981: A preliminary investigation of the interaction of internal gravity waves with a steady shearing motion, J. Fluid Mech., 113, 347-386.

Lindzen, R. S., 1967: Thermally driven diurnal tide in the atmosphere, Quat. J. Roy. Met. Soc., 93, 18-42.

Lindzen, R. S., 1981: Turbulence and stress due to gravity wave and tidal breakdown, J. Geophys. Res., 86C, 9707-9714.

Matsuno, T., 1982: A quasi one-dimensional model of the middle atmosphere circulation interacting with internal gravity waves, J. Meteor. Soc. Japan, 60, 215-226.

Nastrom, G. D., B. B. Balsley, and D. A. Carter, 1982: Mean Meridional winds in the mid and high-latitude summer mesosphere, Geophys. Res. Letters, 9, 139-142.

Orlanski, I. and K. Bryan, 1969: Formation of the thermocline step structure by large-amplitude internal gravity waves, J. Geophys. Res., 74, 6975-6983.

Schoeberl, M. R., D. F. Strobel, and J. P. Apruzese, 1983: A numerical model of gravity wave breaking and stress in the mesosphere, Adv. Earth and Planet. Sci. (this issue).

Theon, J. S., W. Nordberg, L. V. Katchen, and J. J. Horvath, 1967: Some observations on the thermal behavior of the mesosphere, J. Atmos. Sci., 24, 428-438.

Vincent, R. A., and I. M. Reid, 1983: HF Doppler measurements of mesospheric gravity wave momentum fluxes, J. Atmos. Sci., to appear.

Waltersheid, R. L., 1981: Inertio-gravity wave induced accelerations of mean flow having an imposed periodic component: Implications for Tidal observations in the meteor region, J. Geophys. Res., 86, 9698-9706.

Wehrbein, W. M., and C. B. Leovy, 1982: An accurate radiative heating and cooling algorithm for use in a dynamical model of the middle atmosphere, J. Atmos. Sci., 39, 1532-1544.

Weinstock, J., 1982: Nonlinear theory of gravity waves: Momentum deposition, generalized Rayleigh friction, and diffusion, J. Atmos. Sci., 39, 1698-1710.

J. R. Holton and T. Matsuno, Dynamics of the Middle Atmosphere, 117-140.

MULTIPLE "GUST LAYERS" OBSERVED IN THE MIDDLE STRATOSPHERE

Manabu D. Yamanaka* and Hiroshi Tanaka

Institute of Space and Astronautical Science, Tokyo 153, Japan
and Water Research Institute, Nagoya University, Nagoya 464,
Japan

ABSTRACT

Some theoretical aspects and quick results of a balloon observation
of gravity waves and turbulence in the middle stratosphere are shown.
Turbulence layer thickness, previously observed by some workers, is
theoretically related to the distance between the critical level and
breaking level of internal (inertio-) gravity waves. It is predicted
that the microstructure in the layer may involve multiple "gust layers".
The observation was carried out by using "adapted Gill-type" propellor
anemometers which scanned through 600 m in depth under the balloon for
over 30 hours. Quick analyses demonstrate that multiple "gust layers"
of 10-50 m in thickness often appear in the middle stratosphere (altitude: 24-26 km). Vertical shear reaches 10^{-1} s^{-1} in the layers, which
may be enough to generate turbulence. It is suggested that extended
analyses of this observational data will give information on the input
of wave momentum flux into the middle atmosphere.

1. INTRODUCTION

Stratospheric balloon observations have been planned through the
Middle Atmosphere Program (MAP) in order to reveal the origin and nature
of turbulence in the middle atmosphere. Some observations so far demonstrated that there exist multiple thin layers of turbulence in the lower
stratosphere in spite of its strong stratification (Cadet, 1977; Dewan,
1981; Woodman, 1981; Barat, 1982). A possible reason for the origin of
such turbulence layers is critical-level breakdown of upwards propagating internal gravity waves generated in the troposphere (Geller *et al.*,
1975; Tanaka, 1975, 1980, 1982, 1983a; Yamanaka and Tanaka, 1982). However, the following problems have not yet been solved:
  (i) The microstructure and life cycle of the turbulence layer;
  (ii) The role of inertial effect due to the Coriolis force on
      the propagation and breakdown of internal gravity waves;

---

* Graduate student of Nagoya University.

(iii) The climatology of amplitudes and wavelength of upwards
propagating gravity waves at the bottom of the middle atmosphere.
All of these must be solved to clarify the two important roles of turbulence in the dynamics of the middle atmosphere: diffusion (Cadet, 1977;
Dewan, 1981; Barat, 1982) and momentum transport (Lindzen, 1981; Matsuno,
1982; Holton, 1982).

The critical level for steady, non-dissipating, *non-inertial* internal gravity waves is the level where the mean flow takes the same velocity as the wave phase, and it has the following features:

(I) As a wave approaches the critical level, the vertical wavelength tends to zero, and the horizontal velocity perturbation
tends to infinity;

(II) As a wavepacket approaches the critical level, the vertical
group velocity tends to zero;

(III) The momentum flux associated with a wave is independent of
the height except for a gap discontinuity at the critical level.

These imply that internal gravity waves are selected according to their
phase velocity, and that the momentum input at the bottom is added to
the mean flow at the critical *level* (Eliassen and Palm, 1961; Bretherton,
1966; Booker and Bretherton, 1967). In the actual middle atmosphere the
following two ideas on the effective critical-*layer* absorption of internal gravity waves have been presented:

(a) On the basis of (I), the wave *breaks* into turbulence through
some local instabilities near the critical level so as to adjust the flow to the marginal state (Lindzen, 1981; Holton,
1982);

(b) Based on (II), the wavepacket is gradually *dissipated* due to
the viscosity and/or the Newtonian cooling as it approaches the
critical level (Plumb and McEwan, 1978; Matsuno, 1982).

Both theories are equivalent in considering that the wave momentum of
(III) is added effectively to flow in a layer near the critical level
(Dunkerton, 1982). This will be schematically shown in Fig. 2. However,
they are different from each other in the role of turbulence, which is
the result of wave breaking in (a) but is a predominant cause of wave
dissipation in (b). We can find that the critical level turbulence is
inseparable from what works as the eddy viscosity in the middle atmosphere, because both parameterizations (a) and (b) can simulate the general circulation.

The turbulence layer thickness which will be theoretically obtained
in Chapter 2 as the distance between the critical level and the breaking
level is also based on the idea (a) in simpler and more generalized
forms. Suppression of turbulence generation by the molecular viscosity
and the Newtonian cooling were checked by using the same idea as (b)
(Tanaka, 1982, 1983a). The inertial effect due to the Coriolis force
produces "valve"-like absorptions of long-wavelength gravity waves near
Jones' (1967) critical levels, where the wave intrinsic frequency is
equal in magnitude to the Coriolis factor (Yamanaka and Tanaka, 1982).
The turbulence layer predicted by such a steady linear theory is an ensemble of multiple unstable zones along the wavefronts (Geller *et al.*,

1975), which is clear from (I) and (a). In other wards the vertical microstructure of horizontal flow in such a "turbulence" layer is made up of many layers of "gusts", which must be, exactly speaking, incorporated by nonlinear effects (Tanaka, 1975, 1980) and wave transience (Fritts, 1982).

Multiple "gust layers" were actually observed in the middle stratosphere by our balloon observation descrived in Chapter 3. Since our motivation was to clarify the points (i)-(iii) mentioned in the beginning, we planned a balloon instrumentation sophisticated enough to obtain full data on both turbulence and gravity waves (Yamanaka *et al.*, 1983). We achieved an observation duration of 30 hours. Although our analyses have not yet been completed, we can quickly show in Chapter 4 that the "gust layer" provides a key for solving problems of stratospheric turbulence such as its small diffusivity (Cadet, 1977; Barat and Aimedieu, 1981; Barat, 1982) and its diverged inertial subrange (Gage, 1979; Dewan, 1981; Larsen *et al.*, 1982; Balsley and Carter, 1982). Furthermore, one can find from theoretical grounds that analyses of the stratospheric "gusts" lead to quantitative information on the input of wave momentum flux into the whole middle atmosphere, which is the highest aim of our observational study.

2. THEORETICS

2.1. *Local convective instability*

A wave is a series of undurations of the isopycnic surfaces (Fig. 1). Let ζ be the vertical displacement of a parcel, (x, y) the horizontal and vertical coordinates and (k, m) the horizontal and vertical wavenumbers. The inclination of such a material surface to the x-direction is given by $|\partial\zeta/\partial x| = k \cdot |\zeta|$, and that of the wavefront is written by $(2\pi/|m|)/(2\pi/k)$, where we assume k is a positive constant. Local convective instability occurs when the former becomes larger the latter, *i.e.* $|m| > 1/|\zeta|$. From (I) and (a) mentioned in Chapter 1 the wave breaking level is given by the root z of the following equation (Yamanaka and Tanaka, 1982):

$$|m(z) \cdot \zeta(z)| = 1. \qquad (1)$$

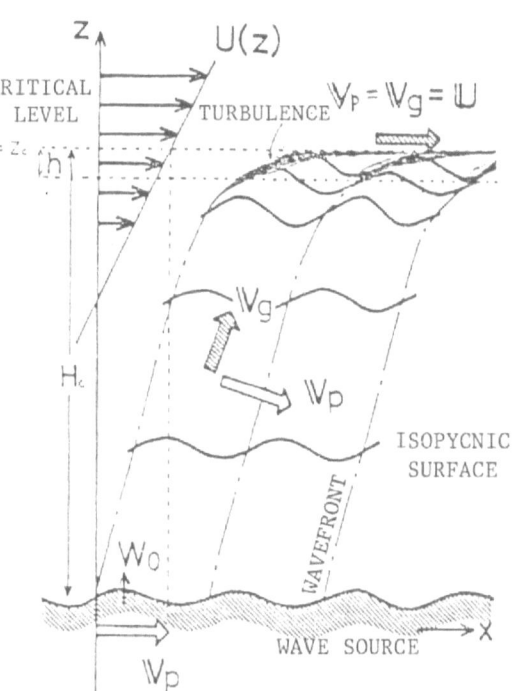

Fig. 1 Propagation and breakdown of internal gravity waves. V_p and V_g denote the phase velocity and the group velocity, respectively.

120

This level is slightly lower than the critical level (Fig. 2). The turbulence layer thickness, h, is obtained from the WKB solution for internal gravity waves in a Boussinesq, non-rotating atmosphere in the same form as derived by Geller et $al.$ (1975):

$$h = \frac{N^{2/3} \cdot W_0{}^{2/3}}{k^{2/3} \cdot |U_z|^{4/3} \cdot H_c{}^{1/3}} , \qquad (2)$$

where N is the Brunt-Väisälä frequency, W_0 the vertical velocity amplitude at the wave source, U_z the basic vertical shear and H_c the height of the critical level measured from the wave source. When we use $N \simeq 2\times 10^{-2}$ s^{-1}, $|U_z| \simeq 2$ m·s^{-1}/km and $H_c \simeq 10$ km, a turbulence layer generated by a wave of $2\pi/k \simeq 50$ km and $W_0 \simeq 1.3$ cm·s^{-1} has thickness estimated by $h \simeq 400$ m. It is found from (2) that h becomes thicker for longer-wavelength, larger-amplitude waves under stronger stratifications.

2.2. Effect of compressibility

For a compressible atmosphere we can transform the variables as $w \rightarrow w^* \cdot \exp(-z/2H)$, where H is the density scale height, assumed constant. Since w^* in the compressible system can be expressed in the same form as w in the Boussinesq system under the WKB approximation (Bretherton, 1966), we finally obtain the marginal state for the compressible atmosphere as follows, in place of (1):

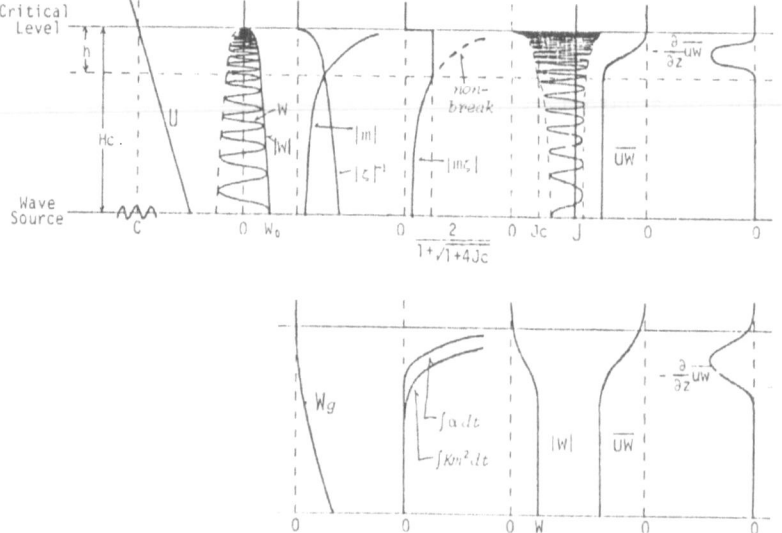

Fig. 2 Effective absorption of internal gravity waves near a critical level. The upper series of figures show the case of breakdown theory, and the lower ones are the case of dissipation theory.

$$\left| m(z) \cdot \zeta(z) \right| \cdot \exp(z/2H) = 1, \tag{3}$$

where m and ζ are quantities derived in the Boussinesq system. Then we have the following expression for turbulence layer thickness (Tanaka, 1982):

$$h = \frac{N^{2/3} \cdot W_0^{2/3}}{k^{2/3} \cdot \left| U_z \right|^{4/3} \cdot H_C^{1/3}} \cdot \exp\frac{H_C}{3H}, \tag{4}$$

which is exactly the same as those derived by Lindzen (1981) and Holton (1982). We can consider that the wave amplification effect due to compressibility is not so important in the stratosphere, where $\exp(H_C/3H)$ may be estimated approximately as unity.

2.3. Local KH (Kelvin-Helmholtz) instability

The local Richardson number $\tilde{J}$ is written as

$$\tilde{J} \equiv \frac{N^2 + \dfrac{\partial \sigma}{\partial z}}{\left(U_z + \dfrac{\partial u}{\partial z} \right)^2} \geq \frac{1 - \left| m\zeta \right|}{\left| m\zeta \right|^2}, \tag{5}$$

where σ is the buoyancy defined by $g \cdot \delta\rho/\bar{\rho}$ ($\bar{\rho}$ and $\delta\rho$: the basic and perturbed density, respectively) and u the perturbed horizontal velocity. We assume in (5) that the mean Richardson number $J \equiv N^2/U_z^2 \gg 1$ and $|u| = (|m|/k) \cdot |w| \simeq N \cdot |\zeta|$. According to the hypothesis (a) mentioned in Chapter 1, we consider that waves should break as $\tilde{J} = J_C$, where J_C is the critical Richardson number for a local stability. Thus the marginal state is expressed by

$$\left| m(z) \cdot \zeta(z) \right| \simeq 2/(\sqrt{1 + 4J_C} + 1). \tag{6}$$

The WKB theory leads to the turbulence layer thickness as follows:

$$h \simeq \frac{N^{2/3} \cdot W_0^{2/3}}{k^{2/3} \cdot \left| U_z \right|^{4/3} \cdot H_C^{1/3}} \cdot \exp\frac{H_C}{3H} \cdot \left(\frac{\sqrt{1 + 4J_C} + 1}{2} \right)^{2/3}, \tag{7}$$

where the effect of compressibility is also considered.

For the local KH instability we put $J_C = 1/4$. We find that breakdown due to the local KH instability begins a little earlier than the local convective instability. Then h in (7) is 1.13 times as thick as predicted from (3), as was suggested by Geller $et\ al.$ (1975). Therefore for estimation of the turbulence layer thickness it is enough to consider only the local convective instability ($J_C = 0$).

2.4. Effects of viscosity and Newtonian cooling

We now discuss the stabilization effects due to molecular viscosity and Newtonian cooling. According to thw wave dissipation theory (b) mentioned in Chapter 1, we may write an attenuation factor as

$$\exp\left(-\left|\frac{dt}{\tau}\right|\right) \simeq \exp\left(-\left|\frac{dz}{W_g}\right|\right) = \exp\left(\left|\frac{dm}{kU_z}\right|\right), \tag{8}$$

where $W_g \equiv \partial\hat{\omega}/\partial m$ is the vertical group velocity. and τ a time constant of attenuation alternatively defined by

$$\tau_{vis} \simeq 1/(\nu m^2) \qquad \text{and} \qquad \tau_{New} \simeq 1/\mu \tag{9}$$

with the molecular kinematic viscosity ν and the Newtonian cooling factor μ. Therefore $|\zeta|$ must be multiplied by the factor (8) in the viscous diabatic atmosphere, which may be important near critical levels because of small W_g and large m in (8) and (9).

On the basis of the considerations §§2.1-4 the turbulence layer thickness with the critical-level breakdown of non-inertial internal gravity waves is generally given by the root h of the following equation (Tanaka, 1982):

$$h \cdot \exp\left\{\frac{2}{9}\left(\frac{h_\nu}{h}\right)^3 + \frac{2h_\mu}{3h} + \frac{h}{3H}\right\}$$

$$= \left(\frac{\sqrt{1+4J_c}+1}{2}\right)^{2/3} \cdot \frac{W_0^{2/3} N^{2/3}}{k^{2/3}|U_z|^{1/3}H_c^{1/3}} \cdot \exp\left\{\frac{2}{9}\left(\frac{h_\nu}{H_c}\right)^3 + \frac{2h_\mu}{3H_c} + \frac{H_c}{3H}\right\}, \tag{10}$$

where

$$h_\nu \equiv \left(\frac{\nu N^3}{k|U_z|^4}\right)^{1/3} \qquad \text{and} \qquad h_\mu \equiv \frac{\mu N}{k|U_z|^2} \ .$$

(7) is a special form of (10) for the inviscid and adiabatic case ($\nu \to 0$ and $\mu \to 0$). We find $h > (2e/3)^{1/3} \cdot h_\nu$ for $\mu = 0$ and $h > (2e/3) \cdot h_\mu$ for $\nu = 0$. Therefore the turbulence layer thickness has a lower limit in the actual atmosphere due to the stabilization effects. In the middle and lower stratosphere this limit may be estimated as about 100 m (Tanaka, 1983a).

2.5. *Inertial effect due to Coriolis force*

In the lower stratosphere long-wavelength ($\sim 10^3$ km), long-period (~ 1 day) gravity waves are often observed (*e.g.* Sawyer, 1961; Cadet and Teitelbaum, 1979)*. It is reasonable to consider that they are modulated by the Coriolis force and to treat them as so-called internal *inertio*-gravity waves. As Jones (1967) has pointed out, if the basic

---

* Cadet and Teitelbaum observed the equatorial stratosphere, where the β-effect may be incorporated. However, in many cases $\beta L/f_0$ (f_0 and β: the values of f and df/dy at a reference latitude; L: a characteristic value of the meridional displacement of parcels with gravity wave) can be small enough to neglect the β-effect (Yamanaka, 1983).

flow is sheared only in the vertical direction, an internal inertio-gravity wave with varying intrinsic frequency $\hat{\omega}$ has critical levels indicated by

$$\hat{\omega} = \pm f , \tag{11}$$

where f is the Coriolis factor assumed to be constant. The critical level for a non-rotating system as studied by Booker and Bretherton (1967) is a special case for $f \to 0$. Propagation and breakdown of inertio-gravity waves are, however, quite different from those of non-inertial waves in the vicinity of the Jones' critical levels (11), as shown in the Appendix.

Two important features of gravity wave propagation affected by the Coriolis force are the *valve effect* and the *turning-level reflection* (Fig. 3). Considering the lower and middle stratosphere in the extra-tropical northern hemisphere, we hereafter set $U_z < 0$, $f > 0$ and $k > 0$.

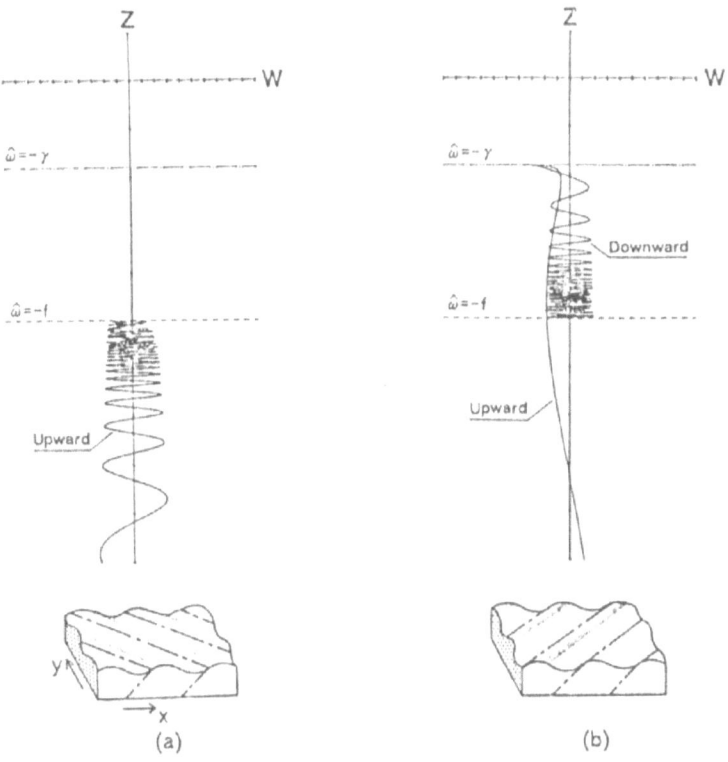

(a)   (b)

Fig. 3 "Valve" effect of a Jones' critical level ($\hat{\omega} = -f$) and relection at a "turning level" ($\hat{\omega} = -\gamma$) for incident inertio-gravity waves: (a) $\ell > 0$; (b) $\ell < 0$.

Then waves with *negative* ℓ, where ℓ is the meridional wavenumber, break down on the lower side of the lower critical level ($\hat{\omega} = -f$), and the turbulence layer thickness is given by

$$h \simeq \frac{|\ell|^{4/5} \cdot W_0^{4/5}}{k^{8/5} \cdot |U_z|^{4/5} \cdot H_C^{3/5}} \ , \tag{12}$$

where we use the simplest breaking-level equation (1) with the aid of the exact expressions for m and ζ in the vicinity of the critical level $\hat{\omega} = -f$ (Yamanaka and Tanaka, 1982). We find that stability of the basic field (N) does not affect the turbulence layer thickness [compare (12) with (2)]. Differences on the indicial characters of parameters in the two thickness expressions result from the respective asymptotic behaviors $|\zeta|$, which is proportionate to $|\hat{\omega} + f|^{-1/4}$ for inertio-gravity waves but to $|\hat{\omega}|^{-1/2}$ for non-inertial waves (see Appendix). On the other hand, waves with *positive* ℓ pass upward through the critical level ($\hat{\omega} = -f$) and reflect at the lower turning level ($\hat{\omega} = -\gamma$). Then they go downward and break down on the upper side of the critical level ($\hat{\omega} = -f$).

All of these striking features due to the inertial effect appear in the following region near the Jones' critical level:

$$\gamma < |\hat{\omega}| < \sqrt{2}f \ , \tag{13}$$

where the conventional dispersion relation of inertio-gravity waves is invalid (see Appendix). We can prove that in the region above the turning level ($\hat{\omega} > -\gamma$) waves coming from the bottom source are evanescent, so that wave absorption between the far upper region ($\hat{\omega} \gg f$) and the lowest bottom ($\hat{\omega} \ll -f$) agrees with that on both sides of the critical level ($\hat{\omega} = 0$) for non-inertial waves. This fact has been pointed out yet by the original study of Jones (1967) and is also clarified by comparing (A3) with (A4) in Appendix. However, in relation to the production of turbulence layers, we should separately consider the non-inertial case, the negative-ℓ inertial case and the positive-ℓ inertial case.

2.6. *Thickness and origin of stratospheric turbulence layers*

The three cases mentioned above are plotted in Fig. 4, where we take a set of typical values for the lower and middle stratosphere: $N \simeq 2 \times 10^{-2}$

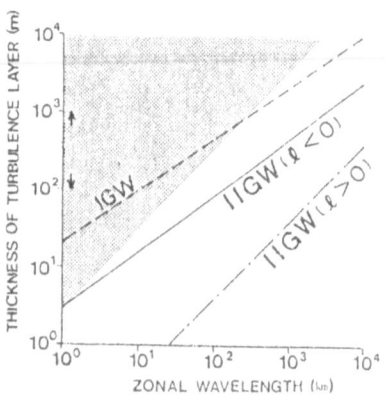

Fig. 4 Turbulence layer thickness estimated for internal gravity waves (IGW) and internal inertio-gravity waves (IIGW) in the stratosphere. The inertial effect is negligible in the shaded area. The most frequently observed range is indicated by arrows.

s^{-1}, $U_z \simeq -2 \times 10^{-3}$ s^{-1}, $H_C \simeq H \simeq 7$ km, $W_0 \simeq 1$ cm·s^{-1} and $|\ell| \simeq k$. We find that the positive-ℓ case is unavailable in the lower and middle stratosphere since h in this case is too thin to overcome the stabilization effect discussed in §2.4. Such waves are considered to propagate almost horizontally near the critical plane ($\hat{\omega} = -f$) until they are entirely dissipated due to the viscosity and the Newtonian cooling. Similarly waves shorter than 10 km in horizontal wavelength cannot produce turbulence layers and must be dissipated. On the other hand for waves longer than about 10^2 km, h predicted from the non-inertial case is included in the region (13), and breakdown of such long gravity waves should be treated by the inertial-wave case. Therefore, in the lower and middle stratosphere, internal gravity waves of 10-10^2 km horizontal wavelength and/or the negative-ℓ inertio-gravity waves of 10-10^4 km can break and produce turbulence layers. Concerning the differences in the turbulence layer thicknesses of the three cases, we draw Fig. 5 based on the local convective instability theory (1).

Observational studies so far made suggest that the thickness of stratospheric turbulence layers is concentrated between 10^2-10^3 m (Dewan, 1981; Woodman, 1981; Barat, 1982). The above theoretical results lead to an explanation of this concentration of the thickness: turbulence layers of 10^2-10^3 m in thickness are produced by both 10^2 km non-inertial waves and 10^3 km inertial waves. Such waves are most frequently observed in the lower stratosphere. It may be very significant that many synoptic-scale wavy disturbances in the extratropical troposphere have negative ℓ's. Another explanation of the thickness concentration can be given from the result that $h \propto k^{-2/3} W_0^{2/3}$ for non-inertial waves and $h \propto k^{-8/5} |\ell|^{4/5} W_0^{4/5}$ for inertial waves [see (2) and (12), respectively]. Tanaka (1983a) points out, based on the quasi-two-dimensional turbulence

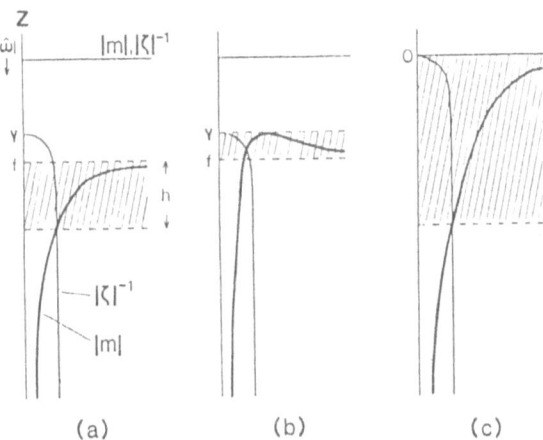

Fig. 5 Three types of generations of turbulence layers. The locally unstable regions predicted by $|m(z)\zeta(z)| > 1$ are hatched: (a) and (b) are inertio-gravity wave cases of $\ell < 0$ and $\ell > 0$, respectively; (c) non-inertial gravity wave case.

theory of Gage (1979) and the energy spectra obtained by Balsley and Carter (1982), that W_0 in the formulae for h decreases as k decreases if the power spectrum of U_0, the horizontal velocity amplitude, obeys $-5/3$ power law of k (*cf*. Chapter 4). This idea implies that W_0 adjusts the magnitude of h with each k so as to concentrate within a somewhat common thickness.

Anyway we can quantitatively understand that the stratospheric turbulence layers so far observed are mainly produced by the critical-level breakdown of internal (inertio-) gravity waves incident from the troposphere. Here we would call attention to three points below in order to apply the theoretical results obtained in this chapter for interpretation of observational facts. First, the "turbulence layer" of h in thickness may be different from the observed one if the anemometric methods are sensitive enough to detect some smaller structures included *within* the layer (see the next section). Second, since h is related to the wave parameters, we can obtain information on the input of gravity-wave momentum into the middle atmosphere by analysing $U_0 W_0$ from a data set of h, k, ℓ, N and U_z with the aid of the continuity equation (see Chapter 4). Last, in view of the results discussed in §§2.5-6, the inertial effect must not be neglected when the horizontal wavelength is longer than 10^2 km. This criterion is also confirmed by Tanaka's (1983b) calculations on the difference of momentum flux divergences between inertio-gravity and internal gravity waves.

2.7. *Structure of turbulence layer and multiple "gusts"*

According to the linear theories treated above, a "turbulence layer" is an ensemble of multiple unstable zones along the wavefronts packed in the vicinity of the critical level (see Fig. 1). Smaller-scale KH-type waves can grow in the unstable zones (see §2.3), and finally break until even smaller scale eddies occupy the "turbulence layer". Therefore, exactly speaking, the dynamics of the turbulence layer should be nonlinear and time-dependent. Some features near the critical level based on nonlinear time-dependent theories are somewhat different from those derived from steady linear theories (*e.g.* Tanaka, 1975, 1980). Furthermore, most waves may be excited in a limited space and/or time domain, *i.e.* as a wavepacket, so that we should consider wave transience (*e.g.* Fritts, 1982). We consider that our observational knowledge is even now insuffisient to clarify the microstructure and life cycle of the stratospheric turbulence layer. Although some turbulence "patches" observed in the lower troposphere and in the upper ocean have been identified as representing some stage of wave breakdown due to KH instabilities (*e.g.* Woods and Wiley, 1972), it seems unclear whether they are related to critical levels or not.

From the above considerations it is clear that both (a) and (b) mentioned in Chapter 1, on so-called *wave-mean flow interaction* theory, are not the exact representations of the "critical layers" but are parameterizations represent their effects on the basic field. However, it is also clear from many theoretical studies on the dynamics of the middle atmosphere that both mechanisms of (a) and (b) are effectively valid. Thus we can consider that both features of turbulence correspond to two extreme stages in the complete life cycle of the "turbulence layer" produced in the vicinity of the critical level: the turbulence in (a) and that in (b) are the starting stage and the final stage, respectively, in the case of a steady forcing of waves (cf. Booker and Bretherton, 1967). Mechanism (a) implies that the perturbed horizontal wind field must be an ensemble of multiple "gusts" from the beginning of breakdown through the decaying of KH billows (viz. Geller et al., 1975; Tanaka, 1975; Fritts, 1982). The thickness of such a "gust layer", say h_g, is approximately given by one half of the vertical wavelength of the breaking wave; e.g. for non-inertial waves

$$h_g \simeq \frac{1}{2} \cdot \frac{2\pi}{|m(z)|}\bigg|_{z=\text{breaking level}} = \frac{\pi}{\sqrt{J}} \cdot h \tag{14}$$

We can estimate $h_g \simeq 30$ m for $h = 100$ m, assuming that the Richardson number of the basic field takes a typical value as $J \equiv N^2/U_z{}^2 = 100$.

Among the observed values of turbulence layer thickness are some thinner values consistent with above estimation of the "gust layer" thickness [see Barat (1982)]. As will be discussed in Chapter 4, in relation to our observational results, such thin structures seem to be independent of the shear intensity of vertical scales larger than 10^3 m, which is the detectable limit of usual balloon trackings. We consider that those are related to "gust layers" not only based on considerations of the anemometric methods but also on the theoretical grounds of the origin of stratospheric turbulence layers. As long as our primary aim is to obtain information on the input of gravity-wave momentum flux into the middle atmosphere (see the last sentence of Chapter 1), measurements which can detect the "gust layers" in which the dynamic range is just intermediate between the balloon motions and the micro-scale turbulence are very important. Thus the next step of our theoretical study will open after the observation of the stratospheric "turbulence layer" as an ensemble of the "gust layers".

3. OBSERVATION

3.1. Balloon instrumentation

We used a zero-pressure balloon of 5000 m³, which was launched from the Sanriku Balloon Center, ISAS, Japan (39°10'N, 141°50'E) at 0702 JST on 20 Sep 1982 (2202 GMT on 19 Sep). Two sets of two-component sensitive propellor anemometers, named "Adapted Gill-type", are borne on two gondolas suspended under the balloon (Fig. 6). The lower gondola could scan 600 m in depth from the upper gondola, which was based on a hint from Cadet et al. (1977). A full description of our instrumentation will be given in Yamanaka et al. (1983).

Observations were performed for 24 hours at the 24-26 km levels and

128

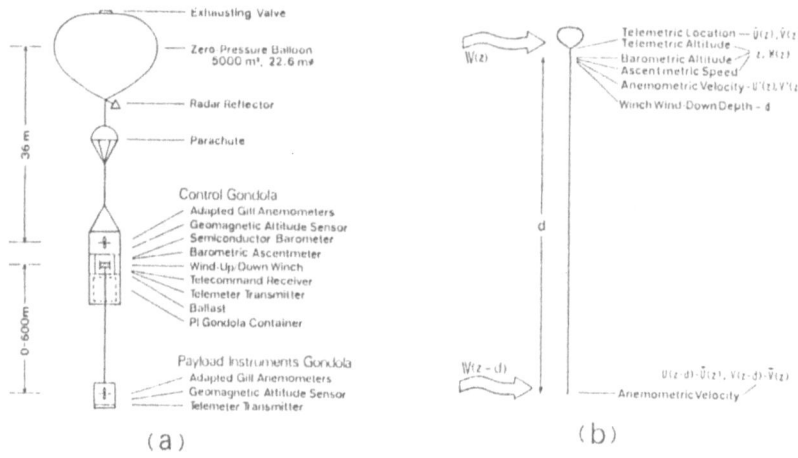

Fig. 6 Balloon instrumentation and observed quantities: (a) positions
of instruments; (b) similarity figure when the lower gondola
would down by 600 m under the upper gondola. $V = (U, V, W)$ is
the vector wind field. Quantities denoted by $\overline{(\ )}$ and $(\ )'$ are
divided by the minimum time scale of balloon motion ($\simeq$ 5 min).

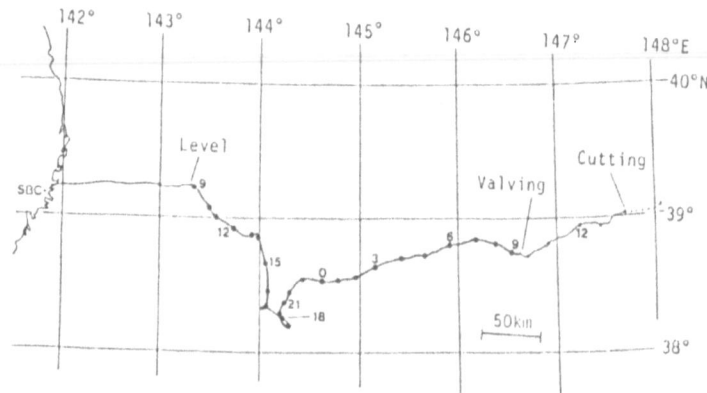

Fig. 7 Balloon trajectory. The level flight at 24–26 km altitudes is
thickened. Locations are indicated every right time of JST.

then for 5 hours at the 20-21 km levels (Fig. 7). The data shown in this paper are only of the telemetric winds and of the anemometric winds at the lower gondola during the former level flight. The winch was telecommanded four times to wind down and up the lower gondola.

3.2. *Sounding errors*

The telemetric sensitivity is 10 m, and its error is estimated as 200 m. The mechanical error of the receiving antennae is 0.05°. We can estimate the error of the telemetric wind data as less than 1 m/s except for the meridional components after 0000 JST on 21 Sep (about 300 km away from Sanriku). The data signals are reliable within 0.1 % of FS, which may not be considered in analyzing the data.

The quasi-adiabatic oscillations of the balloon have periods of about 5 min, but the vertical motion amplitudes are about 50 m or less except for the first hour after arriving at the flight level. This can be confirmed by the realtime data from the barometric sensors (error: 1 μb). The frequency of the balloon quasi-adiabatic oscillation is nearly the highest one which the balloon can follow. Vertical scannings using the winch system (depth error: 10 cm) were carried out in very quiet periods of balloon oscillations, so that phenomena with vertical scales larger than 10 m are observable. The "Adapted Gill" anemometer can detect winds stronger than 0.9 m/s and its distance constant is 42 m (time constant for 2 m/s wind: 21 s) in the middle stratosphere (Yamanaka *et al.*, 1983).

Pendulum motions of the gondolas relative to the balloon can be separately considered as those of the upper gondola relative to the balloon and those of the lower gondola relative to the upper gondola, since the lower gondola ($\simeq$ 8 kg including the wire) is sufficiently lighter than the total weight of the balloon system ($\simeq$ 148 kg). For the former pendulum motions, the oscillation angle was 0.1°, the rotating speed 0.1 rpm and the equivalent relative wind speed 0.5 m/s, so that we can consider that these could not stray into the observational data of anemometers.

On the other hand, for the pendulum motions of the lower gondola to the upper gondola, the following check can be made. Let d be the depth of the lower gondola from the upper one, θ the oscillation angle and g the gravitational acceleration. Then the false wind speeds are given by $d \times \theta \cdot \sqrt{g/d}$. For this value to affect the anemometric data, $\theta \gtrsim 0.1$ rad, *i.e.* the lower gondola should oscillate in magnitudes larger than 5°. Since such an oscillation must be forced by the actual winds, the drag of gondola (drag coefficient: C_D; effective surface: S) to the winds (speed: v) is expressed by $C_D S \rho v^2 / 2$, where ρ is the atmospheric density. Then the gondola oscillation is estimated as $\theta \simeq$ (drag)/(weight) $\sim 10^{-4}$ rad and the gondola rotation as order of 1 rpm, which was actually observed in the data of the attitude sensors. Therefore we can safely confirm that this type of pendulum motion also never appears in the anemometric data for the given sensitivity.

However, the last mentioned oscillations may disturb the wind data through the gondola rotations for the following reason. We observed the horizontal wind vector relative to the balloon by using two-component anemometers and attitude sensors. If the actual wind is a constant vector and the gondola-borne anemometers are rotating, each component anemometer must detect the wind pulsively. Since the anemometer has a time constant of order of 10 sec to follow the wind and hysteresis near the lowest sensitivity, and since the Gill type propellor has a stall region near the rectangular direction (Holmes *et al.*, 1964; Yamanaka *et al.*, 1983), the sounding wind derived by the vector sum of the outputs of two-component anemometers may have a false pulsive feature. Therefore we cannot trust the sounding data for phenomena of scales shorter than the order of 10 sec.

3.3. Basic wind field

The observation was carried out just after the autumnal reversal in the stratosphere, and the mean zonal wind in the middle stratosphere was weak westerly (Fig. 8). The balloon tracking data from the launch to the arrival at the flight level and those through the level flight are analyzed in order to obtain the basic wind field and the wave structures (Figs. 9 and 10). We can find commonplace features of several types of wavelike variations (*cf.* Sawyer, 1961; Cadet and Teitelbaum, 1979).

3.4. Vertical microstructures

The anemometric data of the lower gondola show three types of vertical variations of horizontal wind in the middle stratosphere (Fig. 11):
  (A) homogeneous and heterogeneous wind layers of 100–300 m in thickness;
  (B) multiple "gust layers" of 10–50 m in thickness with the wind amplitudes

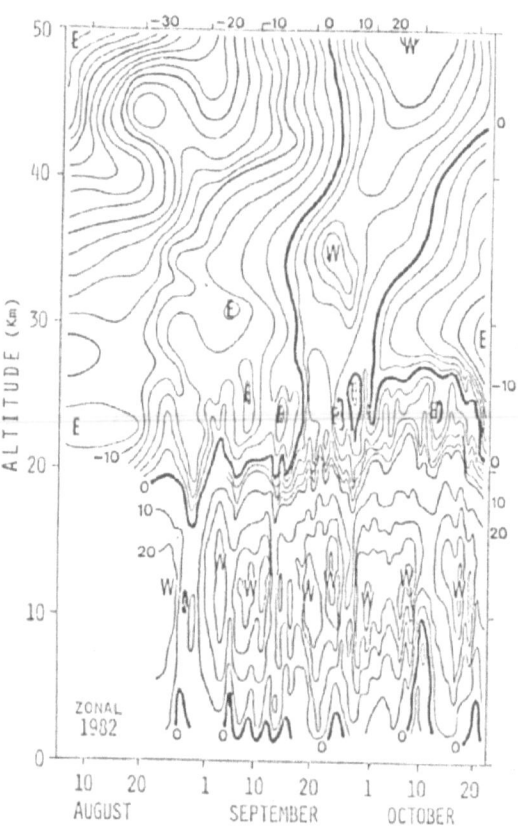

Fig. 8 Time-height section of zonal wind at SBC, Akita (39°43'N, 140°06'E) and Ryori (39° 02'N, 141°50'E). Fluctuations less than 5 km in the vertical direction are filtered out. (Units: m/sec.)

of 1-3 m/s;
    (C) minute variations of a few meters in thickness with wind
    fluctuations of about 0.5 m/s or weaker.
On the basis of the check of sounding errors in §3.2, we consider that
(A) and (C) may be the variations of balloon flight winds and the false
variations due to gondola rotation, respectively. On the other hand,
refering to previous observational results (Dewan, 1981; Woodman, 1981;
Barat, 1982), the scales of (A) and (C) are similar to the typical
"turbulence layer" and the wind fluctuations in the "turbulence layer",
respectively. However, minute strong variations like (B) have never
been reported in the middle stratosphere. We discuss our results mainly
for the "gust layers" of (B) in the next chapter.

4. DISCUSSION

    The variety of stratospheric microstructures shown in Fig. 11 may
be reasonably understood as several stages of wave breakdown, just as
those observed in the upper ocean (see Woods and Wiley, 1972). From the
latter analogy we think that the "gust layers" correspond to the starting stage of wave breakdown. There may be mainly two reasons why such
"gusts" have not been reported in the middle stratosphere: anemometric

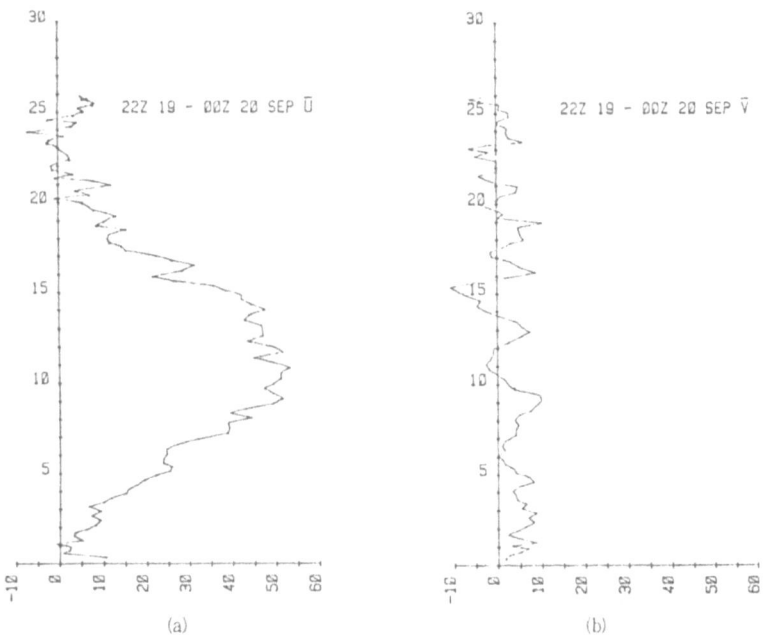

Fig. 9 Vertical wind profiles obtained from the balloon tracking data.
      (a) and (b) are zonal and meridional, respectively.

methods and rare occurrences of the "gusts" themselves. The adapted
Gill anemometers used in our observations are of appropreate dynamic
range to detect them. Rare occurrences of the "gusts", which occur only
for a few percent of the whole observation period, suggest that the input of internal (inertio-) gravity waves into the middle atmosphere is
not so frequent. The maximum value of vertical wind shear in such a
"gust layer" reaches 10^{-1} s^{-1}, so that the "gust" should be "turbulence"
due to the local instability. The multiple structures typically shown
in Fig. 11(b) are very similar to the turbulence layer structures predicted by the *linear* critical layer theory (*viz.* Geller *et al.*, 1975),
which also implies that the "gusts" are the starting stage of wave
breakdown. If the total thickness of an ensemble of multiple "gust

layers" corresponds
to the turbulence
layer thickness dis-
cussed in Chapter 2,
the breaking waves
might have a wave-
length of 10-10^2 km
(for the non-inertial
wave case) or 10^2-10^3
km (for the inertial
wave case).

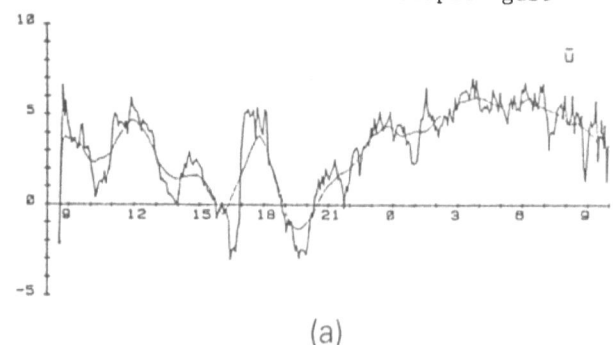

(a)

The *maximum* eddy
diffusivity K_z in the
"gust layer" can be
estimated as follows,
assuming that the
characteristic radius
of the eddy, λ, is
approximated by the
typical thickness of
a "layer", h_g, as the
maximum value:

$$K_z = V \cdot \lambda$$
$$\simeq \sqrt{\overline{\Delta u^2}} \cdot h_g$$
$$\geq 10^5 \, cm^2 s^{-1};$$
$$(15)$$

where V is the char-
acteristic velocity
of eddy and Δu is the
amplitude of wind
speed observed in the
"gust layers ". We
find that in the
"gust layers" the
eddy diffusion can be

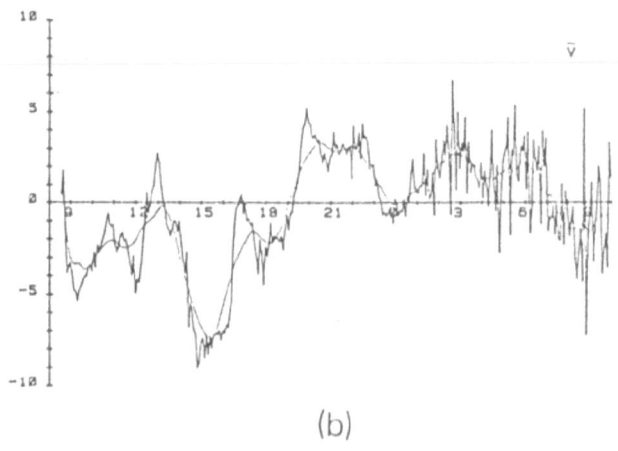

(b)

Fig. 10 Time variations of balloon tracking winds. (a) and
(b) are the zonal and meridional components, respectively. Thin lines are running-mean data over
each 100 min. The data of V after 0000 JST are not
good because of mechanical errors of the antenna.

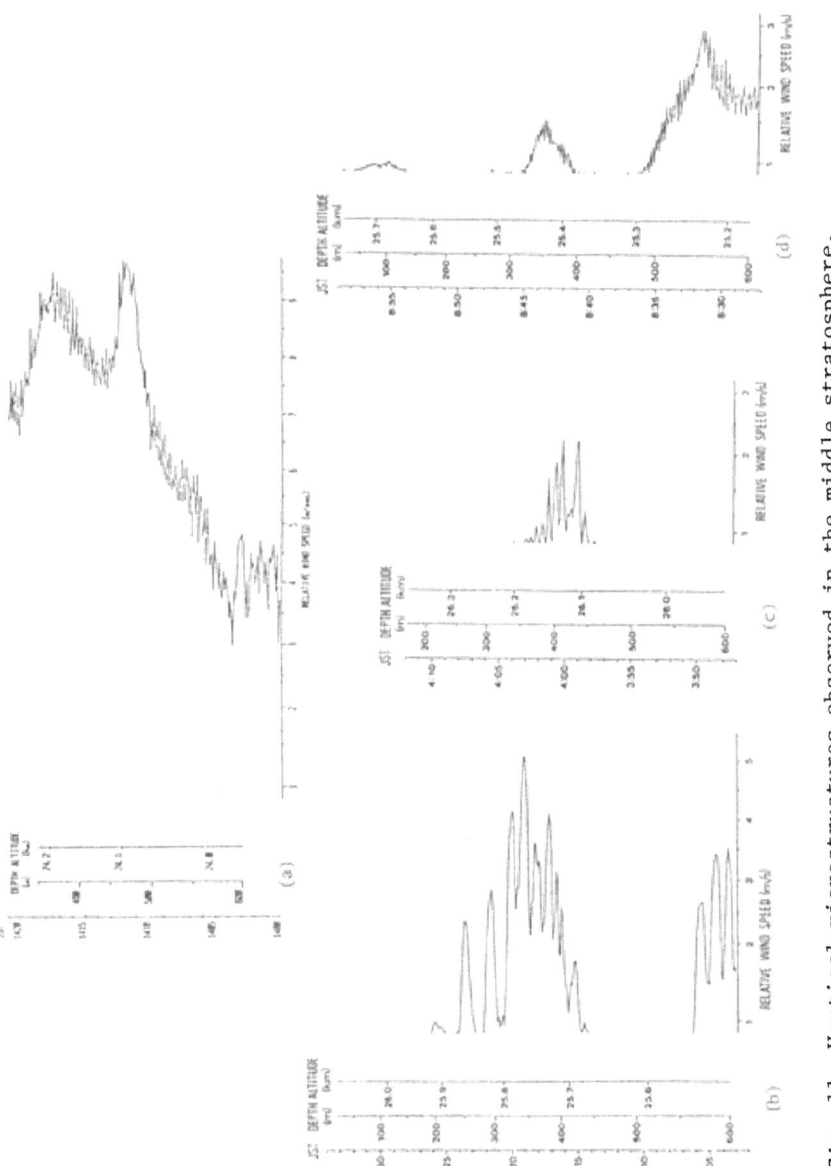

Fig. 11 Vertical microstructures observed in the middle stratosphere.

much larger than the small values ($K_z \simeq 10^2$ cm^{2}s^{-1}) estimated by some observations so far made (Cadet, 1977; Dewan, 1981; Barat, 1982). The present value (15) is almost the same intensity as that estimated for quasi-horizontal mixing due to the Lagrangian motions of planetary waves (see *e.g.* Matsuno, 1980), and is enough to explain the *ad hoc* eddy diffusion ($K_z \simeq 10^3$-10^5 cm^{2}s^{-1}) of the vertical chemical transport in the lower and middle stratosphere (*e.g.* Massie and Hunten, 1981). In point of the fact that chemicals should be transported in every season and in every latitude, the eddy transport due to the "gust layers" seems more powerful than the Lagrangian advection by planetary waves.* As Barat and Aimedieu (1981) and Barat (1982) have been pointed out, eddy diffusivity calculated from wind microstructure data is not always consistent with that based on the microstructure of chemical tracer distribution since the former eddy is assumed to obey local similarity theory. In the case of sporadic and/or skew-spectral eddies, it should be natural that the eddy diffusivity derived by the conventional algorithm does not give the effective value of the chemical transport. The "gust layer" is considered to play the most important role in the transport although it appear only just after the start of wave breakdown.

The "gust layers" are, as mentioned repeatedly, features of internal gravity waves just incident upon their critical levels. They must at last break and induce a turbulence layer of which the top is the critical level. In another words, the critical level behaves like the ground in the case of a turbulent boundary layer. On the other hand, the wind field before the waves arrive may be geostrophic. The dynamics in a frictional layer between the ground and a geostrophic flow can be described by the well-known theory of the *Ekman boundary layer*, for which the depth h_E is effectively given as follows by means of the Coriolis factor f and the eddy viscosity K_z:

$$h_E = \sqrt{2K_z/f}. \tag{16}$$

In the case of the turbulence layer near the critical level, the thickness is given by derived in Chapter 2.** Therefore, for simple internal gravity waves (2), we obtain

---

* As a side view of this fact, the general circulation in the lower stratosphere can be seriously affected by the critical-level breakdown of internal gravity waves, which is just like near the mesopause (see Tanaka and Yamanaka, 1983). Planetary waves are weaker than mountainous waves in the lower stratosphere except for winter high-latitudes.

** Exactly speaking, h_E in (16) is derived as the vertical "e-folding scale" for the ageostrophic motion. However, as parameterized for effects of the planetary boundary layer in the tropospheric circulation studies, h_E can be treated as the "thickness" of a frictional layer. Furthermore, we consider here a situation that waves may arrive at their critical levels without almost any dissipation, which is equivalent to a situation that wave sources exist just below the critical levels and momentum exchanges between the sources and the atmosphere take place in the critical layers (see Fig. 1).

$$K_z = \frac{fh^2}{2} \propto k^{-4/3} \; . \tag{17}$$

We can easily notice that (17) is just the same form in relation to k as that predicted by the similarity theory of homogeneous turbulence within the so-called inertial subrange (see $e.g.$ Townsend, 1976), that is,

$$\text{Energy spectrum} \propto k^{-5/3} \; . \tag{18}$$

We find from (14) that for the "gust layer" thickness h_E the same form of formula as (17) is derived. Hence we can reasonably understand why the -5/3 power law diverges into larger scale motions ($cf.$ Gage, 1979; Dewan, 1981; Balsley and Carter, 1982; Larsen et $al.$, 1982). The above hypothesis implies that waves and turbulence near the critical level are equivalent to local homogeneous eddies in steady conditions.

It is neccessary to develop and use more sensitive anemometers, such as the ionic anemometers (Barat, 1982; Good et $al.$, 1978), to confirm perfectly the nature and life cycle of the "gust" and turbulence induced by the internal (inertio-) gravity waves. However, quantitative analysis of the balloon motion is also needed to utilize data observed by such sensitive anemometers. Especially in this regard the observational studies so far made, we think, must be reexamined.

Since "gust layers" and their multiple structures are actually observed in the middle stratosphere, the theoretical considerations of the critical-level breakdown of internal gravity waves (see Chapter 2) are considered to be a realistic feature in the whole middle atmosphere. In this article the data descriptions and quatitative analyses are very restricted. We are now preparing some papers in which detailed analyses of the whole data obtained by the observation described here are shown. The perfect interpretation of observational facts from a theoretical point of view introduced here should be based on them. We will derive an estimate of the input of gravity-wave momentum flux into the middle atmosphere by connecting the observational results with the theoretical studies in the near future.

5. CONCLUSION

Here we have introduced theoretical results for the turbulence layer thickness and some early looks at our balloon observation on the microstructure of stratospheric winds. To conclude, we shall summarize the relationships between the "gust layers", the turbulence layers and the internal gravity waves on the basis of our theoretical and experimental study:

- A "gust layer" observed in the middle stratosphere has a thickness of 10-50 m and wind variation amplitudes of 1-3 m/s.
- The multiple structure of the "gust layers" is a feature of the "turbulence layer" in its starting stage for which the thickness is observed as 100-300 m.

- The thicknesses of the "gust layer" and of the turbulence layer are theoretically derived as the vertical wavelength of the breaking wave and as the distance between the breaking level and the critical level, respectively.
- The stratospheric turbulence layer of 10^2-10^3 m in thickness is generated by the critical-level breakdown of internal gravity waves of $10-10^2$ km in horizontal wavelength or of inertio gravity waves of 10^2-10^3 km with negative ℓ.
- Internal gravity waves shorter than 10 km and inertio-gravity waves with positive ℓ are dissipated due to the viscosity and Newtonian cooling, so that they cannot produce turbulence.
- The turbulence layer makes eddy diffusion, which is, effectively equivalent to that made by local-similarity eddies in the Ekman boundary layer if the turbulence layer is maintained steadily, but predominant vertical diffusion transport is caused by the "gust-layer" structure as about 10^5 cm^{2}s^{-1}.
- We can estimate the input of wave momentum flux into the middle atmosphere from the data set of the "gust layer" thickness and the basic field quantities.

ACKNOWLEDGEMENTS

We are grateful to Prof. T. Matsuno for criticisms to improve this study, to Prof. J. Nishimura for observational advice and to Prof. A. Ono for his encouragement. Thanks are extended to Prof. J.R. Holton and Dr. Walterscheid for careful readings of the manuscript and to Mr. Y. Matsuzaka and Dr. T. Yamagami for their experimental cooperations. Discussions and correspondence with Profs. J. Barat, I. Hirota, H. Hirosawa, T. Itoh, R. Kimura, S.Kato, M. Uryu, Drs. B.B. Balsley, D. Cadet, S. Fukao, H. Kanzawa, H. Kida, Y. Matsuda, S. Miyahara, K.-K. Tung, Messrs. Y.-Y. Hayashi and K. Mimura are very helpful. This study was supported by the Scientific Research Funds for MAP and for Scientific Ballooning from the Ministry of Education, Science and Culture.

REFERENCES

Balsley, B.B., and D.A. Carter, 1982: The spectrum of atmospheric velocity fluctuations at 8 km and 86 km. *Geophys. Res. Lett.*, 9, 465-468.

Barat, J., 1982: Some characteristics of clear air turbulence in the middle stratosphere. *J. Atmos. Sci.*, 39, 2553-2564.

———— and P. Aimedieu, 1981: The external scale of clear air turbulence derived from the vertical ozone profile: Application to vertical transport measurement. *J. Appl. Meteor.*, 20, 275-280.

Booker, J.R., and F.P. Bretherton, 1967: The critical layer for internal gravity waves in a shear flow. *J. Fluid Mech.*, 27, 513-539.

Bretherton, F.P., 1966: The propagation of groups of internal gravity waves in a shear flow. *Quart. J. Roy. Meteor. Soc.*, 92, 466-480.

Cadet, D., 1977: Energy dissipation within intermittent clear air turbulence patches. *J. Atmos. Sci.*, 34, 137-142.

————, G. Bannerot and B. Brioit, 1977: A two-dimensional sounding technique from a balloon: Application to the study of a stratospheric shallow layer. *J. Appl. Meteor.*, 16, 662-667.

———— and H. Teitelbaum, 1979: Observational evidence of internal inertio-gravity waves in the tropical stratosphere. *J. Atmos. Sci.*, 36, 892-907.

Dewan, E.M., 1981: Vertical transport by small scale turbulence: a critical review. *AFGL-TR-81-0051*, 32pp.

Dunkerton, T.J., 1982: Stochastic parameterization of gravity wave stresses. *J. Atmos. Sci.*, 39, 1711-1725.

Eady, E.T., 1949: Long waves and cyclone waves. *Tellus*, 1, 33-52.

Eliassen, A., and E. Palm: On the transfer of energy in stationary mountain waves. *Geofys. Publ.*, 22(3), 23pp.

Fritts, D.C., 1982: The transient critical-level interaction in a Boussinesq fluid. *J. Geophys. Res.*, 87, 7997-8016.

Gage, K.S., 1979: Evidence for a $k^{-5/3}$ law inertial range in mesoscale two-dimensional turbulence. *J. Atmos. Sci.*, 36, 1950-1954.

Geller, M.A., H. Tanaka and D.C. Fritts, 1975: Production of turbulence in the vicinity of critical levels for internal gravity waves. *J. Atmos. Sci.*, 32, 2125-2135.

Good, R.E., J.H. Brown and G. Harpell, 1978: Development of a corona anemometer for measurement of stratospheric turbulence. *AFGL-TR-78-0070*, 44pp.

Grimshaw, R., 1975: Internal gravity waves: Critical layer absorption in a rotating fluid. *J. Fluid Mech.*, 70, 287-304.

Holmes, R.M., G.C. Gill and H.W. Carson, 1964: A propellor-type vertical anemometer. *J. Appl. Meteor.*, 3, 802-804.

Holton, J.R., 1982: The role of gravity wave induced drag and diffusion in the momentum budget of the mesosphere. *J. Atmos. Sci.*, 39, 791-799.

Jones, W.L., 1967: Propagation of internal gravity waves in fluids with shear and rotation. *J. Fluid Mech.*, 30, 439-448.

Larsen, M.F., M.C. Kelley and K.S. Gage, 1982: Turbulence spectra in the upper troposphere and lower stratosphere at periods between 2 hours and 40 days. *J. Atmos. Sci.*, 39, 1035-1041.

Lindzen, R.S., 1981: Turbulence and stress due to gravity wave and tidal breakdown. *J. Geophys. Res.*, 86, 9707-9714.

————, B. Farrel and K.-K. Tung, 1980: The concept of overreflection and its application to baroclinic instability. *J. Atmos. Sci.*, 37, 44-63.

Massie, S.T., and D.M. Hunten, 1981: Stratospheric eddy diffusion coefficients from tracer data. *J. Geophys. Res.*, 86, 9859-9868.

Matsuno, T., 1980: Lagrangian motion of air parcels in the stratosphere in the presence of planetary waves. *Pure Appl. Geophys.*, 118, 189-216.

————, 1982: A quasi-one-dimensional model of the middle atmosphere circulation interacting with internal gravity waves. *J. Meteor. Soc. Japan*, 60, 215-226.

Olver, F.W.J., 1974: *Asymptotics and Special Functions*. Academic Press, 572pp.

Plumb, R.A., and A.D. McEwan, 1978: The instability of a forced standing

138

wave in a viscous stratified fluid: A laboratory analogue of the quasi-biennial oscillation. *J. Atmos. Sci.*, 35, 1827-1839.

Sawyer, J.S., 1961: Quasi-periodic wind variations with height in the lower stratosphere. *Quart. J. Roy. Meteor. Soc.*, 87, 24-33.

Tanaka, H., 1975: Turbulent layers associated with a critical level in the planetary boundary layer. *J. Meteor. Soc. Japan*, 53, 425-439.

————, 1980: The evolution of a nonlinear critical layer for internal gravity waves. *J. Meteor. Soc. Japan*, 58, 321-326.

————, 1982: Application of WKB theory to turbulence layers in the vicinity of critical levels. *J. Meteor. Soc. Japan*, 60, 1034-1040.

————, 1983a: Turbulence layer thickness in the stratosphere under the presence of viscosity and Newtonian cooling. Submitted to *J. Meteor. Soc. Japan*.

————, 1983b: Momentum flux divergences associated with inertio-gravity and internal gravity waves in the middle atmosphere. Submitted to *J. Meteor. Soc. Japan*.

———— and M.D. Yamanaka, 1983: Atmospheric circulation in the lower stratosphere induced by the mesoscale mountain wave breakdown. (In preparation)

Tokioka, T., 1970: Non-geostrophic and non-hydrostatic stability of a baroclinic fluid. *J. Meteor. Soc. Japan*, 48, 503-520.

Townsend, A.A., 1976: *The Structure of Turbulent Shear Flow*. (2nd Ed.), Cambridge University Press, 429pp.

Woodman, R.F., 1981: Turbulence in the middle atmosphere: a review. *Handbook for MAP*, 2, 293-300.

Woods, J.D., and R.L. Wiley, 1972: Billow turbulence and ocean microstructure. *Deep-Sea Res.*, 19, 87-121.

Yamanaka, M.D., 1983: Critical level shifting with β-effect. (In preparation)

————, Y. Matsuzaka, J. Nishimura and H. Tanaka, 1983: Balloon instrumentation for soundings of stratospheric microstructures, I: Adapted Gill-type propellor anemometer. (In preparation)

———— and H. Tanaka, 1982: Propagation and breakdown of internal inertio-gravity waves near critical levels in the middle atmosphere. Submitted to *J. Meteor. Soc. Japan*.

APPENDIX: *Propagation of internal (inertio-) gravity waves*

The quasi-one-dimensional problem of vertical propagating waves in the case of constant vertical basic shear and constant Coriolis factor is governed by

$$\frac{d^2\tilde{w}}{d\hat{\omega}^2} + \frac{2f[f + i(\ell/k)\hat{\omega}]}{\hat{\omega}(\hat{\omega}^2 - f^2)} \cdot \frac{d\tilde{w}}{d\hat{\omega}} + \frac{J'\hat{\omega} - 2i(\ell/k)f}{\hat{\omega}(\hat{\omega}^2 - f^2)} \tilde{w} = 0, \qquad (A1)$$

where $w \equiv \tilde{w}(z) \cdot \exp[i(kx + \ell y - \omega t)]$ is the perturbed vertical velocity, $\hat{\omega}(z) \equiv \omega - k \cdot U(z)$ the intrinsic frequency and $J' \equiv (N^2/U_z^2) \cdot [1 + (\ell/k)^2]$ the modified Richardson number (Eady, 1949; Jones, 1967; Tokioka, 1970; Grimshaw, 1975; Yamanaka and Tanaka, 1982). Eady showed that solutions around the apparent regular singularity $\hat{\omega} = 0$ contain the baroclinic unstable modes. Jones proved that the critical levels of neutral internal gravity waves exist at the pair of regular singularities $\hat{\omega} = \pm f$. Tokioka solved (A1) numerically and showed that the structures of unstable modes are dependent upon the sign and magnitude of ℓ. Grimshaw studied asymptotic solutions around each singularity of (A1) and pointed out the "valve effect" of the Jones' critical levels and the non-inertial gravity waves being the solutions around a regular singularity $\hat{\omega} = \infty$.

According to analyses by Yamanaka and Tanaka (1982), the exact solutions of (A1) are written by $w = \hat{\omega}^3(\hat{\omega} + f)^{i\ell/k} \cdot F(\hat{\omega})$, where F is the Gauss' hypergeometric function. The leading term is expressed by

$$\tilde{w} \simeq (\hat{\omega} + f)^0 \qquad \text{or} \quad \tilde{w} \simeq (\hat{\omega} + f)^{i\ell/k} \qquad \text{for } \hat{\omega} \simeq -f, \quad \text{(A2)}$$

$$\tilde{w} \simeq \left(\frac{1}{\hat{\omega}}\right)^{-\frac{1}{2}+i\sqrt{J'-(1/4)}} \quad \text{or} \quad \tilde{w} \simeq \left(\frac{1}{\hat{\omega}}\right)^{-\frac{1}{2}-i\sqrt{J'-(1/4)}} \quad \text{for } \hat{\omega} \simeq -\infty. \quad \text{(A3)}$$

Similar results are obtained for positive singularities, $\hat{\omega} = f$ and ∞. In the case of *non-inertial* waves ($f = 0$), Eq. (A1) can be solved easily:

$$\tilde{w} = \hat{\omega}^{\frac{1}{2}\pm i\sqrt{J'-(1/4)}} . \qquad \text{(A4)}$$

It is important that the solutions (A2) near the Jones' critical levels have quite different characters from those of the non-inertial case (A4). This fact leads to all of the strange behaviors of inertio-gravity waves near the Jones' critical levels, as mentioned in §2.5, including the difference between both turbulence layer thicknesses (2) and (12). The solutions (A2) are valid in the region (13) due to the convergence condition of the hypergeometric function. It is clear from (A2) that the behavior of inertio-gravity waves near the Jones' critical levels should be strongly dependent upon the meridional wavenumber ℓ. In physical explanation we consider that this ℓ-dependence arises from the twisting term of the potential vorticity equation (see Yamanaka and Tanaka, 1982).

We summarize below the relation between the solutions and the wave propagation. Each solution can be written in the polar form: $\tilde{w}(z) = |\tilde{w}(z)| \times \exp\{i \cdot \text{Arg}[\tilde{w}(z)]\}$. Then the vertical wavenumber m must satisfy a dispersion relation written as $m \equiv \partial\text{Arg}[\tilde{w}]/\partial z = \text{Im}[\tilde{w}*\tilde{w}_z]/|\tilde{w}|^2$. In general, the quantity $\text{Im}[\tilde{w}*\tilde{w}_z]$ is an invariant if the governing equation takes a *standard* form as $\tilde{w}_{zz} + M(z)\cdot\tilde{w} = 0$ and M is real. If so, the solution is *exactly* expressed by $\tilde{w} = (C/|m|^{1/2})\cdot\exp[i\int mdz]$, where C is a constant, and we have $\tilde{w}_{zz} = -m^2[1 - m^{-3/2}(m^{-1/2})_{zz}]\cdot\tilde{w}$. Hence the *WKB approximation* $m \simeq \pm\sqrt{M}$ is valid if $|M^{-3/4}\cdot(M^{-1/4})_{zz}| \ll 1$ (*cf.* Lindzen *et al.*, 1980). In the vicinity of a regular singularity, say z_c, we have $M \simeq M_0/(z - z_c)^2$, where M_0 is the residue at the pole, so that $M^{-3/4}(M^{-1/4})_{zz} \simeq -(1/4M_0)(z-z_c)^{-2}$. Therefore we should write $m \simeq \pm\sqrt{(4M_0 - 1)/4M_0}\cdot\sqrt{M}$ after the *Liouville-Green method* developed by Olver (1974). We find that the wave amplitude $|\tilde{w}|$ varies with $|m|^{-1/2}$ and the group velocity $W_g \equiv \partial\hat{\omega}/\partial m$ varies with m^{-2}. A kind of wave action density $|\tilde{w}|^2/2\hat{\omega}$ is an invariant. The wave momentum flux can be exactly proved to be proportional to the standard-form invariant $\text{Im}[\tilde{w}*\tilde{w}_z]$. Therefore the wavepacket kinematics is, exactly speaking, valid for the waves described by a standard-form equation.

We can transform Eq. (A1) into a standard form by substituting

$$\tilde{w} \equiv \check{w} \cdot \exp\left[-\int \frac{f\{f + i(\ell/k)\hat{\omega}\}}{\hat{\omega}\,(\hat{\omega}^2 - f^2)}\, d\hat{\omega}\right].\tag{A5}$$

Thus the dispersion relation of internal (inertio-) gravity waves corresponding to the solution of (A1) is written as follows (Yamanaka and Tanaka, 1982):

$$m = \frac{\ell f U_z}{\hat{\omega}^2 - f^2} + \check{m}\;,\tag{A6}$$

where

$$\check{m} \equiv \frac{\text{Im}[\check{w}*\check{w}_z]}{|\check{w}|^2} \simeq \pm \frac{\alpha}{\sqrt{1+\alpha^2}} \cdot \left\{\frac{N^2(k^2 + \ell^2)}{\hat{\omega}^2 - f^2} + \frac{\ell^2 f^2 U_z^2}{(\hat{\omega}^2 - f^2)^2} + \frac{2k^2 f^2 U_z^2}{\hat{\omega}^2(\hat{\omega}^2 - f^2)}\right\}^{1/2}$$

$$= \mp \frac{kU_z\alpha}{\sqrt{1+\alpha^2}} \cdot \left\{J' \cdot \frac{(\hat{\omega}^2 - \gamma^2)(\hat{\omega}^2 + \delta^2)}{\hat{\omega}^2(\hat{\omega}^2 - f^2)^2}\right\}^{1/2}.\tag{A7}$$

α takes ℓ/k for the solutions (A2) and $2\sqrt{J' - (1/4)}$ for (A3) or (A4), and γ and δ are real parameters dependent upon J and ℓ/k. Extremities of (A7) and (A8) for $U_z \to 0$ coincide with the dispersion relation of common use:

$$m^2 = \frac{N^2(k^2 + \ell^2)}{\hat{\omega}^2 - f^2}.\tag{A8}$$

However, the waves corresponding to the solutions (A2) near the Jones' critical level are never contained within the waves discribed by (A8). The first term of (A6) can takes the same order as the second term for the regular (transmitted) mode near the Jones' critical levels, but the former does not exceed the latter in any cases. Thus we can regard the wavepacket velocity based on the dispersion relation (A6) as the energy proceeding velocity. The levels indicated by $\hat{\omega} = \pm\gamma$ in (A7) are called the *turning levels* which reflect waves passing through the *valve* of Jones' critical levels (see Fig. 3).

J. R. Holton and T. Matsuno, Dynamics of the Middle Atmosphere, 141-160.
Copyright © 1984 by Terra Scientific Publishing Company.

INTERNAL GRAVITY WAVE ENHANCEMENT BY THE CHEMICAL HEAT RELEASE DUE TO OXYGEN RECOMBINATION

Yoshiyuki Hayashi and Taroh Matsuno

Geophysical Institute
University of Tokyo

ABSTRACT

The possibility of internal gravity wave enhancement by chemical heating due to atomic oxygen recombination is examined as a cause of large temperature fluctuations in the upper mesosphere and lower thermosphere observed at high latitudes in wintertime.

The problem is treated as a spatial amplification of upward propagating waves in contrast to Leovy's (1966) study, in which temporal growth of waves was discussed as an instability problem.

Downward motion, which is likely to occur in winter at high latitudes, may cause a higher atomic oxygen concentration at lower altitudes as actually observed by Dickinson et al. (1980). By assuming this winter condition as the basic state of the O-concentration, linearized equations for the coupled system are solved. The dispersion relation obtained indicates that upward amplification of gravity waves may occur under favorable conditions. A one dimensional model is used to test the amplification mechanism in a more realistic atmospheric model. Although enhancement is caused in the same amount as expected by the linear theory, an unrealistic temperature increase takes place, which may be due to improper treatment of thermarization processes. Close examination of this point leads to the conclusion that significant wave enhancement is not likely to occur.

1. INTRODUCTION

There is a great difference between the summer and the winter temperature fields in the upper mesosphere and the lower thermosphere at high latitudes. In addition to the general trend that high temperatures (≈ 200 K) occur in winter and low temperatures (≈ 140 K) occur in summer, there is a marked difference in the behavior of temperature variations between those two seasons. The rocket observations of Theon and Smith (1970) performed by the grenade method in the 35 km - 95 km range show that the temperature variations during summer remain within ±10 K, while during winter the variations are twice as large (Fig. 1). Furthermore,

141

an interesting feature is found in winter profiles at about 80 km. Above this height, the temperature fluctuations grow rapidly and reach the extent of ±30 ∿ 40 K. In contrast to this, the fluctuations in summer or in winter below 80km are almost constant regardless of altitude.

The duration and the horizontal extent of the large fluctuations above 80km are not clear; it is not certain whether the temperature variations become large only occasionally or not. The rocket observations (mentioned below) indicate that the large fluctuations lasted at least for a day. The temperature averaged in time, shows the existence of a warm layer at about 80km, where the large fluctuations occur. From a small number of observations available, the high temperature region seems to spread over an area of several thousand kilometres.

The results of each observation are reported in detail by Heath et al. (1974). They trace the phase marching of the temperature disturbance from the data of 6 soundings carried out every three hours from 31th January to 1st February 1967. The general features of those temperature fluctuations indicate that they are internal gravity waves with downward phase propagation. Their figure shows the vertical wavelengths range from 10km to 15km, and the periods are roughly from two to four hours. By the use of the dispersion relation for internal gravity waves, the horizontal wavelength is found to be around 400km. The waves have upward group velocities in the order of 1 m/s. These results suggest that the large temperature fluctuations at altitudes above 80km are due to the amplification of upward propagating gravity waves.

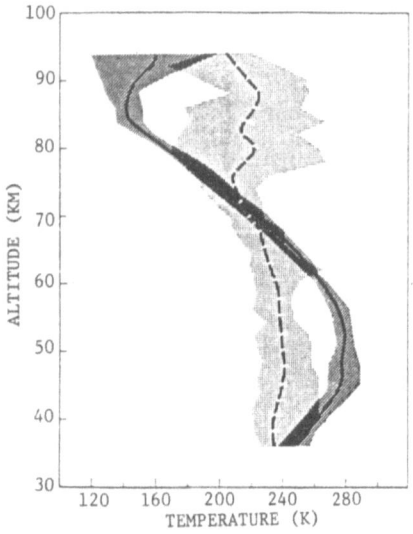

Fig. 1 Seasonal average temperature profiles above Barrow, Alaska (71N). The solid curve is the average of 10 summer soundings and the broken curve the average of 12 winter soundings during 1965-1967. The cross hatched areas surrounding each curve represent the total range of temperatures included in the average. (after Theon and Smith, 1970)

As waves propagate upward in the atmosphere, their amplitudes grow exponentially corresponding to the decrease of the air density unless absorption, reflection and/or diffusion take place. However, such an exponential growth trend is not so obvious in the observations by Theon et al. The temperature variations are almost uniform in summer and also below the altitude of 80km in winter. The behavior at altitudes above 80km in winter seems rather singular. The purpose of this paper is to present some interpretations other than density change concerning those large fluctuations.

Above 80km, atomic oxygen becomes important as a component of the atmosphere. It is especially important as an energy source because chemical energy (5.12eV) is released by its recombination reaction. Leovy (1966) considered the possibility of internal gravity wave amplification by means of this recombination energy. He studied waves with given wavelengths, which are assumed to be generated at this altitude by destabilization due to the oxygen reactions. His result shows that, for instance, it takes 2.6 days at an altitude about 90km to double the amplitude of a wave, the frequency of which is 2×10^{-4} sec (period = 8.7 hour). The growth is certainly caused by atomic oxygen, but the growth time is not short enough. But as Leovy suggested in his paper, if the atomic oxygen density is increased to a certain extent for some reason, a much quicker growth may be expected. In fact, recent observations of the atomic oxygen density by Dickinson et al. (1980) show that the density attains the maximum of 3×10^{12}/cm^3 at the 90km level, which may be sufficient, as will be shown later, to cause amplification.

Actually, in wintertime the general circulation is downward in the high latitudes (e.g. Murgatroyd and Singleton, 1961), while in summer it is upward. Further as the mean zonal wind $\bar{u}$ is westerly in winter, planetary waves are able to penetrate into the mesosphere and may produce strong downward motion. Since the mixing ratio of atomic oxygen increases rapidly with height, if there is a downward flow, the oxygen rich air in the upper atmosphere can be transported downward and an oxygen dense region may be produced. In this way the density increase may occur only in wintertime and become the energy source for gravity waves. It is reasonable to consider atomic oxygen recombination as a reason for the winter amplification.

In the following we examine the possibility of the growth of upward propagating gravity waves by the chemical heat release due to oxygen recombination. Calculation of the growth time as an instability problem (Leovy, 1966) may not be applicable because waves will be amplified only while they are traveling in the oxygen active region. The finiteness of the energy producing region should be taken into account. Hence spatial change of propagating waves will be considered in the presence of oxygen reactions.

2. BASIC EQUATIONS

2.1 Oxygen recombination and heat release

The oxygen reactions which may be important in the region from 80km to 100km are as follows,

sink of O

$O+O+M \rightarrow O_2+M$	$k_1 = 4.8 \times 10^{-33}(T/300)^{-2}$	(cm^6/s)	(2.1a)
$O+O_2+M \rightarrow O_3+M$	$k_2 = 6.2 \times 10^{-34}(T/300)^{-2}$		(2.1b)
$O+OH \rightarrow O_2+H$	$k_3 = 4 \times 10^{-11}$	(cm^3/s)	(2.1c)
$O+HO_2 \rightarrow O_2+OH$	$k_4 = 3.5 \times 10^{-11}$		(2.1d)

sink of O_3

$$O_3 + H \rightarrow O_2 + OH \qquad k_5 = 1.4 \times 10^{-10} \exp(-\frac{470}{T}) \qquad (2.1e)$$

source of HO_2

$$H + O_2 + M \rightarrow HO_2 + M \qquad k_6 = 5.5 \times 10^{-32} (T/300)^{-1.4}. \qquad (2.1f)$$

k_i rate constant (from Chemical Kinetic and Photochemical Data Sheets for Atmospheric Reactions)

The change of oxygen mixing ratio due to reactions (a) - (d) is given as,

$$\partial_t \chi_0 + \underline{V} \cdot \nabla \chi_0 = -\frac{1}{\tau} \chi_0 + S + D\nabla^2 \chi_0, \qquad (2.2)$$

where

$$\frac{1}{\tau} = k_1 \chi_0 [M]^2 + k_2 \chi_{O_2} [M]^2 + k_3 \chi_{OH} [M] + k_4 \chi_{HO_2} [M].$$

τ time constant of O consumption

χ_X mixing ratio of X

$\quad = [X]/[M]$

D diffusion coefficient

S source of O (produced by photon)

The source term S is mainly the production of O by photo-dissociation, which vanishes at night time, or in the polar night region. But there are transports of O by the general circulation, planetary wave motion and so on which maintain the oxygen density even under night conditions, if the life time of O is longer than the transport time. If we assume a standard atmosphere in which the source and sink terms are in balance with the chemical reactions, (2.2) becomes,

$$0 = -\frac{\chi_s}{\tau_s} + S + D\nabla^2 \chi_s. \qquad (2.3)$$

Since the source term for the standard atmosphere does not seem to change greatly, (2.3) may be substituted into (2.2).

$$\partial_t \delta \chi_0 + \underline{V} \cdot \nabla (\delta \chi_0 + \chi_s) = -\frac{1}{\tau} \chi_0 + \frac{1}{\tau_s} \chi_s + D\nabla^2 \delta \chi_0. \qquad (2.4)$$

$\delta \chi_0$ mixing ratio of excess atomic oxygen

$\quad = \chi_0 - \chi_s$

The time constant τ is mostly determined by the recombination (2.1a) above around 95km because of the large value of χ_0. This reaction is most sensitive to the increase of [O] because the reaction speed is proportional to $[O]^2$. If [O] increases significantly, the whole time constant of oxygen consumption will be determined by this reaction. On the other hand, in the lower region where the value of χ_0 is small, the reactions (2.1b), (2.1c) and (2.1d) come to be important. Especially at about 85km the hydrogen catalytic reactions mainly determine the time constant of O. In this case the time constant does not change with the increase of [O]. Since [H] does not increase so rapidly with height as [O], the downdrafts to produce an oxygen dense layer will not cause an increase of hydrogen. Hereafter we shall consider the two extreme cases, in which the reaction (2.1a) becomes either dominant or not. The former case will be referred as "case I" and the latter as "case II". Thus we avoid the complicated calculations of chemical reactions involving components other than atomic oxygen.

The heating due to these reactions is treated by assuming that the released chemical energy should be transformed into the thermal energy immediately and completely. Although the reaction (2.1b) produces O_3, it will be dissociated by the rapid hydrogen reaction (2.1e). The whole reactions are regarded as $O + O \rightarrow O_2$. The equation of temperature change then becomes

$$\frac{d}{dt}T = -\frac{g}{C_p}w + \Delta T[\frac{\chi_O}{\tau} - \frac{\chi_s}{\tau_s}] + \frac{\kappa}{C_p\rho}\nabla^2 T - \frac{1}{\tau_N}T. \tag{2.5}$$

ΔT temperature increase due to oxygen reactions

$$= \frac{\frac{1}{2}\Delta E}{\frac{7}{2}k_B} = 8.45 \times 10^3 K$$

ΔE chemical energy ($= 5.12eV = 8.16 \times 10^{-19}J$)
k_B Boltzman constant ($= 1.38 \times 10^{23} J/K$)
κ thermal conductivity
τ_N Time constant of Newtonian cooling

The energy transformation may be somewhat slow because of the smaller collision frequency, and the chemical energy may be lost by non-thermal radiation. This may be a serious problem and will be briefly discussed later. For the sake of simplicity this non-LTE effect is not considered in the above equations.

2.2 Air motion

For simplicity, the Coriolis term is neglected. As the basic state, we consider an isothermal atmosphere with $T_0 = 239$ K, i.e., scale height $H = 7$km.

As seen from Fig.1 the active layer of gravity waves seems to be within ±10km in thickness around the altitude of 90km. The usual linearized equations with Boussinesq approximation may be used for air motion after separating out the effect of the density change in the form of exponential scaling $\exp(z/2H)$. In the horizontal direction, a wavenumber k is assumed and the variables are separated into wave parts and horizontal means. Then the momentum equations are written as follows,

$$\partial_t u' = -ikp' \qquad - Ku' + \nu\partial_z^2 u' \tag{2.6a}$$

$$\partial_t w' = -\partial_z p' + \frac{g}{T_0}T' - Kw' + \nu\partial_z^2 w'. \tag{2.6b}$$

$$u = Re\{u'(z,t) \times \exp(\frac{z}{2H} + ikx)\}$$

$$w = Re\{w'(z,t) \times \exp(\frac{z}{2H} + ikx)\}$$

$$p = Re\{p'(z,t) \times \exp(-\frac{z}{2H} + ikx)\} \times \rho_0$$

ρ_0 reference density (at z = 95km)
ν molecular viscosity coefficient (= D)

$T = \mathrm{Re}\{T'(z,t) \times \exp(\frac{z}{2H} + ikx)\}$
K   coefficient of the ion drag

The continuity equation is,

$$iku' + \partial_z w' = 0. \tag{2.7}$$

Note that we have neglected the term $-w'/2H$ in the above equation. This approximation is permissible for waves with the vertical wavelength $\lesssim 20$ km. By adopting this Boussinesq approximation, partly non-linear treatment of the problem as described below becomes simpler. In the case that $0 + 0 + M \rightarrow O_2 + M$ is considered as the dominant reaction (case I), the thermodynamic equation becomes,

$$\partial_t T' = -\frac{g}{C_p}w' + \Delta T \frac{2(\bar{\chi}+\chi_s)}{\tau_s\chi_s}\chi' - \frac{T'}{\tau_N} + \frac{\kappa}{C_p\rho_s}\partial_z^2 T' \tag{2.8a}$$

$$\partial_t \bar{T} = -(\frac{g}{C_p} + \partial_z\bar{T})\bar{w} - \exp(\frac{z}{H})\partial_z(\overline{T'w'}) + \Delta T\{\frac{\bar{\chi}^2+2\bar{\chi}\chi_s}{\tau_s\chi_s} + \exp(\frac{z}{H})\frac{\overline{\chi'^2}}{\tau_s\chi_s}\}$$
$$- \frac{\bar{T}}{\tau_N} + \frac{\kappa}{C_p\rho_s}\partial_z^2\bar{T}. \tag{2.8b}$$

$\bar{T}$ horizontal mean temperature difference from T_0
$\bar{w}$ mean vertical velocity
χ perturbation of oxygen mixing ratio

$= \mathrm{Re}\{\chi'(z,t) \times \exp(\frac{z}{2H} + ikx)\}$

$\bar{\chi}$ horizontal mean mixing ratio of atomic oxygen
$\delta\chi_0 = \bar{\chi} + \chi$

On the other hand, if $0 + 0 + M \rightarrow O_2 + M$ is a minor reaction (case II), the thermodynamic equation becomes,

$$\partial_t T' = -\frac{g}{C_p}w' - \Delta T\frac{\chi'}{\tau_s} - \frac{T'}{\tau_N} + \frac{\kappa}{C_p\rho_s}\partial_z^2 T' \tag{2.9a}$$

$$\partial_t \bar{T} = -(\frac{g}{C_p} + \partial_z\bar{T})\bar{w} - \exp(\frac{z}{H})\partial_z(\overline{T'w'}) + \Delta T\frac{\bar{\chi}}{\tau_s} - \frac{\bar{T}}{\tau_N} + \frac{\kappa}{C_p\rho_s}\partial_z^2\bar{T}. \tag{2.9b}$$

We have included the effect of the mean vertical velocity $\bar{w}$, which will be prescribed as an external parameter. In the case I, the continuity equation for atomic oxygen becomes,

$$\partial_t \chi' = -w'\partial_z(\chi_s + \bar{\chi}) - \frac{2(\chi_s+\bar{\chi})}{\tau_s\chi_s}\chi' + D\partial_z^2\chi' \tag{2.10a}$$

$$\partial_t \bar{\chi} = -\bar{w}\partial_z(\chi_s + \bar{\chi}) - \exp(\frac{z}{H})\partial_z(\overline{\chi'w'})$$
$$- \frac{\bar{\chi}^2+2\bar{\chi}\chi_s}{\tau_s\chi_s} - \exp(\frac{z}{H})\frac{\overline{\chi'^2}}{\tau_s\chi_s} + D\partial_z^2\bar{\chi}, \tag{2.10b}$$

and in the case II,

$$\partial_t \chi' = -w'\partial_z(\chi_s + \overline{\chi}) - \frac{\chi'}{\tau_s} + D\partial_z^2\chi' \tag{2.11a}$$

$$\partial_t \overline{\chi} = -\overline{w}\partial_z(\chi_s + \overline{\chi}) - \exp(\frac{z}{H})\partial_z(\overline{\chi'w'}) - \frac{\overline{\chi}}{\tau_s} + D\partial_z^2\overline{\chi} . \tag{2.11b}$$

The effect of the density change remains explicitly in the form of a factor exp(z/H) in the equation for T and χ. The general wind in the horizontal direction is not taken into account. Though the mean temperature change due to recombination is not small as will be shown later, the temperature increase effect is neglected for simplicity. We shall confine ourselves to the effects of the recombination on waves.

2.3 Values of constants

As a standard time constant $\tau_s(z)$, we use the value summarized by G.E. Thomas et al. (1980), which was originally calculated by Park and London (1974). In case I, $\tau_s/2$ is equated to their value, as the linearization of the reaction (2.1a) yields the factor 1/2. In case II, τ_s is set to be the same as their value.

The standard values of [O] are taken from CIRA (1965) by fitting an appropriate function (Fig. 3). Other curves in Fig. 3 are obtained by modifying this as explained below. Fig. 4 shows another series of profiles in which the standard profile is assumed to have greater amount of [O] in the lower region than that of CIRA. Mean downward motion of winter global circulation may cause the higher density in the polar region. It is adequately obtained within the range of observations of Dickinson (1980) (shaded regions in Fig. 3 and Fig. 4) (cf. review of L. Thomas (1980)). The value of [O] are fairly close to those of the winter polar region calculated by Kasting and Roble (1981). We will call this profile the "winter standard profile". Profiles of excess mixing ratio $\overline{\chi}$ are shown in those figures which are derived from the assumption that some additional downdraft (planetary waves, for example) shift the standard mixing ratio conservatively from certain altitudes;

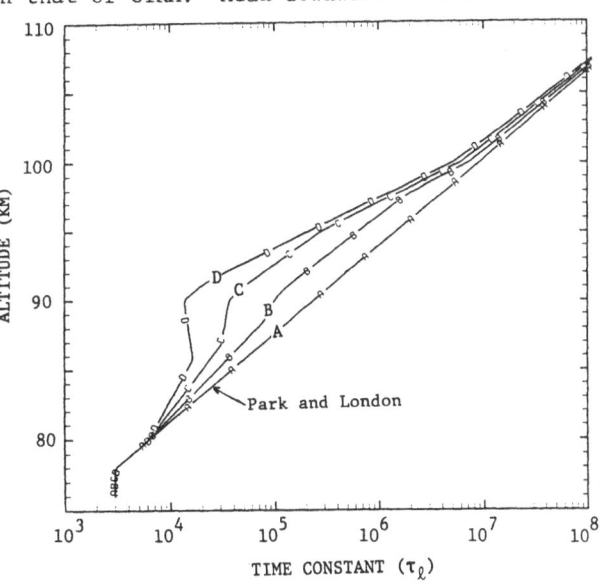

Fig. 2 Height distributions of time constant τ_ℓ. The curve A is the standard profile by Park and London(1974). The curves B, C and D are the reduced time constants in case I by the corresponding increased values of [O] in Fig. 3 produced from CIRA standard [O].

148

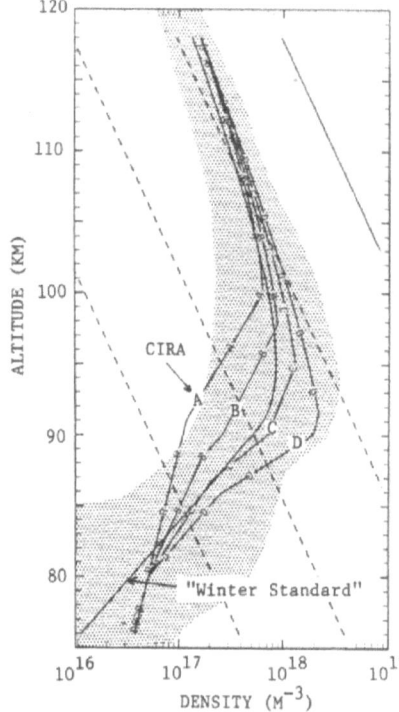

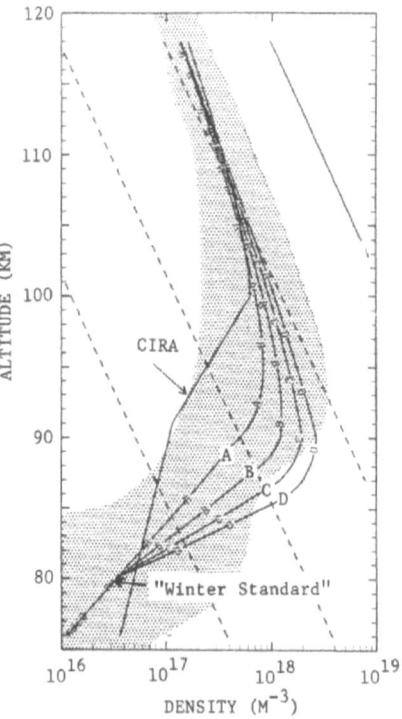

Fig. 3 Profiles of [O] using CIRA as a standard. The curve A is the standard profile (CIRA). B, C and D denote oxygen dense profiles with the maximum lowered length DL = 3km, 6km and 9km respectively. The shadowed area represents the total range of observations by Dickinson (1980). The solid curve is the winter standard profile (cf. Fig. 4). The solid line is [M] and dashed lines correspond to χ = 0.1, 0.01 and 0.001 respectively.

Fig. 4 Profiles of [O] using the winter standard density. The solid curve is the CIRA standard (cf. Fig. 3).

we assumed the following displacement,
$$\bar{\chi}(z) = \chi_s(z') - \chi_s(z)$$

$$z' = \begin{cases} z + \dfrac{z-80km}{90-80km} \cdot DL & \text{for } 80 \leqq z \leqq 90km \\[2mm] z + \dfrac{z-120km}{90-120km} \cdot DL & \text{for } 90 < z \leqq 120km \qquad (2.12) \\[2mm] z & \text{otherwise} \end{cases}$$

DL is the maximum displacement which is assumed to occur at the altitude of 90km. In Fig. 3 and Fig. 4, three cases are shown corresponding to DL = 3 km, 6 km and 9 km.

Since in the upper atmosphere the molecular diffusions (of constituents, heat and momentum) and the so-called ion drag effect become important (e.g. Kato and Matsushita, 1969), the corresponding terms are
included in the equations. The values of molecular viscosity, thermal
conductivity and the ion drag rate are given as follows.

$$D = \nu = 6.2 \times \exp\left(\frac{z-95\text{km}}{H}\right) \qquad (\text{m}^2/\text{s}) \qquad (2.13\text{a})$$

$$\frac{\kappa}{C_p \rho_s} = 8.5 \times \exp\left(\frac{z-95\text{km}}{H}\right) \qquad (\text{m}^2/\text{s}) \qquad (2.13\text{b})$$

$$K = 10^{-6} \times \exp\left(\frac{z-95\text{km}}{2.7\text{km}}\right) \qquad (\text{s}^{-1}) \qquad (2.13\text{c})$$

Newtonian cooling rate is not so definite. Werhbein and Leovy (1982)
show that τ_N ranges from 2 to 10 days at about 80 km according to the
vertical scale of the temperature perturbation. In the present study we
assume τ_N = 5 days. A brief discussion on this value will appear in
secgion 7.

3. LINEAR THEORY

To see how the existence of atomic oxygen alters the structure of
internal gravity waves, we consider the inviscid linearized equations.
From (2.6) $\sim$ (2.11), the inviscid equations are obtained as follows,

$$\partial_t u' = -ikp' \qquad (3.1\text{a})$$

$$\partial_t w' = -\partial_z p' + \frac{g}{T_0} T' \qquad (3.1\text{b})$$

$$iku' + \partial_z w' = 0 \qquad (3.1\text{c})$$

$$\partial_t T' = -\frac{g}{C_p} w' + \Delta T \frac{\chi'}{\tau_\ell} \qquad (3.1\text{d})$$

$$\partial_t \chi' = -w' \partial_z (\chi_s + \bar{\chi}) - \frac{\chi'}{\tau_\ell} \qquad (3.1\text{e})$$

τ_ℓ linearized time constant

$$= \frac{\tau_s \chi_s}{2(\chi_s + \bar{\chi})} \qquad (\text{in case I})$$

$$= \tau_s \qquad (\text{in case II})$$

In case I τ_ℓ is changed from $\tau_s/2$ by considering the excess atomic oxygen $\bar{\chi}$ (cf. Fig. 2). The mean vertical velocity $\bar{w}$ is neglected compared
with the wave advection w' in the above equations. From (3.1d) and (3.
1e) we obtain the relation between T' and w' for a given frequency ω,

$$-i\omega \frac{g}{T_0} T' = -N^2 [1 + \frac{C_p}{g} \Delta T \partial_z (\chi_s + \bar{\chi}) \frac{1+i\omega\tau_\ell}{1+\omega^2\tau_\ell^2}] w'$$

$$\equiv -\tilde{N}^2 w' \qquad (3.2)$$

N Brunt-Vaisala frequency

$$= \frac{g}{\sqrt{C_p T_0}}$$

$\tilde{N}$ complex Brunt-Vaisala frequency

As seen from (3.2) the effect of chemical heating is condensed into the complexity of the Brunt-Vaisala frequency.

The work done by the buoyancy force per unit mass per unit time is given by

$$\frac{g}{T_0}\overline{T'w'} = \text{Im}(\frac{\tilde{N}^2}{\omega})\overline{w'^2} = \frac{g}{T_0}\Delta T \partial_z (\chi_s + \overline{\chi})\frac{\tau_\ell}{1+\omega^2\tau_\ell^2}\overline{w'^2} \tag{3.3}$$

When atomic oxygen exists, the imaginary part of $\tilde{N}^2$ is positive as $\partial_z(\chi_s + \overline{\chi})$ is positive. The correlation between the buoyancy force and the velocity w' arises and wave enhancement will be caused.

In the cases where the time constant τ_ℓ is far from ω^{-1}, the imaginary part can not be large enough to cause significant deformation. In the case $\tau_\ell\omega \gg 1$, the time constant is too long and the situation is the same as that of no atomic oxygen. In the case $\tau_\ell\omega \ll 1$ the time constant is so short that the oxygen fluctuation caused by wave motion immediately adjusts to its surroundings. The highest temperature occurs while an air parcel is in the lowest position from the equilibrium height. So the buoyancy force as a restoring force becomes larger and the real part of $\tilde{N}$ increases. If τ_ℓ is longer than this case, the highest temperature occurs when an air parcel is in its ascending motion. In the real atmosphere, Fig. 2 indicates that τ_ℓ is longer than 10^4 sec and increases exponentially with height so that $\tau_\ell\omega$ is larger than 2.7 even for a 6-hour period wave. In most region, the condition $\tau_\ell\omega \gg 1$ is valid. Hence to obtain a large value of the $\text{Im}(\tilde{N}^2)$, $\partial_z(\chi_s + \overline{\chi})$ should be large to a certain extent or in case I τ_ℓ should become small.

If $\tilde{N}^2$ is regarded as a constant, the following dispersion relation is readily obtained from the system (3.1) with the hydrostatic approximation,

$$m^2 = \frac{\tilde{N}^2}{\omega^2} k^2 \tag{3.4}$$

Then the complex vertical wavenumber m can be obtained for a given set of real ω and k. The spatial growth rate is defined for a wave after traveling a distance L vertically by

$$e^{-\text{Im}(m)\cdot L} \tag{3.5}$$

From (3.4), we see that a large k (short horizontal wavelength) and a small ω (long period) is suitable to give large spatial growth except for the value of $\text{Im}(\tilde{N})$. Leovy (1966) considered wave amplification with time. He obtained complex ω for real k and m. For effective enhancement, ω should be large, since the growth rate is given in the form of

exp(Im(ω)·Δt). This apparent discrepancy is due to the difference
of view point.

In the real atmosphere, various dampings can not be neglected.
A large value of m is favorable
for wave enhancement but also such
a wave suffers from friction.
Hence there will be a certain
range of ω and k within which
waves can be amplified significantly. The dispersion relation
which includes constant diffusion
coefficient ν is obtained by replacing ω in (3.4) by $\omega + i\nu m^2$. The
complex vertical wavenumber m is
iteratively obtained for a given ω
and k. Since τ_ℓ increases exponentially with height, the effect
of Im($\tilde{N}$) is significant only within a relatively thin layer. Hence
L in the definition of spatial
growth rate is set to be 10km.
The diffusion coefficient ν = 30
m/sec.

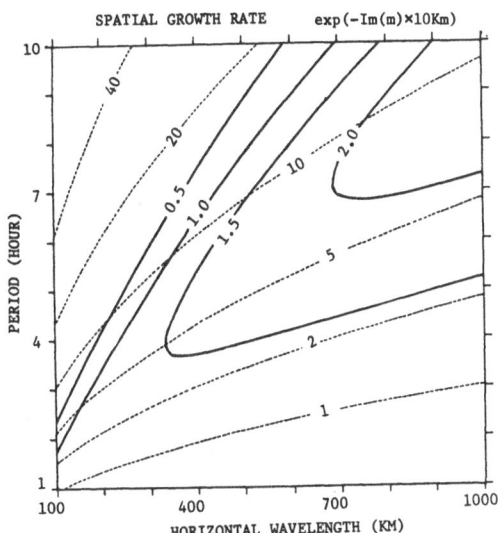

Fig. 5 The values of spatial growth rate exp
(-Im(m)×10km)) with $\partial_z(\chi_s + \overline{\chi}) = 5 \times 10^{-6} m^{-1}$ and
$\tau_\ell = 5 \times 10^4$ sec (the moderate case). Diffusion
constant ν = 30m/s. Dashed contours represent
the periods with which waves propagate through
the length of 10 km with their group velocities.

Fig.5 shows the growth rates
for $\partial_z(\chi_s + \overline{\chi}) = 5 \times 10^{-6} m^{-1}$ and $\tau_\ell = 5 \times 10^4$
sec which may be considered as a
moderate condition (cf. Fig.2 and
Fig.3 or Fig.4). Fig.6 shows
those of an extreme case ($\partial_z(\chi_s + \overline{\chi})$
$= 10^{-5} m^{-1}$ and $\tau_\ell = 2 \times 10^4$ sec). In the
upper left part of these figures,
growth rates tend to be zero because of large values of the real
part of m. The largest growth
rate in the largest O-density case
(Fig.6) is 23 for the wavelength
of 550km and the period of 5.5
hours. In this case, $\omega\tau_\ell \approx 1/6$ and
Im($\tilde{N}/N$)≈O(1), owing to the relatively large gradient of $\chi_s + \overline{\chi}$.
However the value may be too large
because τ_ℓ is around 5 hours so
that downdrafts of realistic magnitude can not bring about such an
amount of excess oxygen to produce
the gradient of $10^{-5} m^{-1}$ at low altitudes against the recombination.

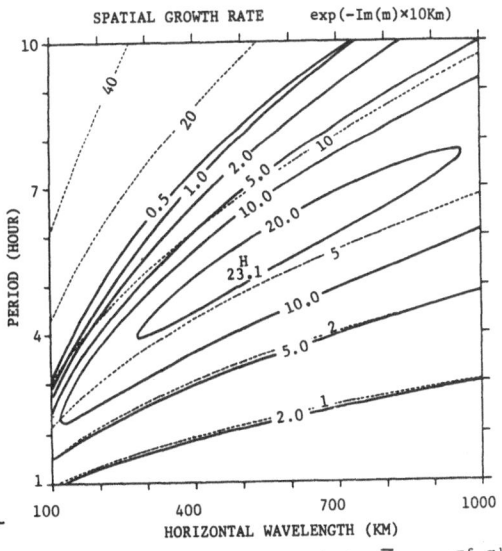

Fig. 6 Same as Fig. 5 except $\partial_z(\chi_s + \overline{\chi}) = 10^{-5} m^{-1}$
and $\tau_\ell = 2 \times 10^4$ sec (the extreme case).

In the moderate case, the waves may be enhanced by a factor of 1.5 $\sim$ 2, which is small compared to the extreme case but may be still significant as an explanation for the observed large temperature fluctuations.

The values of τ_ℓ and $\partial_z(\chi_s + \bar{\chi})$ mentioned above can not be obtained by the standard profiles of Fig.2 and Fig.3 or Fig.4, that is, no enhancement will be caused by these profiles. But there is a possibility that downdrafts cause some increase in [O] at the altitudes where τ_ℓ is sufficiently short, or in case I, some decrease in τ_ℓ at the altitudes where $\partial_z(\chi_s + \bar{\chi})$ is sufficiently large. A considerable enhancement due to recombination is expected by the above discussion. However, there remain some points which require more detailed treatments. The dashed contours of Fig.5 and Fig.6 indicate the times for wave packets to travel through a layer of 10km thickness, which are obtained by the use of pure gravity wave group velocities. In some cases they are of the same order as the time constant τ_ℓ. Except for frictional damping, growth rate increases as the vertical wavenumber m increases, whereas the group velocity is inversely proportional to m and it takes a longer time to propagate the oxygen active layer. In such a case, the change of excess atomic oxygen $\bar{\chi}$ should be taken into account. The steady state considered above may not be realized for those waves. The acceptability of $\bar{\chi}$ profiles should be checked. The profiles of Fig.3 and Fig.4 may not be attained by reasonable down drafts. Moreover the profiles of mean fields $\chi_s + \bar{\chi}$ and τ_ℓ change exponentially with height. It may not be applicable to treat $\tilde{N}$ as a constant coefficient.

4. DESCRIPTION OF NUMERICAL EXPERIMENTS

We use the equations of section 2. The origin of z-axis is chosen at 95km and a layer with a thickness of 30km around this level is taken into consideration (Fig.7). Above this layer, a sponge layer is added where the ion-drag and the molecular diffusion become important. Under the active layer, there is an approach layer, which is taken long enough to pass the initial waves and to avoid the effect of reflection at the lower boundary. Both in the sponge layer and in the approach layer, the scaling rule for the effect of density change is not applied. The factors exp(z/H) in (2.8b) or (2.9b) and (2.10b) or (2.11b) are set to be constant.

Pure gravity waves ($\chi' = 0$) are set free at t = 0 as shown in Fig.7. They are forced at the bottom boundary with given ω and k. The amplitudes of the waves u_0 are 30.0 m/s. At the upper boundary we assumed w = 0 and the diffusion equilibrium be valid for other variables (their vertical gradients vanish).

5. RESULTS OF CASE I (τ DEPENDS ON [O])

5.1 Results for initially given $\bar{\chi}$ profiles

For simplicity's sake, we shall consider the cases where no vertical mean motion exists but the initially given $\overline{\chi}$ profiles have excess oxygen. The calculations were performed for two types of $\overline{\chi}$ profiles produced from CIRA data (Fig.3); the 6km shifted moderate case (line C) and the 9 km shifted extreme case (line D). They are given at t = 0 and relax toward the standard value, while waves with given frequency and wavelength propagate through the layer containing the excess oxygen. The results are summarized in Table I.

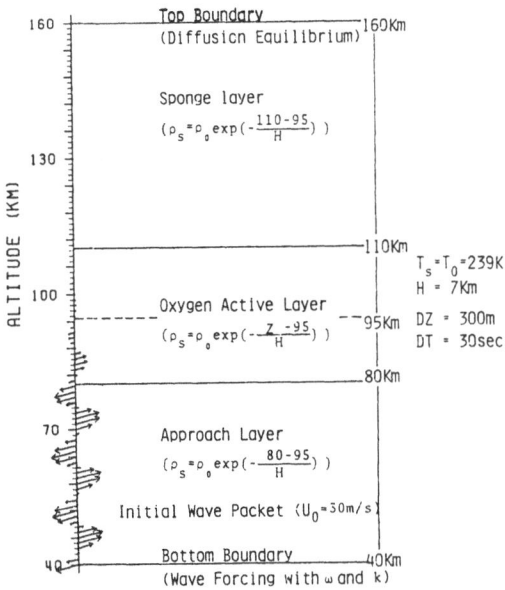

Fig. 7 Schematic representation of numerical experiments.

Fig.8a shows the time-height section for the moderate case with a wavelength of 300km and a period of 4 hours. The growth rate, which is defined in this section as the ratio of the maximum amplitude of u' to the input amplitude u_0 (30m/s), is 1.5, which is almost the same as that of the constant coefficient, linear calculation. The lasting time of the enhancement is roughly determined by the amount of remaining $\overline{\chi}$. It takes about 1600min for $\overline{\chi}$ to reduce by half at 91 km level, which corresponds to the decay of enhancement. As summarized in Table I, those results with the 6km shifted $\overline{\chi}$ profile show roughly the same tendency of the spatial growth rate as that of the linear case (Fig. 5). However, owing to the slow group velocity, the wave with a six-hour period suffers not only from the molecular viscosity but also from the decrease of $\overline{\chi}$. Half of the initial $\overline{\chi}$ has been consumed by the time when the maximum amplitude is attained.

In the case of the 9km shifted $\overline{\chi}$ profile, the spatial growth rate reaches as much as 3.0 for the same period and frequency as those shown in Figs.8. However, because of the rapid consumption of $\overline{\chi}$, the large spatial growth rate obtained in the linearized constant coefficient case (Fig.5) was not realized. A large amount of initially given $\overline{\chi}$ disappeared with a short time constant (400 min at 92 km), without causing the wave amplification.

Fig.8b shows the mean temperature change calculated from (2.8b) for the 6 km shifted case. In this calculation Newtonian cooling is not included. Although the change of the basic temperature field is not included in the wave equations, the temperature increase reaches 160K after about 3 days. A similar calculation excluding gravity waves showed that

154

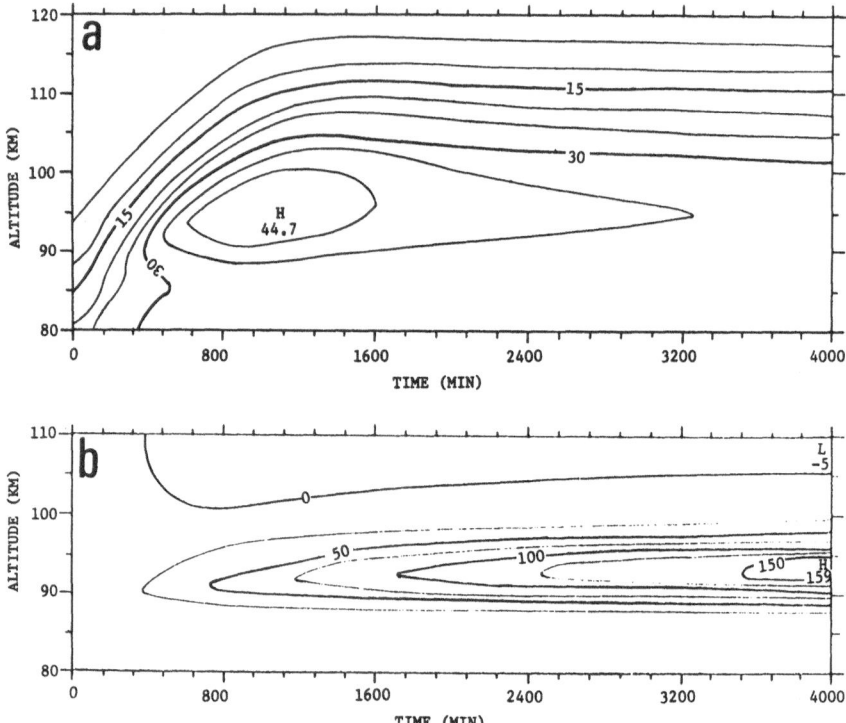

Fig. 8 Time-hight cross section of a) |u'| and b) T̄ with a perild of 4 hours
and a horizontal wavelength of 300km. The amplitude of incident wave is 30
m/s. Profiles A (CIRA) and C of Fig.3 are used as χ_s and $\chi_s + \bar{\chi}$ at t=0. The
contour intervals are 5 m/s in a) and 25K in b).

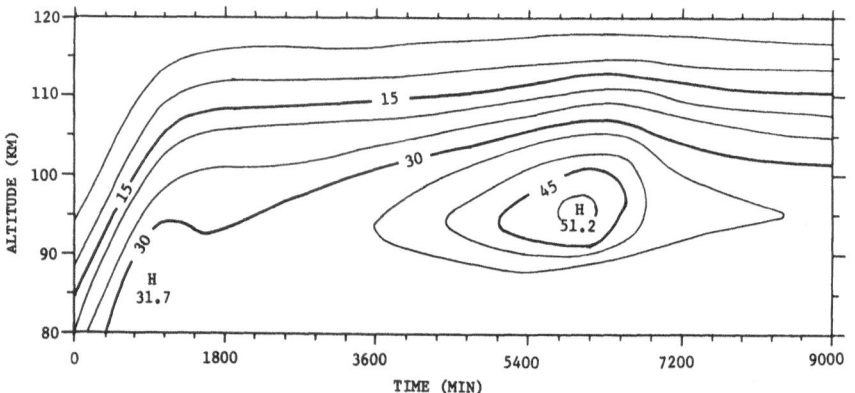

Fig. 9 Time-height cross section of |u'| with $\bar{\chi} = 0$ at t = 0 and the vertical wind
w̄ is switched on at t = 1440min and off at t = 5760min. The incident wave and χ_s
are the same as Figs. 8.

Table I Summary of Numerical Experiments

	χ_s	$\overline{\chi}$		λ_h km	h	λ_z km	C_{gz} m/s	u'max m/s
CASE I	CIRA	$\overline{w}=0$ (No Vertical Motion) $\overline{\chi}$ given at t=0	DL=6km	600	2	26.2	3.6	37.6
					4	13.1	0.9	42.9
					6	8.7	0.4	49.7
				300	2	13.1	1.8	37.6
					4	6.5	0.45	44.7
					6	4.4	0.2	41.2
			DL=9km	600	2	26.2	3.6	54.6
					4	13.1	0.9	90.9
				300	2	13.1	1.8	63.8
					4	6.5	0.45	128.0
		$\overline{w}$ given for 3 days $\overline{\chi}=0$ at t=0		300	4	6.5	0.45	51.2
CASE II	CIRA	$\overline{w}$ given for 3 days $\overline{\chi}=0$ at t=0		300	4	6.5	0.45	36.0
	Winter Standard	$\overline{w}$ given for 3 days $\overline{\chi}=0$ at t=0		300	4	6.5	0.45	46.0
					6	4.4	0.2	60.2
	Winter Standard	$\overline{w}=0$ (No Vertical Motion) $\overline{\chi}=0$ at t=0	DL=0km	300	4	6.5	0.45	34.3
			DL=3km					40.3
			DL=6km					49.0

The input amplitude u_0=30m/s for all cases.

the temperature increase progresses a
little slower than that including the
waves shown in Fig.8b. It reaches 120K
at t=4000min at altitudes of 92-95km.
This result implies that the transport
of O-atoms by gravity waves promotes the
recombination of them. When Newtonian
cooling (τ_N=5 days) is included, the maximum temperature of 90K occurs at t=4320
min and z=94km for this no wave case.
This value may be still too large for a
realistic model of the atmosphere. There
arises a question whether the profiles of
$\overline{\chi}$ might not be realistic; they may be too
large.

5.2 With a mean vertical wind $\overline{w}$

In this section, a mean downward
motion $\overline{w}$ is included to check whether the
initially given $\overline{\chi}$-profiles of Fig.3 and
the resultant wave amplifications are
reasonable or not. $\overline{\chi}$ is set to be 0 at
t=0 and it will be generated from the
standard oxygen value (CIRA data) by the
following vertical wind,

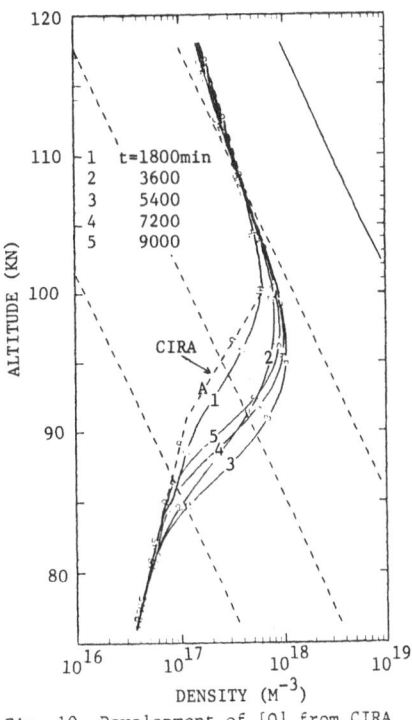

Fig. 10 Development of [O] from CIRA
standard profile for the case of Fig. 9.

$$\bar{w} = \begin{cases} -3.0 \ \cos[\pi \cdot \dfrac{z - 90km}{110 - 70km}] & cm/s & 70 \leqq z \leqq 110 \ km \\ 0 & & otherwise \end{cases} \quad (5.1)$$

As the time constant ranges from 2 to 3 days at about 90km, the value of 3cm/s was chosen so that air mass can descend 9km in 3 days. Newtonian cooling is also included, since the temperature increase is expected to reach a large amount.

Fig.9 shows the time-height section of the amplitudes of u' with the same parameter values as the wave shown in Figs.8. The vertical motion $\bar{w}$ is switched on at t=1day and off at t=4days. The obtained growth rate of the wave is about the same as one in Fig.8a. The development of [0] profile is shown in Fig.10, which indicates that the downdraft of 3cm/s is sufficient to produce a distribution similar to the 6km-shifted profile of Fig.3 within 2 to 3 days. The temperature increase reaches as high as 192K after 3 days, though about 100K is expected from adiabatic change due to $g/C_p \cdot \bar{w} \times 3days$. Thus we can confirm that the results obtained by prescribing initial $\bar{\chi}$-profiles may be acceptable if a downward motion of 3cm/s does exist. Then we have to solve the problem of unrealistically high temperatures.

So far we have assumed that the three body reaction (2.1a) is the main recombination process. Actually the reaction is dominant above 95 km so that the reduction of the time constant in our model is reliable only above this level. On the other hand the enhancement obtained in the calculations so far is caused by the change of time constant at altitudes from 85km to 95km where $\partial_z(\chi_s + \bar{\chi})$ is still sufficiently large. The time constant at this region, however, is determined by the reactions (2.1b)-(2.1d) rather than (2.1a). The extreme temperature increase may be due to the improper treatment of oxygen reactions.

6. RESULTS OF CASE II (τ DOES NOT DEPEND ON [0])

6.1 With the CIRA standard profile

The situation is the same as 5.2 except for the time constant and the equations of $\bar{T}$ and $\bar{\chi}$. The vertical motion $\bar{w}$ is switched on at t=0 and off at t=3days. The obtained growth rate is about 1.2 after 3days which may not be so remarkable. The temperature increase becomes a little moderate (about 100K after 3 days), since the time constant at about 95km is much longer than case I. The altitude of the maximum amplitude is 89km which is lower than case I. (cf. Fig.9)

The reason for insignificant enhancement is that oxygen could not come down to the region (80km-90km) where the time constant is short enough for the wave-chemistry coupling. The oxygen profile of CIRA has a peak at 100km and decreases rapidly below this height (Fig.3). To obtain a large $\partial_z(\chi_s + \bar{\chi})$ in the region where the coupling can take place oxygen must be lowered more than 9km, which is impossible by a moderate downdraft.

6.2 With the winter standard profile

If there is more atomic oxygen in the lower layer, 3cm/s downdraft may be able to cause a large $\partial_z(\chi_s + \overline{\chi})$ at about 85km. By considering the peculiarity of the winter polar region, we may consider another profile which contains much more oxygen in these region. Noted that the CIRA model is intended to represent the global and seasonal average state. From now on, we shall use a new standard profile (Fig.4) as χ_s, which is supposed to be suited to winter high latitude conditions. The oxygen peak is located lower than that of CIRA profile. In Fig.3 this winter standard profile is also inserted. A similar numerical experiment to the one described previously was carried out.

When the vertical wind (5.1) is switched on at t=0 and off at t=3days, an amplification of waves by factor 1.5 is obtained. Fig. 11 shows the time change of [O]. This figure indicates that the value of $\partial_z(\chi_s + \overline{\chi})$ reaches 5×10^{-6} at 85km where τ_0 is about 5×10^4 sec. The obtained maximum amplitude (46m/s) is explained by the linear theory with these values.

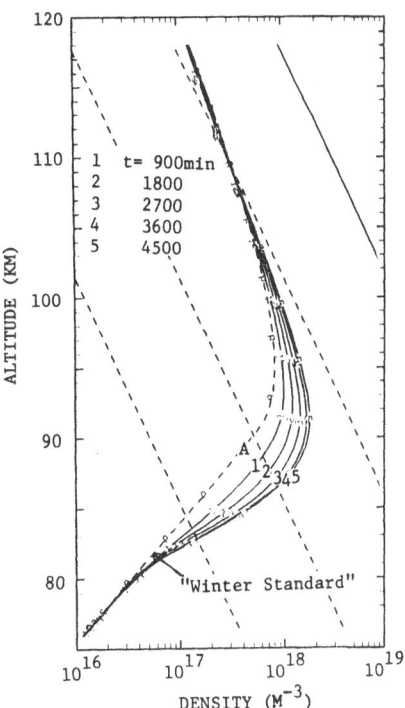

Fig. 11 Development of [O] from the winter standard profile. $\chi = 0$ at $t = 0$ and the vertical wind is switched on at $t = 0$ and off at $t = 4320$min. The incident wave is the same as Figs. 8 but the curve A of Fig. 5 (the winter standard) is used as χ_s.

The temperature increase is similar to that of case I. It reaches about 190K after 3 days. An increase of $\overline{\chi}$ necessarily gives rise to a rapid heating as [O] declines toward its standard value in the short time constant region. "Realistic" oxygen profiles which contain a small amount of oxygen at 80-85km inevitably cause unrealistic temperature increase whenever the background downward motion is introduced.

If we regard the 6km shifted oxygen from the winter standard profile (line C of Fig.4) as χ_s and do not incorporate $\overline{w}$, the wave can be enhanced to reach 50m/s after 1000min. Since $\overline{w}=0$ in this case, all the temperature increase is due to the wave effect. The amount of temperature increase reached 30K after 3 days at 89km. Even though these results are interesting, the assumed χ_s seems to be unrealistic in that [O] exceeds 10^{12}cm^{-3} at 86km and the maximum reaches 2×10^{12}cm^{-3}.

A peculiarity of the temperature behavior in this case is that there occurs a temperature decrease under the high temperature region, which can not be seen in the previous calculations, since it is masked by the large increase due to $\bar{w}$. It reaches about -10K after 3 days. By a brief analysis of the heat transport property it can be shown that growing waves transport heat upward, just the opposite process to that discussed by Waltersheid and Schoeberl in this Seminar.

7. DISCUSSION AND CONCLUSION

The linear theory of section 3 indicates that there is a possibility of wave enhancement due to oxygen under certain conditions assuming dense [O] in the winter polar regions. The growth rates may be large enough to explain the temperature variations around the mesopause observed by Theon and Smith. The numerical experiments including the vertical and time changes of [O] also show the validity of the linear theory. The increase of $\partial_z(\chi_s + \overline{\chi})$ and τ_ℓ with altitude cancel out to result in a nearly constant $\tilde{N}$ so that a growth rates of section 3 with constant coefficient equations become valid. However, they also reveal some difficulties.

The first is that a too large temperature increase about 200K per 3 days is caused by the excess oxygen as it reduces to its standard value rapidly. The necessary [O] profile required by the linear theory is rather easily obtained by a downward motion of 3cm/s within 3 days from the winter standard profile which we think is not so unrealistic. There may be some inconsistencies between oxygen density and the time constant. And in this paper it is attributed to the implicit catalytic cycles, that is, the density of hydrogen. We assumed restoration to standard profiles to skip the complicated calculations for catalytic actions. There is a possibility that the time constant is too small for the given profile of oxygen. Or, conversely, the excess oxygen is too large for the given profile of the time constant. If the obtained oxygen profile or the time constant profile is not applicable, the conditions required by the linear theory can not be realized.

If the profiles of τ_s and χ_s are reasonable and the value of downdraft (3cm/s) is acceptable, we should examine the thermodynamical equation in order to search for a cause of the high temperature. One of the uncertain processes is Newtonian cooling. We used τ_N =5 days in our experiments. But if τ_N is around a day, the heat may be removed without any influence on the wave enhancement. Werbein and Leovy (1982) showed that Newtonian cooling coefficients due to CO_2 and O_3 reach about 2 days at 80 km for thin limit temperature disturbance of 5 km thickness and 5 days for boxcar limit (25km thickness). However the vertical extents of the high temperature layers in our experiments are not small enough to accept 2 days as a value of τ_N. Further even if this is accepted, the effect is not strong enough to counteract the rapid temperature increase (200 K/3 days).

At thermospheric levels (above 90km), thermal radiation due to the

63 μ lines of atomic oxygen becomes important. However, the magnitude of the cooling rate is estimated to be about 6 K day^{-1} for the case of 1% oxygen mixing ratio. This cooling is too weak to diminish the temperature rise. At these levels non-LTE effect becomes important which tends to decrease the cooling due to the CO_2 and O_3 bands (Wehrbein and Leovy, 1982). Thus, the anomalous temperature increase can not be removed by ordinary thermal radiation.

Another uncertain process in the thermodynamic equation is the assumption of high thermarization efficiency of chemical energy. The released energy (5.1eV) can not be directly transformed into kinetic energy. It is supposed that a considerable fraction of this high energy may be transferred into the vibrations of N_2 and O_2 molecules. These molecules easily excite the vibrations of CO_2 (ν_3) and H_2O (ν_2). If there are a sufficient number of collisions, those vibration energy will be converted into kinetic energy, as we assumed in section 2. However, even at 60km, the radiative life time of CO_2 (ν_3) (about 2.4×10^{-3}) becomes shorter than collisional deactivation time (e.g. Houghton, 1969) so that a considerable amount of 5.1eV may be lost by CO_2 radiation through the nonthermal process. In the Newtonian cooling non-LTE has an effect on the exitation of radiation levels, while in the energy cascading process it intercepts the thermalization of high energy quanta. This is a most likely candidate to resolve the problem of the anomalous high temperature. Ironically if the high temperature is removed by this effect, it also diminishes the wave enhancement.

In either case (unrealistic profiles of excess oxygen or unrealistic efficiency of energy conversion), the energy which waves can utilize will be greatly reduced, which means that the chemical induced wave enhancement is not likely to occur in the atmosphere, except in the case that χ_s is fairly large from the beginning at low altitudes where τ_s is sufficiently small and collision frequency is sufficiently large.

Acknowledgements

We would like to express our gratitudes to Drs. C.B.Leovy and J.R. Holton for their careful reading of the manuscript, and to Drs. T. Ogawa and K. Fukuyama for valuable informations. We are also grateful to Dr. K. Gambo and other members of the Meteorological Laboratory of Tokyo University for useful discussions. Last but not least we thank Mrs. K. Kudo for her skilfull typing.

References

U.S.Department of Transportation, 1980: Chemical kinetic and photochemical data sheets for atmospheric reactions. Federal Aviation Administration, Washington D.C.

CIRA, 1965: North-Holland Publ. Co.

Dickinson, P.H.G., L. Thomas, E.R. Williams, D.B. Jenkins and N.D. Twiddy, 1980: The determination of the atomic oxygen concentration and associated parameters in the lower ionosphere. Proc. R. Soc. Lond., A.369, 379-408.

Heath, D.F., E. Hilsenrath, A.J. Kruger, W. Nordberg, C. Prabhakara and J.S. Theon, 1974: in "Structure and dynamics of the upper atmosphere", F. Verniani (ed.), Elsevier Scientific Publ. Co., 131-198.

Houghton, J.T., 1969: Absorption and emission by carbon-dioxide in the mesosphere. Quart. J. Roy. Meteor. Soc., 95, 1-20.

Kasting, J.F. and R.G. Roble, 1981: A zonally averaged chemical-dynamical model of the lower thermosphere. J. Geophys. Res., 86, 9641-9653.

Kato, S. and S. Matsushita, 1969: A consideration on the tidal wave transmission through the ionized atmosphere. J. Geomag. Geoelectr., 21, 471-478.

Leovy, C.B., 1966: Photochemical destabilisation of gravity waves near the mesopause. J. Atmos. Sci., 23, 223-232.

Murgatroyd, R.J. and F. Singleton, 1961: Possible meridional circulations in the stratosphere and mesosphere. Quart. J. Roy. Meteor. Soc., 87, 125-135.

Park, J.H. and J. London, 1974: Ozone photochemistry and radiative heating of the middle atmosphere. J. Atmos. Sci., 31, 1898-1916.

Theon, J.S. and W.S. Smith, 1970: Seasonal transitions in the thermal structure of the mesosphere at high latitudes. J. Atmos. Sci., 27, 173-176.

Thomas, G.E. et al., 1980: Scientific objectives of the solar mesosphere explorer mission. Pure and Applied Geophys., 118, 591-615.

Thomas, L., 1980: The composition of the mesosphere and thermosphere. Phil. Trans. R. Soc. Lond., A296, 243-260.

Wehrbein, W.M. and C.B. Leovy, 1982: An accurate radiative heating and cooling algorithm for use in a dynamical model of the middle atmosphere. J. Atmos. Sci., 39, 1532-1544.

TIDES AND FREE OSCILLATIONS

J. R. Holton and T. Matsuno, Dynamics of the Middle Atmosphere, 163-172.

AN OVERVIEW ON TIDAL OBSERVATIONS

Susumu Kato, Takehiko Aso,
Radio Atmospheric Science Center, Kyoto University
and
Robert A. Vincent
Departmemt of Physics, University of Adelaide

ABSTRACT

While various theoretical works on atmospheric tides have been done considering realistic atmospheric models, systematic tidal observation has remained in a very rudimental situation until quite recently. However, it is encouraging to find now that meteor-, partial reflection- and MST radars are supplying data of the middle atmosphere tides about which we were almost ignorant in the past. The present paper shows some highlights of recent observation by these radars, suggesting new problems.

1. INTRODUCTION

Since the establishmemt of classical tidal theory in the 1960s, much research (e.g. Forbes, 1982a, b) has been done using realistic atmospheric models which include such effects as dissipation and interactions with zonal winds. These are factors which are neglected in the classical theory. Contrary to various results in these theoretical or computational approaches, tidal observations have remained in a rather rudimentary state until quite recently.

Atmospheric tides have two outstanding characteristics, one being global as known from the wavenumber of 1, 2 etc. along longitudes and the pole-to-pole extension along meridians, and the other being vertically extensive as from the surface to the thermosphere. Disappointingly, so far, systematic observation has been limited to pressure tides on the surface, and geomagnetic tides and tides in the meteor-wind zone (80-100 km) in the upper atmosphere. No reliable data were available on tides in the middle atmosphere which plays a crucial role in determining tidal excitation and propagation. Now, however, we are reaching a novel observational stage. MST-type radars (MSTR, hereafter) are opening a new era together with lidars and satellite remote sensors.

The radars will supply data which are continuous in time, making it possible to produce vertical profiles of the wind velocity, extending from a few km above the ground over much of the middle atmosphere.

Note, however, that the 30-60 km height region is still difficult to observe due to the very weak radar returns. Another drawback is that the mesosphere is observed only in daylight hours; diurnal tides are thus observed only incompletely. Partial reflection radars (PRR, hereafter) are similar to MSTR, in capacity, in observing tides in the mesosphere (e.g. Vincent and Ball, 1981). These radars which can measure winds in the 60-100 km region by day and the 80-100 km region by night, are now in operation at several locations over the world. Global coverage by MSTR and PRR is, as yet, far from being complete, but new facilities are under construction and in planning at various other locations.

Lidars are also supplying vertical profiles of atmospheric constituents, but the profiles are restricted in height range and duration in observation so that the facilities can be used for tidal observation at some future time. Satellite remote sensing is unique in being able to supply temperature data on global scale averaged over a few months. Note, however, satellite infrared radiometer observations have successfully detected the lunar temperature tides at 80 km (e.g. Schlapp, 1981).

Whilst these technical innovations will give a significant improvement of tidal observations, the tides, global and vertically extensive, will only be fully elucidated through global cooperation using both novel and conventional observing techniques. The patient accumulation of data over many years also remains invaluable. We believe that the observation of tides, together with other atmospheric waves, will see a significant advancement in the foreseeable future. What follows will give a few highlights suggesting such an advancement. As in other field of science, the new experiments would not only solve some conventional problems but also raise other ones.

2. VERTICAL STRUCTURE OF TIDES IN THE MIDDLE ATMOSPHERE

MSTR and PRR are now providing information on the vertical structure which defines the basic behavior of tides in the middle atmosphere. Two successful results will be presented below; one is an equatorial observation at Jicamarca (Perú, 12.0°S) and the other on a high-latitude observation at Saskatoon (Canada, 52°N), Poker Flat (Alaska, 65°N) and Mawson (Antarctica, 68°S, 67°E).

At Jicamarca, during 19-21 November, 1981, the Kyoto University group (Aso et al., 1983) carried out an observation with the large Jicamarca VHF radar which was originally built as an incoherent scatter radar to observe ionospheric plasma, but was used in the present experiment as a MSTR i.e. a coherent radar. The stratosphere (15-30 km) and the mesosphere (60-90 km) were observed simultaneously, but only in daylight hours (0600-1800 local time). The vertical and westward velocities were obtained simultaneously using two radar beams. The data, averaged over the three consecutive days with respect to local time, gave tidal components on the assumption that the wind velocity consists of a prevailing, plus diurnal or semi-diurnal components. This assumption seems reasonable as discussed by Countryman and Dolas (1982) for mesospheric tidal winds over Jicamarca. Thus for the mesosphere, semidiurnal component is inferred from the daytime data. For the stratosphere, on the other hand, diurnal component is reasonably fitted to the observation. During the same period, Roettger et al. of the Max Planck Institute for Aeronomy (Roettger, private communication, 1982) made observations mainly of the stratosphere at Arecibo. Their result shows agreement for the diurnal component with Jicamarca stratosphere, which is a further demonstration of consistency of the radar observations. One can compare the observed results with Forbes' calculation (1982a, b) which is based on realistic atmospheric dynamic models. A fairly good agreement is found in general (Figs. 1 and 2).

Mesospheric tides were observed in high latitudes, not simultaneously but in the same season between 1979-1982, by MSTR at Poker Flat and by PRR at Saskatoon and Mawson. The diurnal tide profile, thus obtained, is illustrated in Fig. 3. The Poker Flat and Saskatoon data are discussed by Carter and Balsley (1982), while the Mawson data were provided by R. MacLeod (private communication). It is impressive that all results including the Antarctic observation are in an excellent agreement, suggesting that the diurnal-tidal wind blows towards the pole, most strongly at local noon in summer. As in Figs. 1 and 2, Forbes' calculation is shown in Fig. 3. Whilst an excellent agreement is found in phase, the calculated amplitude is much smaller than that observed. The stratospheric data during this period at these sites were not available.

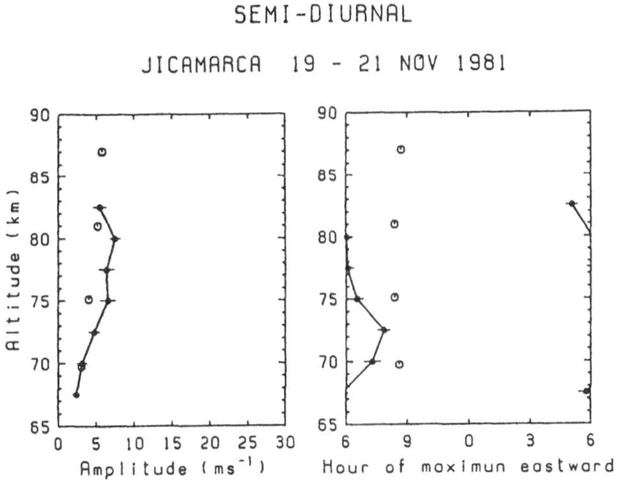

SEMI-DIURNAL

JICAMARCA 19 - 21 NOV 1981

Figure 1. Semidiurnal component of zonal wind in the mesosphere over Jicamarca. The bars for each point give the error. The open circle gives Forbes' calculation for 12°S at the December solstice.

DIURNAL

JICAMARCA 19 - 21 NOV 1981

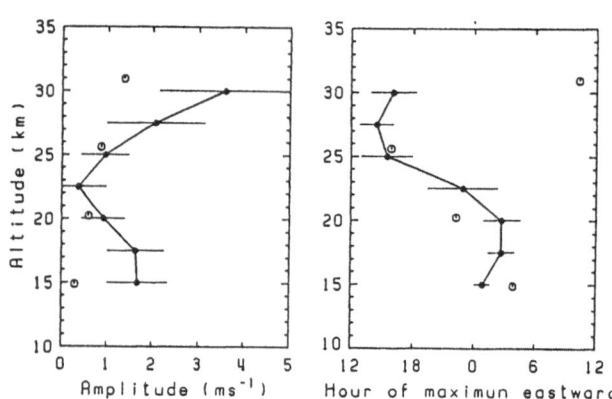

Figure 2. Similar to Fig. 1 except for the diurnal component in the stratosphere.

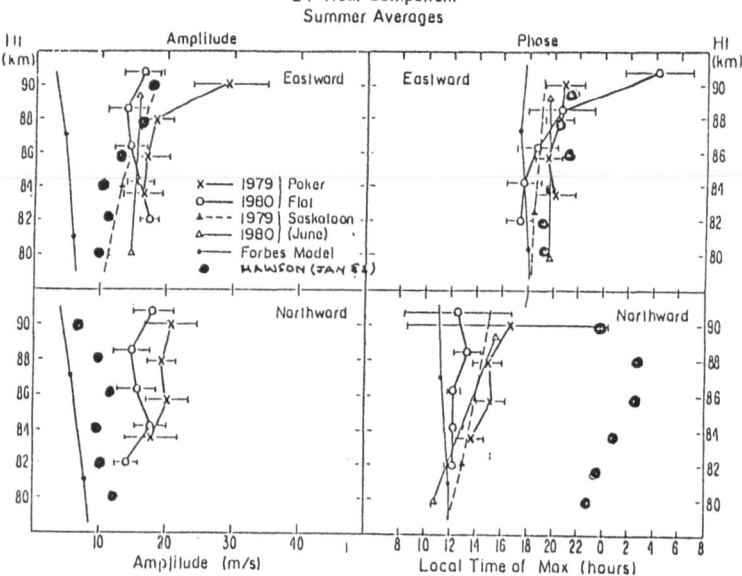

Figure 3. Diurnal mesospheric tidal winds in high latitudes. On the diagram by Carter and Balsley (1982) an observational result taken at Mawson during January, 1982, by R. MacLeod (private communication, 1982) is plotted as the thick circles. The thick line gives Forbes' calculation for 66°N.

However, it is interesting that Wallace and Hartranft (1969) obtained the diurnal tidal wind in the stratosphere by balloon radiosonde experiments. Annual mean 12 hr wind differences observed are computed from monthly mean wind statistics at 27 levels between the surface and 10mb (about 30 km). Fig. 4 shows the result where one finds a remarkable similarity to Fig. 3 in that the calculated tidal amplitude are also much smaller than those observed. Note that the observed phase was, however, in agreement, showing dominance of the first negative mode of the diurnal tide.

Here, we face a clear-cut problem, very probably, in calculation. The important heat source for exciting tides is the middle atmosphere ozone, absorbing solar ultraviolet, especially the ozone on the topside of the distribution. The Nimbus 4 BUV ozone data (Borucki and Eberstein, 1980/1981) show an anomalous distribution of ozone with latitude. The anomaly is more pronounced on the topside, with an enhanced density increase at high latitudes in summer. Forbes' calculation (1982a, b) was based on a simple distribution of ozone.

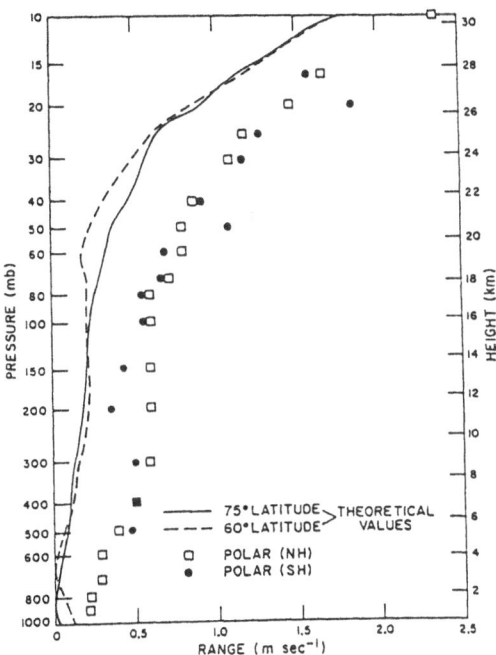

Figure 4. Diurnal stratospheric tidal winds in high latitudes (Wallace and Hartranft, 1969). Not the amplitude but the range (2 x amplitude) is shown. Theoretical values are from Lindzen's tide tables (1967). Note that a remarkable similality is found between Figs. 3 and 4; theory gives weaker amplitudes than are observed.

3. LUNAR TIDES AT METEOR HIGHTS; WHAT HAVE ACCUMULATED DATA SHOWN?

Lunar tides are interesting phenomena because of their well-known excitation mechanism despite their small amplitude. They may be informative on atmospheric parameters which affect their propagation. Lunar tides have successfully been detected from meteor wind data by the Kyoto University meteor radar (Tsuda et al., 1981). The period, for which the data were obtained, is two years 1979 and 1980, the total number of days being 140 with about 1,500 meteor echoes per day. It is interesting to compare the present result with that calculated by Forbes (1982b). Since meteor heights are those at which the calculated phase varies rapidly, comparisons can often be in error. Nevertheless, one can conclude that there is a considerable discrepancy between calculation and observation (Fig. 5). Note that the observed seasonal variation of lunar tides in Fig. 5 is consistent with the lunar temperature tides observed by the TIROS-N series of satellites (Brownscome and Schlapp, 1983). Also lunar geomagnetic tides seem to be consistent with the finding that lunar tides are stronger in winter than in summer (Shiraki, 1981).

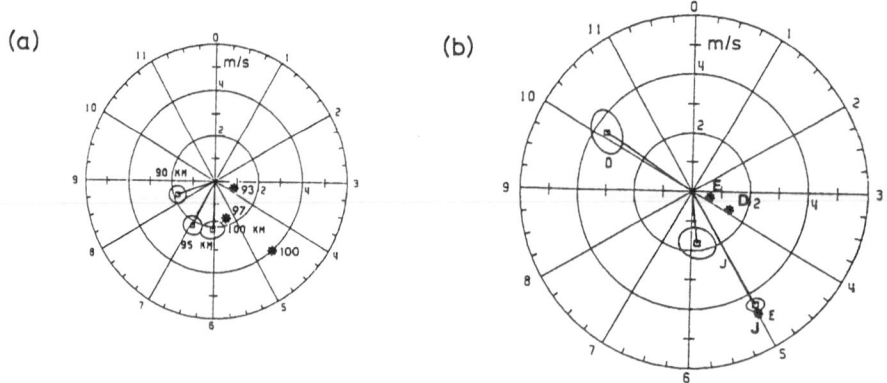

Figure 5. Lunar M_2-tidal winds (northward) at meteor heights adapted from Tsuda et al., 1981.
(a) Averaged state over all seasons at 90, 95, 100km. Forbes' calculation (1982b) arveraged over equinox, June & December solstices at 36° N is indicated for the heights 93, 97 and 100 km by * in the harmonic dial.
(b) Seasonal state averaged over heights between 90 and 100 km. Forbes' calculation averaged over three heights 93, 97 and 100 km is plotted as in (a).

How the discrepancy can be resolved belongs to future study. Observationally, it is desirable to accumulate enough data for detecting lunar tides at other locations.

4. DISCUSSION

Classical tidal theory assumes, among other things, a static horizontally uniform and non-dissipative atmosphere. Thus, for a period of the mean solar day or its submultiples, tidal modes are given in terms of Hough functions which define their amplitude distribution with colatitude. The vertical structure is different for each mode, and this gives a characteristic which is widely used in observation for distinguishing between modes. Because of the linear models used, each mode of the tides, once excited by a heat source of the same mode, remains independent as it propagates vertically. The heat source distribution determines which tidal modes are present at any height. New modes are neither produced nor destroyed.

However, such a simple situation is far from being realistic and observations have also shown the need for a more realistic treatment. The atmosphere, in reality, is non-uniform and disspative. These complications have been considered in recent work (e.g. Lindzen and Hong, 1974; Forbes 1982a, b; Aso et al., 1981; Walterscheid et al., 1980). An important point in this work is consideration of the zonal winds which require corresponding nonuniform temperature and density distribution. This approach shows that many modes are produced by non-linear interactions among modes as the original tidal waves propagate. As is noticed in Section 2 and even in Section 3, the approach seems fairly successful in helping to interpret observations.

In Section 3 a considerable discrepancy is shown between theory and observation. Since lunar tides have a definite excitation, the discrepancy, if any, is due to the transmission dependent on the atmospheric model used in the theory. Though inconclusive, as yet, the discrepancy cannot disappear by slight modification of the adopted models which cover various zonal wind intensities. Introduction of meridional winds may be another possibility which has not been considered. Rather, one must face more complicated situation e.g. a non-stationary background atmosphere in the presence of waves other than tidal ones. It is well-known that tidal winds are often observed to vary from day to day (Fig. 6), suggesting the existence of such waves. Whilst an attack on the non-stationary problem is desirable, observations should supply more definite information than now available of wave behavior. These waves would modulate tides both systematically (in the case of planetary wave types) and randomly (in the case of gravity wave types, e.g. Walterscheid, 1981). In order to resolve this new question, radar observations will also be of vital importance.

As to the accumulation of observational data, non-migrating tides could be an important subject besides lunar tides. Whereas systematic data analysis was done for surface pressure tides (Haurwitz, 1965), little is known observationally about non-migrating tides both in the middle atmosphere and the thermosphere (Kato et al., 1982).

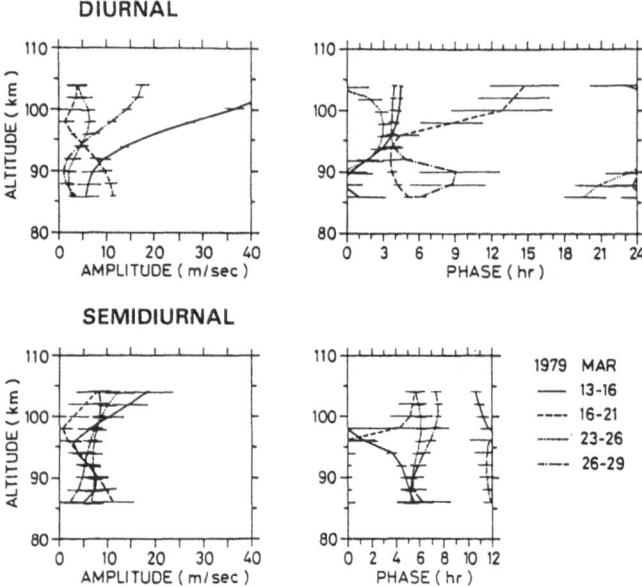

Figure 6. Tidal wind fluctuation from day to day (Tsuda, 1982). The observation was made by the Kyoto University meteor radar at Shigaraki (35°N, 136°E). The northward tidal wind is shown; the phase indicates the time of maximum northward wind.

ACKNOWLEDGEMENT

The authors thank Mr. Rod MacLeod, University of Adelaide for permitting them to show his observation of tides at Mawson. Their thanks are due to Dr. J. M. Forbes, Department of Physics, Boston College, for sending them the table of his tidal calculations. Thanks are also due to the American Geophysical Union and the American Meteorological Society for permitting them to reproduce the figures in this paper. Part of the work has been supported by the Nissan Science Foundation.

REFERENCES

Aso, T., T. Nonoyama and S. Kato, 1981: Numarical simulation of semi-diurnal atmospheric tides. J. Geophys. Res., 86, 11388-11400.

Aso, T., J. Roettger, Y., Maekawa, P. Czechowsky, R. Ruester, G. Schmidt, I. Hirota, R. F. Woodman and S. Kato (to be published in 1983).

Borucki, W. and I. J. Eberstein, 1980/81: Comparison of the Nimbus-4 BUV ozone data with Ames two-dimensional model. Pure and Applied Geophys., 119, 26-749.

Brownscome, J. L. and D. M. Schlapp, 1983: The lunar semi-diurnal tide observed by stratospheric sounding units on the TIROS-N series of satellites. J. Atmos. Terr. Phys., 45, 27-32.

Carter, D. A. and B. B. Balsley, 1982: The summer wind field between 80 and 93 km observed by the MST radar at Poker Flat, Alaska (65°N). J. Atmos. Sci., 39, 2905-2915.

Countryman, I. D. and P. M. Dolas, 1982: Observations on tides in the e-quatorial mesosphere. J. Geophys. Res., 87 (C2), 1336-1342.

Forbes, J. M., 1982a: Atmospheric tides 1, model description and results for the solar diurnal component. J. Geophys. Res., 87 (A7), 5222-5240.

Forbes, J. M., 1982b: Atmospheric tides 2, the solar and lunar semidi-urnal components. J. Geophys. Res., 87 (A7), 5241-5252.

Haurwitz, B., 1965: The diurnal surface-pressure oscillation. Arch. Mete-orol. Geophys. Biokl., A14, 361-379.

Kato S., T. Tsuda and F. Watanabe, 1982: Thermal excitation of non-migrating tides. J. Atmos. Terr. Phys., 44, 131-146.

Lindzen, R. S., 1967: Thermally driven diurnal tide in the atmosphere. Quart. J. Roy. Met. Soc., 93, 18-42.

Lindzen, R. S. and S. Hong, 1974: Effects of mean wind and horizontal temperature gradients on solar and lunar tides in the atmosphere. J. Atmos. Sci., 31, 1421-1446.

Schlapp, D. M., 1981: Lunar tides in the stratosphere and mesosphere from NIMBUS 6 data, J. Atmos. Terr. Phys., 43, 205-207.

Shiraki, M., 1981: Seasonal dependence of lunar daily geomagnetic varia-tions in different regions of the world. J. Geomag. Geoelectr., 33, 467-501.

Tsuda. T., 1982: Kyoto Meteor Radar and its application to observation of atmospheric tides. Ph. D. Thesis , Department of Electronics, Kyoto University.

Tsuda, T., J. Tanii, T. Aso and S. Kato, 1981: Lunar tides at meteor heights. Geophys. Res Lett., 8, 191-194.

Vincent, R. A. and S. M. Ball, 1981: Mesospheric winds at low- and mid-latitudes in the southern hemisphere. J. Geophys. Res., 86, 9159-9169.

Wallace, J. M. and F. R. Hartranft, 1969: Diurnal wind variations, sur-face to 30 kilometers. Month. Wea. Rev., 97, 446-455.

Walterscheid, R. L., 1981: Inertio-gravity wave induced accelerations of

172

mean flow having an imposed periodic component: Implications for tidal observations in the meteor region. J. Geophys. Res., 86, 9698-9706.

Walterscheid, R. L., J. G. DeVore and S. V. Venkateswaran, 1980: Influence of mean zonal motion and meridional temperature gradients on the solar semidiurnal atmospheric tides: A revised spectral study with improved heating rates. J. Atmos. Sci., 37, 455-470.

J. R. Holton and T. Matsuno, Dynamics of the Middle Atmosphere, 173-180.
Copyright © 1984 by Terra Scientific Publishing Company.

LINEARIZED STEADY CALCULATIONS OF
SEMIDIURNAL TIDES IN THE MIDDLE ATMOSPHERE

Takehiko Aso and Susumu Kato

Radio Atmospheric Science Center
Kyoto University, Uji 611, Japan

ABSTRACT

Linearized steady modelings of the solar semidiurnal atmospheric
tides in the middle atmosphere have been worked out. In this work,
latitudinal temperature gradient and associated mean zonal wind are
reconsidered based on previous works. It is shown that the results
seem to reasonably delineate radar observations at meteor heights.

1. INTRODUCTION

The present work deals with linearized modeling of semidiurnal
atmospheric tides in the middle atmosphere with a latitudinal
temperature gradient and associated mean zonal wind. The primitive
equations are invoked here as governing equations to describe the
linearized, steady state response due to semidiurnal tidal forcings.
In our previous paper (Aso et al, 1981), a single partial differential
equation for the quantity corresponding to perturbation geopotential has
been used to look into details of the sensitivity of tidal structures at
meteor heights to possible changes in forcing, latitudinal temperature
structure and associated mean zonal wind. This work follows Lindzen
and Hong (1974)'s original work to whom a credit should be given on this
problem. In these works, the latitudinal temperature variation T_1 and
the mean zonal wind V are assumed to be small quantities relative to the
mean equatorial temperature and the phase velocity of tides,
respectively. Namely, a small parameter ϵ conforms to the ratios of
these quantities and terms of the order of ϵ^2 or higher have been
neglected.

In the present work, revisions to this small ϵ approximation are
taken into account by retaining all the terms relevant to T_1 and V in
deriving the aforementioned single second order partial differential
equation. Intuitively, some change in the calculated results is
expected in the higher latitude region where small ϵ approximation
becomes slightly less valid. To retain all the terms relevant to T_1
and V, the derivation of a single partial differential equation has
again been carried out through algebraic manipulation using symbolic

173

computation software. The results, comparisons with earlier results
and some observations will be given with regard to solar semidiurnal
atmospheric tides.

2. METHOD OF NUMERICAL MODELING

The equation system used is the linearized set of equations of
motion, continuity equation, equation of state and energy equation for a
shallow, rotating spherical atmosphere which takes into consideration
the meridional temperature gradient and associated mean zonal wind.
Basic approximations pertinent to the classical tidal theory (Chapman
and Lindzen, 1970) are implicitly assumed. As is mentioned in the
introduction, the small ϵ approximation which conforms to the earlier
work is removed and a more generalized single partial differential
equation is derived by the REDUCE-II program at the Kyoto University
Data Processing Center. Numerical integration of this equation is
performed using the algorithm given by Lindzen and Kuo (1969). The
temperature and background wind model adopted throughout the work is
formulated by applying the Lindzen and Hong type analytic representation
(Lindzen and Hong, 1974) to the model based mainly on the CIRA (1972)
and on Murgatroyd (1957). The dissipation is introduced in terms of
Newtonian cooling and Rayleigh friction. This is adopted from
Schoeberl and Strobel(1978)'s work on the middle atmosphere circulation;
above about 110 - 120 km, it tends to act as a spongy layer to absorb
perturbation energy from below. The upper boundary condition is thus
specified by letting the solution approach a constant value at higher
altitude. These and all other details are identical to earlier
works. It must be noted that these approximations confine the present
analysis to be valid below about 110 km.

3. RESULTS

We will model the linearized steady response of the solar
semidiurnal atmospheric tides to the conventional (2,2) + (2,3)
(solstice only) + (2,4) water vapor and ozone heatings given by Lindzen
and Hong (1974) and Hong and Lindzen (1976). The phase here is
expressed in hours relative to 0300 LT throughout. Figure 1 compares
three dimensional displays of amplitude and phase of the northerly and
westerly wind components under June solstitial condition calculated by
(a) the original version and (b) the present revised version. In the
figure, amplitude and phase are plotted versus altitude (2 - 110 km) and
colatitude (30 - 150°) with amplitude in logarithmic scale. The
colatitude 0 - 90° refers to the northern hemisphere. The upper
boundary is set at 150 km and vertical and latitudinal grid size is 1 km
and 5°, respectively. In Figure 1(a), as was originally mentioned in
Lindzen and Hong (1974), the mean wind induces higher order modes.
Specifically, amplitude tends to be larger at higher latitudes due to
the predominance of higher order modes excited by the strong westerly
jet in the winter hemisphere (colatitude 90 - 180°). It must be noted
that the small ϵ approximation becomes less valid especially below about
80 km of the winter higher latitude region, which might possibly induce

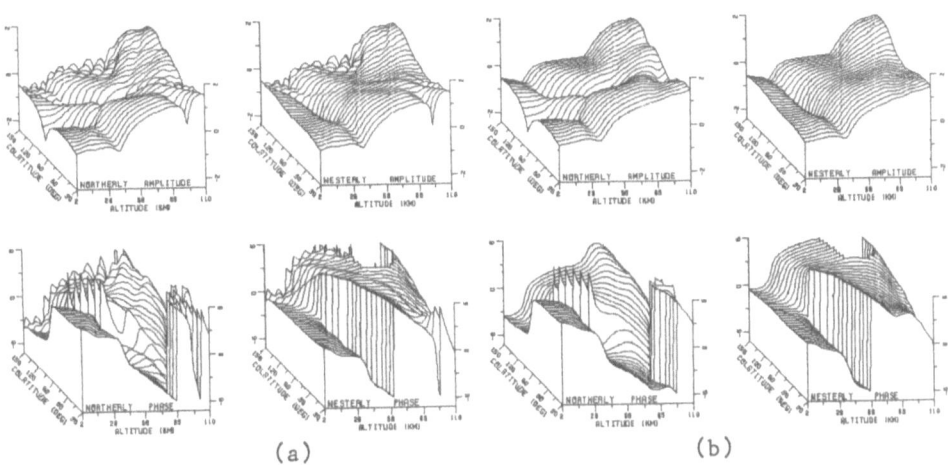

(a)   (b)

Figure 1. Amplitude and phase of northerly (left) and westerly winds
for the solar semidiurnal tide at solstice, plotted versus altitude
and colatitude.   (a) Original version, and (b) revised version.

undulating, irregular structures with some loss of accuracy.   This
contrasts with the newly obtained result shown in Figure 1(b) where the
small ϵ approximation is not taken into consideration.   It is clearly
seen that though the basic features of amplitude and phase are
unchanged, complicated structures at higher latitudes in Figure 1(a)
almost disappear and the present generalization renders the calculated
results of the revised version more accurate and convincing.

We have examined the numerical stability of the present model based
on the revised code.   Combination of the height of the upper boundary
z_{max} at 250 km and 150 km and latitudinal grid size of 5°, 4.5° and
3.6°, respectively, induces no appreciable change in tidal fields below
about 110 km where present calculations are valid.   The difference
between those for vertical grid sizes of 1 km and 500 m turns out to be
small.   Also, a non-uniform grid is introduced for the colatitude which
involves finer grid size around critical latitudes corresponding to the
denominator of the classical Laplace tidal equation.   In the present
case, this corresponds to the poles.   A grid size of 1° within 5
degrees of both poles and 5° otherwise is tested, and in this case the
calculated results show no significant change either.   Hence we use an
upper boundary at 150 km, a uniform grid size of 1 km in altitude and 5°
in latitude throughout unless otherwise noted.

Figure 2 specifically compares altitude profiles of northerly and
westerly wind components at solstice at 35°N.   Solid line refers to a
no mean wind regime, and dashed and dotted lines to the solstitial
temperature and mean wind condition for the revised and original

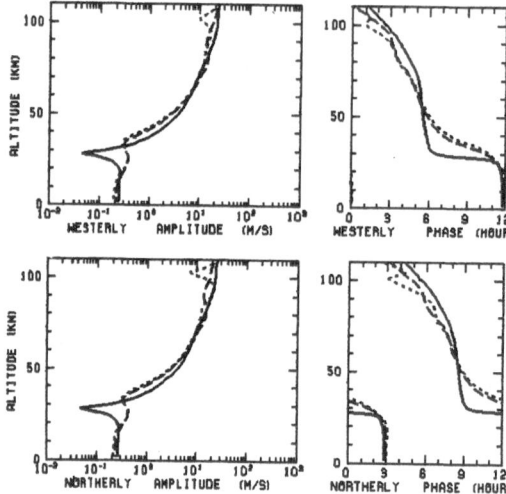

Figure 2. Altitude profile of northerly (lower) and westerly wind components at 35°N at solstice for no wind (solid) and wind (dashed: revised, dotted : original) regimes.

versions, respectively. Two versions give identical results for the no wind condition as they should. When a background wind is included, wind-induced higher order modes generally contribute to produce shorter vertical wavelengths. The dotted line representing the original small ϵ version for solstitial wind shows, by and large, irregular structure

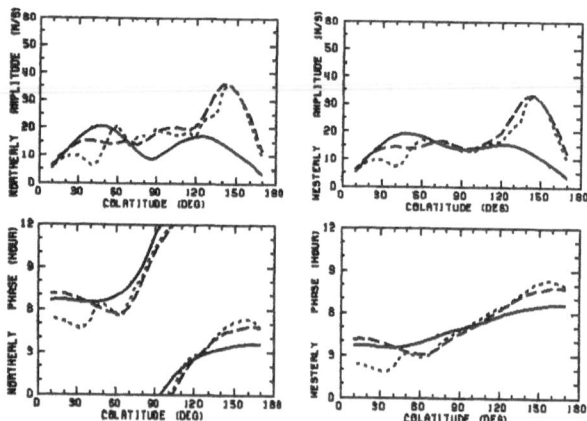

Figure 3. Latitudinal profile of northerly (left) and westerly wind components at 90 km. Lines are as in Figure 2.

at around 100 km compared to a more smooth variation of the revised version. Although the difference at lower altitudes is comparatively slight at this latitude, the amplitude minimum and associated phase change due to mixing of induced higher order modes reveal some discrepancies between the two at meteor heights. A comparison with respect to latitudinal profiles at 90 km is illustrated in Figure 3. Each line represents the same condition as in Figure 2. For the no wind case, antisymmetric (2,3) forcing is responsible for the asymmetry in the tidal fields. For the solstitial background wind, the discrepancy seems to be significant at middle to high latitudes of the summer hemisphere. Amplitudes of the zonal and meridional winds show differences of less than 10 m/sec and phases of about 2 hours or less. In the winter hemisphere, results do not so much deviate as in the summer hemisphere at this altitude.

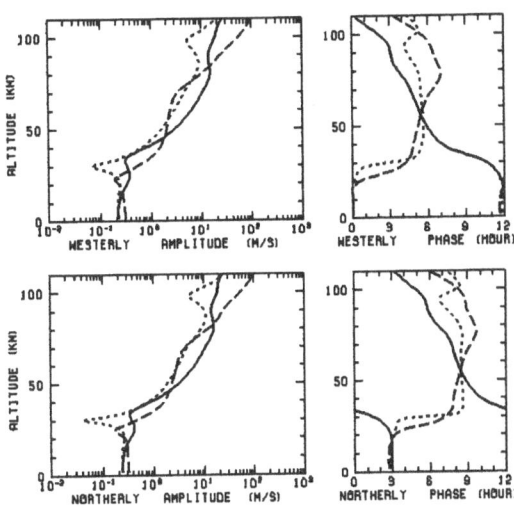

Figure 4. Seasonal variation of altitude profile of northerly (lower) and westerly wind components at 35°N for June solstice (solid), December solstice (dashed) and equinox (dotted).

Next we will survey briefly the seasonal change of semidiurnal tides at equinox and June and December solstices. In Figure 4, we show altitude profiles of wind velocities at 35°N as in Figure 2. In the figure, amplitude and phase are plotted for equinoctial, June and December solstitial conditions. It is evident that significant seasonal change exists at all heights. At December solstice, the amplitude tends to become larger and phase tends to lag compared to equinox or June solstice in the meteor region. The corresponding latitudinal profile is shown in Figure 5 at the altitude of 90 km. At solstices, due to the effect of asymmetric background temperature and zonal wind along with antisymmetric (2,3) mode excitation, the amplitude of zonal and meridional winds is larger in the winter hemisphere by about a factor of two than in the summer hemisphere at middle to high latitudes; phase varies considerably from simple in-phase (zonal) and anti-phase (meridional) characteristics between the two hemispheres. At equinox, the background wind model is rather symmetric which gives an almost symmetric structure even in the windy condition. It must be noted that these might depend on assumed forcing and wind models.

178

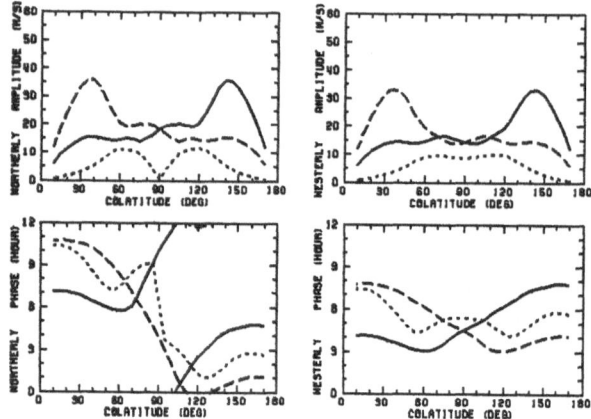

Figure 5. Seasonal variation of latitudinal profile of northerly (left) and westerly wind components at 90 km. Lines are as in Figure 4.

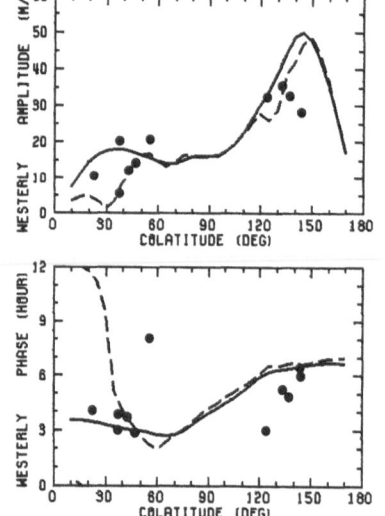

Figure 6. Comparison of eastward wind component at 95 km with radar wind data. Solid line refers to the revised, and dashed line to the original versions, respectively. Note that phase is in hours relative to 0300 or 1500.

Now we describe the comparison of the present tidal modeling with some of the observations reported hitherto. In Figure 6, a latitudinal plot of observed amplitude and phase of the eastward wind is compared with calculated results of the revised as well as original versions. These data are based on the radar observations carried out at various periods (including CTOP campaign in 1974 - 1976) during summer and winter in the northern hemisphere (Müller, 1966; Fellous et al., 1975; Roper and Salah, 1978; Glass et al., 1978; Clark, 1978). The wintertime data are plotted in the colatitude of 90 - 180°. The altitude approximately corresponds to 95 km. The calculated results are basically consistent with the observed amplitude and phase. It should be emphasized that for the summer hemisphere, rapid phase change at around 60°N predicted by the original version does not appear, and the prsent result seems to delineate the observed phase change more reasonably. Amplitude asymmetry found in the solstitial observations is reproduced by both models. In November 1981, an international cooperative observation of tides was carried out. Latitudinal profiles of amplitude and phase of the meridional component at 90 km from several stations (R. A. Vincent, 1982, private communication) compare well in phase structure with the present calculation for the December solstitial condition. The observed amplitude is relatively small compared to the present numerical model. If we alter the forcing and/or the background wind models, we could anticipate a more rigorous coincidence. However, non-steadiness of atmospheric tides should also be taken into account for comprehensive agreement between short-term observation and theory as is pointed out by Forbes(1982, private communication).

4. CONCLUSION

The linearized steady calculation of solar semidiurnal atmospheric tides in the middle atmosphere with background temperature gradient and mean zonal wind is reformulated following the works of Lindzen and Hong (1974) and Aso et al. (1981). In the present scheme, the small parameter approximation which neglects higher order quantities associated with latitudinal temperature variation and background mean zonal wind is removed. Calculated results indicate that amplitude and phase of zonal and meridional wind components show more organized structures in the middle to high latitude region. Amplitude and phase show revisions of less than 10 m/sec and 2 hours at these latitudes between the original and revised values of the horizontal wind component at 90 km for the present wind model. These numerical results are found to be in reasonable agreement with some of the observational data at meteor heights.

ACKNOWLEDGEMENT

The authors thank Profs. R. S. Lindzen and J. R. Holton for carefully reading the manuscript. Part of the present work was supported by the Grant-in-Aid for Scientific Research in 1982 by the Ministry of Education, Science, and Culture of Japan. Computations were carried out at the Data Processing Center of Kyoto University, and also at the Computer Center of the Institute of Plasma Physics, Nagoya University and the Institute of Space and Astronautical Science as part of the collaboration projects at both the Institutes. REDUCE-II program was made by Dr. Y. Kanada of Tokyo University.

REFERENCES

Aso, T., T. Nonoyama, and S. Kato, 1981: Numerical simulation of semidiurnal atmospheric tides. J. Geophys. Res., 86(A13), 11388-

11400.

Chapman, S., and R. S. Lindzen, 1970: Atmospheric Tides. D. Reidel Pub. Co., Dordrecht-Holland, 200pp.

CIRA, 1972: COSPAR International Reference Atmosphere. Akademic Verlag, Berlin, 450pp.

Clark, R. R., 1978: Meteor wind data for global comparisons. J. Atmos. Terr. Phys., 40, 905-911.

Fellous, J. L., R. Bernard, M. Glass, M. Massebeuf, and A. Spizzichino, 1975: A study of the variations of atmospheric tides in the meteor zone. J. Atmos. Terr. Phys., 37, 1511-1524.

Glass, M., R. Bernard, J. L. Fellous, and M. Massebeuf, 1978: The French meteor radar facility. J. Atmos. Terr. Phys., 40, 923-931.

Hong, S.-S., and R. S. Lindzen, 1976: Solar semidiurnal tide in the thermosphere. J. Atmos. Sci., 33, 135-153.

Lindzen, R. S., and S.-S.Hong, 1974: Effects of mean winds and horizontal temperature gradients on solar and lunar semidiurnal tides in the atmosphere. J. Atmos. Sci., 31(5), 1421-1446.

Lindzen, R. S., and H.-L. Kuo, 1969: A reliable method for the numerical integration of a large class of ordinary and partial differential equations. Mon. Wea. Rev., 97(10), 732-734.

Müller, H. G., 1966: Atmospheric tides in the meteor zone. Planet. Space Sci., 14, 1253-1272.

Murgatroyd, R. J., 1957: Winds and temperatures between 20 km and 100 km - a review. Quart. J. Roy. Met. Soc., 83, 417-458.

Roper, R. G., and J. E. Salah, 1978: Preliminary results from the URSI/IAGA cooperative tidal observations program (CTOP). J. Atmos. Terr. Phys., 40, 879-885.

Schoeberl, M. R., and D. F. Strobel, 1978: The zonally averaged circulation of the middle atmosphere. J. Atmos. Sci., 35, 577-591.

J. R. Holton and T. Matsuno, Dynamics of the Middle Atmosphere, 181-197.

ZONAL MEAN WINDS INDUCED BY SOLAR DIURNAL TIDES IN THE LOWER
THERMOSPHERE FOR A SOLSTICE CONDITION

Saburo Miyahara*

Department of Physics, Faculty of Science,
Kyushu University, Fukuoka 812, Japan

ABSTRACT

    The generation of mean winds in the lower thermosphere for a
solstice condition due to momentum transport by dissipating solar
diurnal tidal waves is discussed. The calculated results show that
even for a solstice condition the induced mean zonal wind is almost
symmetric about the equator in the upper mesosphere and the lower
thermosphere. The distribution of the induced mean zonal winds is
similar to observed results so far as the wind direction is concerned.
    The present results confirm that dissipating solar diurnal tidal
waves contribute greatly to the cinfiguration of the mean zonal winds
in the upper mesosphere and the lower thermosphere.

1. INTRODUCTION

    The general circulation of the lower thermosphere (at about 100km
height) seems to be considerably different from that in the middle
atmosphere and the upper thermosphere. Although the observed mean
zonal winds distribution is not definitive yet, an easterly wind
exists in the low latitude region all through the year, and westerly
winds exist in the middle latitude regions in both summer and winter
hemispheres (CIRA,1972, Groves,1969, 1972, Iljichev and Portnyagin,
1977). It seems unlikely that this symmetric zonal wind system is
produced by thermal forcing due to the absorption of solar insolation.
    The present author (Miyahara,1981; hereafter referred to as M),
discussed the generation of zonal mean winds due to momentum transport
by solar diurnal tides in the lower thermosphere for an equinox
condition, and showed that dissipating solar diurnal tidal waves
produce a mean zonal wind system in the lower thermosphere which is
similar to that observed. In the present study the calculation is

---

*Currently at Geophysical Fluid Dynamics Program, Princeton University,
Princeton, New Jersey 08540, U.S.A.

extended to solstice conditions. In M, the effects of small scale
motions and tidal breakdown in the mesosphere and the lower
thermosphere were parameterized in two ways; i.e. eddy diffusion and
Rayleigh friction. That study showed no remarkable difference due to
the difference of the parameterizations so far as the behavior of the
solar diurnal waves is concerned, and the calculated results in both
parameterizations seemed to explain the overall feature of the
observed diurnal tidal waves. On the other hand, the magnitude of the
induced mean zonal winds in that model greatly depended on the form of
parameterization; but both results were acceptable.

Recently Lindzen(1981) and Matsuno(1982) revealed that the
physical origin of the Rayleigh friction which is used in numerical
model studies of the middle atmosphere (e.g. Leovy,1964, Schoeberl and
Strobel,1978, Holton and Wehrbein,1980) can be attributed to the
momentum transport by upward propagating internal gravity waves.

In the present model the effects of small scale motions on mean
fields are parameterized by a prescribed vertical eddy diffusion
alone, and the deceleration effects of internal gravity waves are not
included. Those effects greatly depend on the distribution of the
mean zonal winds throughout the region where the internal gravity
waves are propagating, so that we must take into account the effect of
background winds due to the general circulation of the middle
atmosphere to estimate the deceleration effects. This problem is
discussed in a companion paper (Miyahara,1983).

Although the effects of internal gravity waves on tidal waves are
an interesting problem (Walterscheid,1981), it is complicated to take
those effects into account. In the present model it is assumed that
tidal waves are dissipated by the prescribed vertical eddy diffusion
alone.

In Section 2 we describe the outline of the dynamical model. In
Section 3 the behavior of solar diurnal tidal waves for a solstice
condition is discussed. In Section 4 the zonal mean fields induced by
solar diurnal tidal waves are presented. Section 5 is devoted to
conclusions and remarks.

2. THE DYNAMICAL MODEL

a) Basic equations

The primitive equation system is separated into a zonally
averaged equation system and a perturbation equation system. With the
log-pressure coordinate, the basic equations are written as follows:
The zonally averaged equation system:
the eastward momentum equation

$$\frac{\partial \bar{u}}{\partial t} + \frac{1}{a \sin\theta} \frac{\partial \bar{u}\bar{v}\sin\theta}{\partial \theta} + \frac{1}{P} \frac{\partial P\bar{u}\bar{w}}{\partial Z} + 2\Omega\cos\theta\,\bar{v} = \frac{g}{P}\frac{\partial}{\partial Z}\left(\frac{\mu}{H_0}\frac{\partial \bar{u}}{\partial Z}\right)$$

$$+ \frac{1}{P}\frac{\partial}{\partial Z}\left(\frac{P K_{EV}}{H_0^2}\frac{\partial \bar{u}}{\partial Z}\right) - D_{\lambda\lambda}\bar{u} + \bar{F}_u \,, \qquad (2.1)$$

the southward momentum equation

$$\frac{\partial \bar{V}}{\partial t} - 2\Omega \cos\theta \, \bar{u} = -\frac{1}{a}\frac{\partial \bar{\Phi}}{\partial \theta} + \frac{g}{P}\frac{\partial}{\partial Z}\left(\frac{\mu}{H_0}\frac{\partial \bar{V}}{\partial Z}\right) + \frac{1}{P}\frac{\partial}{\partial Z}\left(\frac{PK_{EV}}{H_0^2}\frac{\partial \bar{U}}{\partial Z}\right)$$
$$- D_{\theta\theta}\bar{V} + \bar{F}_V \, , \tag{2.2}$$

the hydrostatic equation

$$\frac{\partial \bar{\Phi}}{\partial Z} = \frac{R}{m}\,T \, , \tag{2.3}$$

the thermodynamic equation

$$\frac{\partial \bar{T}}{\partial t} + \frac{1}{a\sin\theta}\frac{\partial \bar{T}\bar{V}\sin\theta}{\partial \theta} + \frac{1}{P}\frac{\partial P\bar{W}\bar{T}}{\partial Z} + N^2\bar{W} = \frac{g}{P}\frac{\partial}{\partial Z}\left(\frac{K}{C_PH_0}\frac{\partial \bar{T}}{\partial Z}\right)$$
$$+ \frac{1}{P}\frac{\partial}{\partial Z}\left[\frac{PK_{EC}}{H_0^2}\left(\frac{\partial \bar{T}}{\partial Z} + \kappa\bar{T}\right)\right] - \alpha_T\bar{T} + \bar{F}_T \, , \tag{2.4}$$

the continuity equation

$$\frac{1}{a\sin\theta}\frac{\partial \bar{V}\sin\theta}{\partial \theta} + \frac{1}{P}\frac{\partial P\bar{W}}{\partial Z} = 0. \tag{2.5}$$

The perturbation equation system:
the eastward momentum equation

$$\frac{\partial u'}{\partial t} + 2\Omega \cos\theta \, v' = -\frac{1}{a\sin\theta}\frac{\partial \Phi'}{\partial \lambda} + \frac{g}{P}\frac{\partial}{\partial Z}\left(\frac{\mu}{H_0}\frac{\partial u'}{\partial Z}\right)$$
$$+ \frac{1}{P}\frac{\partial}{\partial Z}\left(\frac{PK_{EV}}{H_0^2}\frac{\partial u'}{\partial Z}\right) - D_{\lambda\lambda}u' \, , \tag{2.6}$$

the southward momentum equation

$$\frac{\partial v'}{\partial t} - 2\Omega \cos\theta \, u' = -\frac{1}{a}\frac{\partial \Phi'}{\partial \theta} + \frac{g}{P}\frac{\partial}{\partial Z}\left(\frac{\mu}{H_0}\frac{\partial v'}{\partial Z}\right)$$
$$+ \frac{1}{P}\frac{\partial}{\partial Z}\left(\frac{PK_{EV}\partial v'}{H_0^2\partial Z}\right) - D_{\theta\theta}v' \, , \tag{2.7}$$

the hydrostatic equation

$$\frac{\partial \Phi'}{\partial Z} = \frac{R}{m}T' \, , \tag{2.8}$$

the thermodynamic equation

$$\frac{\partial T'}{\partial t} + N^2W' = \frac{g}{P}\frac{\partial}{\partial Z}\left(\frac{K}{C_PH_0}\frac{\partial T'}{\partial Z}\right) + \frac{1}{P}\frac{\partial}{\partial Z}\left[\frac{PK_{EC}}{H_0^2}\left(\frac{\partial T'}{\partial Z} + \kappa T'\right)\right]$$
$$- \alpha_T T' + \frac{J}{C_P} \, , \tag{2.9}$$

the continuity equation

$$\frac{1}{a\sin\theta}\frac{\partial u'}{\partial \lambda} + \frac{1}{a\sin\theta}\frac{\partial v'\sin\theta}{\partial \theta} + \frac{1}{P}\frac{\partial PW'}{\partial Z} = 0 \, . \tag{2.10}$$

In these expressions the upper bar denotes the zonal mean, the prime denotes a wave disturbance, and subscript 0 denotes the basic quantity.

The notations used are

t	:	time
λ	:	longitude
θ	:	co-latitude
z	:	a measure of height $=-\ln(p/p_0)$
g	:	acceleration of gravity
R	:	universal gas constant
m	:	mean molecular weight of the air
p	:	pressure
p_0	:	a constant reference pressure
H_0	:	scale height of basic state
$\bar{u},\bar{v}$	:	eastward and southward component of zonal mean velocity, respectively
$\bar{w}$	:	a measure of zonally averaged vertical motion $(=dz/dt)$
u',v'	:	perturbation of eastward, southward velocity, respectively
w'	:	a measure of perturbation vertical motion
N^2	:	a measure of static stability $(=\partial T_0/\partial z +\kappa T_0)$
T_0	:	temperature of the basic state
$\bar{T}$	:	zonally averaged departure of local temperature from the basic state
T'	:	perturbation of departure of local temperature
$\bar{\Phi}$	:	zonally averaged departure of local geopotential from the basic state
ϕ'	:	perturbation of local geopotential
	:	angular velocity of the earth's rotation
a	:	radius of the earth
C_p	:	specific heat at constant pressure
κ	:	ratio of gas constant to specific heat at constant pressure (R/mc)
J	:	tide-generating heating
$\bar{F}_u,\bar{F}_v,\bar{F}_T$	:	zonally averaged divergence of eastward, southward eddy momentum flux and thermal eddy flux, respectively
$D_{\lambda\lambda},D_{\theta\theta}$	:	coefficients of ion drag terms
K_{EU}	:	coefficient of eddy viscosity
K_{EC}	:	coefficient of eddy conductivity
μ	:	coefficient of molecular viscosity
K	:	coefficient of molecular conductivity
d_T	:	coefficient of Newtonian cooling.

The explicit form of zonally averaged eddy flux terms are as follows:

$$\bar{F}_u = -\frac{1}{a}\frac{\partial \overline{u'v'}}{\partial \theta} - \frac{2\cot\theta}{a}\overline{u'v'} - \frac{1}{P}\frac{\partial P\overline{u'w'}}{\partial z}, \qquad (2.11)$$

$$\bar{F}_v = -\frac{1}{a}\frac{\partial \overline{v'v'}}{\partial \theta} + \frac{\cot\theta}{a}\overline{u'u'} - \frac{\cot\theta}{a}\overline{v'v'} - \frac{1}{P}\frac{\partial P\overline{v'w'}}{\partial z}, \qquad (2.12)$$

$$\bar{F}_T = -\frac{1}{a}\frac{\partial \overline{v'T'}}{\partial \theta} - \frac{\cot\theta}{a}\overline{v'T'} - \frac{1}{P}\frac{\partial P\overline{T'w'}}{\partial z}. \qquad (2.13)$$

We have neglected the nonlinear terms in (2.2) since these terms are negligibly small as shown in M. We have also neglected the feedback effect of the induced mean winds on the wave as well as the effect of the background wind due to the general circulation as assumed in M. Unlike planetary waves the phase velocity of the diurnal tidal wave is much faster than the velocity of the background zonal winds in the low to middle latitude region where our attention is focused, so that the neglect of the effects of mean zonal winds may not result in crucial error (Lindzen and Hong,1974, Miyahara,1975, Walterscheid et al.,1980).

b) Tide-generating heating and infrared cooling

We use the tide-generating heating calculated for a solstice condition. The tide-generating heating by H_2O is calculated by the same method used by Forbes and Garrett(1978). For O_3 heating, we use the model of ozone concentration at the solstices given by Groves(1982) and evaluate the heating rate using the parameterization for ozone absorption of Lacis and Hansen(1974). The SSMIN (sunspot minimum) heating at solstice calculated by Forbes and Garrett(1976) is used for the heating by UV and EUV in the thermosphere. The tide-generating heating distribution used in the present model is shown in Fig.1.

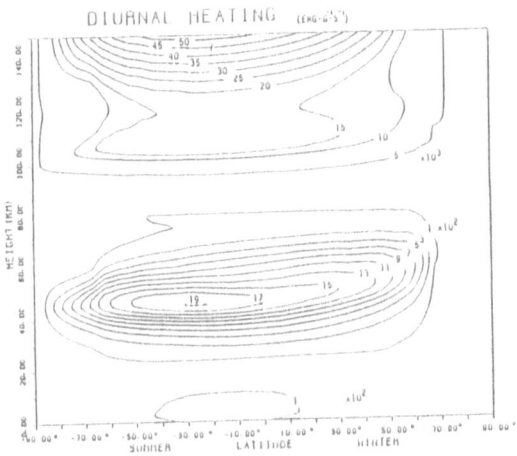

Fig.1 Meridional cross section of tide-generating heating for solstice.

The infrared cooling is parameterized in the form of Newtonian cooling and the coefficient used in the present model is shown in Fig.2.

c) The model atmosphere and numerical model

The model atmosphere is the same one used in M except for the vertical profile of the coefficient of eddy diffusion. In the present model we assume that the effective Prandtl number K_{EV}/K_{KC} is equal to one as in M. The assumed height dependence of the coefficient of vertical eddy diffusion used in the present model is shown in Fig.2. The magnitude of the coefficient below 50km height and above 130km height is smaller than that used in M. The distribution in the lower thermosphere is almost similar to the equinoctial one calculated by Alcayde et al.(1979).

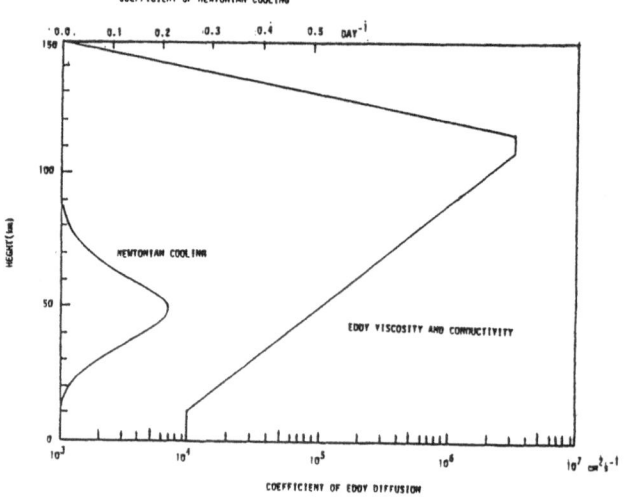

Fig.2 Height dependence of the coefficients of eddy diffusion and Newtonian cooling.

The lower boundary is the earth's surface. The upper boundary is set at 21.5 scale height (the real height is about 170km). The meridional extent of the present model is from the winter pole to the summer pole. The boundary conditions are same as those used in M except that the conditions at the equator are removed in the present model.

The finite difference and the time integration schemes are the same as those in M, and $\Delta z=0.1$, $\Delta\theta=5°$ and $\Delta t=300$ sec.

3. WAVE SOLUTIONS

The wave solution becomes steady about 20 days after the starting time of the time integration. The height distributions of amplitude and phase of u' and v' at several latitudes at 22.6 days after the starting time are shown in Figs.3 and 4. As is seen from these figures, the vertical wave length is about 30km, the amplitude grows with height below 90km height, and the distributions of amplitude and phase are almost symmetric about the equator in low to middle latitude regions. These results show that the wave mainly consists of the symmetric $S_{1,1}$ mode in these regions below 90km height. The phase at ±60° is almost constant with height below 80km height. However, the symmetry about the equator is not so good as in the low to middle latitude region. The reason for the difference of symmetry with latitude is as follows. The antisymmetric component of tide-generating heating mainly consists of the $S_{1,1}$ mode which has

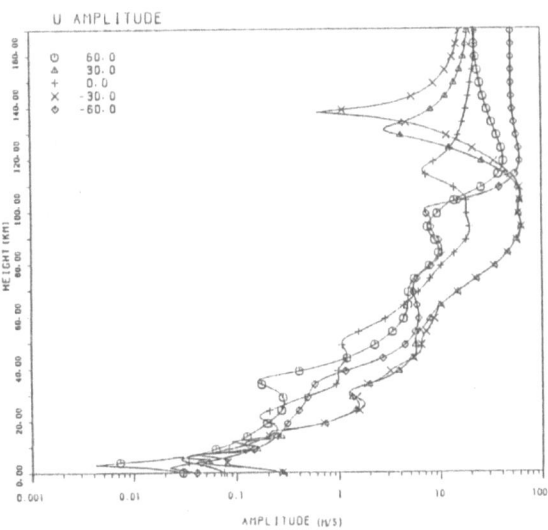

Fig.3a Height distributions of amplitude of u' at -60° (◇), -30 (X) , 0° (+), 30° (△) and 60° (☉). Minus sign shows winter hemisphere.

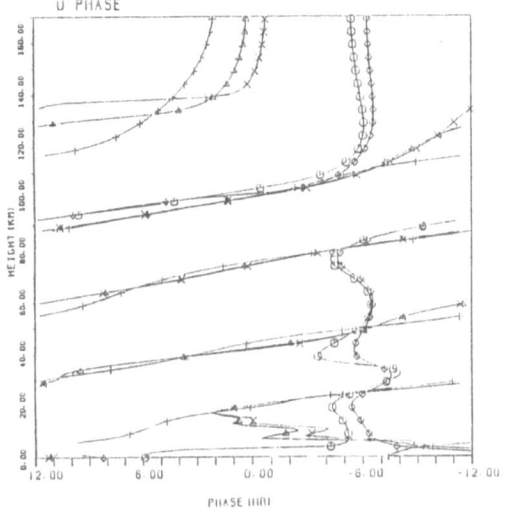

Fig.3b As in Fig.3a except for phase of u'.

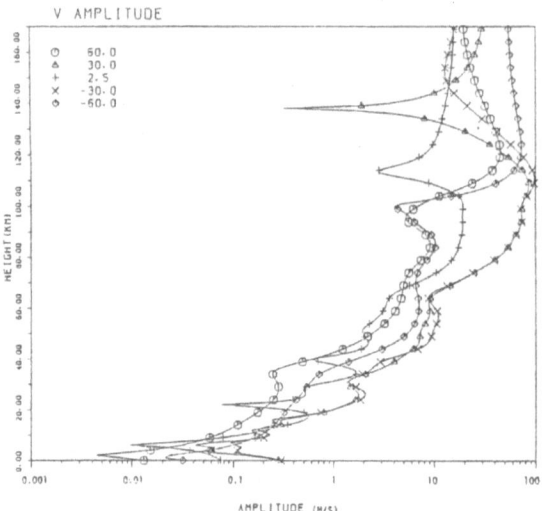

Fig.4a Height distributions of amplitude of v' at –60° (◊),
–30° (✗), 2.5° (+), 30° (△) and 60° (⊘). Minus sign shows
winter hemisphere.

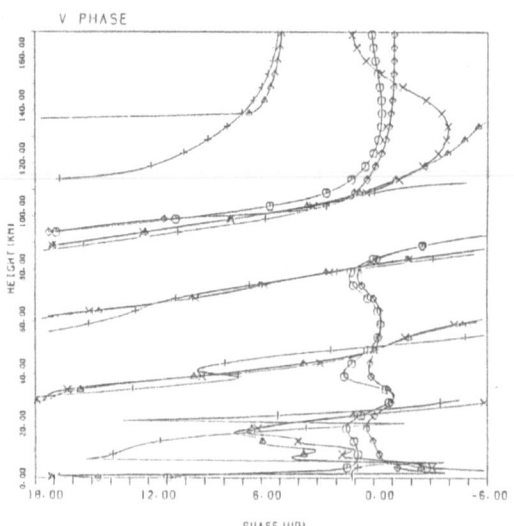

Fig.4b As in Fig.4a except for phase of v'.

large amplitude at high latitudes, and the vertically propagating
antisymmetric modes which have shorter vertical wave lengths than the
S$_{1,1}$ mode are dissipated by the prescribed eddy viscosity. The
vertical profiles of phases at ±60° and 80 to 100km height where
tide-generating heating is weak show that the wave consists of the S$_{1,1}$

mode propagating from below even at high latitudes. The growth of amplitude with height almost ceases at about 90km height due to the prescribed eddy diffusion of the order of $10^6 cm^2 s^{-1}$.

The lower thermosphere is a transition region where propagating tidal waves are replaced by an in situ tide. In the region above 160km, diffusive equiliblium is realized; the amplitude and the phase are almost constant with height.

The behavior of the wave solutions below about 100km height calculated in the present model is almost the same as that of M except for small asymmetry, and the present results seems to explain the overall features of the diurnal tidal waves below about 100km height. However, recently the day-to-day variation and the existence of antisymmetric modes of solar tide were reported (e.g. Salar et al.1977, Aso and Kato,1980, Aso and Vincent,1982). These features may be caused by the asymmetry and variability of the background wind and temperature fields (e.g. Lindzen and Hong, 1974, Walterscheid et al.,1980, Aso and Kato,1980), and/or they may be generated locally by, for instance, effects of internal gravity waves (Walterscheid,1981). To take these effects into account in a solar diurnal tide simulation is a problem for the future.

4. INDUCED ZONAL MEAN FIELDS

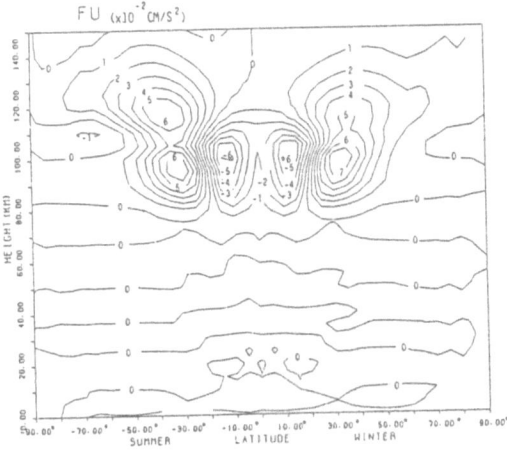

Fig.5 Meridional cross section of mean zonal forcing $\overline{F}u$.

Figs. 5,6 and 7 show the meridional cross sections of mean zonal forcing $\overline{F}u$, mean meridional forcing $\overline{F}v$ and mean heating $\overline{F}_T$, respectively, calculated from the tidal wave solution obtained in Section 3. Reflecting the symmetry of the tidal wave solution, $\overline{F}u, \overline{F}v$ and $\overline{F}_T$ are almost symmetric about the equator except in the thermosphere where the symmetry is broken by the effect of the in situ tide. The distributions of these terms in the region below about 100km height are almost the same as those calculated in M except for

190

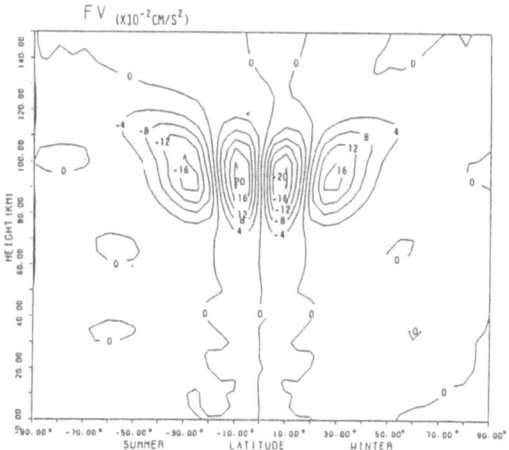

Fig.6 As in Fig.5 except for mean meridional forcing $\bar{F}v$.

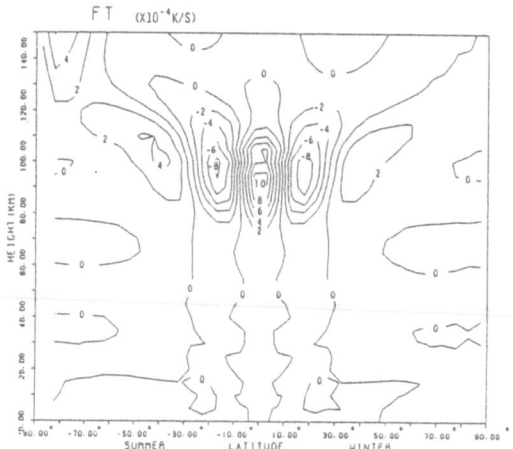

Fig.7 As in Fig.5 except for mean heating $\bar{F}_T$.

somewhat larger magnitude. The heating rate due to the tidal wave at high latitudes in the summer thermosphere is larger than that for the equinox condition calculated in M. However, the heating rate is two orders smaller than the mean heating rate due to the absorption of UV and EUV in the summer thermosphere (Dickinson et al.,1977). In the winter high latitude region of the thermosphere, although the mean solar heating is very small (Dickinson et al.,1977),it is still larger than the mean heating due to the tidal waves. Thus, even at the solstices the mean heating due to the tidal waves in the thermosphere (from 140 to 150km height region) is negligible.

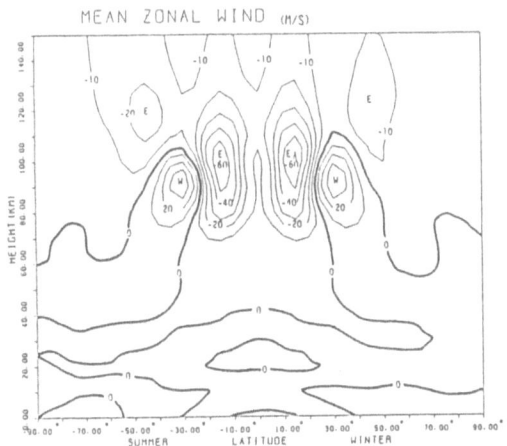

Fig.8 Meridional cross section of induced mean zonal wind.

Fig.8 shows the meridional cross section of the induced mean zonal wind. Easterly winds, with maximum speed of 60 ms⁻¹ at ±15° around 100km height, are produced in low latitudes and westerly winds, with maximum speed of 35ms⁻¹ at ±30° at 95km height, are produced in middle latitudes. Although the effects of the Coriolis term and the nonlinear terms in the zonally averaged eastward momentum equation are important as shown in M, the induced mean winds in the upper mesosphere and the lower thermosphere are in the same direction as that of $\bar{F}u$. On the other hand, the direction of the induced mean zonal wind above the lower thermosphere does not coincide with that of $\bar{F}u$. The easterly winds at ±45° around 120km height are induced by the Coriolis force acting on the equatorward winds (see Fig.9) which are produced by the mean heating and cooling at around ±45°(see Fig.7). This difference in the mechanism of mean zonal wind generation is due to the difference in the structure of tidal waves in the two regions. In the former case, the mean wind generation is mainly due to the dissipating S₁,₁ mode; in the latter case it is due to the in situ tide.

Reflecting the symmetry of the tidal wave solution, the induced mean zonal winds are almost symmetric about the equator except for the middle to high latitude region in the thermosphere.

The calculated distribution of the induced mean zonal winds in the lower thermosphere is similar to the observed results (CIRA,1972, Groves,1969, 1972, Iljichev and Portnyagin,1977) so far as the wind directions are concerned. The distributions of easterly winds in the lower thermosphere in the present model and in M are somewhat different from those in the previous papers (Miyahara,1978b, 1980) in which the induced easterly wind has its maximum over the equator. As the observed mean zonal wind distribution in the lower thermosphere is not definitive, we cannot decide which latitudinal distribution is preferable. This problem is closely related to the dissipation of tidal waves. In the present model we assume ad hoc profiles of eddy viscosity and eddy conductivity. However, the latitudinal distribution of induced mean zonal winds may be grearly affected by the relative importance of mechanical and thermal dissipation (Andrews and McIntyre, 1976b, Teitelbaum and Vial,1981). We need more careful discussion of the dissipation and breakdown of the solar diurnal tide (Lindzen,1981).

In his study of the semiannual oscillation, Hirota(1978) found a mesopause semiannual oscillation with a $-40ms^{-1}$ wind in the easterly phase of the oscillation at Ascension Island ($8°S$). The calculated mean zonal wind at the corresponding position in the present model is about $-30ms^{-1}$. The present result also suggest that the dissipating diurnal tide is an important candidate for the easterly momentum source of the mesopause semiannual oscillation (Dunkerton,1982).

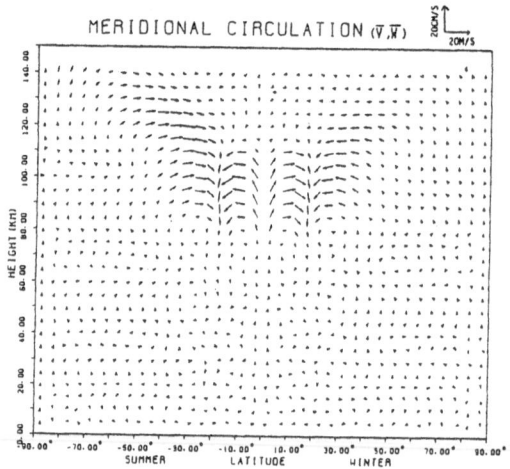

Fig.9 Eulerian-mean meridional circulation.

Fig.9 shows the Eulerian-mean meridional circulation calculated for the present model. In the upper mesosphere and the lower thermosphere, there is a four-cell circulation, with upward motions over the equator and in middle to high latitudes and downward motions at around 20 of latitude. In the thermosphere there is a two-cell circulation, with upward motions in high latitudes and downward motion in low latitudes. This circulation system is quite similar to that calculated in M except that the magnitude is somewhat larger.

Figs.10 and 11 show the residual mean circulation and the Lagrangian-mean meridional circulation, respectively (Andrews and McIntyre,1976a, 1978). In the present calculation, the Stokes correction is calculated up to the second order of the wave amplitude, and the effect of the induced mean wind shear is not included. Except for the region where the wave is excited, these two circulations coincide with one another provided that the wave is steady and nondissipative (Andrews and McIntyre,1976a, 1978). In the present case these conditions are not fulfilled, so that these two

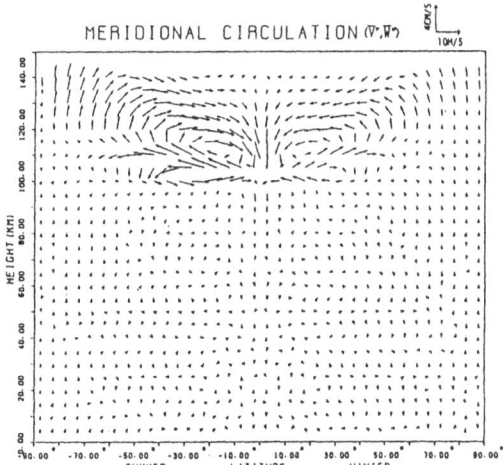

Fig.10 Residual mean meridional circulation.

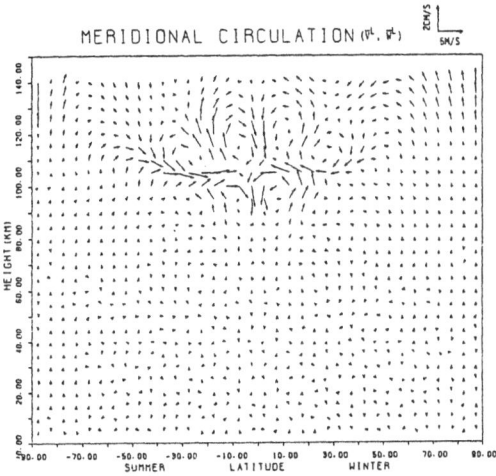

Fig.11 Lagrangian-mean meridional circulation.

circulations do not coincide. Needless to say, these circulations do not coincide with Eulerian one. However, all three circulations are similar in the middle to high latitude region in the thermosphere, where the motion is directed upward and equatorward. The circulation in this region is actually a convective one induced by the mean heating due to the heat transport by the in situ tide as mentioned earlier. It is for this reason that these three different meridional circulations are similar.

The residual and the Lagrangian-mean circulations in the upper mesosphere and the lower thermosphere are completely different from the Eulerian one, and the magnitude of these circulations is much

194

smaller than the Eulerian one because of the tendency of the cancellation between the Eulerian-mean winds and eddy heat flux convergence or Stokes drifts, respectively.

The time scales of the vertical diffusion and the vertical transport by the Lagrangian-mean motion are same order in the upper mesosphere and the thermosphere. Thus, in the present case we cannot consider that the Lagrangian-mean circulation represents the actual mass transport.

Concerning the mean zonal winds, the Eulerian mean almost coincides with the Lagrangian mean for the solar diurnal tide. However, this is not the case for semidiurnal tide, because of the long vertical wave length of the $S_{2,2}$ mode (Nakamura,1976, Miyahara, 1978a).

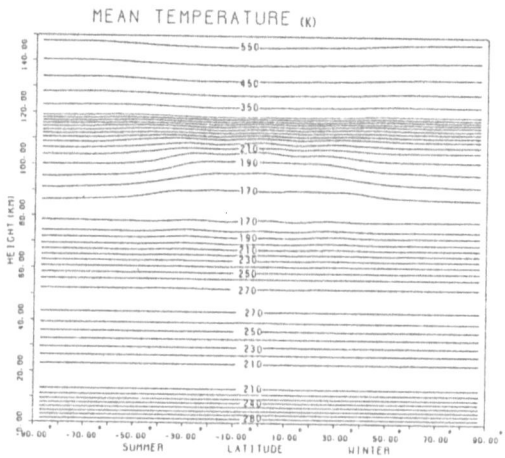

Fig.12 Meridional cross section of zonal mean temperature.

Fig.12 shows the calculated temperature profile. Although the anomalies of the order of 10K are seen in equatorial regions in the lower thermosphere, the effect of the tidal waves on the mean temperature is not so large as on the mean wind distribution.

5. CONCLUSIONS AND REMARKS

In the present paper, we have discussed the behavior of the solar diurnal tidal waves and their induced mean fields for a solstice condition. The results show that dissipating solar diurnal tidal waves can greatly contribute to the configuration of the mean winds in the upper mesosphere and the lower thermosphere.

The dissipative effect due to small scale motions and/or tidal wave breakdown itself is parameterized by the prescribed eddy diffusion. The present result seems to explain the overall features of the observed diurnal tidal waves. However, day-to-day variations in the lower thermosphere seen in the observed results (e.g. Salah et al.,1977) are not simulated in the present model. In order to simulate these variations, it may be necessary to take into account transient variations of the background mean wind or tide-generating heating, or the effects of internal gravity waves (Walterscheid, 1981). It is a problem for the future to take into account these effects.

The calculated distribution of the induced mean zonal winds in the lower thermosphere is similar to the observed results (CIRA,1972, Groves,1969, 1972, Iljichev and Portnyagin,1977) so far as the wind directions are concerned. However, the latitudinal distribution of the induced mean zonal winds may be greatly affected by the relative importance of mechanical and thermal dissipations on the wave (Andrews and McIntyre,1976b, Teitelbaum and Vial,1981), so that more precise treatment of the dissipation and breakdown of the solar diurnal tide (Lindzen,1981) may be required to discuss the problem.

It is also found that the magnitude of the induced mean zonal winds around the mesopause is sufficiently strong to explain the easterly phase of the mesopause semiannual oscillation found by Hirota(1978).

Further discussion of the role of solar diurnal tidal waves in the middle atmosphere circulation is developed in the companion paper (Miyahara,1983).

ACKNOWLEDGEMENTS

The author wishes to express his thanks to Prof. M. Uryu for his discussions and criticisms. The auther also thanks Prof. J. R. Holton and Dr. M. R. Schoeberl for help in clarifying the manuscript. Thanks are extended to Miss Motomura for typing. This work was financially supported in part by MAP of Japan. The computations were performed by the use of the FACOM M-200 computer at the Computer Center of Kyushu University.

REFERENCES

Alcayde, D.,J.Fontanari, G.Kockarts, P.Bauer and R.Bernard, 1979: Temperature, molecular nitrogen and turbulence in the lower thermosphere inferred from incoherent scatter data. Ann. Geophy., 35, 41-51.
Andrews, D.G., and M.E. McIntyre, 1976a: Planetary waves in horizontal and vertical shear; The generalized Eliassen-Palm relation and mean zonal acceleration. J. Atmos. Sci.,33,2031-2048.
Andrews, D.G., and M.E. McIntyre, 1976b: Planetary waves in horizontal and vertical shear; Asymptotic theory for equatorial waves in weak shear. J. Atmos. Sci.,33,2049-2053.
Andrews,.D.G., and M.E. McIntyre, 1978: An exact theory of nonlinear waves on a Lagrangian-mean flow. J. Fluid Mech.,89,609-646.
Aso, T., and S.Kato, 1980: Simulation of atmospheric tides. J. Meteorol. Soc. Japan, 58,286-291.

Aso T., and R.A. Vincent, 1982: Some direct comparisions of
  mesospheric winds observed at Kyoto and Adelaide. J. Atmos. Terr.
  Phys.,44,267-280.
CIRA,1972: COSPAR International Reference Atmosphere (CIRA).
  Akademie-Verlag, Berlin.
Dickinson, R.E., et al., 1977: Meridional circulation in the
  thermosphere, II, Solstice conditions. J. Atmos. Sci.,34,178-192.
Dunkerton, T., 1982: Stochastic parameterization of gravity wave
  stress. J. Atmos. Sci.,39,2325-2333.
Forbes, J.M., and H.B. Garrett, 1976: Solar diurnal tide in the
  thermosphere. J. Atmos. Sci.,33,2226-2241.
Forbes, J.M., and H.B. Garrett, 1978: Thermal excitation of
  atmospheric tides due to insolation absorption by O_3 and H_2O.
  Geophys. Res. Lett.,5,1013-1016.
Groves, G.V., 1969: Wind models from 60 to 130km altitude for
  different months and latitudes. J. Brit. Interplan. Soc.,22,
  285-307.
Groves. G.V., 1972: Annual and semiannual zonal wind components and
  corresponding temperature and density variations, 60-130km.
  Planet. Space Sci.,20,2099-2112.
Groves, G.V. 1982: Hough components of ozone heating. J. Atmos. Terr.
  Phys.,44,111-121.
Hirota, I., 1978: Equatorial waves in the upper stratosphere and
  mesosphere in relation to the semi-annaual oscillation of the
  zonal wind. J. Atmos. Sci.,35,714-722.
Holton, J.R., and W.M. Wehrbein, 1980: A numerical model of the
  zonal mean circulation of the middle atmosphere. Pure Appl.
  Geophys.,118,284-306.
Iljichev, Yu.D., and Yu. I. Portnyagin, 1977: Schemes of the global
  height-latitude cross sections of the wind field up to 100km.
  IAMAP,CMUA collection of extended summaries of contributions
  presented at CMUA sessions, IAGA/IAMAP Joint Assembly,1977,
  Seatle, p33-1-p33-5.
Lacis, A.A., and J.E. Hansen, 1974: A parameterization for the
  absorption of solar radiation in the earth's atmosphere.
  J. Atmos. Sci.,31,118-133.
Leovy, C.B., 1964: Simple models of thermally driven mesospheric
  circulation. J. Atmos. Sci.,25,327-341.
Lindzen, R.S., 1981: Turbulence and stress owing to gravity wave
  and tidal breakdown. J.G.R.,86,9707-9714.
Lindzen, R.S., and S.S. Hong, 1974: Effects of mean winds and
  horizontal temperature gradients on solar and lunar semidiurnal
  tides in the atmosphere. J. Atmos. Sci.,31,1421-1445.
Matsuno, T., 1982: A quasi one-dimensional model of the middle
  atmosphere circulation interacting with internal gavity waves.
  J. Meteorol. Soc. Japan,60,215-226.
Miyahara,S., 1975: The effects on the atmospheric lunar tide of the
  mericional temperature gradient and the zonal winds. J. Meteorol.
  Soc. Japan, 53,55-68.
Miyahara, S., 1978a: Zonal mean winds induced by vertically
  propagating atmospheric tidal waves in the lower thermosphere.

J. Meteorol. Soc. Japan,56,86-97.

Miyahara, S., 1978b: Zonal mean winds induced by vertically propagating atmospheric tidal waves in the lower thermosphere: Part II. J. Meteorol. Soc. Japan,56,548-558.

Miyahara, S., 1980: Solar diurnal tides and the induced zonal mean flows. J. Meteorol. Soc. Japan,58,302-306.

Miyahara, S., 1981: Zonal mean winds induced by solar diurnal tides in the lower thermosphere. J. Meteorol. Soc. Japan,59,303-319.

Miyahara, S., 1983: A numerical simulation of the zonal mean circulation of the mimddle atmosphere including effects of solar diurnal tidal waves and internal gravity waves;solstice condition. In this issue.

Nakamura, K., 1976: On the "Wave momentum" of deep internal gravity waves. J. Meteorol. Soc. Japan,54,331-333.

Salar, J.E., et al., 1977: Comparision of simultaneous tidal observations by incoherent scatter radars. Ann. Geophys.,33, 95-102.

Schoeberl, M.E., and D.F. Strobel, 1978: The zonally averaged circulation of the middle atmosphere. J. Atmos. Sci.,35,577-591.

Teitelbaum, H., and F. Vial, 1981: Momentum transfer to the thermosphere by atmospheric tides. J.G.R.,9693-9697.

Walterscheid, R.L., 1981: Inertio-gravity wave induced accelerations of mean flow having an imposed periodic component: Implications for tidal observations in the meteor region. J.G.R., 86,9698-9706.

Walterscheid, R.L., J.G. Devore, and S.V. Venkateswaran, 1980: Influence of mean zonal motion and meridional temperature gradients on the solar semidiurnal atmospheric tide: A revised spectral study with improved heating rates. J. Atmos. Sci.,37, 455-470.

J. R. Holton and T. Matsuno, Dynamics of the Middle Atmosphere, 199-213.

NORMAL MODE ROSSBY WAVES OBSERVED IN THE UPPER STRATOSPHERE

Isamu Hirota and Toshihiko Hirooka

Geophysical Institute
Kyoto University

ABSTRACT

A global analysis is made of large-scale traveling planetary waves
in the upper stratosphere by the use of stratospheric height and thickness data up to the 1 mb level derived from the Stratospheric Sounding
Unit (SSU) on TIROS-N and NOAA-A satellites for the period December 1979
through November 1981.

The results of space-time power and cross spectral analyses show
the global existence of a westward traveling wave of zonal wavenumber 1
with a period of about 5 days and a westward traveling wave of zonal
wavenumber 2 with a period of about 4 days.

The structure and behavior of these waves are investigated by using
the components of these period bands separated by a numerical band-pass
filter. The 5-day wave is very similar to the (1,1) mode, which is the
first symmetric wavenumber 1 Rossby normal mode in an isothermal atmosphere. Similarly, the 4-day wave appears to correspond to the (2,1)
mode. Vertical structure of both waves coincides well with the theoretical expectation of free external waves.

1. INTRODUCTION

Since the earlier studies of Deland(1964), Eliasen and Machenhauer
(1965) and others, there have been many observational studies of planetary-scale westward traveling waves in the troposphere and lower stratosphere. The longest waves, i.e., zonal wavenumber s=1, are classified
into two categories in terms of their periodicity: one is "5-day waves"
and the other is "16-day waves".

Madden and Julian(1972, 1973) performed cross spectral and composite
wave analyses for the 500 mb height and sea-level pressure data, and
showed that the 5-day wave can be identified with the first symmetric
s=1 normal mode Rossby wave for an isothermal atmosphere. Madden(1978)
also made a cross spectral analysis of the sea-level pressure data and
the geopotential height data for several pressure levels up to the lower

199

stratosphere (30 mb).

However, for the region above the level of conventional balloon observations, our knowledge of the traveling planetary waves is fragmentary. Concerning the upper stratosphere, Rodgers(1976) reported that s=1 temperature waves have a spectral peak in the range 4.5-6.2 days, using radiance data derived from the Selective Chopper Radiometer (SCR) on Nimbus 5. Concerning the region between 80 km and 100 km, some studies using wind data observed by meteor radar and UHF radar have reported wind fluctuations with a period of 5-6 days, which are considered to be associated with the traveling planetary waves (Salby and Roper, 1980; Massebeuf et al., 1981; Hirota et al., 1983). It was also reported that the coherence between ionosonde measurements and stratospheric temperatures indicates a significant peak near a period of 5 days (Fraser and Thorpe, 1976; Fraser, 1977). Hence, the 5-day wave seems to appear in regions as high as the ionosphere.

In this connection, it will be of interest to investigate the structure and behavior of large-scale normal mode Rossby waves in the upper stratosphere. The main purpose of the present study is to expand the height region as well as the seasonal time span with the aid of global satellite observations. Attention is mainly paid to the 5-day wave in this paper.

2. DATA AND ANALYSIS METHOD

The data used in this study are the global stratospheric geopotential thickness and height analyses at 1200 GMT which are derived from radiances measured by the Stratospheric Sounding Unit (SSU), High-resolution Infra-Red Sounder (HIRS), and Microwave Sounding Unit (MSU) instruments on TIROS-N and NOAA-A (See Miller et al., 1980). These data were obtained through the British Meteorological Office. We use the data in the global form with 5x5° latitude-longitude grid-spacing during the period December 1979 through November 1981. The pressure levels available are 100, 20, 10, 5, 2 and 1 mb. Missing data in time series are obtained by linear interpolation.

In order to study the nature of traveling planetary waves, we intend to filter the waves of some specified frequency bands which have prominent peaks in the energy spectra.

The primary process through which we attempt to achieve this aim is the space-time spectral analysis. The technique we use is that of the Maximum Entropy Method (MEM) after Hayashi(1977, 1981, 1982). The advantage of the MEM is to give space-time spectra with fine frequency resolution even with data of a short time record length. The procedure of this spectral analysis is the following: The grid-point data are expanded into zonal Fourier harmonics at each latitude and each level for each day. The resulting time dependent complex Fourier coefficients of wavenumber s [$F_s(t) = C_s(t) - iS_s(t)$] are treated as separate time series, where $C_S(t)$ and $S_S(t)$ are the cosine and sine coefficients, respectively. The $F_S(t)$ time series is divided into four seasonal segments and the seasonal averages are removed. The four seasonal segments are December, January, February (DJF), March, April, May (MAM), June, July, August (JJA), and September, October, November (SON). A spectral analysis is performed for the $F_S(t)$ time series of the geopotential height fields for each season. For the cross spectral analysis, cross spectra between the $F_S(t)$ of 40°N and that of other latitudes are calculated. 40°N is selected here as the reference latitude, because the latitudinal structure of the normal mode Rossby waves of interest has their maximum amplitude near 40°N.

Next, a numerical band-pass filter is used to separate the dominant frequency components. The frequency band and bandwidth of the filter are determined by the results of the spectral analysis. The numerical filter used is the Ormsby filter (Ormsby, 1961). This filter is characterized by having a sharp cut-off response. Seasonal variations and the structure of the traveling waves are investigated using the filtered components.

3. RESULTS

3.1 5-day Wave of s=1

(a) Spectral Analysis   We calculated space-time spectra smoothed by

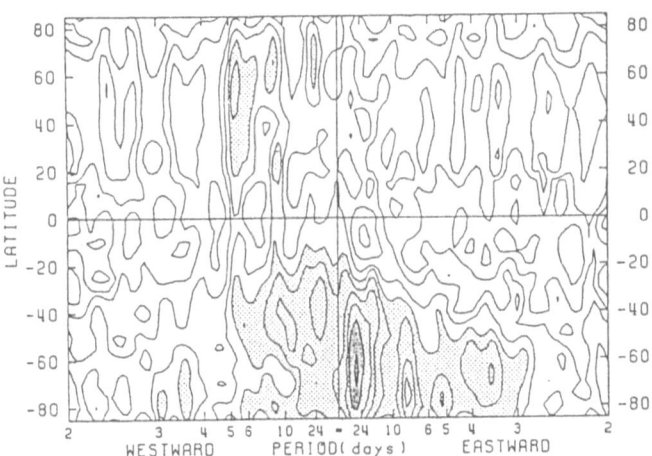

Fig.1 Latitude-period section of the space-time power spectral density ($m^2 \cdot day$) of s=1 components at 1 mb for a period June 1 through August 31 in 1980. The shading denotes regions greater than 10^4 $m^2 \cdot day$. Contours are 3x10, 10^2, $3x10^2$, 10^3, $3x10^3$, 10^4, $3x10^4$, 10^5, $3x10^5$ and 10^6 $m^2 \cdot day$.

averaging over a frequency bandwidth of 1/120 day^{-1}, choosing 20 as the
length of the prediction error filter. To begin with, we present the
results of the power spectral analysis. Fig.1 shows a latitude-period
plot of contours of the power spectra at 1 mb for the period June 1
through August 31 (JJA) in 1980. It can be seen that clear predominant
variance of the westward 5-6 day period band runs from the Northern
Hemisphere (NH) to the Southern Hemisphere (SH). The same characteristic also can be recognized in the MAM and SON seasons, but such pre-
dominancy cannot be recognized in the DJF season. We also performed
the same spectral analysis for the lower altitude pressure levels, and
the results indicate that the spectral power is more predominant as the
altitude becomes higher.

Since we may identify such predominant variance with the 5-day wave,
we next performed the cross spectral analysis choosing 40°N as a reference latitude for the same season as that in Fig.1. Fig.2 shows a
latitude-period plot of contours of the coherence squared. Surprisingly,
the strong coherence belt, whose period band is the same as that in
Fig.1, reaches to 50°S. Moreover, its region is in-phase with 40°N.
Therefore, we consider that this wave of 5-6 day period, i.e., the 5-day
wave, is identified with the (1,1) mode. This is also confirmed by the
global pattern as will be shown later (See Figs. 6 and 7).

To make sure of the predominance of westward traveling waves, it is
interesting to see the time change of the geopotential height fields in
August 1980, which is a month of this interesting season. Fig.3 shows a
longitude-time section of the deviation of 1 mb height from the time and

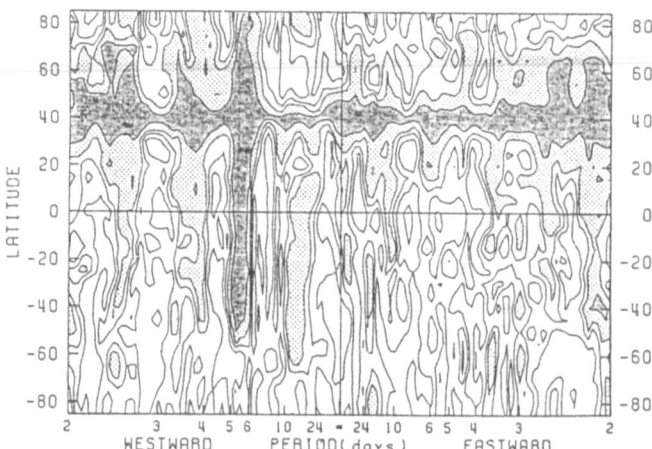

Fig.2   Latitude-period section of the coherence squared of s=1 components
        at 1 mb. The reference latitude is 40°N. Light shade denotes
        regions greater than 0.8, and dark shade, greater than 0.95.
        Contours are 0.3, 0.6, 0.8 and 0.95.

zonal average at 40°N. It is easily seen that there appears a s=1 westward traveling wave with a period of 5-6 days. It is also noteworthy
that a 4-6 day period zonal wind fluctuation in the upper mesosphere and
lower thermosphere was observed by the Arecibo(18°N) UHF radar for the
same period as that in Fig.3 (Hirota et al., 1983). This wind fluctuation was probably the geostrophic wind variation associated with the
passage of the 5-day wave.

(b) Structure and Seasonal Variation   Next, we investigated further
characteristics of the 5-day wave. We first passed the Fourier components of s=1 through a numerical band-pass filter of about 5 days. The
following analyses were performed on the components obtained by such a
procedure. It will be necessary, however, to pay attention to the fact
that not only the westward traveling 5-day wave but also the eastward
moving '5-day wave' is obtained by this filtering method.

Fig.4 shows the seasonal variation of the 1 mb 5-day wave at 40°N.
The upper part denotes the amplitude and the lower part expresses the
daily values of the phase differences between 40°N and other latutudes
(20°N, Equator, 20°S and 40°S). If four lines of the phase differences
gather near 0 degree, we can say that the wave has an in-phase structure
in the North-South direction. It is found from this figure that except
for the DJF season, the periods of large amplitude coincide with the
in-phase periods, and that the periods of large amplitude of the 5-day
wave roughly 1.5 months long occur irregularly.

Fig.5 shows a 1 mb 5-day feature of the amplitude calculated from

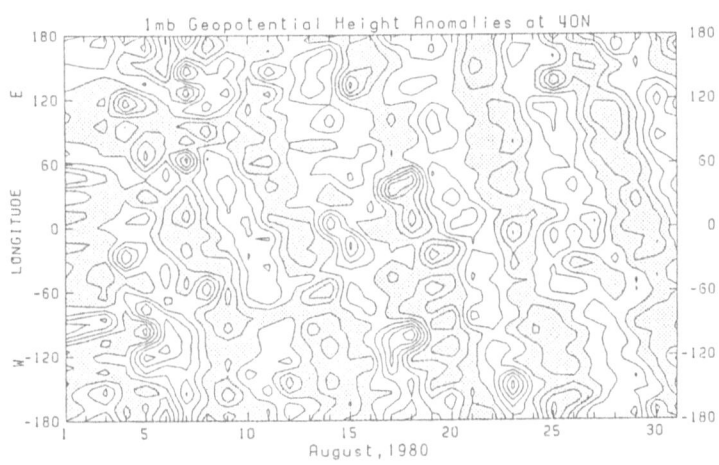

Fig.3   Longitude-time section of the deviation of 1 mb geopotential
        height from the time and zonal average at 40°N for August 1980.
        Shaded regions denote negative anomalies. The contour interval
        is 50 m.

the filtered components. This figure shows a spatially regular pattern
of near equatorially symmetric structure with the maximum amplitude near
40-50 degrees and a spatially irregular pattern which appears in the
winter season of both hemispheres with the maximum amplitude near 70
degrees. The regular pattern coincides with the 5-day wave, and it

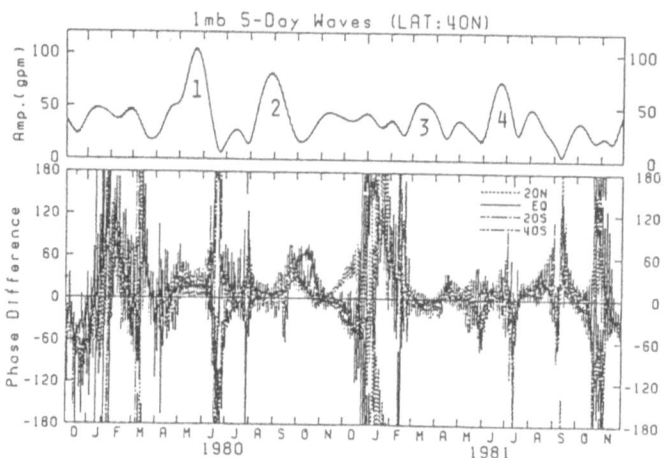

Fig.4 Seasonal variation of the 1 mb 5-day wave at 40°N. Upper: Amplitude. Lower: Relative phase differences (degrees) between 40°N
and other latitudes (20°N, Equator, 20°S and 40°S). Numbers '1',
'2', '3' and '4' denote the large amplitude periods (See the
text).

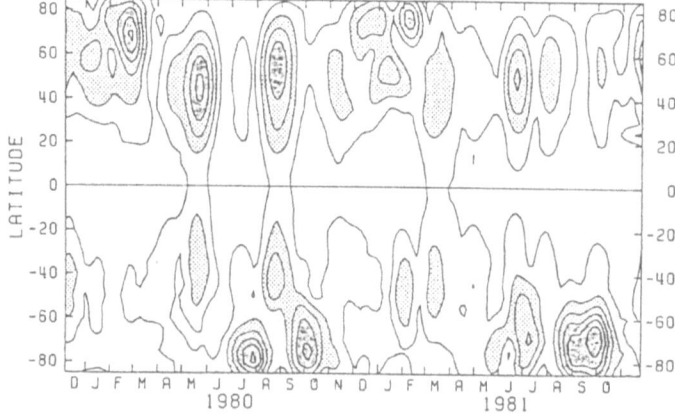

Fig.5 Latitude-time section of the filtered wave amplitude at 1 mb
averaged over 10 days. Light shade denotes regions greater than
40 m, and dark shade, greater than 80 m. The contour interval
is 20 m.

should be pointed out that the amplitude in the NH is always greater
than that in the SH. On the other hand, the irregular pattern turns
out to coincide with the eastward moving wave. This was found by check-
ing the daily phase change. This eastward moving wave is a part of the
eastward moving variance with a wide period range of 4-10 days (See
Fig.1). In the NH winter, the irregular pattern dominates over the
whole hemisphere, whereas in the SH winter the irregular pattern co-
exists with the regular pattern near 50°S. This difference of the two
hemispheric winters corresponds to the fact that the spectral power is
not predominant in the DJF season as mentioned before.

Fig.6 illustrates the latitudinal structure of the 5-day wave during the predominance period (August 1980). In this figure, the eastward
moving component is dominant in higher latitudes of the SH. This figure
illustrates that the typical latitudinal structure of the 5-day wave is
quite similar to the (1,1) mode, even though the amplitude in the NH is
greater than that in the SH. This hemispheric asymmetry is consistent
with Geisler and Dickinson(1976) and Salby(1981a,b).

The traveling pattern of the 5-day wave is shown in Fig.7, which is
for the 6 successive days of August 28 through September 2 in 1980.
These figures show the regular westward traveling pattern over the region of 85°N through 50°S and the irregular pattern to the south of 50°S.

Next, in order to examine the vertical structure of the 5-day wave,
we chose the large amplitude periods of the wave. The periods were determined by the appropriate critical values of the amplitude and the
relative phase differences, and are indicated by numbers '1', '2', '3'
and '4' in Fig.4. Period 1 is from May 3 to June 6 in 1980, Period 2 is
from August 8 to September 12 in 1980, Period 3 is from February 27 to
April 1 in 1981, and Period 4 is from June 7 to June 30 in 1981. We can

1mb 5-Day Wave

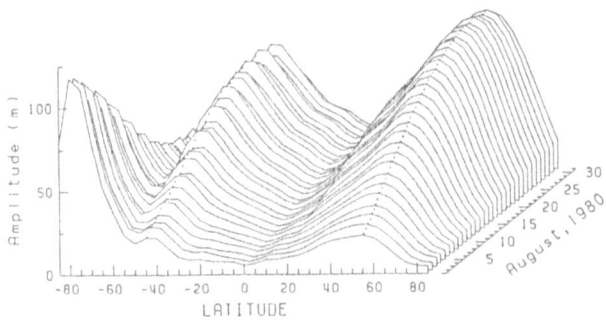

Fig.6   Three-dimensional plot of the 1 mb 5-day wave amplitude versus
        latitude and time for August 1980.

obtain an average amplitude for each period at each pressure level.
These average amplitudes are plotted versus pressure levels in Fig.8
with the theoretical line for the Lamb mode. Every line is in good
agreement with the theoretical line. We also calculated average relative phase differences between the 1 mb level and other five levels for
the same periods as those in Fig.8. As is seen from the results shown
in Fig.9, there is small westward tilt with height in each case. These
characteristics of the vertical structure indicate that the 5-day wave
has the nature of free waves. Here, we will refer to the seasonal

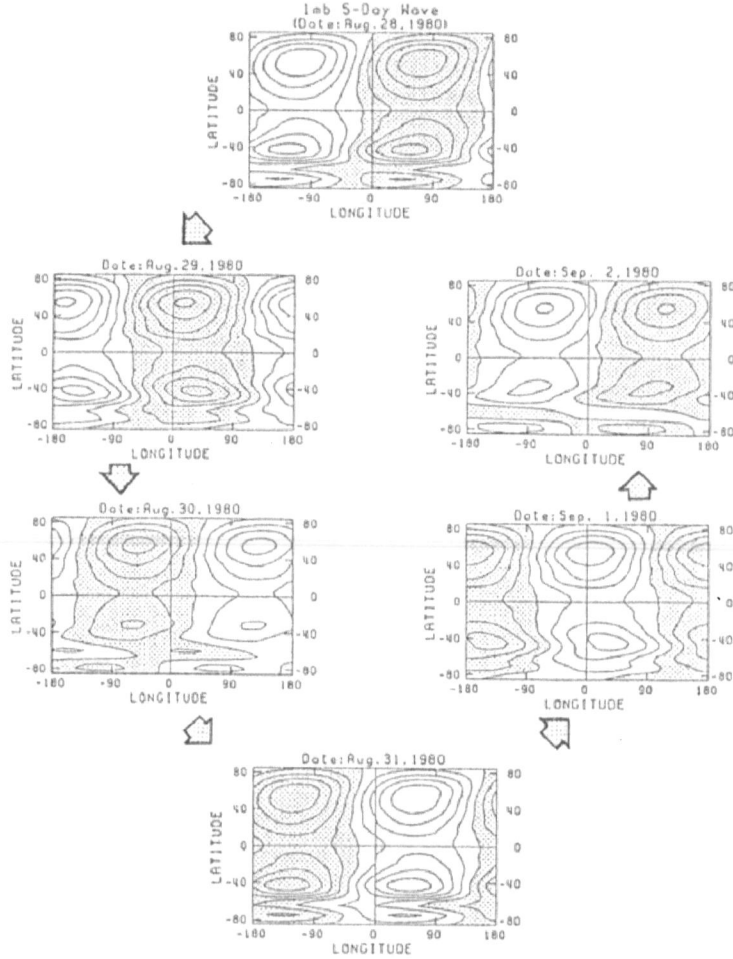

Fig.7   The traveling pattern of the 1 mb 5-day wave for the 6 successive
days of August 28 through September 2 in 1980. Shaded regions
denote negative anomalies. The contour interval is 20 m.

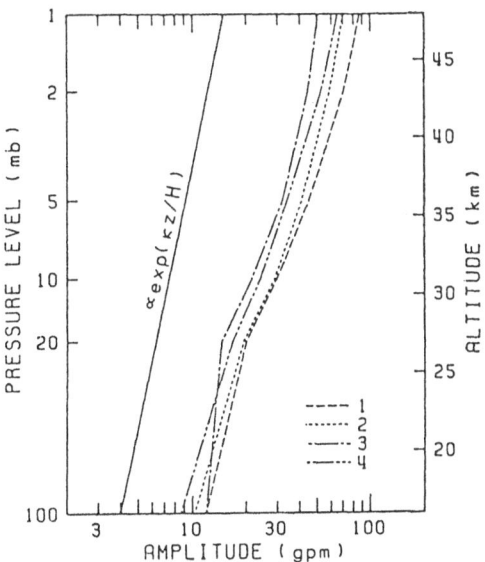

Fig.8 Amplitude of the 5-day wave at 40°N for the large amplitude periods. Numbers '1', '2', '3' and '4' coincide with those in Fig.4. The solid line denotes the theoretical expectation for the Lamb mode.

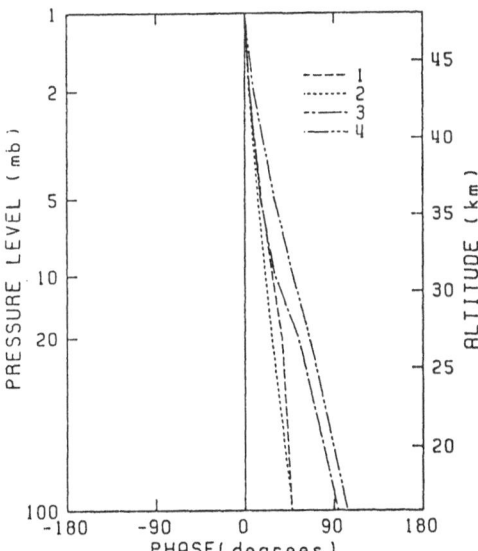

Fig.9 The relative phase differences between the 1 mb level and other five levels for the same periods as those in Fig.8.

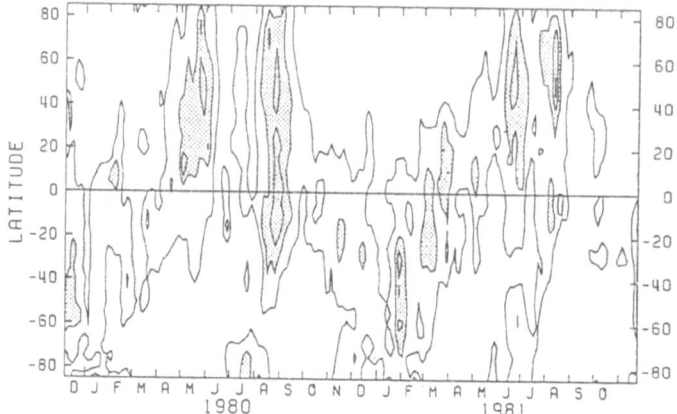

Fig.10 Latitude-time section of the ratio of the filtered wave amplitude
to 10-day root-mean-square wave amplitude of s=1 at the 1 mb
level. Shade denotes regions greater than 0.4. The contour
interval is 0.2.

change of the vertical structure by comparing with the results of Salby's
numerical model (1981a,b). Period 1 and 4 are considered to approximate
the solstice conditions, while Period 3 is considered to approximate the
equinox condition. The vertical growth rate of the amplitude seems to
be slightly enhanced in the solstice conditions consistent with Salby's
results. For the phase tilt, however, there appears no significant difference between the two groups in contrast to Salby's results.

Finally, it is interesting to see what portion of the total s=1
disturbance is occupied by the 5-day wave. Fig.10 shows a latitude-time
section of the ratio of the filtered wave amplitude to root-mean-square
amplitude of s=1 at the 1 mb level. It is found that the ratio changes
seasonally and takes large values reaching to 0.7 in the summer hemisphere. This is because the 5-day wave is predominant in the summer
hemisphere, while in the winter hemisphere other waves, mainly stationary
waves, have much greater ampliutde so that the 5-day wave contributes
much less to the total s=1 variance.

3.2 4-day Wave of s=2

For the s=2 westward traveling planetary wave with a period of 4-5
days, there were a few studies based on spherical harmonic analyses in
the latter half of the 1960's (Eliasen and Machenhauer, 1965, 1969;
Deland and Lin, 1967). In the last decade, however, there have been few
reports dealing with the s=2 traveling wave, and the structure and behavior of the wave are little known. Hence, we tried to analyze the s=2
traveling wave with a period of 4-5 days using the same methods as we
used for the 5-day wave analysis.

(a) Spectral Analysis   We calculated space-time power spectra for each season. The predominant power spectra are not recognized so readily in this period range as for the 5-day wave of s=1, but significant power can be seen in Fig.11, which illustrates the power spectra at 1 mb for the period March 1 through May 31 (MAM) in 1981. This spectral characteristic is similar to that of the 5-day wave except for the period band of about 4 days.

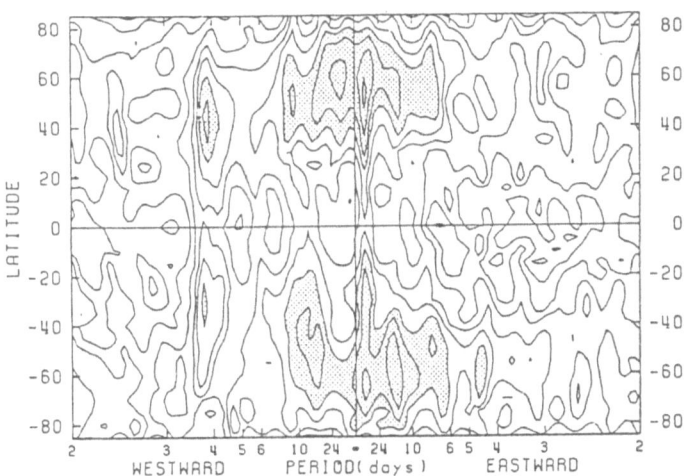

Fig.11   Same as Fig.1 except for s=2 components during March 1 through May 31 in 1981.

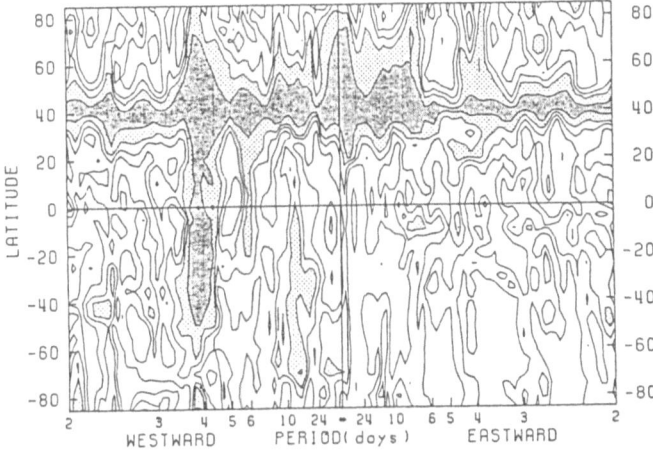

Fig.12   Same as in Fig.2 except for s=2 components in the same season as that in Fig.11.

210

Fig.12 shows a latitude-period section of the coherence squared, whose reference latitude is 40°N, in the same season as that in Fig.11. Again, a strong coherence belt is found, whose region is in-phase with 40°N. Therefore, this wave of 4 day period (the 4-day wave) is also likely to be identified with the (2,1) mode.

(b) Structure and Seasonal Variation   To see the nature of the 4-day wave in more detail, we again passed the Fourier components of s=2 through a numerical band-pass filter.

Fig.13 is the same figure as Fig.4 except for the 4-day wave and for using values at 30°S instead of 40°S. It is again found that the

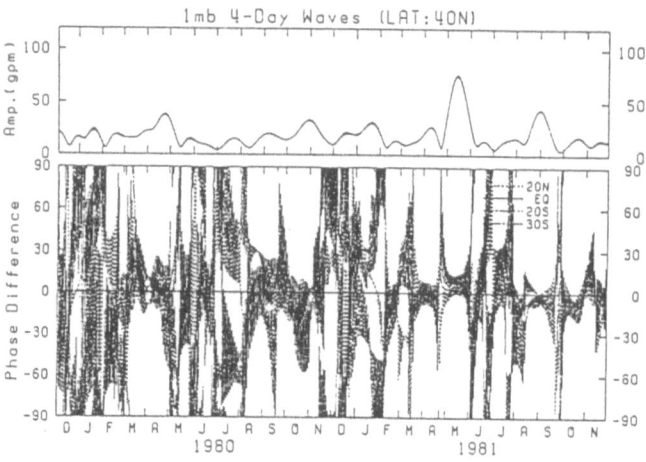

Fig.13   Same as in Fig.4 except for the 4-day wave and for using values at 30°S instead of 40°S.

1mb 4-Day Wave

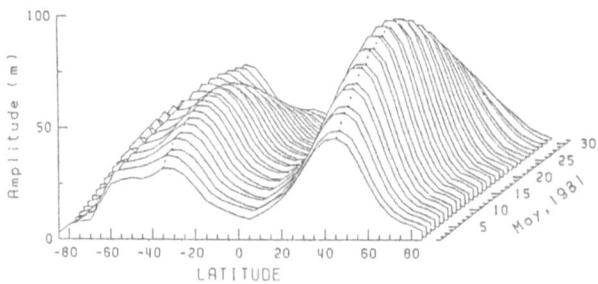

Fig.14   Same as in Fig.6 except for the 4-day wave during May 1981.

periods of large amplitude coincide with the in-phase periods but that
the 4-day wave is not predominant as repeatedly as was the 5-day wave.
However, the 4-day wave often appears except in the DJF season, even
though its amplitude is not large. It is noteworthy that the predominant periods of the 4-day wave never overlap those of the 5-day wave.

Fig.14 illustrates a latitudinal structure of the 4-day wave during
a predominance period (May 1981), indicating that the latitudinal structure of the 4-day wave is similar to that of the (2,1) mode.

Fig.15 illustrates the traveling pattern of the 4-day wave during
the period May 10 through May 13 in 1981. These figures show the regular westward traveling pattern over the region of 85°N through 40°S.
There also exists eastward moving component to the south of 40°S, and
the pattern changes with a time scale of 2 days by superposition of the
westward and eastward waves.

For the vertical structure of the wave, we calculated an average
amplitude and an average phase difference in the same way as we did for

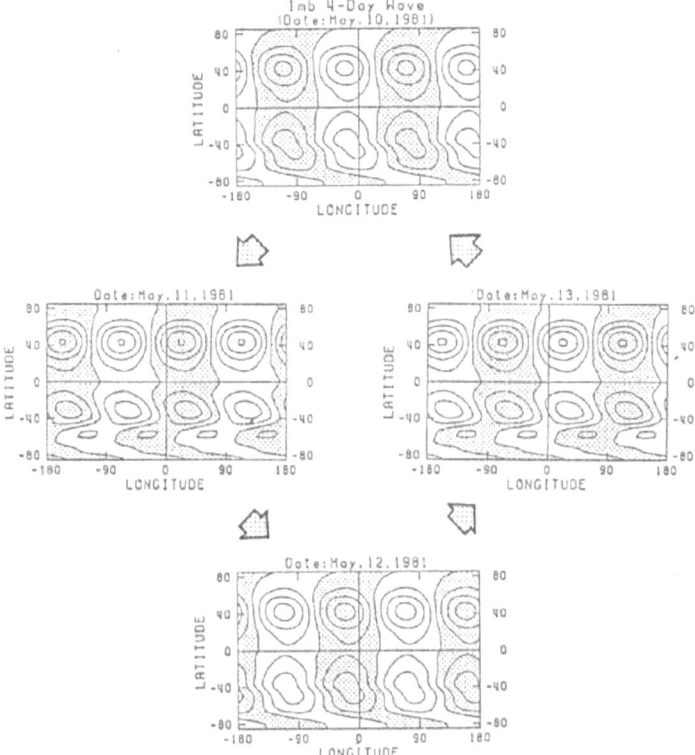

Fig.15 Same as in Fig.7 except for the 4-day wave during a period May
10 through May 13 in 1981.

the 5-day wave. The characteristics of the 4-day wave are the same as those of the 5-day wave, and suggest that the 4-day wave is also a free wave.

4. CONCLUDING REMARKS

In the present analysis, the existence of normal mode Rossby waves whose structure is identified with the (1,1) and (2,1) modes has been found in the upper stratosphere. The vertical structure of these waves corresponds to that of the free waves predicted theoretically. However, there still remain some problems to be investigated about the normal mode Rossby waves. The first is the structure of the waves above the stratopause, since the results of theoretical models (Geisler and Dickinson, 1976; Salby, 1981a,b) indicate large change of the wave structure in this region. The next is the problem of the waves of other modes with smaller meridional scales (e.g., the 16-day wave). It will be very interesting to see whether these waves exist or not in the upper stratosphere. Therefore, further observational studies are needed for the understanding of free Rossby waves in the middle atmosphere.

ACKOWLEDGMENTS

We are grateful to Dr. S. A. Clough of the British Meteorological Office for his courtesy in supplying us with the TIROS-N/NOAA-A SSU data. Thanks are also due to Drs. M. A. Geller and M. L. Salby for their helpful comments.

REFERENCES

Deland, R. J., 1964: Travelling planetary waves. Tellus, 16, 271-273.

———, and Y-J. Lin, 1967: On the movement and prediction of traveling planetary-scale waves. Mon. Wea. Rev., 95, 21-31.

Eliasen, E., and B. Machenhauer, 1965: A study of the fluctuations of the atmospheric planetary flow patterns represented by spherical harmonics. Tellus, 17, 220-238.

———, and ———, 1969: On the observed large-scale atmospheric wave motions. Tellus, 21, 149-166.

Fraser, G. J., 1977: The 5-day wave and ionospheric absorption. J. Atmos. Terr. Phys., 39, 121-124.

———, and M. R. Thorpe, 1976: Experimental investigations of ionospheric/ stratospheric coupling in southern mid latitudes-1. Spectra and cross-spectra of stratospheric temperatures and the ionospheric f-min parameter. J. Atmos. Terr. Phys., 38, 1003-1011.

Geisler, J. E., and R. E. Dickinson, 1976: The five-day wave on a sphere with realistic zonal winds. J. Atmos. Sci., 33, 632-641.

Hayashi, Y., 1977: Space-time power spectral analysis using the maximum entropy method. J. Meteor. Soc. Japan, 55, 415-420.

———, 1981: Space-time cross spectral analysis using the maximum entropy method. J. Meteor. Soc. Japan, 59, 620-624.

———, 1982: Space-time spectral analysis and its applications to atmospheric waves. J. Meteor. Soc. Japan, 60, 156-171.

Hirota, I., Y. Maekawa, S. Fukao, K. Fukuyama, M. P. Sulzer, J. L. Fellous, T. Tsuda and S. Kato, 1983: Fifteen-day observation of mesospheric and lower thermospheric motions with the aid of the Arecibo UHF radar. J. Geophys. Res. (in press).

Madden, R. A., 1978: Further evidence of traveling planetary waves. J. Atmos. Sci., 35, 1605-1618.

———, and P. R. Julian, 1972: Further evidence of global-scale 5-day pressure waves. J. Atmos. Sci., 29, 1464-1469.

———, and ———, 1973: Reply. J. Atmos. Sci, 30, 935-940.

Massebeuf, M., R. Bernard, J. L. Fellous and M. Glass, 1981: Simultaneous meteor radar observations at Monpazier (France, 44°N) and Punta Borinquen (Puerto Rico, 18°N). II-Mean zonal wind and long period waves. J. Atmos. Terr. Phys., 43, 535-542.

Miller, D. E., J. L. Brownscombe, G. P. Carruthers, D. R. Pick and K. H. Stewart, 1980: Operational temperature sounding of the stratosphere. Phil. Trans. Roy. Soc. London, A296, 65-71.

Ormsby, J. F. A., 1961: Design of numerical filters with applications to missile data processing. J. Association for Computing Machinery, 8, 440-466.

Rodgers, C. D., 1976: Evidence for the five-day wave in the upper stratosphere. J. Atmos. Sci., 33, 710-711.

Salby, M. L., 1981a: Rossby normal modes in nonuniform background configurations. Part I: Simple fields. J. Atmos. Sci., 38, 1803-1826.

———, 1981b: Rossby normal modes in nonuniform background configurations. Part II: Equinox and solstice conditions. J. Atmos. Sci., 38, 1827-1840.

———, and R. G. Roper, 1980: Long-period oscillations in the meteor region. J. Atmos. Sci., 37, 237-244.

LARGE-SCALE WAVES AND WAVE, MEAN-FLOW INTERACTION

J. R. Holton and T. Matsuno, Dynamics of the Middle Atmosphere, 217-251.
Copyright © 1984 by Terra Scientific Publishing Company.

THE QUASI-BIENNIAL OSCILLATION

R. Alan Plumb*
CSIRO Division of Atmospheric Research, Aspendale 3195,
Australia

ABSTRACT

The quasi-biennial oscillation manifests itself most clearly in
the equatorial stratosphere, as a periodic reversal of the zonal wind
with an average period of about 28 months. Other observations
indicate a similar oscillation in stratospheric temperature and ozone,
and perhaps in certain tropospheric elements. These observations are
reviewed briefly but the main emphasis is directed toward a
theoretical treatment of the oscillation. The theory of Holton and
Lindzen (1972), that the zonal wind oscillation in the equatorial
stratosphere is driven by upward propagating equatorial waves, is
reviewed in detail; supporting evidence from observational analyses
and a laboratory experiment is discussed. Amongst other things, the
theory suggests why the quasi-biennial oscillation has not yet been
found in general circulation models.

1. INTRODUCTION

The search for periodicities in atmospheric data has a long
history and still receives considerable attention in the literature.
Apart from the obvious diurnal and annual cycles and their harmonics,
success has been limited and is at best controversial – any signal
that may exist is usually difficult to separate from the noise. There
is, however, one remarkable exception. Figure 1 shows a time series
of 50 mb and 30 mb monthly mean zonal winds at Singapore (1°20'N).
This station being close to the equator, the annual cycle is small and
the series is dominated by a clear, if slightly irregular, oscillation
with an average period of a little over 2 yr. This pheomenon, which
has come to be known as the quasi-biennial oscillation (QBO), was
first reported by Reed (1960) and by Veryard and Ebdon (1961). The
discovery put an end to confusion that had arisen from attempts to
interpret early stratospheric wind observations on a steady-state
basis.

---

*Present address: Geophysical Fluid Dynamics Program, Princeton University.

218

The zonal wind oscillation illustrated in Figure 1 is strongly
concentrated in the equatorial stratosphere. Nevertheless an
oscillation of similar period has been revealed in many tropospheric
elements, in the high latitude stratosphere and in the mesosphere.
These observational studies have been reviewed extensively by
Landsberg (1962), Newell et.al. (1974) and, for the tropical
stratosphere, by Wallace (1973). This being the case it is not
intended in this review to discuss the observations in detail, but
rather to highlight the major observed properties of the QBO.

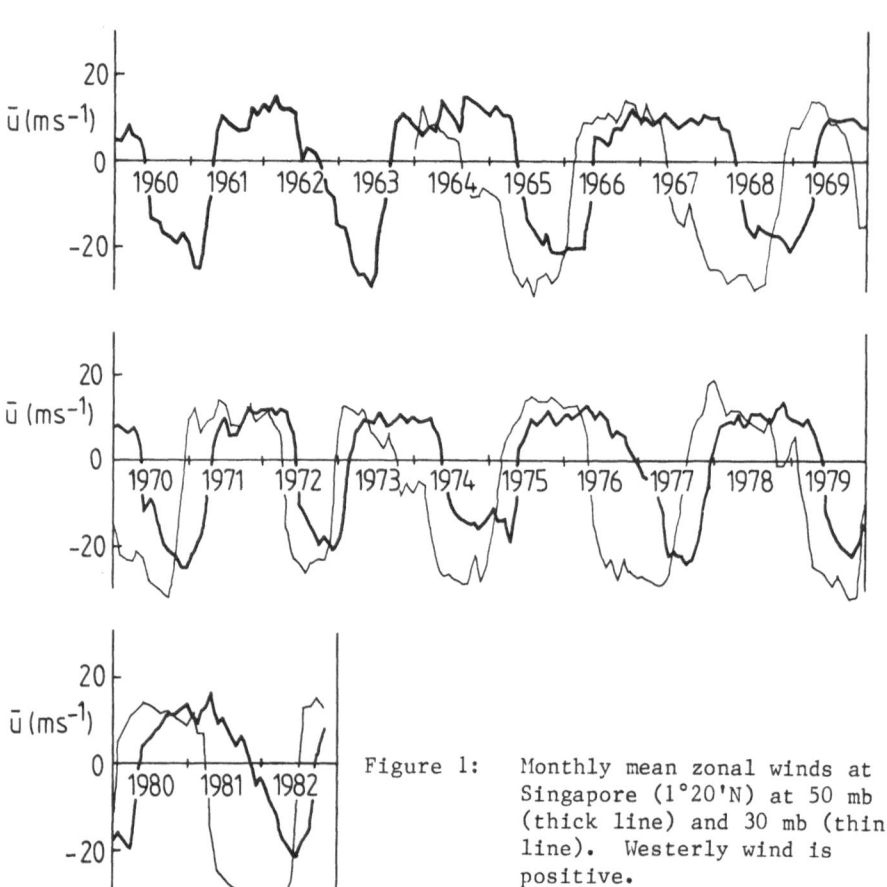

Figure 1: Monthly mean zonal winds at
Singapore (1°20'N) at 50 mb
(thick line) and 30 mb (thin
line). Westerly wind is
positive.

The major emphasis here will be on a theoretical discussion of the phenomenon and possible generating mechanisms. While several suggested external and internal causes of the period of the QBO have been put forward, only the theory of Holton and Lindzen (1972) has successfully explained the structural features of the phenomenon in the equatorial stratosphere. This theory will be discussed in detail and supporting evidence will be presented.

2. OBSERVATIONS

2.1 THE STRATOSPHERIC ZONAL WIND OSCILLATION

The time series shown in Figure 1 is typical of stratospheric wind records from all near-equatorial stations. Inspection of such data reveals that the amplitude and phase of the QBO is independent of longitude. Its structure is most clearly revealed by plotting contours of monthly zonal wind anomalies (thus removing the steady background flow as well as the annual cycle and its harmonics) as a function of height and time. Such a section for stations near 9°N is shown in Figure 2. (This figure was originally produced by Wallace (1973) and updated by Coy (1979)). A number of features are apparent.

i)   The period is irregular, varying from a minimum of 22 months to a maximum of 34 months with an average of about 28 months. The periods of successive cycles (measured between the transitions from easterly to westerly flow at 30 mb) are shown in Figure 3. On all but two of the twelve cycles, the period exceeds 24 months. It is clear that the oscillation is not biennial, contrary to the views expressed by some authors (e.g. Newell et al. 1974). Nevertheless, Figs. 1 and 2 (despite the fact that the latter has the annual cycle removed) contain evidence that the oscillation does interact with the annual cycle, at least to the extent that there appears to be a tendency for certain phases of the QBO to synchronize with the calendar. For example in the Singapore record of Fig. 1 all eight of the W-E wind reversals at 30 mb occur during the half-year between November and April.

ii)   The amplitude of the oscillation near 9°N is about 20 ms^{-1} above 50 mb, decreasing rapidly below this level. (The oscillation is barely discernible at 100 mb).

iii) The phase of the oscillation descends at a rate of about 1 km per month.

iv)   The oscillation exhibits east-west asymmetry. As noted by Wallace (1973), the maximum westerly acceleration at a given level is consistently larger than the maximum easterly acceleration, and the corresponding vertical shears are stronger as the westerly phase is becoming established.

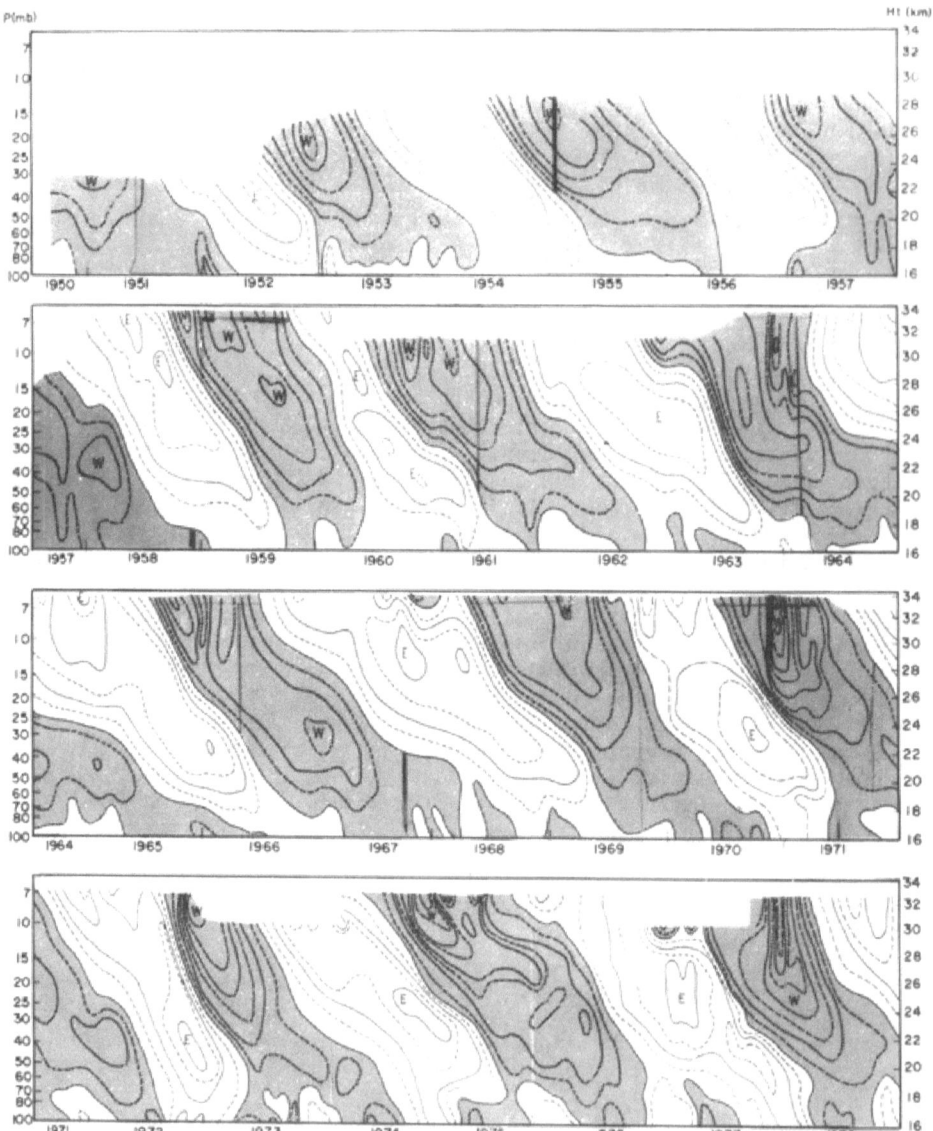

Figure 2: Time-height section of the zonal wind shear 9°N (Canal Zone until 1970, Kwajalein July 1970 onward). The 15-year average of monthly means were subtracted to remove the annual and semi-annual cycles. Solid isotachs are at intervals of 10 ms^{-1}. Westerlies are shaded. (After Coy, 1979).

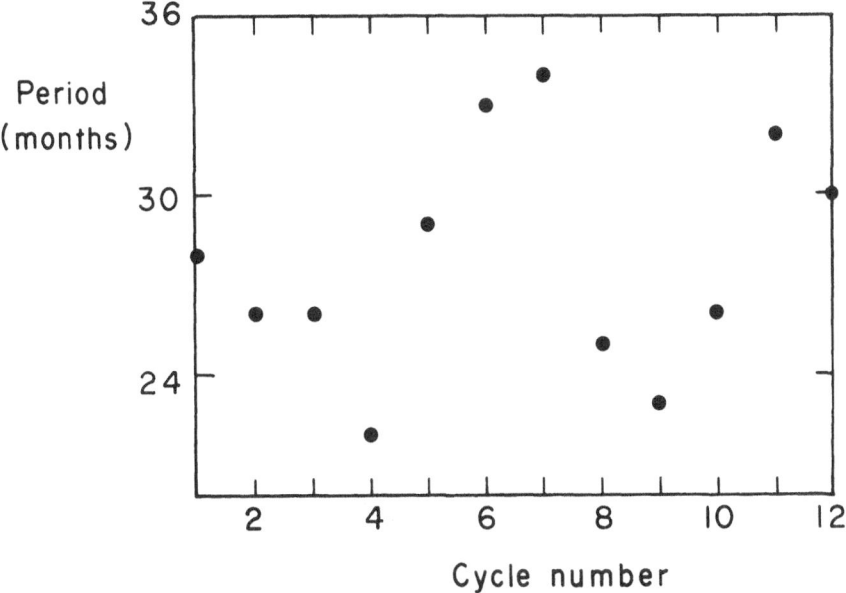

Figure 3. Period of successive cycles of the QBO at 30 mb, measured between the times of transition from easterly to westerly flow. Cycle 1 begins August 1952 (cf. Quiroz 1981).

In the meridional plane, the oscillation decays rapidly away from the equator; its amplitude decreases to one-half of the equatorial value within about 15° latitude and to very small values in the subtropics (see Fig. 6 of Wallace, 1973 and Fig. 5 of Tucker, 1979). There is evidence, however, that this oscillation reappears at higher latitudes(Tucker 1979, Holton and Tan 1980,1982). Holton and Tan composited 16 years of monthly mean geopotential data for the Northern Hemisphere with respect to the direction of the equatorial QBO. An example of their results – the height difference between their January composites – is shown in Fig. 4. The dominant (and statistically significant) feature, which was found to persist in all the months November-March, is an enchancement of the polar night cyclone during periods of equatorial westerlies, but above normal heights at lower latitudes (around 50°N). This implies an enhancement of the high-latitude westerly jet (by about 5 ms^{-1}) in phase with the equatorial oscillation, and a weaker oscillation with opposite phase at around 30-40° latitude.

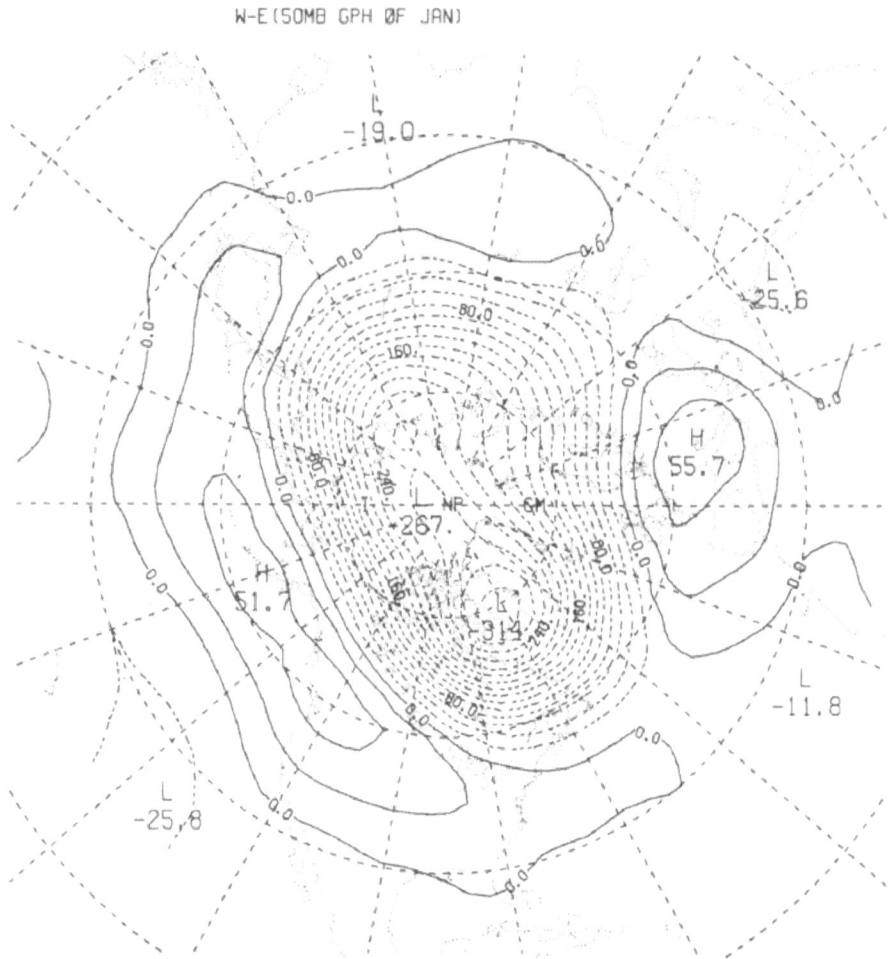

W-E(50MB GPH 0F JAN)

Figure 4: The quasi-biennial oscillation at 50 mb. Height differences (gpm) between January composites (westerly minus easterly) based on the direction of zonal wind at 50 mb on the equator. (After Holton and Tan, 1980).

2.2 OTHER STRATOSPHERIC MANIFESTATIONS

Quasi-biennial variations have been observed in a number of other stratospheric parameters, in particular temperature and ozone. Veryard and Ebdon (1961) first reported a stratospheric temperature QBO at Canton Island (3°S). Although the QBO is largely an equatorial phenomenon, its period is so long that one expects it to be in geostrophic balance almost everywhere (the inverse of the Coriolis parameter equals 2 years at latitude 0.03°). Therefore the QBO should be in approximate thermal wind balance. This implies a temperature maximum just below the westerly wind maximum or, in a single-station time series, temperature maximum before westerly wind maximum. Reed (1962, 1964) found the phase of the temperature oscillation to be consistent with this picture but with a phase reversal in the subtropics. Newell et al. (1974) show an oscillation of maximum amplitude 2K whose phase is also approximately consistent with thermal wind balance, while Angell and Korshover (1978) give an amplitude of about 1K in equatorial regions, falling off to 0.3K in middle latitudes. They found maximum equatorial temperatures at or slightly preceding the time of west wind maximum; Northern Hemisphere middle latitudes were approximately in phase with the tropics but the phase lagged with increasing latitude in the Southern Hemisphere.

A quasi-biennial fluctuation in total ozone was noted by Funk and Garnham (1962). Newell et al. (1974) show the oscillation at several stations; the QBO is detectable (at least during certain intervals) at all latitudes. Pittock (1968) analysed vertical distributions of ozone during 1965-67 at Aspendale (38°S) and observed that phase changes in ozone density descend at about 1 km per month. Later, however, (Pittock 1977) he was unable to find evidence of a QBO during 1965-73. Hasebe (1980) describes a QBO in zonal-mean total ozone during the years 1962-76, the amplitude maximizing in mid-latitudes of the Northern Hemisphere (see the article by Hasebe (1983) in this volume).

A possible QBO in stratospheric water vapour has also been identified; Hyson (1983) has found large fluctuations in H_2O mixing ratio in the lower stratosphere over Australia, which appear to correlate well with stratospheric wind QBO at Singapore.

2.3 THE TROPOSPHERIC QBO

Evidence for a QBO in tropospheric data has been accumulating for some years; an extensive discussion of these results is given in Landsberg (1962) and Newell et al. (1974), to which the reader is referred for references.

Landsberg lists some 24 claims of quasi-biennial periodicity in various tropospheric elements which appeared in the literature before the discovery of the zonal wind oscillation in the stratosphere. Indeed, he attributes the discovery of the QBO to Clayton (1885) who found a periodicity of 25 months in certain surface data at a number of locations in the United States (some of Clayton's data are reproduced by Newell et al. 1974).

Like the stratospheric oscillation, most authors place the period of the tropospheric phenomenon at slightly longer than 2 years. Bugayev et al. (1972) suggested that the two oscillations might be related. This led Ebdon (1975) to seek differences in Northern Hemisphere surface and 500 mb fields between mid-season months in which the 30 mb equatorial winds were respectively easterly or westerly. In January he found differences over large parts of the hemisphere, and claimed that the number of significant points "probably indicates significance at greater than the 1 percent level". He also found apparently significant differences in July but not in April or October. Ebdon's results appear to be supported by Holton and Tan (1980) who extended their compositing study to 1000 mb. Their results, shown in Fig. 5, give a structure similar to that found at 50 mb (Fig. 4), viz., decreased heights at high latitudes and increased heights in middle latitudes during period of westerly equatorial wind at 50 mb. (The amplitude is however, less than that at 50 mb, so the signal is not barotropic). These results are similar to those of Ebdon and, since only 6 of 16 years of data used by Holton and Tan are common to both studies, this fact enhances confidence in the results. They are also qualitatively consistent with the finding by Angell and Korshover (1977) of a tendency for the 300 mb north circumpolar vortex to be contracted at the time of 50 mb equatorial west wind maximum.

Variability of quasi-biennial period in certain indices of the Southern Hemisphere circulation was identified by Trenberth (1979, 1980); however, he noted that some aspects of this variability do not appear to correlate with the QBO in stratospheric wind. This raises the possibility that at least some of the claimed QBO in the troposphere may in fact be unrelated to the stratospheric phenomenon. Indeed, Holton and Tan (1982) observed that, because of the similarity in period of the QBO and the Southern Oscillation, the time series used in these studies may be too short to permit separation of the two signals. Since it seems intrinsically more likely that the tropospheric variability is related to the Southern Oscillation, rather than the stratospheric QBO, it is difficult on present evidence to accept the tropospheric signal as a manifestation of the latter. In fact, similar uncertainty may apply to some of the stratospheric results. For example (Van Loon et. al. (1981) found correlations in the circulation of the mid-latitude stratosphere with the Southern Oscillation; their results are in some ways similar to Holton and Tan's QBO composites.

225

W-E(1000MB GPH ØF JAN)

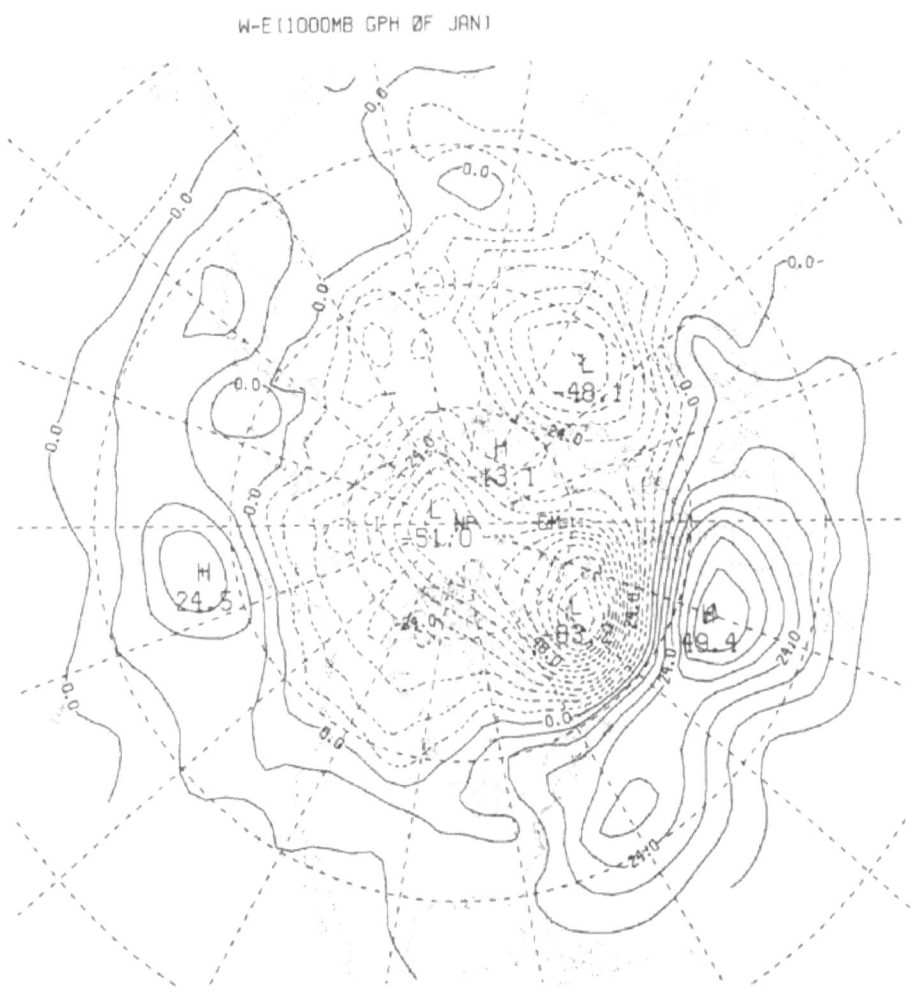

Figure 5:   The quasi-biennial at 1000 mb.   Otherwise as Figure 4.
(After Holton and Tan, 1980).

3. THEORY

3.1 GENERAL

Attempts to provide a theoretical basis for understanding the QBO have largely fallen into two categories. These are attempts to explain (i) the periodicity of the oscillation or (ii) the observed structure and required momentum budget of the zonal wind oscillation in the equatorial stratosphere.

The first category includes claims of a relationship with solar activity (e.g. Shapiro and Ward 1962, Probert-Jones 1964, Sugiura 1976). Others have suggested that the period is internally controlled. Landsberg (1962) proposed that a biennial oscillation could arise if the atmospheric circulation should somehow become out of balance one year and subsequently (i.e. the following year) overcompensate. Similar arguments were put forward by Newell et. al. (1974), Nicholls (1978) and, more generally, by Brier (1978). At first sight it might appear that such processes would necessarily produce a two-year cycle, but Brier argued that the interaction between the annual cycle and negative feedback would give cycles of two or (probably less often) three or more years. It was noted in Section 2.1, however, that the duration of successive QBO cycles shows no particular preference for 24 (much less 36) months. More fundamentally, none of these theories provides an explanation for the structure of the observed oscillation, nor of the nature of its generating mechanism. Therefore they have not gained wide acceptance.

Viewing the observational evidence it is difficult to avoid the impression that the centre of action of the QBO lies in the equatorial stratosphere. It is here that the oscillation most clearly manifests itself, explaining 90% of the zonal wind variance* (Newell et al. 1974) This being the case it is not surprising that the second approach is the route via which the most promising theory has arisen.

In fact, consideration of the angular momentum budget of the equatorial stratosphere immediately places some constraints on acceptable theories. At certain phases of the QBO a westerly acceleration occurs at the equator while there is a westerly (spatial) maximum there. The equator is then a region of maximum absolute angular momentum and, since angular momentum is conserved in inviscid axisymmetric motion, it seems likely nonaxisymmetric motions must be responsible for the observed acceleration. Therefore it proves necessary to seek eddy motions capable of providing the required driving of the zonal flow.

---

*This of course arises partly ecause of the vanishing of the annual cycle near the equator; nevertheless the QBO amplitude has a clear maximum here.

Horizontal transfer by stratospheric planetary waves would at
first sight appear to be the most likely process. Tucker (1965) and
Wallace and Newell (1966) found evidence of quasi-biennial variations
in the equatorward flux of zonal momentum, but these studies did not
give any clues to the origin of the periodicity. Detailed
consideration of the structure of the equatorial winds (in particular
the downward phase propagation) makes it unlikely that transfer by
planetary waves can be the generating mechanism of the QBO (see the
discussion in Wallace 1973). This does not, however, rule out the
possibility that planetary waves contribute to the momentum budget of
the equatorial lower stratosphere (in view of their dominance in the
winter stratosphere, it would be surprising if they did not); their
possible influence on the QBO will be addressed in Section 5.2, below.

Lindzen and Holton (1968) suggested instead that the required
momentum sources could be provided through vertical transfer by upward
propagating equatorial waves. Such waves are predicted theoretically
(e.g. Matsuno 1966, Lindzen 1967) as equatorially trapped solutions of
Laplace's tidal equation. Waves observed in the lower stratosphere
(Yanai and Maruyama 1966, Wallace and Kousky 1968) are identifiable
with Kelvin waves and mixed Rossby-gravity waves of the theory (see
Wallace 1973 for a discussion of the agreement between observation and
theory). A summary of the observed properties of these waves is given
in Table 1. It seems clear that they are of tropospheric origin.
Holton (1972) argued that stratospheric equatorial waves could be
generated as a response to random tropospheric forcing; on the basis
of a series of control experiments with a general circulation model,
Hayashi and Golder (1978) found tropospheric latent heat release to be
an essential generation mechanism. Detailed discussion of this
important and not fully resolved question is beyond the scope of this
review; for our purposes, the essential fact is that these
upward-propagating waves do exist in the equatorial stratosphere.

3.2 THE HOLTON-LINDZEN THEORY

Equatorial waves transfer momentum upward and are thus
potentially capable of generating mean zonal motion within the
stratosphere. Steady waves cannot, however, force mean flow
accelerations unless they are dissipated, either via mechanical or
thermal damping or via absorption at a critical level where the

TABLE 1

Typical data for stratospheric equatorial waves

(adapted from Wallace 1973)

	Kelvin wave	Rossby-gravity wave
Zonal wavenumber	1-2	4
Period (ground level)	10-20 days	4-5 days
Phase speed (ground based)	25 ms^{-1}	-23 ms^{-1}
Equatorial symmetry	SYMMETRIC	ANTISYMMETRIC
Latitudinal scale * L	1000 km	1000 km
Vertical wavelength	6-10 km	4-8 km
Amplitudes:		
Zonal wind	8 ms^{-1}	$2-3 \text{ ms}^{-1}$
Meridional wind	0	$2-3 \text{ ms}^{-1}$
Vertical velocity	$1-2 \text{ mm s}^{-1}$	$1-2 \text{ mm s}^{-1}$
Temperature	2-3 K	1 K
Geopotential height	4 m	30 m

*According to equatorial beta-plane theory (Lindzen 1967), the Kelvin wave geopotential has lateral structure $\exp(-y^2/2L_K^2)$ while that of the mixed Rossby-gravity wave has the form $(y/L_{RG}) \exp(-y^2/2L_{RG}^2)$.

Doppler-shifted phase speed (relative to the local mean flow) vanishes (e.g. Holton 1975, Andrews and McIntyre 1976a, McIntyre 1980). Where this dissipation occurs the waves force a mean flow acceleration; if for present purposes we neglect the effects of the mean meridional circulation (q.v.), this acceleration may be written

$$\frac{\partial \overline{u}}{\partial t} = - \sum_i \frac{\partial F_i}{\partial z} \tag{1}$$

where F_i is the effective momentum flux (the negative of the Eliassen-Palm flux) associated with the ith wave component. Now the Kelvin wave (which has a westerly zonal component of phase velocity) has $F_K > 0$ and so, from (1), forces westerly acceleration where the wave is dissipated (i.e. where F_K decreases with height). For the Rossby-gravity wave $F_{RG} < 0$ (Lindzen 1971) and its absorption induces easterly mean acceleration. Thus the observed waves are able to provide sources of both easterly and westerly momentum (and, as will be seen in Section 4.2, their magnitudes are sufficiently large to account for the observed acceleration in the QBO). The interaction between the mean flow and these two opposing sources of momentum forms the basis of the Holton-Lindzen theory.

Initially, Lindzen and Holton (1968) suggested that the waves are absorbed at critical surfaces where $\overline{u} - c = 0$ (where c is the wave phase velocity). This presented problems, partly because a continuous spectrum of waves is required (which is not borne out by observations) but mainly because critical surfaces do not seem to exist, especially for the Kelvin wave (c.f. Table 1 and Figure 1; the mean zonal wind velocity is always much smaller than 25 m s^{-1} throughout the record). Later, however, they (Holton and Lindzen 1972) showed that wave absorption via radiative damping in the stratosphere could provide efficient wave-mean flow interaction and produced on this basis a model that reproduced the major features of the QBO.

Before presenting details of this model, it is useful first to consider a simpler theoretical study of Plumb (1977). This study does not attempt to model the stratosphere, being concerned with the interaction between the mean flow and internal gravity waves, but qualitatively its properties are the same and its greater simplicity allows clearer conceptual insight. Further, it provides the basis for an experimental validation of the theory (q.v., Section 4.1). Consider a vertically unbounded, non-rotating, stratified fluid subjected to a standing wave forcing of the form cos kx . cos kct on its lower boundary. This wave may of course be regarded as the sum of two equal but opposite travelling waves of the form Re $(....e^{ik(x \pm ct)})$. These waves radiate upward into the fluid where they are dissipated by (say) a Newtonian cooling process at a rate α. If the mean flow is slowly varying in height and time then one may use a two-scaling (WKB) technique to derive the profile of the wave momentum fluxes

$$F_i \propto \exp - \int_o^z g_i(z').dz' \qquad (2)$$

where the attenuation rate is

$$g_i(z) = \frac{\text{Dissipation rate}}{\text{(vertical group velocity)}} = \frac{\alpha}{k(\overline{u}-c_i)^2/N}; \qquad (3)$$

N is the buoyancy frequency. The dependence of group velocity on a positive power of $(\overline{u} - c)$ is crucial to the theory. Just as for equatorial waves, the internal wave of positive ("westerly") phase speed has $F_+ > 0$, while $F_- < 0$ for the easterly wave and, since the wave amplitudes are equal on the lower boundary, $F_+ + F_- = 0$ everywhere and, by (1), there is no net acceleration. Therefore the state $\overline{u} = 0$ is an equilibrium solution, which is obvious by reasons of symmetry). This state is, however, unstable. To see that this is so, consider Figure 6(a), in which a westerly mean flow perturbation has been introduced. In regions of westerly flow, the Doppler-shifted phase speed $c - \overline{u}$ of the westerly wave is reduced; by (3), therefore, its vertical group velocity is reduced and the wave is attenuated more rapidly in the vertical. Conversely the easterly wave is attenuated less rapidly, since $c + \overline{u}$ is increased. It then follows from (1) that westerly acceleration dominates in this region, as depicted in Figure 6(a). As a result of this selective attenuation of the westerly wave, however, at high levels the easterly wave dominates and therefore the net acceleration becomes easterly there. For reasons discussed in Plumb (1977), the level of maximum acceleration is below the level of maximum $\overline{u}$; consequently the level of maximum $\overline{u}$ must descend with time. Therefore a finite mean flow develops from the instability in the manner shown in Figure 6(b).

The subsequent evolution of the flow is depicted schematically in Figure 7. The shear zone separating the low-level westerly regime from the easterly flow above becomes increasingly narrow with time (Fig. 7(a)); eventually this must break down, e.g. by momentum diffusion. This has the effect of destroying the westerly regime from above and therefore "switching" the low-level flow into an easterly regime (Fig. 7(b)). When this occurs the westerly wave is no longer attenuated at low levels and penetrates to greater heights. Consequently a high level westerly acceleration of the mean flow takes place. Eventually (Fig. 7(c)) a westerly regime develops there which

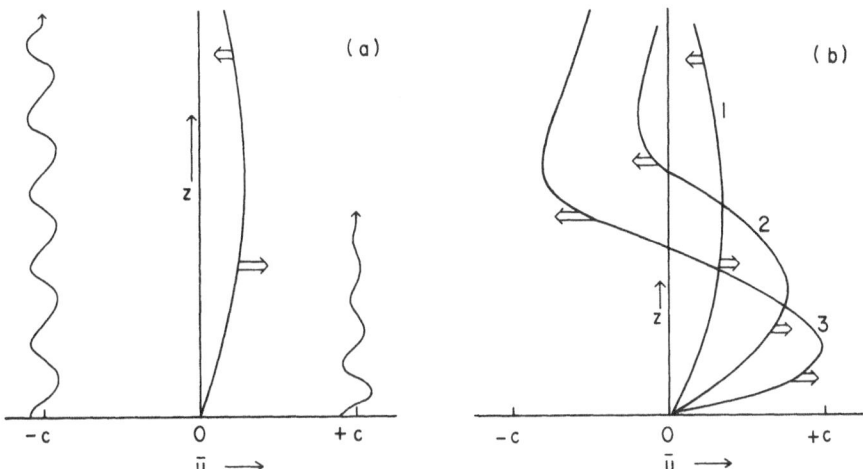

Figure 6: Schematic representation of the instability of zonal flow
in a stratified fluid with standing wave forcing applied
at a lower boundary. Zonal mean velocity $\bar{u}$ vs. height z.
(After Plumb 1982b).
(a) Onset of instability from a small zonal flow
   perturbation.
(b) Early stages of subsequent evolution.
Double arrows: Approximate location and direction of
maximum acceleration.
Wavy lines: Schematic representation of penetration of
wave components.

moves downward until the interior shear layer again becomes narrow
enough for diffusion to act (Fig. 7d). At this stage, the flow has
been reversed in sign (cf. Fig. 7(a)); subsequently the evolution is
identical with (but opposite in sign to) that described above until
the original structure is re-established (Fig. 7(d) - (e) - (f) -
(a)). Thus the interplay between the mean flow profile and the feedback via selective attenuation of the wave motions generates an
oscillation with descending regimes of alternate easterly and westerly
flow.

With this in mind we now turn to Holton and Lindzen's (1972)
model of the equatorial stratosphere. Strictly, the equatorial wave
problem is a three-dimensional one but this was simplified by considering an equivalent two-dimensional system, essentially by a latitudinal averaging. The equivalence is not exact but the major
features of the problem should not be in serious error. Then the
attenuation rates in (2) become

232

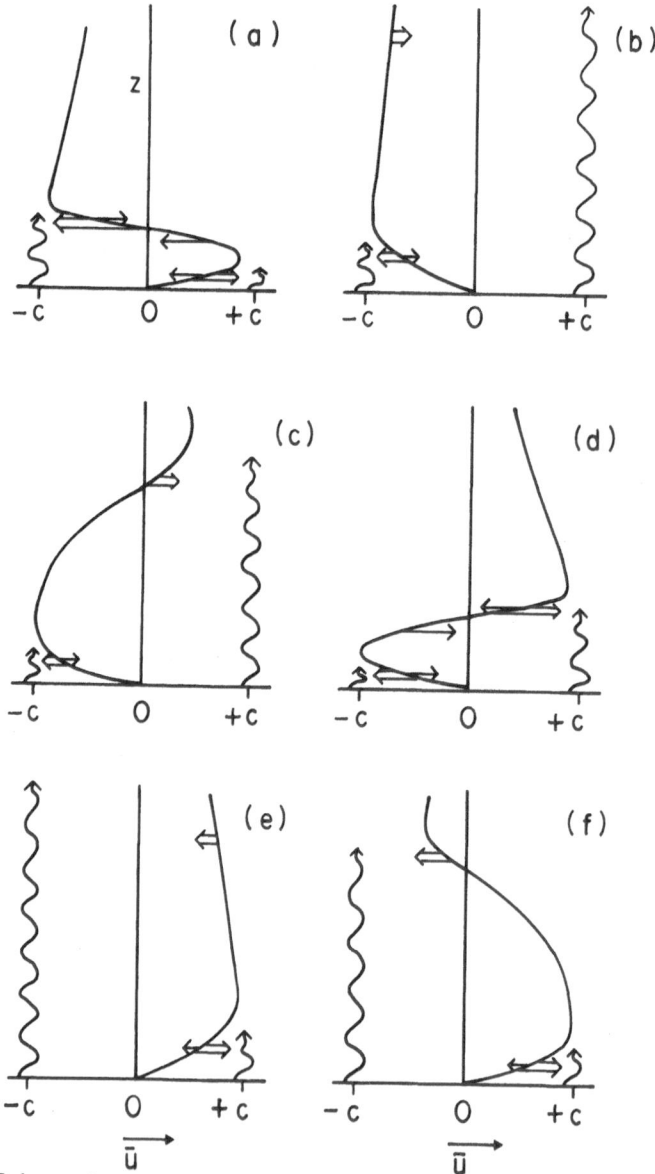

Figure 7: Schematic representation of evolution and structure of fully developed flow (following on from Figure 6). Six stages in one complete cycle of the mean flow oscillation.
Single arrows: acceleration arising from viscous forces.
Double arrows: acceleration arising from wave attenuation.
Wavy lines: Schematic representation of penetration of wave components

Kelvin wave:
$$g_K = \frac{N\alpha}{k_K \overline{(u - c_K)}^2}$$

(4)

Rossby-gravity wave:
$$g_{RG} = \frac{N\alpha}{k_{RG}\overline{(u - c_{RG})}^2} \left\{ \frac{\beta}{k^2_{RG}\overline{(u - c_{RG})}} - 1 \right\}$$

(where β is the equatorial beta parameter) which are respectively the appropriate relations of Kelvin and Rossby-gravity waves derived from a WKB analysis of equatorial waves in vertical shear (Lindzen 1971). The Kelvin wave attentuation rate is identical with that appropriate to internal gravity waves; the functional form of g_{RG}, however, is different. Qualitatively, this does not matter since the fundamental property that g varies as a negative power of $|\bar{u} - c|$ is retained (and in fact, because of the cubic term, is enhanced). A consequence of the difference in the functional dependence of g_K and g_{RG} on $(\bar{u} - c)$ is the introduction of an easterly-westerly asymmetry into the flow.

Then, using (2), Holton and Lindzen integrated Eq. (1) into which two additional terms were introduced. First, a term $K\partial^2\bar{u}/\partial z^2$ was included to allow for momentum diffusion which, as discussed above, is necessary to permit switching from one regime to another. The diffusion coefficient used was $K = 0.3$ m^2s^{-1}. Second, a high-level semi-annual forcing was imposed in the model in order to drive a semi-annual oscillation similar to that observed above 30 km in the tropics (Reed 1966). However Plumb (1977) argued on theoretical grounds that high-level effects can have no significant influence on the flow beneath and confirmed this by repeating Holton and Lindzen's calculation without the semi-annual forcing. Results of this calculation are shown in Figure 8.

Comparing Figures 8 and 2, it is clear that the Holton-Lindzen model has captured the major features of the QBO, in particular the downward propagation of the alternating regimes is well simulated. The period is about 3 years. In the simpler model of Plumb (1977), the oscillation period was found to be proportional to the timescale

$$T = k\hat{\hat{c}}^3/N\alpha\hat{F}$$

(5)

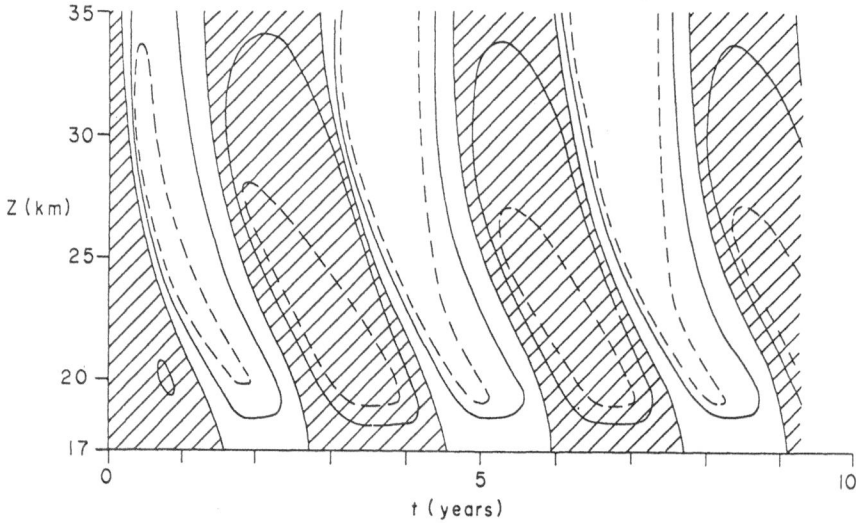

Figure 8: Theoretical evolution of mean zonal flow $\bar{u}$ at the equator
according to the Holton-Lindzen model, with data as given
in Table 2. Solid contours are at intervals of 15 ms^{-1},
dashed contours represent ± 22.5 ms^{-1}. Westerlies are
shaded. (After Plumb, 1977). Model data as follows:

Buoyance frequency $N = 2.16 \times 10^{-2}$ s^{-1} Newtonian cooling rate:

$$\alpha = 0.55 + 0.56 \left(\frac{z-17 \text{ km}}{6.62 \text{ km}}\right) \times 10^{-6} \text{ s}^{-1}, \quad z < 30 \text{ km};$$

$$\alpha = 1.65 \times 10^{-6} \text{ s}^{-1}, \quad z > 30 \text{ km};$$

Wave data:

	Kelvin Wave	Rossby-gravity wave
c	30 ms^{-1}	-30 ms^{-1}
k	1.57 x 10^{-7} m^{-1}	6.28 x 10^{-7} m^{-1}
	(zonal wavenumber 1)	(zonal wavenumber 4)
F (z = 17 km)	4 x 10^{-3} m^2 s^{-2}	-4 x 10^{-3} m^2 s^{-2}

where $\hat{k}$, $\hat{c}$ and F are typical values appropriate to the forcing waves. However, Dunkerton (1981b) has suggested that, in a compressible atmosphere, a dependence

$$T = cH/F \qquad (6)$$

may be more appropirate, where H is the density scale height. The uncertainty in the parameters in (5) and (6) implies a corresponding uncertainty - probably a factor of about 2 - in the theoretically predicted period (e.g. using the value c = 25 ms^{-1} in (5), rather than 30 ms^{-1}, would yield a period of 25 months). By the same token, the observed irregularity in the wave period could therefore be consistent with the theory if there is a corresponding variation in wave forcing.

There are some notable differences, in vertical structure and in the asymmetry of the oscillation. The vertical penetration of the oscillation, particularly of the easterly regime, is greater than observations suggest (cf Fig. 2); however Hamilton (1981) shows that agreement with observation is improved if more realistic dissipation rates are used. In Figure 8 the easterly regime appears more rapidly than the westerly whereas it was noted in Section 2.1 that the opposite asymmetry obtains in reality. This could perhaps be rectified by varying the relative magnitudes of the two waves (within observational limits) - cf. the calculations of Plumb (1977) and of Dunkerton (1981b) with asymmetrical forcing. However, while a two-dimensional theory simulates the major features of the QBO, it might be expected that a more rigorous calculation is required to reporduce observed structural details.

3.3 THE LATITUDINAL STRUCTURE

Introduction of latitudinal wind shear into theoretical models of the interaction of equatorial waves with the zonal flow greatly complicates the problem, which can no longer be treated by the simple procedures described above. Nevertheless, Boyd (1978a,b) has developed a technique that allows the two-scaling method to be used even in the presence of strong latitudinal shear; Plumb and Bell (1982a) confirmed the accuracy of Boyd's method in realistic zonal wind configurations. However, this approch has yet to be successfully applied to a model of the QBO.

The alternative to a two-scaling approach is explicit resolution of the equatorial waves. This immediately presents problems because the small vertical wavelength of the waves (cf Table 1) then demands small vertical model resolution which in turn places severe constraints on the time step permitted in numerical integration. Holton (1979) used a primitive equation model with vertical resolution 1 km and in which either a Kelvin wave or mixed Rossby-gravity wave

236

forcing was specified at a lower boundary to investigate the structure
of the mean flow generated by the waves. The model was integrated
for 60 days, during which time downward moving shear layers were
generated within about 10° of the equator, much as in the
corresponding phase of the QBO.

An explicit simulation of the QBO in a beta-plane model was
performed by Plumb and Bell (1982b). They circumvented the need for
explicit time-resolution of the waves by adopting in part the
procedure followed in the 1-D models, viz., by assuming that the mean
flow is slowly varying in time (but not in space). This assumption,
which is in reality questionable (Dunkerton 1981a,b), allowed a fairly
long time step (1d) so that despite the fine vertical resolution of
500 m integration over several years became practicable. Results of
such a calculation are shown in Fig. 9. The structure of the model
QBO is similar to that observed, one significant improvement over the
1-D calculation being that the east-west asymmetry is now in the
correct sense. The amplitude of the oscillation is, however, about
one-half of that observed (it was in fact possible to generate
stronger zonal winds by reducing the model viscosity but this caused
resolution problems, even with the fine resolution used). The
latitudinal structure, exemplified

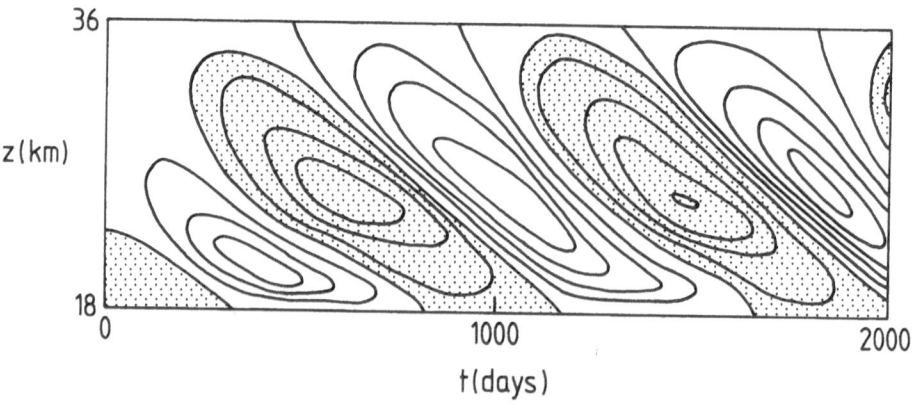

Figure 9: Time height plot of the mean zonal wind at the equator
in the model of Plumb and Bell (1982b). Contours at
intervals of 2 ms^{-1}; easterlies are shaded.

in Fig. 10, is also in qualitative agreement with observations
although, apart from the amplitude differences, the latitudinal scale
is rather too small. The thermal structure (Fig. 10b) is in
approximate thermal wind balance with the wind field and is maintained
against thermal dissipation by a Lagrangian-mean meridional
circulation (Fig. 10c). As discussed in detail by Plumb and Bell
(1982b), this circulation makes a significant contribution to the
momentum budget (Fig. 11) in two ways. Very close to the equator,
vertical advection retards the descent of easterly shears (cf. Fig.
10c and Fig. 11a) and accelerates the descent of westerly shear. This
appears to be the main mechanism responsible for the observed
asymmetry of the QBO, although Dunkerton (1982) has argued that other,
more subtle, effects play a role. A few degrees off the equator, the
Coriolis acceleration associated with the mean meridional flow becomes
important. As can be seen in the example of Fig. 10, its role in the
momentum budget is largely dissipative (Fig. 11b) and, in fact, this
acts to reduce the latitudinal scale of the equatorial jets and to
drive the reversed-phase oscillation beyond $y = 1200$ km. However the
latitudinal extent of this circulation (Fig. 10c) is too small to
drive any significant oscillation beyond about 2000 km from the
equator.

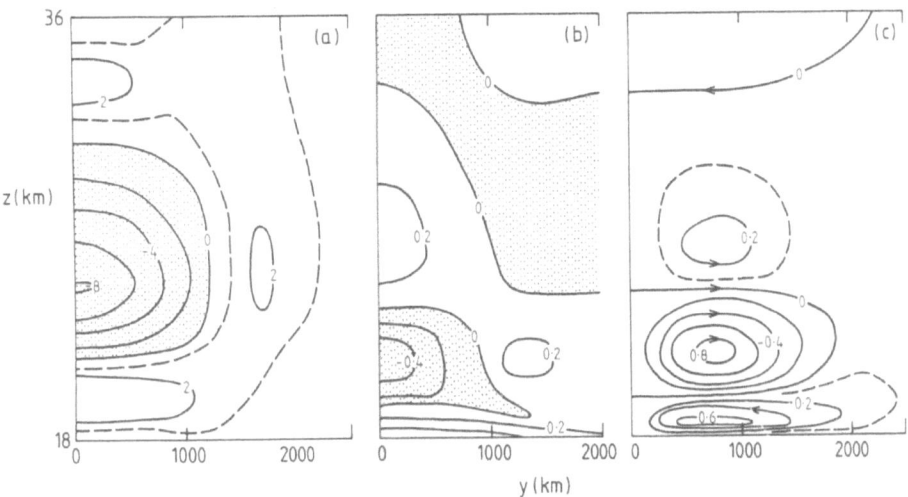

Figure 10: Height/latitude structure of the man state in the model of
            Plumb and Bell (1982b) at the time of maximum easterlies
            (1500d; cf Fig. 9). (a) Zonal wind (ms^{-1}); (b) Potential
            temperature departure from initial state (K): (c) Mass
            streamfunction of the Lagrangian-mean meridional circulation
            (m^2s^{-1}).

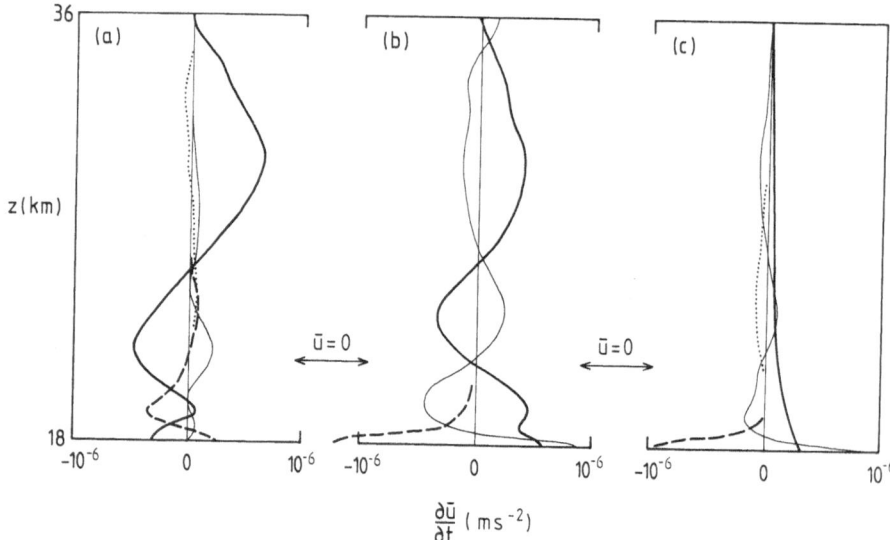

Figure 11: Contribution to the momentum budget at t = 1500d (time of maximum easterlies) in the model of Plumb and Bell (1982b) at (a) 100 km, (b) 900 km and (c) 1700 km from the equator. Heavy solid: wave-induced acceleration. Light solid: contribution from Lagrangian mean meridional circulation. Dashed/dotted: diffusion terms. (The diffusion terms are not plotted where their amplitudes do not exceed 5×10^{-8} ms^{-2}).

4. EVIDENCE SUPPORTING THE HOLTON-LINDZEN THEORY

4.1 A LABORATORY ANALOGUE

The generation of an oscillatory mean flow by internal waves was demonstrated in a laboratory experiment by Plumb and McEwan (1978). This experiment was a very close realisation of the internal wave study of Plumb (1977), described above. An annulus filled with salt-stratified water was subjected to a standing-wave forcing imposed on a flexible lower boundary. The only major difference between the laboratory system and the theoretical model is that in the experiment the upward propagating waves are damped by viscosity (internally and in sidewall boundary layers) rather than by radiation. In reality, however, the precise mechanism of wave dissipation is qualitatively unimportant. In any case it was possible to modify the theory to include viscous dissipation, and thereby to provide a basis for comparison with the experimental results.

Provided the forced wave amplitude was large enough to overcome the stabilizing effects of viscosity, a strong mean flow, which exhibited a long period oscillation (with a period of typically 1 hr, cf. about 20s period of the wave forcing), developed spontaneously in the fluid. The structure and evolution of this flow was found to be in good agreement with the theory; an example of such a comparison is shown in Figure 12. Further, the variation of wave penetration properties at different stages of the cycle (as illustrated in Figure 7) were observed in the experiment and noted to be in accord with the theory. Thus the experiment provides confirmation of the basic theoretical mechanism described in Section 3.2.

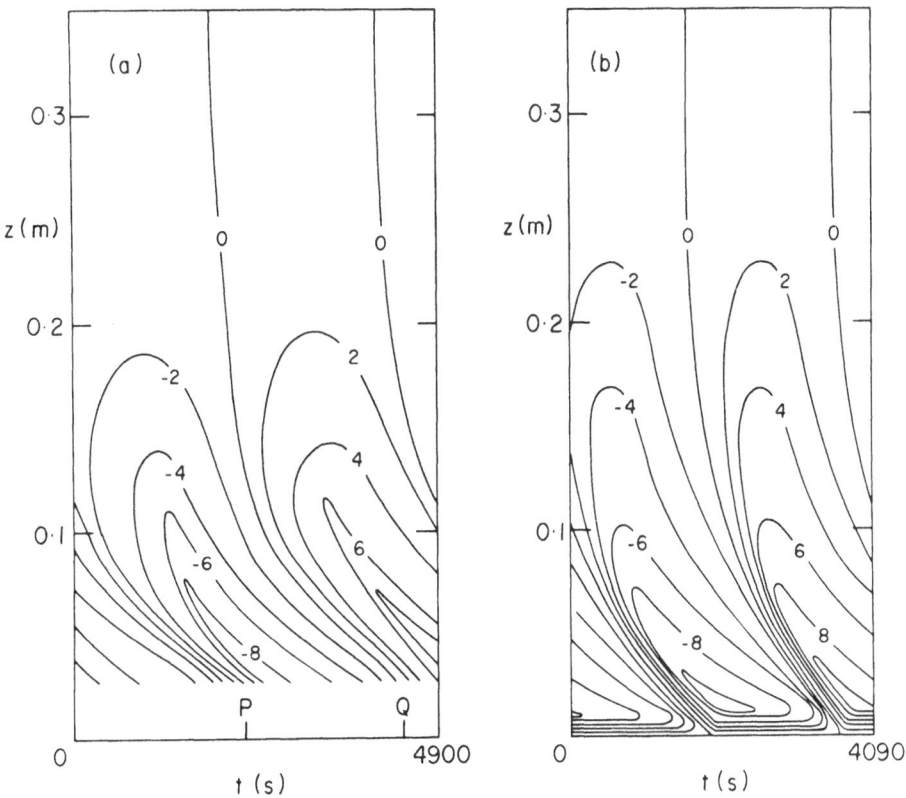

Figure 12: Comparison of (a) experimental measurements and (b) theoretical predictions for the fully developed zonal flow in an annulus of salt-stratified water with standing wave forcing applied on a flexible lower boundary. Contours of mean azimuthal velocity (mm s^{-1}) vs. height and time. (No measurements were taken in the lowest few milimeters of the fluid.) One complete cycle of the mean flow oscillation is shown in each case. (After Plumb and McEwan, 1978).

4.2 OBSERVATIONS OF WAVE DRIVING IN THE EQUATORIAL STRATOSPHERE

The importance of equatorial wave driving in the momentum budget of the equatorial stratosphere has been demonstrated by several observational studies. Even the early observations of equatoral waves provided qualitative confirmation of selective wave attenuation. In a westerly wind regime, only easterly waves are observed (Maruyama 1967, 1979) while only the westerly Kelvin waves are observed during easterly phases of the QBO (Wallace and Kousky 1968, Maruyama 1979), in agreement with the schematic depiction in Figure 7. Estimates of zonal mean acceleration resulting from observed attenuation of the Kelvin waves (Wallace and Kousky 1968, Maruyama 1979) show that these waves are of sufficiently large amplitude to account for the westerly acceleration observed in the QBO.

A more detailed quantitative analysis of observational data was undertaken by Lindzen and Tsay (1975). They analysed three months of data from an interval when a westerly wind regime in the lower stratosphere was being replaced by descending easterlies above. The monthly mean zonal wind profiles and the three-monthly mean are shown in Figure 13(a). The structure of observed wave motion in the period range 4-6 days during this interval was well fitted by the sum of three modes: a Kelvin wave (wavenumber 3, period 5 days) a Rossby-gravity wave (wavenumber 3, period 4.4 days) and also an equatorially symmetric, easterly Rossby wave of wavenumber 4 and period 5 days. Using the observations to calibrate the wave amplitudes, Lindzen and Tsay calculated the wave-induced mean flow acceleration from equations of the form of Eqs. (1), (2) and (4), using the observed $\bar{u}$ profiles and a constant Newtonian cooling rate of $(10 \text{ days})^{-1}$. Results are compared with the observed mean acceleration over the three-month period in Figure 13(b). In the altitude range 24-27 km there is strong easterly acceleration arising from the wave driving. (Lindzen and Tsay in fact found that the Rossby wave made a major contribution to this forcing). The calculated acceleration based on the three-month mean zonal flow has a single peak at about 25 km, where the wave is, according to theory, strongly absorbed by the three-month profile. An alternative calculation based on the three individual monthly mean winds gave three peaks between 23 km and 27 km, being the sum of three such single-peaked distributions. It is likely that the "correct" three-month mean acceleration, the time integral of the instantaneous accelerations in a continuously changing flow, would appear as a smoothed version of the latter profile. This would be in good agreement with the measured $\partial\bar{u}/\partial t$ above 23 km. It may therefore be deduced from Figure 13 that the driving associated with the observed equatorial waves is indeed largely responsible for the observed mean accelerations above 23 km during the period of Lindzen and Tsay's analysis.

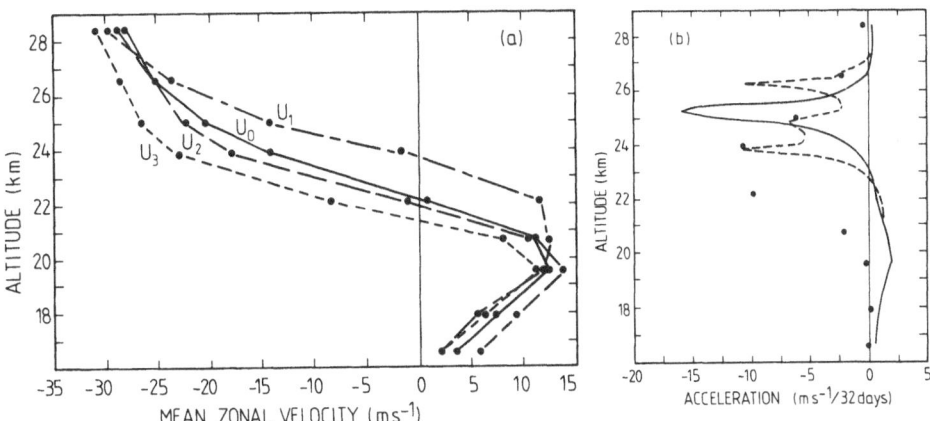

Figure 13: Mean flow and wave driving over the Marshall Islands, 1
April-1 July, (a) Mean zonal velocity. U_0 represents the
3-month mean while U_1, U_2 and U_3 are monthly means for the
three successive months. (b) Observed (dots) acceleration
of the zonal wind (ms^{-1}/32 days) over the 3-month interval.
Solid line: wave-induced acceleration based on the profile
U_0. Dashed line: wave-induced acceleration calculated as an
average of the result for U_1, U_2 and U_3. (After Lindzen
and Tsay, 1975).

This is not the case, however, below 23 km where the observed
acceleration is strongly easterly but the calculations predict a small
westerly forcing. (A similar but larger deficit was found at 6.71°N).
Lindzen and Tsay suggested that the observed acceleration could be
generated by an easterly gravity wave whose vertical wavelength would
be too small to be resolved by the observations. Andrews and McIntyre
(1976a) argued that this wave could not explain the deficit away from
the equator and suggested that long-period Rossby waves were more
likely to be responsible.

Referring back to Section 3.2 and the model of Plumb and Bell
(1982b), however, a different interpretation may be placed on the
acceleration deficit. In order that an oscillatory mean flow be
generated, the theory requires that some process in addition to wave
driving contribute to the momentum budget. In the 1-D models (Section
3.2) this takes the form of diffusion across the shear zone separating
the westerly and easterly regimes. In Plumb and Bell's (1982b) model
this also occurs, but, off the equator, angular momentum advection by
the Lagrangian mean meridional circulation performs much the same
function (cf. Fig. 11). Thus an easterly acceleration must be induced

by processes other than wave driving in the lower portion of the shear
layer - which is precisely the region where Lindzen and Tsay found the
deficit. It therefore seems possible that their results are quite
consistent with theory, even below 23 km, and that there may be no
need to invoke other, unobserved sources of easterly momentum. That
is not to say, however, that Lindzen and Tsay's results rule out the
importance of such sources; see Section 5.2, below.

5. REMAINING PROBLEMS

5.1 WAVE DISSIPATION PROCESSES

In the 1-D models, the precise mechanism of wave dissipation
appears to be unimportant - only thus could the laboratory experiment
described in Section 4.1 be regarded as an analogue of the
stratospheric QBO. Andrews and McIntyre (1976a,b) found, however, on
the basis of an asymptotic theory for weak shear, that the latitudinal
structure of the easterly acceleration induced by the mixed
Rossby-gravity wave is sensitive to the nature of the wave dissipation
(i.e. thermal vs. mechanical damping). They showed that the
acceleration peaks off the equator for a steady Rossby-gravity wave
dissipated purely by thermal damping,* but on the equator if
mechanical dissipation of comparable intensity is also present. The
latter aspect was confirmed by Holton (1979), who noted that
observations indicte that the easterly acceleration in the
stratospheric QBO does indeed appear to maximize very close to the
equator, while Plumb and Bell (1982a) confirmed Andrews and McIntyre's
prediction in the presence of realistic shear. However, Holton (1979)
found that, over 40 days of integration, the mean acceleration tended
to peak at the equator even when much of his mechanical dissipation
was suppressed; it appeared that this results from the effect of wave
transience, i.e. the non-steadiness of wave amplitude, which in fact
may become quite large in the descending shear layers of the QBO
(Dunkerton 1981a,b).

However, it is clear from generalized Lagrangian-mean theory
(Andrews and McIntyre 1978; see Plumb and Bell 1982a) that permanent
mean flow changes (i.e. those that outlive the locally temporary wave
amplification) on the equator still require mechanical damping. At
first sight, this seems difficult to explain, since the time scale for
mechanical damping of the large-scale flow much be much longer than
the radiative decay time (typically 10 days in the stratosphere) -

---

*In fact, they showed that thermally-damped Rossby-gravity waves would
tend to amplify any pre-existing cross-equatorial shear and hence
generate an equatorially-asymmetric easterly wind regime, apparently
contrary to observations. However, apart from the symmetrizing
effects of mechanical dissipation noted here, any cross-equatorial
shear would probably be destroyed by inertial instability (Dunkerton
1981c, Stevens 1983).

otherwise no QBO could ever develop. The problem could be resolved, however, if the equatorial waves break in the regions of strong shear - as indeed they may (Kousky and Koermer 1974). Then the dissipation may become in part mechanical via a cascade to small scales.

5.2 THE ROLE OF PLANETARY WAVES IN THE FORCING OF THE EASTERLIES

The possibility that planetary waves of mid-latitude origin may contribute to the momentum budget of the easterly phase of the QBO has been raised by Dunkerton (1983). He found that if the Rossby-gravity wave forcing of the mean flow is removed from the Holton-Lindzen model and replaced by a simple representation of planetary wave forcing, (the Kelvin wave forcing is retained) the model still predicts QBO - like evolution of the zonal wind. In principle, therefore, there may be no need to invoke Rossby-gravity waves as a source of easterly momentum. In practice, however, it still seems likely that the Rossby-gravity waves are important. For one thing, Lindzen and Tsay's analysis discussed above shows that these waves make a substantial contribution to the momentum budget of the easterlies; also, the changeover from westerlies to easterlies sometimes occurs around the equinoxes (see Figs. 1 and 2), when the planetary waves are usually weak. On the other hand, a contribution from planetary waves might help to explain the apparent interaction of the QBO with the annual cycle; it was noted in Section 2.1 that the easterlies are more likely to appear around Northern winter - precisely when the planetary waves are most active. Probably, we must await further observational studies (or the successful simulation of the QBO in a general circulation model) to resolve this question.

5.3 THE EXTRATROPICAL STRATOSPHERIC QBO

None of the theoretical models discussed above provide an explanation of the observed high-latitude component of the stratospheric QBO. In the beta-plane model of Plumb and Bell, the meridional circulation is not sufficiently extensive in latitude to provide the required communication; the influence of spherical geometry is unlikely to alter this result (Plumb 1982a). In fact, energy considerations argue against any such direct coupling. Since the high latitude QBO is not barotropic in structure (Angell and Korshover 1978, Holton and Tan 1980) it is subject to relatively efficient radiative dissipation. (This contrasts with the equatorial oscillation, for which the available potential energy is a very small component; see the discussion in Plumb and Bell, 1982b). It is difficult to see how the long-period equatorial QBO could survive the drain of energy which such a direct linkage would imply.

The alternative, then, is that the extratropical QBO is driven by some other source which is in some way modulated by the equatorial winds. Tucker (1979) and Holton and Tan (1982) suggested that planetary waves may perform this function and, indeed, Holton and Tan

found evidence of a QBO in the amplitude of stratospheric planetary
waves, but this statistical relationship was not sufficiently
well-defined to suggest what the causal link might be. However,
recent developments in the theory of stratospheric warmings (see
McIntyre 1982, Plumb 1982b) lead us to believe that the interaction of
planetary waves with the subtropical "zero wind" surface where the
zonal wind vanishes is an important precursor to stratospheric major
warmings in high latitudes. The location of this surface is strongly
influenced by the phase of the equatorial QBO and it therefore is
quite conceivable that planetary wave coupling could thus generate a
quasi-biennial modulation at high latitudes. This is clearly an
avenue for future study.

Progress in our understanding of such coupling should also lead
to an appreciation of the mechanisms involved in the observed QBO in
ozone and perhaps water vapour. Planetary waves and the meridional
circulation may play a role here; in the case of water vapour Hyson
(1983) suggested that it may be necessary to invoke a quasi-biennial
modulation of the troposphere – stratosphere exchange processes in the
tropics.

5.4   THE TROPOSPHERIC QBO

In Section 2.3 we discussed the evidence of tropospheric
manifestations of QBO. At this stage it is difficult to do more than
speculate on possible causes. Of these, modulation of the ultra-long
planetary waves by the equatorial stratospheric winds is again a
possibility. Theoretical studies indicate the possibility of changes
in the tropospheric planetary wave structure being produced by changes
in the wind structure of the mid-latitude lower stratosphere (Bates
1977, Geller and Alpert 1980) but the possible response to variability
of the tropical lower stratosphere is at present unknown. The
possibility of such coupling is intriguing and worthy of further
investigation. If the existence of a tropospheric QBO, even of small
amplitude, is confirmed by future studies, this should tell us much
about the ability of the atmosphere to transmit information in both
latitude and height.

5.5   GENERAL CIRCULATION MODELLING

Despite the theoretical progress of the past fifteen years and
despite the development of multilevel general circulation models of
the middle atmosphere, none of these models has yet reproduced a QBO.
Equatorial waves are generated in these models (Hayashi and Golder
1978, Hayashi et al. 1983) and their interactions with the mean state
is simulated well enough to produce a realistic semi-annual
oscillation in the GFDL "SKYHI" model (Mahlman et al. 1983).
Nevertheless the theory reviewed above suggest at least two reasons
why present models fail to produce a quasi-biennial oscillation. The
first of these is a matter of resolution. The vertical wavelength of

the forcing waves is typically 5-10 km (Table 1), although this is
reduced in the dynamically important regions where the waves'
Doppler-shifted phase speed become small. It is probably reasonable
to state that a vertical resolution of about 1 km is required in the
lower strosphere in order to resolve adequately the (possibly
breaking) equatorial waves. (Note that the Kelvin waves responsible
for the westerly phase of the semi-annual oscillation have greater
zonal phase speed and larger vertical wavelength than those
responsible for the QBO; therefore resolution requirements are less
severe for the semi-annual oscillation). A second requirement is that
dissipation must be weak enough to allow the QBO to develop. Plumb
and McEwan (1978) found theoretically and experimentally that a
laboratory "QBO" developed only if the wave forcing of the mean flow
was strong enough to overcome mean flow dissipation. Quantitatively,
this requires that the mean flow dissipation time be longer than about
one "QBO" period.

It appears that these criteria are not met in current general
circulation models. For example, the GFDL "SKYHI" model discussed in
this volume by Mahlman et al. (1983) has a vertical grid spacing of
about 2 km in the lower stratosphere, a little larger than the
criterion suggested above. Further, the model is probably
over-dissipative. For example, the horizontal diffusion is
represented by a nonlinear viscosity (see Andrews et al. 1983)

$$K = 0.0133 \ \Delta_y^2 \ D$$

where $\Delta_y \simeq 500$ km and D is the horizontal deformation. With $D \simeq \bar{u}/L$
$\simeq 30$ ms$^{-1}/10^3$ km, this implies a mean flow dissipation time $L^2/K \simeq 100$
days, much less than the QBO period (cf. a dissipation time scale of
around 600 days in Holton's (1979) model).

It therefore appears than that we require models which are less
dissipative and of better resolution in the lower stratosphere.
General circulation models have been steadily improving in both
respects and one might hope that these problems will soon be overcome.
The ability to simulate the QBO is an important prerequisite for any
satisfactory model of stratospheric climatology.

ACKNOWLEDGEMENTS

This paper was presented at the US/Japan Seminar of "Dynamics of
the Middle Atmosphere: in Honolulu, November 1982. I am grateful to
the organisers, Drs. James Holton and Taroh Matsuno, for their
invitation to attend. My participation was supported in part by a
grant from the Ian Potter Foundation, Melbourne, which is gratefully
acknowledged.

I also thank T.J. Dunkerton, J.R. Holton, J.D. Mahlman, M.E.
McIntyre and M. Takahashi for discussions on the subject matter of
this review and for comments on an earlier version of the paper.

REFERENCES

Andrews, D.G., J.D. Mahlman and R.W. Sinclair (1983): Eliassen-Palm diagnostics of wave, mean flow interaction in the GFDL "SKYHI" general circulation model. Submitted to J. Atmos. Sci.

Andrews, D.G. and McIntyre, M.E. 1976a: Planetary waves in horizontal and vertical shear: the generalised Eliassen-Palm relation and the mean zonal acceleration. J.Atmos.Sci., 33, 2031-2048.

Andrews, D.G. and McIntyre, M.E. 1976b: Planetary waves in horizontal and vertical shear: asymptotic theory for equatorial waves in weak shear. J.Atmos.Sci., 33, 2049-2053.

Andrews, D.G. and McIntyre, M.E. 1978: An exact theory of nonlinear waves on a Lagrangian-mean flow. J.Fluid Mech. 89, 609-646.

Angell, J.K. and Korshover, J. 1977: Variation in size and location of the 300 mb north circumpolar vortex between 1963 and 1975. Mon. Wea. Rev., 105, 19-25.

Angell, J.K. and Korshover, J. 1978: Estimate of global temperature variations in the 100-30 mb layer between 1958 and 1977. Mon. Wea. Rev., 106, 1422-1432.

Bates, J.R. 1977: Dynamics of stationary ultra-long waves in middle latitudes. Quart. J. R. Met. Soc., 103, 397-430.

Boyd, J.P. 1978a: The effects of latitudinal shear on equatorial waves, Part I. Theory and methods. J. Atmos. Sci., 35, 2236-2258.

Boyd, J.P. 1978b: the effects of latitudinal shear on equatorial waves. Part II. Application to the atmosphere. J. Atmos. Sci., 35, 2259-2267.

Brier, G.W. 1978: The quasi-biennial oscillation and feedback processes in the atmosphere-ocean-Earth system. Mon. Wea. Rev., 106, 938-946.

Bugayev, V.A., Kats, A.L. and Ugryumov, A.I. 1972: The two-year cycle in atmospheric circulation. In problems of the general circulation of the atmosphere, Pogosyan, Kh.P. (ed.), Leningrad, Gidrometeoizdat.

Clayton, H.H. 1885: A lately discovered meteorological cycle. Amer. Meteorol. J. 1, 130.

Coy, L. 1979: An unusually large westerly amplitude of the quasi-biennial oscillation. J. Atmos. Sci. 35, 174-176. (See also the Corrigendum, J. Atmos. Sci. 36, p. 913).

Dunkerton, T. 1981a: Wave transience in a compressible atmosphere. Part I: Transient internal wave, mean-flow interaction. J. Atmos. Sci., 38, 281-1297.

Dunkerton, T. 1981b: Wave transience in a compressible atmosphere. Part II: Transient equatorial waves in the quasi-biennial oscillation. J. Atmos. Sci., 38, 298-307.

Dunkerton, T.J. 1981c: On the inertial stability of the middle atmosphere. J. Atmos. Sci., 38, 2354-2364.

Dunkerton, T. 1982: Shear zone asymmetry in the observed and simulated quasi-biennial oscillations. J. Atmos. Sci., 38, 461-469.

Dunkerton, T.J. 1983: Laterally-propagating planetary waves in the easterly phase of the quasi-biennial oscillation. Atmosphere-Ocean, 21, 55-68.

Ebdon, R.A. 1975: The quasi-biennial oscillation and its association with tropospheric circulation patterns. Meteor. Mag., 104, 282-297.

Farkas, E. 1970: Variation of ozone and stratospheric temperatures over high latitudes of the Southern Hemisphere. New Zealand Sci.J., 13, 386-409.

Funk, J.P. and Garnham, G.J. 1962: Australian ozone observations and a suggested 24-month cycle. Tellus, 14, 378-382.

Geller, M.A. and Alpert, J.C. 1980: Planetary wave coupling between the troposphere and the middle atmosphere as a possible sun-weather mechanism. J. Atmos. Sci., 37, 1197-1215.

Hamilton, K. 1981: The vertical structure of the quasi-biennial oscillation: Observations and theory. Atmosphere-Ocean, 19, 236-250.

Hasebe, F. 1980: A global analysis of the fluctuation of total ozone II: Non-stationary annual oscillation, quasi-biennial oscillation and long-term variations in total ozone. J. Met. Soc. Japan, 58, 104-117.

Hasebe, F. 1983: The global structure of the total ozone fluctuations observed on the time scales of two to several years. (This volume).

Hayashi, Y. and Golder, D.G. 1978: The generation of equatorial transient planetary waves: control experiments with a GFDL general circulation model. J. Atmos. Sci., 35, 2068-2082.

Hayashi, Y., Golder, D.G. and J.D. Mahlman (1983): Stratospheric and mesopheric Kelvin waves simulated by the GFDL "SKYHI" general circulation model. Submitted to J. Atmos. Sci.

Holloway, J.L. and Manabe, S. 1971: Simulation of climate by a global general circulation model. I. Hydrologic cycle and heat balance. Mon. Wea. Rev., 99, 335-370.

Holton, J.R. 1972: Waves forced in the equatorial stratosphere generated by tropospheric heat sources. J. Atmos. Sci., 29, 368-375.

Holton, J.R. 1975: The dynamic meteorology of the stratosphere and mesosphere. Meteor. Monographs., 15, No. 37, 216 pp.

Holton, J.R. 1979: Equatorial wave-mean flow interaction: a numerical study of the role of latitudinal shear. J. Atmos. Sci., 36, 1030-1040.

Holton, J.R. and Lindzen, R.S. 1972: An updated theory for the quasi-biennial oscillation of the tropical stratosphere. J. Atmos. Sci., 29, 1076-1080.

Holton, J.R. and Tan, H.-C. 1980: The influence of the equatorial quasi-biennial oscillation at 50 mb. J. Atmos. Sci., 37, 2200-2208.

Holton, J.R. and Tan, H.-C. 1982: The quasi-biennial oscillation in the Northern Hemisphere lower stratosphere. J. Met. Soc. Japan, 60, 140-148.

Hyson, P. 1983: Stratospheric water vapour over Australia. Quart. J.R. Met. Soc., 109, 285-294.

Kousky, V.E. and Koermer, J.P. 1974: The nonlinear behaviour of atmospheric Kelvin waves. J. Atmos. Sci., 31, 1777-1783.

Labitzke, K. 1965: On the mutual relation between stratosphere and troposphere during period of stratospheric warmings in winter. J. App. Met., 4, 91-99.

Landsberg, H.E. 1962: Biennial pulses in the atmosphere. Beitr. Phys. Atmos., 35, 184-194.

Lindzen, R.S. 1967: Planetary waves on beta-planes. Mon. Wea. Rev., 95, 441-451.

Lindzen, R.S. 1971: Equatorial planetary waves in shear. Part I. J. Atmos. Sci., 28, 609-622.

Lindzen, R.S. and Holton, J.R. 1968: A theory of the quasi-biennial oscillation. J. Atmos. Sci., 25, 1095-1107.

Lindzen, R.S. and Tsay, C.Y. 1975: Wave structure of the tropical stratosphere over the Marshall Islands area during 1 April - 1 July 1958. J. Atmos. Sci., 32, 2008-2021.

Mahlman, J.D., and L.J. Umscheid 1983: Dynamics of the middle atmosphere -success and problems of the GFDL "SKYHI" general circulation model. (This volume).

Manabe, S. and Mahlman, J.D. 1976: Simulation of seasonal and interhemispheric variations in the stratospheric circulation. J. Atmos. Sci., 33, 2185-2217.

Maruyama, T. 1967: Large-scale disturbances in the equatorial lower stratosphere. J. Met. Soc. Japan, 45, 391-408.

Maruyama, T. 1979: Equatorial wave intensity over the Indian Ocean during the years 1968-1972. J. Met. Soc. Japan, 57, 39-52.

Matsuno, T. 1966: Quasi-geostrophic motions in the equatorial area. J. Met. Soc. Japan, 44, 25-43.

McIntyre, M.E. 1980: An introduction to the generalized Lagrangian-mean description of wave, mean-flow interaction. PAEGOPH, 118, 152-176.

McIntyre, M.E. 1982: How well do we understand stratospheric warmings? J. Met. Soc. Japan, 60, 37-65.

Newell, R.E., Kidson, J.W., Vincent, D.G. and Boer, J.G. 1974: The general circulation of the tropical atmosphere. Vol. 2. The MIT Press, 371 pp.

Nicholls, N. 1978: Air-sea interaction and the quasi-biennial oscillation. Mon. Wea. Rev., 106, 1505-1508.

Pittock, A.B. 1968: Seasonal and year-to-year ozone variations from soundings over South-Eastern Australia. Quart. J. R. Met. Soc., 94, 563-575.

Pittock, A.B. 1977: Climatology of the vertical distribution of ozone over Aspendale (38°S 145°E). Quart. J. R. Met. Soc., 103, 575-584.

Plumb, R.A. 1977: the interaction of two internal waves with the mean flow: implications for the theory of the quasi-biennial oscillation. J. Atmos. Sci., 34, 1847-1858.

Plumb, R.A. and McEwan, A.D. 1978: The instability of a forced standing wave in a viscous stratified fluid: a laboratory analogue of the quasi-biennial oscillation. J. Atmos. Sci., 35, 1827-1839.

Plumb, R.A. 1982a: Zonally symmetric Hough modes and meridional circulations in the middle atmosphere. J. Atmos. Sci., 38, 983-991.

Plumb, R.A. 1982b: The circulation of the middle atmosphere. Aust. Met. Mag., 30, 107-121.

Plumb, R.A. and Bell, R.C. 1982a: Equatorial waves in stready zonal shear flow. Quart. J. R. Met. Soc., 108, 313-334.

Plumb, R.A. and Bell, R.C. 1982b: A model of the quasi-biennial oscillation on an equatorial beta-plane. Quart. J. R. Met. Soc., 108, 335-352.

Probert-Jones, J.R. 1964: An analysis of the fluctuation in the tropical stratospheric wind. Quart. J. R. Met. Soc., 90, 15-26.

Quiroz, R.S. 1981: Period modulation of the stratospheric quasi-biennial oscillation. Mon. Wea. Rev., 109, 665-674.

Reed, R.J. 1960: The structure and dynamics of the 26-month oscillation. Paper presented at the 40th anniversary meeting of the Amer.Met.Soc. Boston.

Reed, R.J. 1962: Evidence of geostrophic motion in the equatorial stratosphere. Quart. J. R. Met. Soc., 88, 324-327.

Reed, R.J. 1964: A tentative model of the 26-month oscillation in tropical latitudes. Quart. J. R. Met. Soc., 90, 441-466.

Reed, R.J. 1966: Zonal wind behaviour in the equatorial stratosphere and lower mesosphere. J. Geophys. Res., 71, 4223-4233.

Shapiro, R. and Ward, F. 1962: A neglected cycle in sunspot numbers. J. Atmos. Sci., 19, 506-508.

Stevens, D.E. 1983: On symmetric stability and instability of zonal mean flows near the equator. J. Atmos. Sci., 40, 882-893.

Sugiura, M. 1976: Quasi-biennial geomagnetic variation caused by the sun. Geophys. Res. Lett., 3, 643-646.

Trenberth, K.E. (1979) Interannual variability of the 500 mb zonal mean flow in the Southern Hemisphere. Mon. Wea. Rev., 107, 1515-1524

Trenberth, K.E. (1980) Atmospheric quasi-biennial oscillations. Mon. Wea. Rev., 108, 1370-1377.

Tucker, G.B. 1965: The divergence of horizontal eddy flux of momentum in the lower equatorial stratosphere. Quart. J. R. Met. Soc., 91, 356-359.

Tucker, G.B. 1979: The observed zonal wind cycle in the southern hemisphere stratosphere. Quart. J. R. Met. Soc., 105, 263-273.

Van Loon, H., Zerefos, C.S., and Repapis, C.C. 1981: Evidence of the Southen Oscillation in the stratosphere. Publication No. 3., Academy of Athens, Research Centre for Atmospheric Physics and Climatology, Athens, Greece.

Veryard, R.G. and Ebdon, R.A. 1961: Fluctuations in tropical stratospheric winds. Meteor. Mag., 90, 125-143.

Wallace, J.M. and Kousky, V.E. 1968: Observational evidence of Kelvin waves in the tropical statosphere. J. Atmos. Sci., 25, 900-907.

Wallace, J.M. and Newell, R.E. 1966: Eddy fluxes and the biennial stratospheric oscillation. Quart. J. R. Met. Soc., 92, 481-489.

Yanai, M. and Maruyama, T. 1966: Stratospheric wave disturbances propagating over the equatorial Pacific. J. Met. Soc. Japan, 44, 291-294.

J. R. Holton and T. Matsuno, Dynamics of the Middle Atmosphere, 253-269.

A 2-DIMENSIONAL NUMERICAL MODEL OF
THE SEMI-ANNUAL ZONAL WIND OSCILLATION

Masaaki Takahashi

Department of Physics, Faculty of Science,
Kyushu University, Fukuoka 812, Japan

ABSTRACT

The semiannual oscillation of the mean zonal wind in the
equatorial upper stratosphere and mesosphere is investigated using a
two-dimensional numerical model. In the model, the following
mechanisms are included: Easterlies of the semiannual component just
after the solstices at the equator are caused by nonlinear advection
of the easterlies in the summer hemisphere which is forced by the
solar heating. Westerlies just after the equinoxes are caused by the
momentum deposition due to Kelvin waves. But, easterly acceleration of
the mean zonal wind due to meridionally propagating extratropical
planetary waves is excluded from the present model.
Numerical simulations show a semiannual oscillation of the mean
zonal wind whose amplitude is nearly 20 m/s at the stratopause level
over the equator. But the altitude of the oscillation is slightly
higher than that of the observed oscillation. Further, constant
westerlies blow in the equatorial lower stratosphere (z=20–40km),
because of westerly acceleration due to Kelvin waves. And there is an
upward propagation of the westerly wind at the equinox. This is
because easterly acceleration due to nonlinear advection is small in
this region. These results suggest that the observed easterly winds in
this region may be produced by extratropical planetary waves.

1. INTRODUCTION

There have been many observational studies of the semiannual
oscillation of the mean zonal wind in the equatorial upper
stratosphere and mesosphere (Quiroz and Miller, 1967; Angell and
Korshover, 1970; Belmont and Dartt, 1973; Hopkins, 1975; Hirota, 1978;
etc.), since the discovery by Reed (1965, 1966). The characteristic
features of the semiannual oscillation are as follows: The maximum
easterlies(westerlies) appear 20-30 days following the
solstices(equinoxes), and the phase of the westerly wind propagates
downward. The position of the maximum amplitude with about 25 m/s
exists near the stratopause level in the equatorial regions (cf. Reed,

1966, Fig. 1).

There have also been several theoretical studies of the semiannual zonal wind oscillation. Meyer (1970) studied this phenomenon using a two-dimensional axisymmetric numerical model which included solar heating as the only external force, to test the suggestion by Reed (1966) that the semiannual oscillation is driven by the twice-yearly passage of the sun across the equator and the strong absorption of solar radiation in the upper stratosphere and mesosphere. However, because of the oversimplified heating function, he was not able to obtain large enough easterlies (nearly -30 m/s) at the solstices over the equator. He then concluded that a semiannual varying eddy momentum source is required to drive the semiannual zonal wind oscillation.

Recently, Holton and Wehrbein (1980a) simulated a semiannual oscillation of the mean zonal wind using a two-dimensional model based on the primitive equations, in which the diabatic heating by ozone absorption of the solar radiation is explicitly calculated. In their model, they obtained a semiannual oscillation with -30 m/s 30 days following the solstices and -5 m/s 30 days following the equinoxes over the equator. However, they have failed to obtain a westerly wind at the equinox.

One of the most important problems in the phenomenon of the semiannual oscillation is to explain the generation of the westerly momemtum, just as in the quasi-biennial oscillation of the mean zonal wind in the tropical lower stratosphere. It has been already suggested that the westerly wind component may be caused by the eddy momemtum flux convergence due to Kelvin waves, whose period is shorter than that of the Kelvin waves involved in the quasi-biennial oscillation (cf. Holton, 1975).

In this connection, Hirota (1978, 1979) analyzed equatorial disturbances in the upper stratosphere and mesosphere by the use of meteorological rocket and satellite data, and identified these disturbances as Kelvin waves, noting that:

1. The vertical wavelength is 15-20 km.
2. The zonal wavenumber 1 is dominant.
3. The period is 4-9 days.
4. This wave is observed predominantly during
   the easterly phase of the semiannual oscillation.

On the basis of this work, using a one-dimensional model which includes Kelvin waves dissipated by Newtonian cooling only, Dunkerton (1979) was able to simulate a semiannual oscillation considerably well. But his model assumed _a priori_ a Rayleigh-friction-like easterly forcing which restored the zonal wind to -25 m/s during 90 days of each cycle (=180 days). Thus, the cause of the easterly phase of the semiannual oscillation is still open to question. Part of the easterly acceleration may be due to the mean meridional wind producing nonlinear advection of the easterlies in the summer hemisphere forced by the solar heating. And part of the easterly acceleration may be due to easterly acceleration by planetary waves propagating from the extratropical latitudes (cf. Hirota, 1980).

In order to simulate the semiannual oscillation of the mean zonal wind, we construct a two-dimensional numerical model which contains the following mechanisms: The generation of easterlies at the solstices over the equator is due to nonlinear advection of the easterlies in the summer hemisphere by the meridional wind forced by the solar heating (cf. Holton and Wehrbein, 1980a; Mahlman and Sinclair, 1980), while that of westerlies at the equinoxes is due to the acceleration by Kelvin waves which propagate upward from the troposphere and are dissipated by Newtonian cooling and vertical eddy viscosity in the tropical upper stratosphere (cf. Dunkerton, 1979).

As mentioned above, the generation of the easterly wind of the semiannual oscillation may be partly due to the easterly acceleration by planetary waves (Hopkins, 1975; Hirota, 1976). But the present model does not include the effect of planetary waves.

2. MODEL AND THE BASIC EQUATIONS

The present two-dimensional (zonally averaged) numerical model for semiannual oscillation of the mean zonal wind in the equatorial stratosphere includes the following processes: The easterly wind at the solstice is due to the mean meridional wind producing nonlinear advection of the easterly wind in the summer hemisphere forced by the solar heating (e.g., Holton and Wehrbein, 1980a). The westerly wind at the equinox is produced by the acceleration by Kelvin waves which propagate upward from the tropical troposphere, and are dissipated by Newtonian cooling and vertical eddy viscosity in the tropical upper stratosphere and lower mesosphere (e.g., Dunkerton, 1979). In the present paper, we include the zonal wind acceleration due to dissipated Kelvin waves alone because of two-dimensionality of the model.

For simplicity, we shall make the following assumptions:

1. The quasi-geostrophic and the quasi-hydrostatic approximations are assumed for zonal mean flow fields.

2. The terms of time derivative and nonlinear advection of temperature in the zonal mean thermodynamic equation are omitted.

The assumption 2 may be allowed for the following reasons: We are interested in the semiannual oscillation at the stratopause level in the equatorial regions. At that level, the magnitude of the Newtonian cooling coefficient is nearly 2.5×10^{-6} /s. On the other hand, the frequency of the semiannual oscillation is $2\pi/180$ /days $\fallingdotseq 4 \times 10^{-7}$ /s. Then, we may neglect the time derivative term, $\partial \overline{T}/\partial t \dot{=} 0$. Next, from observations (cf. CIRA, 1965), $\partial \overline{T}/\partial y \fallingdotseq 0$, in the regions with which we are concerned. Then, we may neglect the nonlinear advection of temperature (Meyer, 1970; Holton, 1975).

The following notations are used unless otherwise mentioned.

t : time
θ : latitude
p : pressure
z : a measure of 'height' $\equiv -H \ln(p/p_s)$
H : scale height (=7km)
p_s : a constant reference pressure

ρ : mean density
$\bar{u}$: zonal mean eastward velocity
$\bar{v}$: zonal mean northward velocity
$\bar{w}$: a measure of 'vertical velocity' $\equiv \overline{dz/dt}$
$\bar{\varphi}$: zonal mean geopotential
Ω : angular velocity of the earth
a : radius of the earth
N : buoyancy frequency $= 2 \times 10^{-2}$/s
dy : northward distance increment $= a \cdot d\theta$
ν : vertical eddy viscosity coefficient which is assumed to be dependent only on height
α_R: Rayleigh friction coefficient which is assumed to be dependent only on height
α_N: Newtonian cooling coefficient which is assumed to be dependent only on height
Q_R: cooling for the reference state
R : gas constant for dry air
C_p: specific heat of dry air at constant pressure
κ : ratio of gas constant to specific heat at constant pressure $= R/c_p$

a. Basic equations

With the above assumptions, we get the following zonally averaged equations;

$$\frac{\partial \bar{u}}{\partial t} + \bar{v}\frac{\partial \bar{u}}{\partial y} + \bar{w}\frac{\partial \bar{u}}{\partial z} - \left(2\Omega + \frac{\bar{u}}{a\cos\theta}\right)\bar{v}\sin\theta = \frac{1}{\rho}\frac{\partial}{\partial z}\left(\rho\nu\frac{\partial \bar{u}}{\partial z}\right) + F_k - \alpha_R\bar{u}, \quad (2.1)$$

$$\left(2\Omega + \frac{\bar{u}}{a\cos\theta}\right)\bar{u}\sin\theta = -\frac{\partial\bar{\varphi}}{\partial y}, \quad (2.2)$$

$$N^2\bar{w} = \frac{\kappa J}{H} - \left(Q_R + \alpha_N\frac{\partial\bar{\varphi}}{\partial z}\right), \quad (2.3)$$

$$\frac{1}{\cos\theta}\frac{\partial}{\partial y}(\bar{v}\cos\theta) + \frac{1}{\rho}\frac{\partial}{\partial z}(\rho\bar{w}) = 0, \quad (2.4)$$

where J is the diabatic heating rate per unit mass due to absorption of the solar radiation by ozone, and F_k represents the westerly acceleration due to Kelvin waves (see, following subsections). These two terms may produce the semiannual oscillation at the equatorial stratopause level.

In Fig. 1, shown is the vertical profile of ν (solid line) used in the present model. The functional form is as follows,

$$\nu = 4 \times 10^6 \times 10^{(z-110\,km)/33\,km} \quad cm^2/s . \quad (2.5)$$

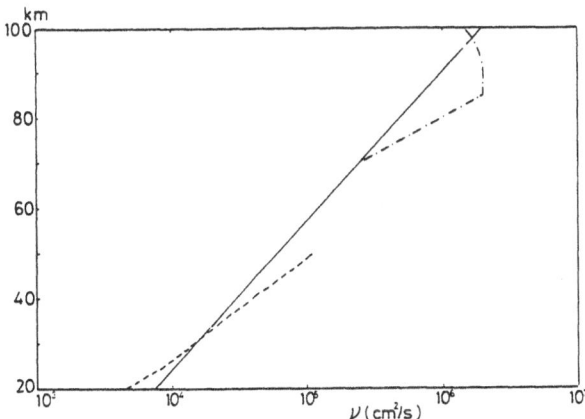

Fig. 1: Vertical profile of the vertical eddy
viscosity coefficient.

The dashed line in Fig. 1 is quoted from Massie and Hunten (1981).
Their vertical eddy viscosity coefficient was obtained so as to fit
the calculated vertical distribution of the minor constituents in the
stratosphere to the observed one. The chain line shows the vertical
eddy viscosity coefficient produced by the breaking of the diurnal
tide which was calculated by Lindzen (1981). The value of ν which we
adopt in this model agrees with, to some extent, those obtained by
above authors.

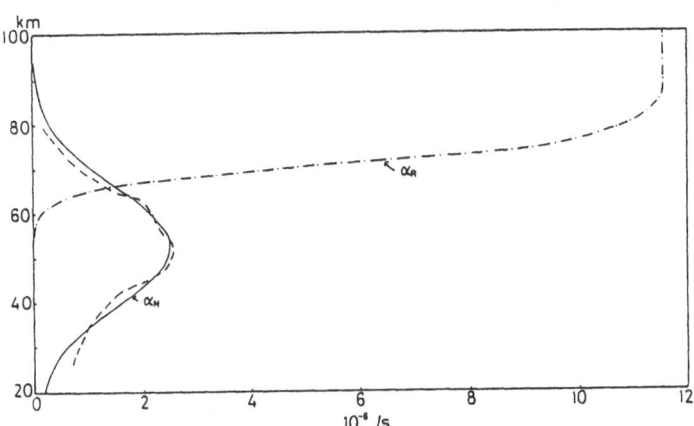

Fig. 2: Vertical profiles of Newtonian cooling
coefficient (solid line) and Rayleigh
friction coefficient (chain line).

The solid line in Fig. 2 shows the vertical profile of the
Newtonian cooling coefficient α_N in the present model whose

functional form is as follows,

$$\alpha_N = 2.5 \times 10^{-5} \cdot \exp\left(-\left((z - 52.5\,\text{km})/19\,\text{km}\right)^2\right) \quad /s. \quad (2.6)$$

The dashed line in Fig. 2 is α_N which was obtained by Dickinson (1973). The value of α_N in the present model is slightly smaller than that obtained by Dickinson (1973) at the heights of 20-30 km.

As to Rayleigh friction coefficient α_R, it has been noticed that some braking mechanism is necessary to decelerate the mean zonal wind in the mesosphere since the work of Leovy (1964). Recently, Lindzen (1981) and Matsuno (1982) show independently that the deceleration mechanism expressed by Rayleigh friction can be resulted from the breaking effects by gravity waves and/or tidal waves. But our main concern in the present paper is not the semiannual oscillation at the mesopause level (Hirota, 1978), but the oscillation at the stratopause level (cf. Dunkerton, 1982a, b). Thus, for simplicity, we do not adopt an explicit formula for acceleration by gravity waves, but use the Rayleigh friction parameterization in order to obtain a realistic mean zonal wind distribution in the mesosphere. The chain line in Fig. 2 shows the vertical profile of α_R used in the present model. The functional form ia as follows,

$$\alpha_R = 1/(2\,\text{days}) \cdot \left(1 + \tanh\left((z - 71\,\text{km})/5\,\text{km}\right)\right). \quad (2.7)$$

The value is almost 0 in the stratosphere, and the damping time in the upper mesosphere and lower thermosphere (near the top boundary) is 1 day. The value near the top boundary is twice that used in Holton and Wehrbein (1980a, b).

b. Solar heating

First, we should note that the easterly wind in the equatorial upper stratosphere and lower mesosphere at the solstice is very sensitive to the meridional distribution of the solar heating rate J. For example, in the case of $J/c_p = 8 \cdot \sin^3\theta \cdot \exp\left(-\left((z - 50\,\text{km})/17\,\text{km}\right)^2\right)$ ($^\circ$K/day), the easterly wind at 55 km over the equator at the solstice is -4 m/s, without the westerly acceleration due to Kelvin waves. On the other hand, in the case of $J/c_p = 8 \cdot \sin\theta \cdot \exp\left(-\left((z - 50\,\text{km})/17\,\text{km}\right)^2\right)$ ($^\circ$K/day), the easterly wind at the same time and place is -78 m/s. For the solar heating, we adopt the parameterization of Lacis and Hansen (1974), and the daily averaged solar heating is calculated by using approximation 1 of Cogley and Borucki (1976).

In this paper, for simplicity, we assume an ozone distribution $u(z,\theta)$ (cm, NTP) as follows,

$$u(z,\theta) = \frac{d}{1 + \exp\left((z - f)/g\right)} \quad (2.8)$$

where,

$$d = 0.26 + \frac{0.15}{0.5\pi} \cdot |\theta| \quad cm , \qquad (2.9)$$

$$f = 25 - \frac{5}{0.5\pi} \cdot |\theta| \quad km , \qquad (2.10)$$

$$g = 4.4 + \frac{0.3}{0.5\pi} \cdot |\theta| \quad km . \qquad (2.11)$$

Fig. 3 shows the latitude-height distribution of the ozone number density $n(z, \theta) = -6 \times 10^{23}/2.24 \times 10^4 \cdot \partial u/\partial z (/cm^3)$ used in the present model.

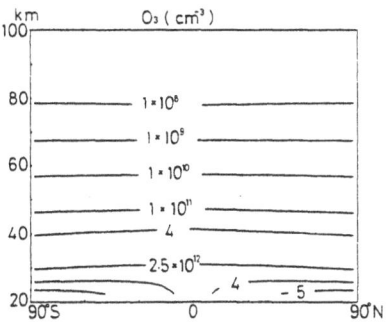

Fig. 3: Latitute-height section of the ozone number density (cm^{-3}) used in the present model

The distribution is similar to that of the observed ozone in the lower and upper stratosphere (cf. Park and London, 1974). Figs. 4a, b show the daily averaged solar heating rate J/C_p at the solstice and equinox, when the above expression is used in the ozone distribution.

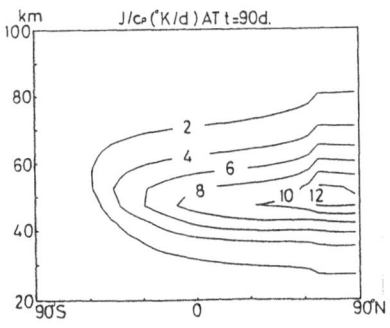
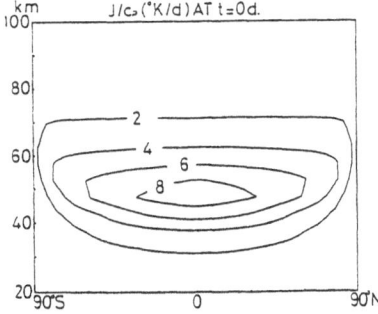

Fig. 4a: Calculated daily averaged solar heating rate $(°K/day)$ at the solstice.

Fig. 4b: Same as Fig. 4a but for equinox.

The magnitude of J/C_p at the stratopause level in the summer polar region is slightly smaller than those of previous authors (cf. Park and London, 1974). This is because the ozone amount in that region used in the present model is slightly smaller than the observed values.

c. Zonal wind acceleration due to Kelvin waves

In the present paper, the disturbance equations for Kelvin waves are not solved explicitly. Instead, we estimate the zonal wind acceleration due to Kelvin waves using a two-scaling formalism (Lindzen, 1971; Andrews and McIntyre, 1976) under the following assumptions.

  1. The effects of latitudinal shear of the basic
     mean zonal wind are small.
  2. The acceleration is due to $1/\rho \cdot \partial(\overline{\rho u'w'})/\partial z$ only.

Concerning assumption 1, it is first noted that the effect of the latitudinal shear on Kelvin waves seems to be small (cf. Boyd, 1978a, b; Holton, 1979). Next, concerning the effect of the latitudinal shear on the acceleration by Kelvin waves, Holton (1979) has shown by numerical simulation that the acceleration due to Kelvin waves tends to reduce the pre-existing shear. On the other hand, Andrews and McIntyre (1976) have pointed out that Kelvin waves with weak thermal dissipation in a weak horizontal shear flow can magnify the pre-existing shear. The difference between these two results could be due to the effects of mechanical damping and wave transience which are included in the former. In other words, it could depend on the ratio of the thermal damping effect to   mechanical damping whether the pre-existing horizontal shear would be magnified or not. In the real atmosphere, the horizontal shear could not become so large that some barotropic instability could be caused, though Andrews and McIntyre (1976) expected. Thus, we do not consider the effect of the latitudinal shear of the basic mean zonal wind on Kelvin waves. That is, we assume the normal mode structure in the meridional direction.

The assumption 2 may be allowed because v' is small and $\overline{v'\psi'_z}$ is small in the case of Kelvin waves (cf. Takahashi and Uryu, 1981; Plumb and Bell, 1982).

From the above assumptions, we can estimate the momemtum flux due to Kelvin waves which are dissipated by Newtonian cooling and vertical eddy viscosity, following Lindzen (1971) and Andrews and McIntyre (1976) as follows,

$$\rho \overline{u'w'} = \frac{B}{(c - \bar{u}_{eq}(z,t))^{1/2}} \cdot \exp\left[-\int_0^z \left(\frac{N\alpha_N}{(c - \bar{u}_{eq})^2 k}\right.\right.$$
$$\left.\left. + \frac{N^3 \nu}{(c - \bar{u}_{eq})^4 k}\right] dz\right] \cdot \exp\left[-\frac{2\Omega a}{(c - \bar{u}_{eq})} \cdot \theta^2\right] , \qquad (2.12)$$

where c is the phase velocity of the Kelvin waves, k = s/a is the zonal wave number of the Kelvin waves, and $\bar{u}_{eq}(z,t)$ is the mean zonal flow at the equator.

Here, we use

$$\beta = 0.31293 \, ,$$ (2.13)

this value corresponds to $\int_{-\infty}^{\infty} \rho \, \overline{u'w'} \cdot d\theta = 13.3 (cm^2/s^2) \cdot \rho\,(17 km)$
(cf. Dunkerton, 1979). And we use

$$C = 60 \, m/s \quad , \quad S = 1 \quad ,$$ (2.14)

in the present paper.

d. Initial and boundary conditions

   Numerical integration is initiated from a motionless state, and
the initial time of the numerical simulation t=0 is taken to be the
northern hemisphere vernal equinox.
   As the boundary conditions at the bottom and the top, we impose

$$\overline{u} = 0 \quad at \quad z = 20 \, km \quad ,$$ (2.15)

$$\frac{\partial \overline{u}}{\partial z} = 0 \quad at \quad z = 100 \, km \quad .$$ (2.16)

The meridional boundary conditions are as follows,

$$\overline{\chi} = 0 \quad at \quad \theta = \pm \frac{\pi}{2} \quad ,$$ (2.17)

where $\overline{\chi}$ is mean meridional mass stream function which is defined as
follows,

$$\frac{\partial \overline{\chi}}{\partial z} = - \rho \overline{v} \cos\theta \quad , \quad \frac{\partial \overline{\chi}}{\partial y} = \rho \overline{w} \cos\theta \quad .$$ (2.18)

e. Finite-difference scheme

   The numerical procedure is as follows. First, $\overline{\chi}$ is solved at each
altitude using the equation for the y-derivative of (2-3),

$$N^2 \frac{\partial}{\partial y} \left(\frac{1}{\rho \cos\theta} \frac{\partial \overline{\chi}}{\partial y} \right) = \frac{\partial}{\partial y} \left(\frac{\kappa J}{H} \right) + \alpha_N \frac{\partial}{\partial z} \left(2 \Omega \sin\theta \, \overline{u} + \frac{\overline{u}^2 \tan\theta}{a} \right) .$$ (2.19)

Next, using this $\overline{\chi}$ and the acceleration due to Kelvin waves, we
integrate (2-1) numerically (Euler scheme) in order to obtain $\overline{u}$ at the
next step. An implicit scheme is adopted for dissipative terms.
   The grid intervals are taken as $\Delta\theta = 5°$ and $\Delta z = 5$ km. The time
increment Δt is set at 1.5 hours. For convenience the year is taken to
be 360 days. The model run is extended for a total period of 450 days,
and the discussions of the results are for the interval 90-450 days.

3. NUMERICAL RESULTS

a. Wind distributions at the solstice and equinox

In this subsection, we discuss the zonal mean circulation at the solstice and equinox.

Fig. 5a shows the mean zonal flow at the solstice (t=270days, northern hemisphere winter solstice). The maximum westerly exists at (30°N, 55km) with 97 m/s, while the maximum easterly is -55 m/s at (20° S, 60km).

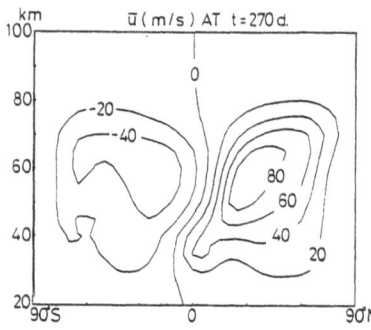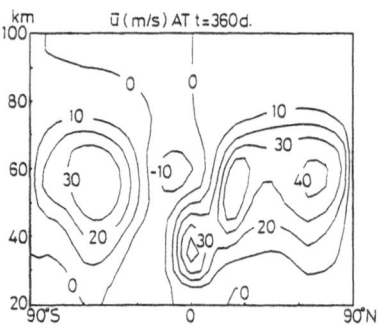

Fig. 5a: Latitude-height section of the computed mean zonal wind at the solstice

Fig. 5b: Same as Fig. 5a but for equinox

These positions may be more equatorward and at lower altitudes than those of the observations (cf. CIRA, 1965). However, according to Reed (1966), there is some evidence in which the easterly jet core is located at (13°, 42km), and hence this distribution of the mean zonal wind does not seem to be so unrealistic. The zonal wind in the upper mesosphere and lower thermosphere is smaller than that in the model of Holton and Wehrbein (1980a), because we choose a larger value of Rayleigh friction coefficient than that of Holton and Wehrbein. The easterly flow at 50 km level over the equator has a magnitude of -15 m/s. The zonal wind in the equatorial lower stratosphere (z=20-40km) is westerly (42m/s at (0°,35km)). This westerly is due to the acceleration by Kelvin waves. On the other hand, as is seen in observations (cf. CIRA, 1965), the easterly wind blows in the equatorial stratosphere, and extends to 20° in the winter hemisphere. This discrepancy may be due to the easterly acceleration near the critical surface by planetary waves which is excluded from the present model. In the upper mesosphere over the equator, weak easterly wind blows in the model, while westerly wind is seen in observations (cf. CIRA, 1965).

Fig. 5b shows the latitude-height distribution of the mean zonal flow at the equinox (t=360days, northern hemisphere vernal equinox). The westerlies have three maxima in the northern hemisphere at (0°, 35km), (25°N, 55km), and (65°N, 55km). These magnitudes are 46m/s, 55m/s, and 42m/s, respectively. There is not the easterly wind in the equatorial lower stratosphere which is seen in the observations (cf. CIRA, 1965). There is the easterly wind in low latitudes except in the equatorial lower stratosphere. On the other hand, the easterly wind

does not exist in the equatorial upper stratosphere and lower
mesosphere in the observations at the equinox (cf. CIRA, 1965).

Figs. 6a, b show the latitude-height sections of the mean
meridional wind at the solstice and equinox, respectively. The
corresponding mean vertical motions are shown in Figs. 7a, b.

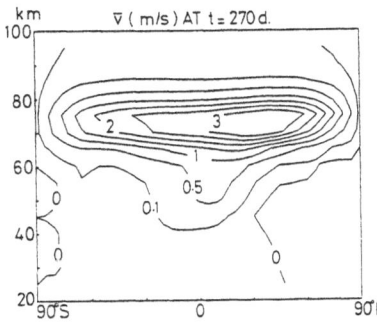

Fig. 6a: Latitude-height section of
the mean meridional wind
at the solstice

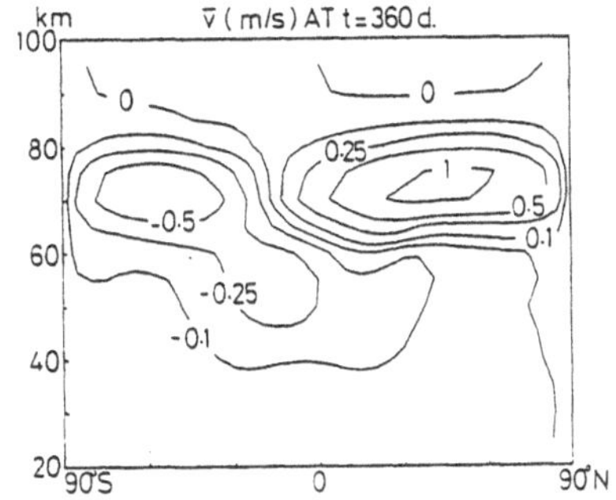

Fig. 6b: Same as Fig. 6a
but for equinox

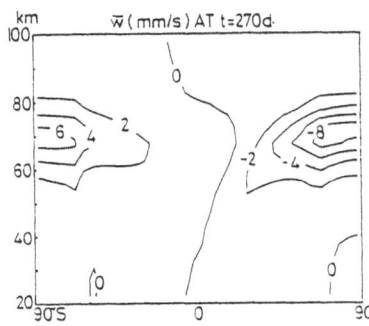
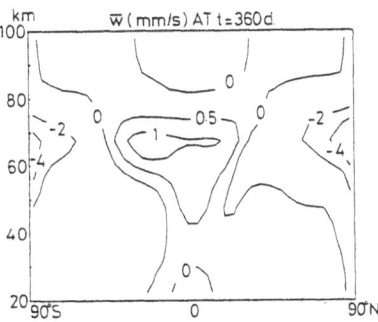

Fig. 7a: Same as Fig. 6a but
for vertical wind

Fig. 7b: Same as Fig. 7a
but for equinox

At the solstice, the mean meridional wind is directed from the summer
hemisphere to the winter hemisphere. The maximum velocity is 3.5 m/s
at (30° N, 70km). The velocity at 50 km over the equator is 53 cm/s.
The resulting easterly acceleration by nonlinear advection
at this height over the equator is -37 m/s/month. This magnitude is
sufficient for an acceleration of the semiannual oscillation = 50
m/s/3months. Thus, the easterly wind at the stratopause level over the
equator at the solstice(Fig. 5a) in the present model is certainly
produced by the nonlinear advection of the easterlies in the summer
hemisphere. The corresponding vertical motion is upward(downward) in
the summer(winter) hemisphere. The maximum upward velocity is 6.5 mm/s

at (82.5°S, 67.5km), and the maximum downward velocity is -8.9 mm/s at (87.5° N, 72.5km). At the equinox, the mean meridional motion in the mesosphere is a poleward motion. The corresponding vertical motions are upward motions in the equatorial regions and downward motions in the polar regions.

b. Semiannual oscillation

Fig. 8a shows the time-height section of the "raw" mean zonal wind at the equator, while Fig. 8b shows that with time-mean and annual cycle removed. There is a semiannual oscillation whose amplitude is 20 m/s at 50 km. The maximum wind exists 20-30 days following the solstice(equinox).

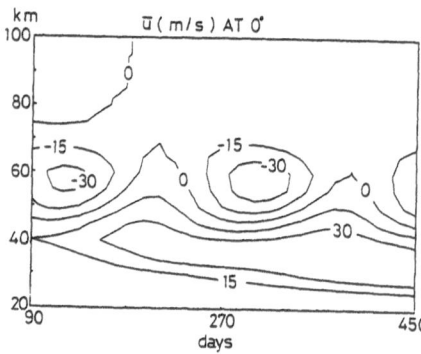

Fig. 8a: Time-height section of the "raw" mean zonal wind at the equator

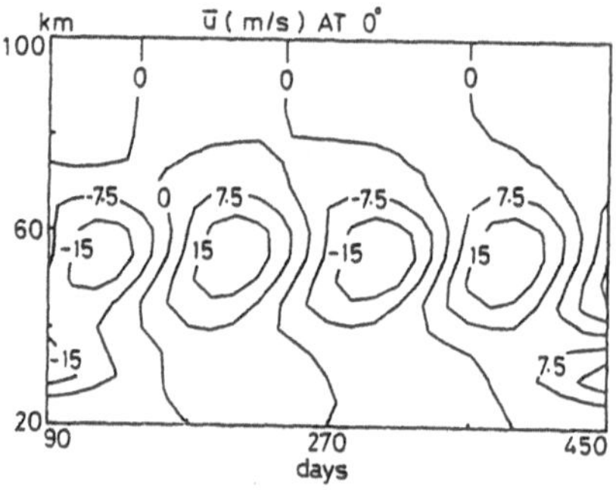

Fig. 8b: Same as Fig. 8a but for time-mean and annual cycle removed

Thus, we can explain, to some extent, the observed semiannual oscillation using this model. But, there are several features different from the observed semiannual oscillation (cf. Reed, 1966; Hirota, 1978). First, the height of the calculated semiannual oscillation is slightly higher than that of the observed oscillation which is 40-60 km. Second, the semiannual zonal wind oscillation at 45km over the equator is always westerly and that at 60km is almost easterly. On the other hand, the former is westerly at the equinox and easterly at the solstice, and the latter is mainly westerly in the real atmosphere (Reed, 1966). Another feature different from the observations is the constant westerlies in the lower stratosphere. For example, the magnitude of the westerly wind is about 35 m/s at 35km over the equator. On the other hand, as is seen in Reed (1966, Fig. 4), the constant easterly (-10 - -15 m/s) blows at 20-50 km levels in the equatorial regions. Finally, the computed westerly wind just after the equinox propagates upward. On the other hand, the westerly wind propagates downward in the observations (cf. Reed, 1966).

Fig. 9 shows time-height section of the mean zonal flow over the equator, for the case in which the westerly acceleration due to Kelvin

waves is excluded. Fig. 10 shows the vertical profile of the mean
zonal wind over the equator at the times of 300 days (near the
solstice) and 210 days (near the equinox), for the case in which the
acceleration due to Kelvin waves is excluded. As seen from these
figures, there are large easterlies near the stratopause level whose
magnitudes are -51m/s at 300 days, and -21m/s at 210 days at 55km,
respectively. But in the equatorial lower stratosphere, there are very
small easterlies. For this reason, when we include the acceleration
due to Kelvin waves in the model, the large constant westerlies are
produced in the lower stratosphere by the westerly momentum of Kelvin
waves.

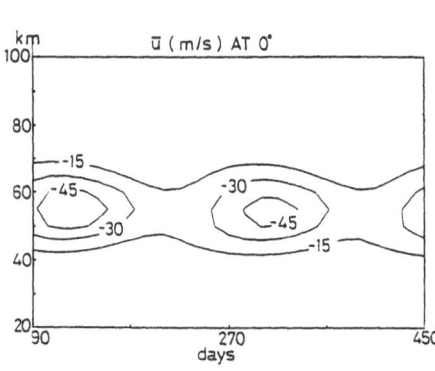

Fig. 9: Time-height section of
the mean zonal flow
at the equator for the
case in which the accel-
eration due to Kelvin
waves is excluded.

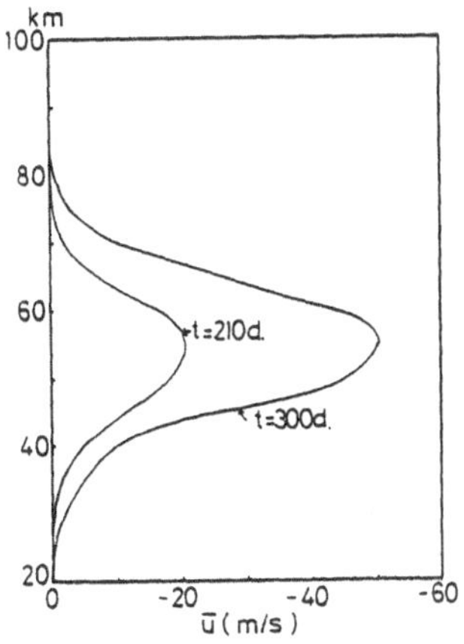

Fig. 10: Vertical profiles of
the mean zonal wind
over the equator for
the case in which the
acceleration due to
Kelvin waves is excluded.

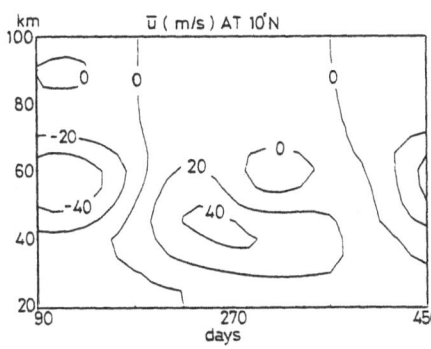

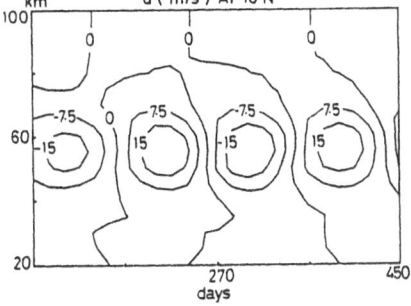

Fig. 11a: Same as Fig. 8a but
for 10°N.

Fig. 11b: Same as Fig. 8b but
for 10°N.

Figs. 11a, b show time-height section of the "raw" mean zonal wind at 10°N, and that with time-mean and annual cycle removed. There is mainly the annual component(33m/s) of the mean zonal wind, and the easterly wind at the winter solstice is only slightly seen in Fig. 11a. This result shows that the easterly winds in the summer hemisphere are not sufficiently advected into this latitude. On the other hand, the easterly wind is seen at 20°N in the real stratosphere (cf. CIRA, 1965). These results suggest that the easterly acceleration due to meridionally propagating extratropical planetary waves near the critical surface is important in obtaining a more realistic semiannual zonal wind oscillation.

4. CONCLUSIONS AND REMARKS

We have investigated the semiannual oscillation of the mean zonal wind in the equatorial upper stratosphere and mesosphere using a two-dimensional numerical model. The mechanisms of the oscillation which are included in the present model are: The easterlies just after the solstices are caused by nonlinear advection of the easterlies in the summer hemisphere which is produced by the solar heating. The westerlies just after the equinoxes are caused by vertically propagating Kelvin waves which are produced in the troposphere, and dissipated by Newtonian cooling and vertical eddy viscosity in the stratosphere and mesosphere. The westerly acceleration due to Kelvin waves has been estimated using the WKB approximation following Lindzen (1971) and Andrews and McIntyre (1976). The present model has excluded the easterly acceleration due to meridionally propagating extratropical planetary waves.

We have obtained a semiannual oscillation of the mean zonal wind whose amplitude is nearly 20 m/s at 50km over the equator using the present model. But the height of the oscillation is slightly higher than that of observations, and there is a constant westerly wind in the equatorial lower stratosphere. On the other hand, there is a constant easterly wind in the observations (cf. Reed, 1966). Also, the westerly wind at the equinox propagates upward in the present model, while the westerly wind propagates downward in the observations (cf. Reed, 1966). These results suggest that a larger easterly acceleration added to the easterly acceleration due to the solar heating is needed in the equatorial lower stratosphere.

The cause of the larger easterly acceleration, which is needed in order to explain the observed semiannual oscillation, may be an acceleration due to planetary waves meridionally propagating from the mid- and high-latitudinal troposphere to equatorial stratosphere.

Finally, there is another semiannual oscillation at the equatorial mesopause level, in which the easterly(westerly) wind of the oscillation appears just after the equinox(solstice) (Hirota, 1978). The present model cannot explain this oscillation because we do not include the explicit acceleration due to gravity waves and/or tidal waves (Lindzen, 1981; Matsuno, 1982), but include a Rayleigh friction parameterization instead. In another paper (Takahashi, 1983), the author develops a model which includes the acceleration due to gravity waves using a simplified version of Matsuno's (1982) method, instead of a Rayleigh friction parameterization. In that model, to some extent, we can obtain the semiannual oscillation at the mesopause level as well as the oscillation at the stratopause level over the equator (cf. Dunkerton, 1982a, b).

ACKNOWLEDGMENTS

The author is particularly grateful to Profs. M. Uryu and S. Miyahara, Kyushu University, for their continuing guidances and encouragements and for their many suggestions and stimulating discussions throughout this work. The author would also like to thank Drs. J. R. Holton and J. D. Mahlman for their helpful comments on the manuscript. He is also grateful to a organiser of this seminar, Prof. T. Matsuno, for his invitation to attend.
This work was financially supported in part by MAP of Japan. The numerical computations were made by the use of FACOM M-200 computer at Kyushu University.

REFERENCES

Andrews, D.G. and M.E. McIntyre, 1976: Planetary waves in horizontal and vertical shear: Asymptotic theory for equatorial waves in weak shear. J. Atmos. Sci., 33, 2049-2053.

Angell, J.K. and J. Korshover, 1970: Quasi-biennial, annual, and semiannual zonal wind and temperature harmonic amplitudes and phases in the stratosphere and low mesosphere of the northern hemisphere. J. Geophys. Res., 75, 543-550.

Belmont, A.D. and D.G. Dartt, 1973: Semiannual variation in zonal wind from 20 to 65 kilometer at $80^\circ - 10^\circ$. J. Geophys. Res., 78, 6373-6376.

Boyd, J.P., 1978a: The effects of latitudinal shear on equatorial waves. Part I : Theory and methods. J. Atmos. Sci., 35, 2236-2258.

-----, 1978b: The effects of latitudinal shear on equatorial waves. Part II : Applications to the atmosphere. J. Atmos. Sci., 35, 2259-2267.

CIRA, 1965: COSPAR International Reference Atmosphere, North-Holland, Amsterdam, 313pp.

Cogley, A.C. and W.J. Borucki, 1976: Exponential approximation for daily averaged solar heating or photolysis. J. Atmos. Sci., 33, 1347-1356.

Dickinson, R.E., 1973: Method of parameterization for infrared cooling between altitudes of 30 and 70 kms. J. Geophys. Res., 78, 4451-4457.

Dunkerton, T., 1979: On the role of the Kelvin wave in the westerly phase of the semiannual zonal wind oscillation.

268

J. Atmos. Sci., 36, 32–41.

———, 1982a: Stochastic parameterization of the gravity
wave stresses. J. Atmos. Sci., 39, 1711–1725.

———, 1982b: Theory of the Mesopause Semiannual Oscillation.
J. Atmos. Sci., 39, 2681–2690.

Hirota, I., 1976: Seasonal variation of the planetary waves
in the stratosphere observed by the Nimbus 5 SCR.
Quart. J. Meteor. Soc., 102, 757–770.

———, 1978: Equatorial waves in the upper stratosphere and
mesosphere in relation to the semiannual oscillation of
the zonal wind. J. Atmos. Sci., 35, 714–722.

———, 1979: Kelvin waves in the equatorial middle atmosphere
observed by Nimbus 5 SCR. J. Atmos. Sci., 36, 217–222.

———, 1980: Observational evidence of the semiannual oscillation
in the tropical middle atmosphere - A review.
Pure Appl. Geophys., 118, 217–238.

Holton, J.R., 1975: The dynamic meteorology of the stratosphere
and mesosphere. Academic Press, New York, 319pp.

———, 1979: Equatorial wave-mean flow interaction:
A numerical study of the role of latitudinal shear.
J. Atmos. Sci., 36, 1030–1040.

——— and W.M. Wehrbein, 1980a: A numerical model of the zonal
mean circulation of the middle atmosphere.
Pure Appl. Geophys., 118, 284–306.

———, 1980b: The role of forced planetary waves in the
annual cycle of the zonal mean circulation of the
middle atmosphere. J. Atmos. Sci., 32, 1968–1983.

Hopkins, R.H., 1975: Evidence of polar-tropical coupling
in upper stratospheric zonal wind anomalies.
J. Atmos. Sci., 32, 712–719.

Lacis, A.A. and J.E. Hansen, 1974: A parameterization
for the absorption of solar radiation in the earth's
atmosphere. J. Atmos. Sci., 31, 118–133.

Leovy, C., 1964: Simple models of the thermally driven
mesospheric circulation. J. Atmos. Sci., 21, 327–341.

Lindzen, R.S., 1971: Equatorial planetary waves in shear:
Part I. J. Atmos. Sci., 28, 609–622.

———, 1981: Turbulence and stress owing to gravity wave
and tidal breakdown. J. Geophys. Res., 86, 9707–9714.

Mahlman, J.D. and R.W. Sinclair, 1980: Recent results from
the GFDL troposphere-stratosphere-mesosphere general
circulation model. Collection of Extended Abstracts
Presented at ICMUA Sessions and IUGG Symposium 18
XVII IUGG General Asembly, 11–18pp.

Massie, S.T. and D.M. Hunten, 1981: Stratospheric eddy
diffusion coefficients from tracer data.
J. Geophys. Res., 86, 9859–9868.

Matsuno, T., 1982: A quasi one-dimensional model of the
middle atmosphere circulation interacting with
internal gravity waves. J. Meteor. Soc. Japan,
60, 215–226.

Meyer, W.D., 1970: A diagnostic numerical study of the
    semiannual variation of the zonal wind in the
    tropical stratosphere and mesosphere.
    J. Atmos. Sci., 27, 820-830.
Park, J.H. and J. London, 1974: Ozone photochemistry and
    radiative heating of the middle atmosphere.
    J. Atmos. Sci., 31. 1898-1916.
Plumb, R.A. and R.C. Bell, 1982: A model of the quasi-biennial
    oscillation on an equatorial beta-plane.
    Quart. J. R. Meteor. Soc., 108, 335-352.
Quiroz, R.S. and A.J. Miller, 1967: Note on the semi-annual
    wind variation in the equatorial stratosphere.
    Mon. Wea. Rev., 95, 635-641.
Reed, R.J., 1965: The quasi-biennial oscillation of the
    atmosphere between 30 and 50 km over Ascension
    Island. J. Atmos. Sci., 22, 331-333.
-----, 1966: Zonal wind behavoir in the equatorial
    stratosphere and lower mesosphere.
    J. Geophys. Res., 71, 4223-4233.
Takahashi, M., 1983: A numerical model of the semiannual
    oscillation. J. Meteor. Soc. Japan (to be submitted).
--- and M. Uryu, 1981: The Lagrangian-mean motions
    forced by steady, dissipating equatorial waves:
    Part I. J. Meteor. Soc. Japan, 59, 781-800.

J. R. Holton and T. Matsuno, Dynamics of the Middle Atmosphere, 271-287.

A NUMERICAL SIMULATION OF THE ZONAL MEAN CIRCULATION OF THE MIDDLE ATMOSPHERE INCLUDING EFFECTS OF SOLAR DIURNAL TIDAL WAVES AND INTERNAL GRAVITY WAVES; SOLSTICE CONDITION

Saburo Miyahara*

Department of Physics, Faculty of Science,
Kyushu University, Fukuoka 812, Japan

ABSTRACT

The general circulation of the middle atmosphere thermally driven by solar insolation for a solstice condition is discussed. Mechanical damping is provided by the parameterization of the drag force due to internal gravity waves proposed by Matsuno(1982) and the acceleration due to dissipating diurnal tidal waves.
The calculated mean zonal winds and temperatures are quite similar to those observed. The present results show that dissipating internal gravity waves and solar diurnal tidal waves contribute greatly to the dynamics of the general circulation of the mesosphere and the lower thermosphere.

1. INTRODUCTION

The general circulation of the middle atmosphere is mainly driven by the absorption of solar insolation in the ozone layer. Leovy(1964) studied the general circulation of the zonal mean middle atmosphere, using a linearized two dimensional model. He found that realistic mean zonal wind and zonal mean temperature distributions around the mesopause (weak mean zonal wind, the highest temperature at the winter pole and the lowest temperature at the summer pole, hereafter referred to as inverse temperature gradient, e.g. CIRA, 1972) are obtained by including a Rayleigh friction and a Newtonian cooling with damping rates of the order of $10^{-6}s^{-1}$. Recently, Schoeberl and Strobel(1978) and Holton and Wehrbein(1980) studied the general circulation of the middle atmosphere, and they obtained realistic mean zonal wind and zonal mean temperature distributions by assuming a strong Rayleigh friction (relaxation time 0.5 day^{-1}) around the mesopause.
It has been speculated that the physical origin of the required large Rayleigh friction in the mesosphere may be the breakdown of internal gravity waves and the solar diurnal tide due to exponential

* Currently at Geophysical Fluid Dynamics Program, Princeton University, Princeton, New Jersey 08540, U.S.A.

271

growth of their amplitudes with height (Houghton,1978, Holton and Wehrbein,1980). However, no explicit discussion of this hypothesis was carried out.

Recently Lindzen(1981) and Matsuno(1982) revealed that the physical origin of the Rayleigh friction can be attributed to the momentum transport by upward propagating internal gravity waves. In their models the mechanism is treated explicitly as a problem of wave-mean flow interaction by making use of the WKB method. In the former model the momentum deposited in the mean field is attributed to the breakdown of the waves, on the other hand in the latter model it is attributed to the attenuation of the waves by a prescribed vertical eddy diffusion.

The weak mean zonal winds and the inverse temperature gradient around the mesopause are well simulated by introducing the above mentioned effect of internal gravity waves into a quasi one-dimensional model of the middle atmosphere circulation (Matsuno, 1982). A similar result was also obtained by Holton(1982) using a β -channel model of the middle atmosphere circulation including the effect of internal gravity waves as formulated by Lindzen(1981).

The mean zonal wind distribution in the lower thermosphere (at about 100km height) seems to be considerably different from that in lower layers. An easterly wind exists in low latitude region all through the year and a westerly wind exists in middle latitude region in both summer and winter hemispheres (CIRA,1972, Groves, 1969, 1972, Iljichev and Portnyagin,1977). It seems to be difficult to produce this symmetric zonal wind system by means of the thermal forcing due to the absorption of the solar insolation. Indeed, this symmetric zonal wind system in the lower thermosphere is not reproduced by thermal forcing in numerical models of the thermosphere circulation (e.g. Dickinson et al.,1977, Schoeberl and Strobel,1978).

The present author has shown that the symmetric zonal wind system in the lower thermosphere may be caused by the momentum transport by solar diurnal tides, and that the dissipating solar diurnal tides produce a mean zonal wind system in the lower thermosphere which is similar to that obserbed (Miyahara,1981, 1983, hereafter referred to as M1, M2).

In the present study, taking into account the effect of internal gravity waves formulated by Matsuno(1982) and the effect of dissipating solar diurnal tidal waves, the zonal mean circulation of the middle atmosphere and the thermosphere is discussed using the two dimensional numerical model used in M2.

2. THE DYNAMICAL MODEL

a) Basic equations

The model is based on the zonally averaged primitive equation system in the log-pressure coordinate, and the basic equations are written as follows:

the eastward momentum equation

$$\frac{\partial \bar{u}}{\partial t} + \frac{1}{a \sin \theta} \frac{\partial \overline{u \bar{v}} \sin \theta}{\partial \theta} + \frac{1}{P} \frac{\partial P \overline{u w}}{\partial z} + 2\Omega \cos \theta \, \bar{v} = \frac{g}{P} \frac{\partial}{\partial z} \left(\frac{\mu}{H_0} \frac{\partial \bar{u}}{\partial z} \right)$$
$$+ \frac{1}{P} \frac{\partial}{\partial z} \left(\frac{P K_E}{H_0^2} \frac{\partial \bar{u}}{\partial z} \right) - D_{22} \bar{u} + \bar{F}_u + \bar{F}_{IGW} , \qquad (2.1)$$

the southward momentum equation

$$\frac{\partial \bar{v}}{\partial t} - 2\Omega \cos \theta \, \bar{u} = -\frac{1}{a} \frac{\partial \bar{\Phi}}{\partial \theta} + \frac{g}{P} \frac{\partial}{\partial z} \left(\frac{\mu}{H_0} \frac{\partial \bar{v}}{\partial z} \right) + \frac{1}{P} \frac{\partial}{\partial z} \left(\frac{P K_E}{H_0^2} \frac{\partial \bar{v}}{\partial z} \right)$$
$$- D_{\theta\theta} \bar{v} + \bar{F}_v , \qquad (2.2)$$

the hydrostatic equation

$$\frac{\partial \bar{\Phi}}{\partial z} = \frac{R}{m} \bar{T} , \qquad (2.3)$$

the thermodynamic equation

$$\frac{\partial \bar{T}}{\partial t} + \frac{1}{a \sin \theta} \frac{\partial \overline{T \bar{v}} \sin \theta}{\partial \theta} + \frac{1}{P} \frac{\partial P \overline{w T}}{\partial z} + N^2 \bar{w} = \frac{g}{P} \frac{\partial}{\partial z} \left(\frac{K}{C_P H_0} \frac{\partial \bar{T}}{\partial z} \right)$$
$$+ \frac{1}{P} \frac{\partial}{\partial z} \left[\frac{P K_E}{H_0^2} \left(\frac{\partial \bar{T}}{\partial z} + \kappa \bar{T} \right) \right] - \alpha_T \bar{T} + \bar{F}_T + \frac{J}{C_P} , \qquad (2.4)$$

the continuity equation

$$\frac{1}{a \sin \theta} \frac{\partial \bar{v} \sin \theta}{\partial \theta} + \frac{1}{P} \frac{\partial P \bar{w}}{\partial z} = 0 \qquad (2.5)$$

where $\bar{F}_{IGW}$ denotes the divergence of the zonal momentum flux due to internal gravity waves, and $\bar{F}_u$, $\bar{F}_v$ and $\bar{F}_T$ are divergence of eastward, southward momentum flux and thermal flux due to solar diurnal tides, respectively. J is zonal mean heating due to the absorption of solar insolation, K_E is coefficient of eddy diffusion, and α_T is coefficient of Newtonian cooling. The profiles of K_E and α_T are shown in Fig.2 of M2. We have neglected nonlinear terms in (2.2) since the mean zonal wind may be almost in geostrophic balance. The other notations are same as used in M2.

b) Zonal momentun flux due to internal gravity waves

In the present model it is assumed that gravity waves are attenuated by the prescribed vertical eddy diffusion. By making use of the WKB approximation, the zonal momentum flux associated with internal gravity waves is given as follows (Plumb and McEwan,1978, Matsuno,1982);

$$F(z) = \sum_i F_i(z) = \sum_i \overline{P u_i' w_i'} = \sum_i F_{i0}(0) \exp \left[-\int_0^z \frac{2 K_E N^3 k_i^3}{\hat{\omega}_i^4} dz' \right], \qquad (2.6)$$

where the subscript i denotes each gravity wave mode, and $F_{i0}(0)$ is the momentum flux of a component of internal gravity waves at the bottom of the model atmosphere, $\hat{\omega}_i$ and k_i are the Doppler-shifted frequency and horizontal wave number of the internal gravity wave, respectively. The dissipation rate is inversely proportional to the fourth power of the Doppler-shifted frequency.

In the actual situation not only the mean zonal wind but also large scale wave motions, for instance tidal waves and planetary waves may cause the Doppler-shift (Walterscheid,1981). However, it is a very complicated problem to take into account the Doppler-shift due to large scale wave motions in the present model, so that we consider the Doppler-shift only due to the mean zonal wind. In this case, the waves which propagate opposite in direction to the background mean zonal wind can propagate into upper levels and then be dissipated by the prescribed eddy diffusion and decelerate the background mean zonal wind at that level. On the other hand, the waves which propagate same direction as the mean zonal wind are dissipated in the lower layers by eddy diffusion.

The spectrum of internal gravity waves in the atmosphere may greatly depend on the distributions of the source (Lindzen,1981). However, we have no concrete information about the source of internal gravity waves in the atmosphere, so we adopt a very simplified ad hoc spectrum in the present model. Two waves, with zonal phase velocities $c = \pm 16 \, ms^{-1}$ and horizontal wavelength 200km, are considered. The magnitude of the zonal momentum flux is assumed to be

$$F_{\pm 0}(0) = \pm 2.4 \times 10^{-1} \, cm^2 s^{-2} \times 1013 \, mb \, . \tag{2.7}$$

These values are the same order as those adopted by Matsuno(1982). We also assume that $F_{\pm 0}(0)$ are independent of latitude and time.

Then, the drag force due to internal gravity waves is given as

$$\overline{F}_{IGW}(z) = -\frac{1}{P} \frac{\partial}{\partial z} \left[F_+(z) + F_-(z) \right] \, . \tag{2.8}$$

c) Mean heating due to the absorption of solar insolation

The mean heating rate at a solstice condition is calculated. For the mean heating due to the absorption by O_3, we use the model for

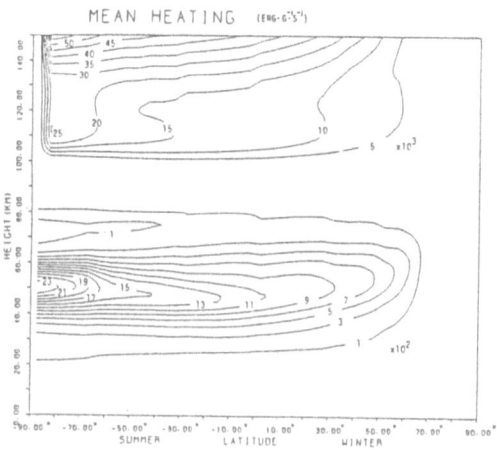

Fig.1 Meridional cross section of mean heating rate.

ozone concentration at the solstice given by Groves(1982), and for evaluating the heating rate we use the parameterization proposed by Lacis and Hansen(1974). We calculate the mean heating rate at the solstice by UV and EUV in the thermosphere following Forbes and Garrett(1976). The height-latitude cross section of mean heating rate used in the present model is shown in Fig.1.

The model atmosphere and the integration scheme are the same as used in M2. We neglect the effect of the mean wind system in the troposphere, so we assume $\overline{U}=\overline{V}=\overline{T}=0$ at the bottom (earth's surface) of the model atmosphere.

In the present study, we only consider the steady state solution for a solstice condition obtained by time integration. However, if we used a time dependent heating rate, the present model would be capable of time dependent simulation of the mean fields.

3. RESULTS AND DISCUSSIONS

In the present study, numerical integrations were conducted for the following two cases:
A) The acceleration effect due to tidal waves is not taken into account.
B) The acceleration effect due to tidal waves is taken into account.

In both cases time integrations are conducted until the steady state is obtained.

Case A)
The profile of the calculated mean zonal wind is shown in Fig.2. The calculated zonal winds in the stratosphere and the mesosphere are westerly in the winter hemisphere and easterly in the summer

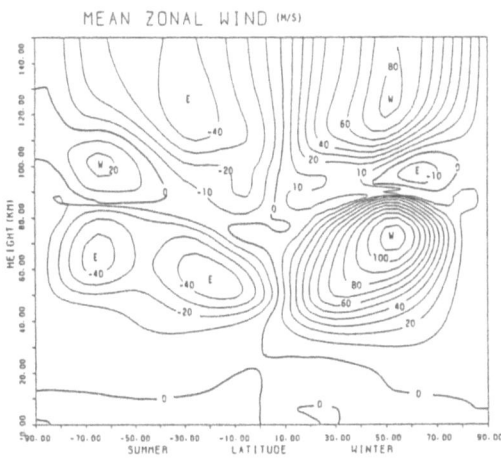

Fig.2 Meridional cross section of the calculated mean zonal wind.

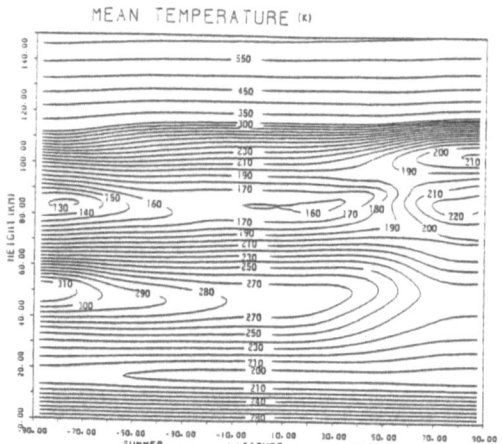

Fig.3 Meridional cross section of the calculated zonal mean
   temperature.

hemisphere. The calculated westerly maximum is 110ms^{-1} at 55° of
latitude and 70km in height, and the easterly maxima are about –40ms^{-1}
at 20° of latitude and 60km in height and 65° of latitude and 65km in
height. These zonal winds decrease rapidly with height and become
almost zero at about 70km height in low latitude region and at about
85 to 90km at high latitudes. Zonal winds with opposite direction
appear above these zero levels. Another zonal wind system is produced
by UV and EUV heating in the thermosphere. These zonal wind systems
calculated in the present study are qualitatively coincident with the
observations (e.g. CIRA,1972).
     The profile of the calculated mean temperature is shown in Fig.3.
The winter to summer temperature gradient around the stratopause and
the inverse temperature gradient around the mesopause are
qualitatively simulated in the present model. However, the temperature
differences between the summer and the winter poles in both regions
are somewhat larger than obsered (CIRA,1972). The cause of the large
temperature difference around the stratopause may be due to the large
heating rate around the summer pole used in the present model. The
cause around the mesopause is closely related to the unrealistically
strong vertical shear of the induced mean zonal wind around the
mesopause (especially in the winter hemisphere, $Ri < 1/4$ at 40° to 50°N)
calculated in the present model. This tendency is also seen in
Matsuno's(1982) result. This strong shear zone is not seen in
Holton's(1982) result in which the internal gravity wave-braking
parameterization proposed by Lindzen (1981) is used.
     The profile of calculated mean meridional wind is shown in Fig.4.
Meridional winds directed from the summer hemisphere to the winter one

the mesopause and in the thermosphere. The meridional winds around the mesopause have maxima at about 50° to 60° of latitude and the magnitude is 10ms⁻¹ in the summer hemisphere and it is 20ms⁻¹ in the winter hemisphere. This meridionl wind system is larger in magnitude and the position of the maximum wind is different from that calculated by numercial models using a Rayleigh friction parameterization (e.g. Holton and Wehrbein,1980). In those models the meridional wind has its maximum at about 30° of latitude in the winter hemisphere just as the meridional winds of the thermosphere calculated in the present model. This difference is due to the effect of internal gravity waves as shown later. The obseved result for the meridional winds around the mesopause, which is similar to the present result, is reported by Groves (1969).

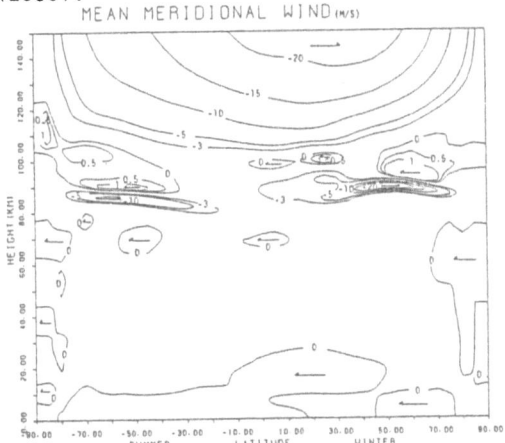

Fig.4 Meridional cross section of the calculated mean meridional wind.

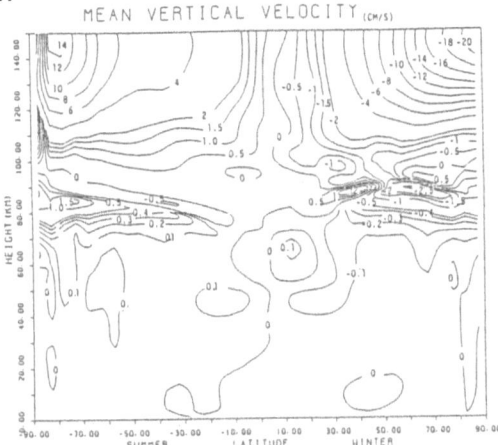

Fig.5 Meridional cross section of the calculated mean vertical motion.

278

Fig. 5 shows the profile of the calculated mean vertical motion. The magnitudes of the upward and downward motions in the high latitude regions around the mesopause are 1.5cms^{-1} and 2.5cms^{-1} , respectively. These values are also much stronger than those calculated by two dimensional numerical models using a Rayleigh friction parameterization (e.g. Holton and Wehrbein,1980).

Fig. 6 shows the meridional circulation. We can see two thermally direct circulations driven by O$_3$ heating in the middle atmosphere and UV and EUV heating in the thermosphere.

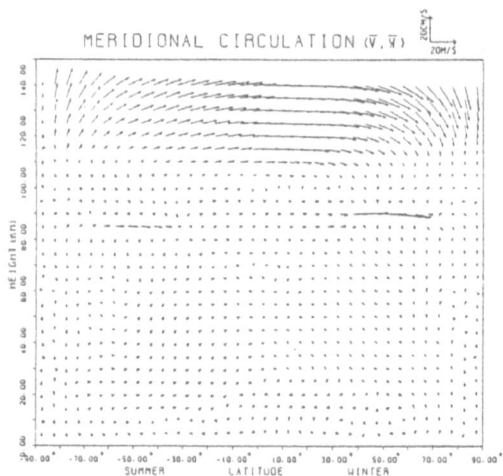

Fig.6 Distribution of the calculated meridional circulatons.

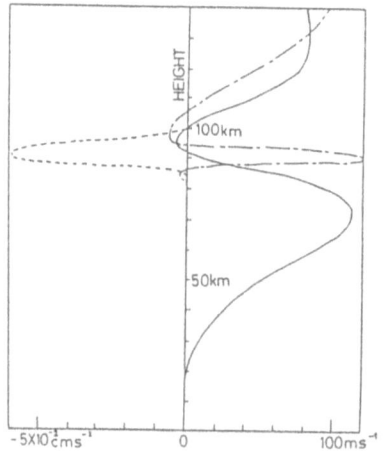

Fig.7 Vertical distributions of $\bar{u}$(——),$\bar{v}$ x 5(—·—) and $\overline{v_{16}w}$ (----) at 55° of latitude in the winter hemisphere.

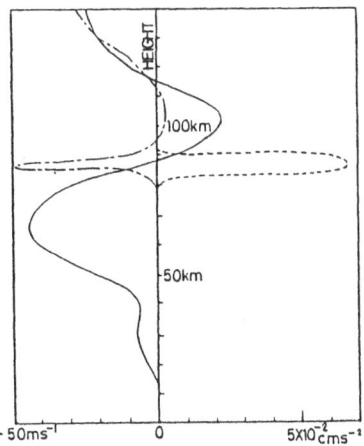

Fig.8 As in Fig.7 except for $\bar{u}$(——), $\bar{v}$ x 5(—·—) and
$\bar{F}_{IGW}$ (-----) at 65° of latitude in the summer hemisphere.

Figs.7 and 8 show vertical distributions of $\bar{u}$, $\bar{v}$ and $\bar{F}_{IGW}$ at 55°
of latitude in the winter hemisphere and at 65° of latitude in the
summer hemisphere, respectively. Around the mesopause there appears a
strong $\bar{F}_{IGW}$ which has the opposite direction to the thermally induced
mean zonal winds in the lower layers. Due to this strong forcing, the
thermally induced mean zonal wind is strongly decelerated around the
mesopause and an inverse direction wind is generated above that level.
At the level of strong $\bar{F}_{IGW}$ there appears a strong meridional wind
which balances with $\bar{F}_{IGW}$ through the Coriolis force.

As mentioned earlier the vertical shear of the induced mean zonal
wind around the mesopause is unrealistically strong, and it may be
unstable. This tendency is also seen in Matsuno's(1982) result. This
strong shear zone is not seen in Holton's(1982) result in which the
parameterization of the forcing due to internal gravity waves proposed
by Lindzen(1981) is used. On the other hand, although the present
result and Matsuno's(1982) result simulate the reverse mean zonal wind
above the mesopause, it is not simulated by Holton's(1982) results.

Fig.9 shows that internal gravity waves play a similar role in
the low latitude region. However, the thermally induced mean zonal
wind is weak in this region, so that the Doppler-shift effect is small
and the peak of $\bar{F}_{IGW}$ appears at lower heihgt than the case of high
latitude regions.

In the present model a weak westerly is produced at about 75 km
height. This westerly is induced by the forcing due to the eastward
propagating internal gravity wave. This result shows that the
semi-annual mean zonal wind oscillation at about 80km height reported
by Hirota(1978) may be produced by internal gravity waves proposed by
Dunkerton(1982). However, the westerly is much smaller than observed.
Stronger internal gravity waves than those used in the present model

or other westerly sources, for instance Kelvin waves, may be required.

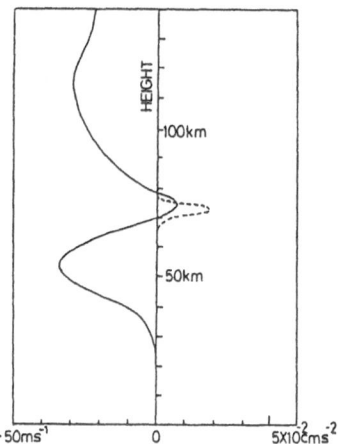

Fig.9 Vertical distribution of $\bar{u}$(——) and $\bar{F}_{low}$ (----) at 5° of latitude in the summer hemisphere.

Case B)

As shown in M1 and M2, solar diurnal tidal waves contribute greatly to the configuration of the mean winds in the upper mesosphere and the lower thermosphere. In these studies, however, the mean winds generated by solar diurnal tidal waves alone were discussed. In the present study the effects of solar diurnal tidal waves are taken into account to the above mentioned numerical model of the middle atmosphere circulation.

The effect of the mean winds on the solar diurnal tidal waves is neglected as in case of M2. As we neglect the effect of mean winds, we may be able to use the same forcing terms due to the dissipating solar diurnal tidal waves calculated in M2.

The profile of the calculated mean zonal wind is shown in Fig. 10. The easterlies in the low latitude region and the westerlies in the middle latitude region in the upper mesosphere and the lower thermosphere are stronger than those of case A in which tidal effect is not included. We may say qualitatively that the present result is a linear combination of the previous results of M2 and case A.

The calculated mean zonal wind distribution is quite similar to that of CIRA,1972 (Fig.11) except in the troposphere and the low latitude region of the stratosphere and the mesosphere. The difference in the troposphere is a natural result of the simplicity of the present model. The easterly winds in the low latitude region of the lower stratosphere of the CIRA,1972 model show the easterly phase of the quasi-biennial oscillation which is induced by equatorial waves (Holton and Lindzen,1972). The effects of equatorial waves are not included in the present model, so that these easterlies are not reproduced in the present model. The westerlies in the low latitude

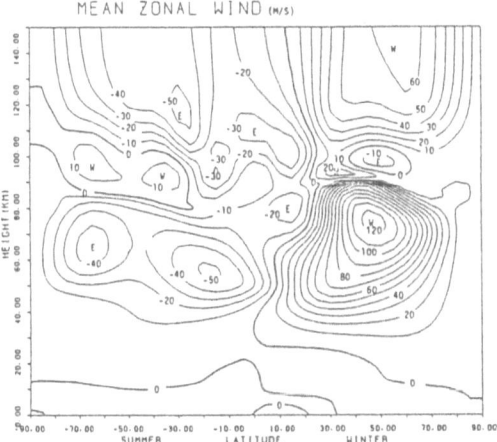

Fig.10 Meridional cross section of the calculated mean zonal winds.

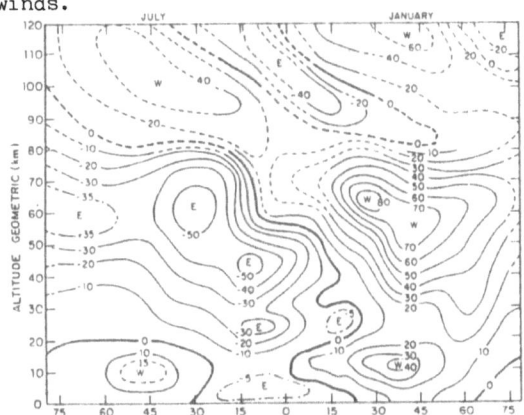

Fig.11 Meridional cross section of mean January and July zonal winds (CIRA,1972).

region of the mesosphere of the CIRA,1972 model may show the westerly phase of the semiannual oscillation around the mesopause reported by Hirota(1978). The westerly phase of the semiannual oscillation can be reproduced by the present model, if we use a stronger $\overline{F}_{low}$ in the low latitude region than that used in the present calculation (Dunkerton,1982) or include appropriate Kelvin waves.

On the other hand, the easterlies in the low latitude region calculated by the present model suggest that the dissipating solar diurnal tide is an important candidate for the source of the easterly phase of the semiannual oscillation at 80km height (Dunkerton,1982).

The present result also suggests that the easterlies observed in the low latitude region of the lower thermosphere (CIRA,1972, Groves,1969,1972, Iljichev and Portnyagin,1977) are produced by dissipating solar diurnal tidal waves.

Fig.12 shows the profile of calculated mean temperature. As shown in M2, the effect of the solar diurnal tide on the mean temperature field is small, so the temperature distribution is similar to that of case A (Fig.3).

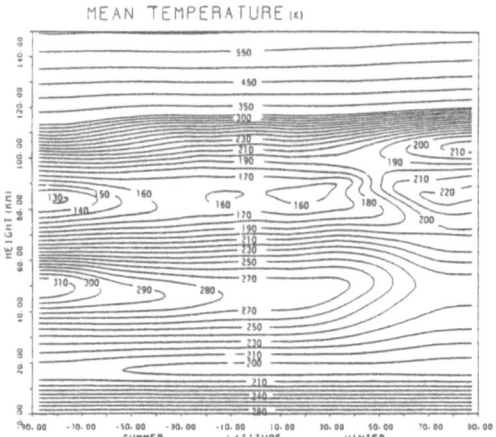

Fig.12 Meridional cross section of the calculated zonal mean temperature.

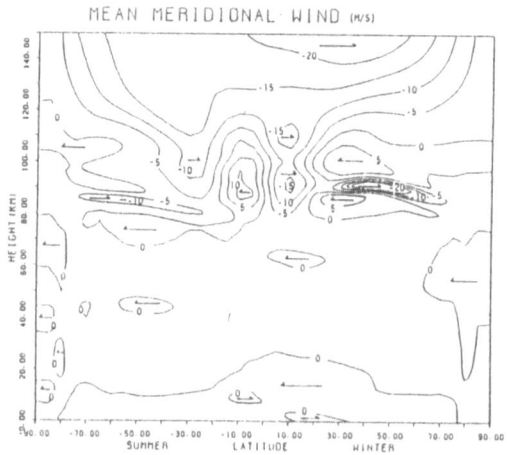

Fig.13a Meridional cross section of the calculated mean meridional wind.

Figs.13a, 13b and 13c show the profiles of the calculated mean meridional wind, mean vertical wind, and mean meridional circulation, respectively. We may say that qualitatively the present result is a linear combination of the results of case A and M2.

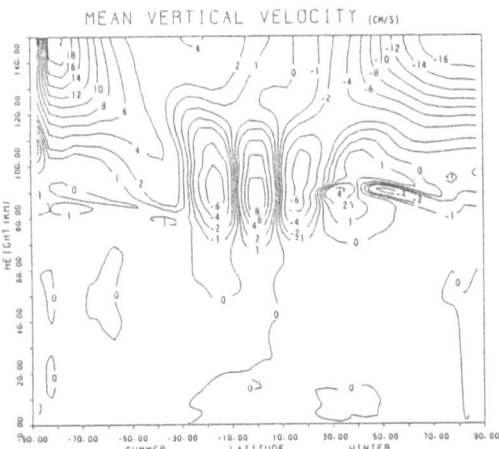

Fig.13b Meridional cross section of the calculated mean vertical
        motion.

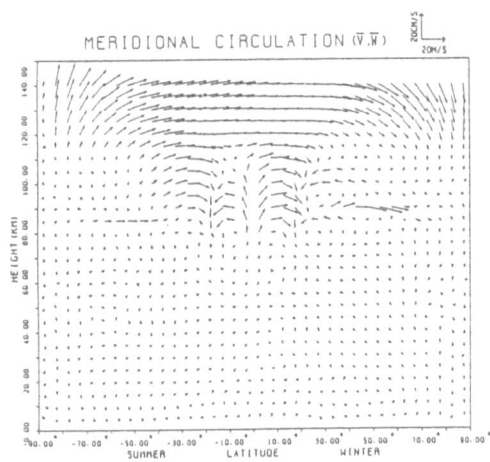

Fig.13c Distribution of the calculated meridional circulation.

Fig.14 shows the profile of the calculated residual mean
meridional circulation (Andrews and McIntyre,1976, 1978a). The
residual mean vertical velocity has a tendency to balance the diabatic
heating (Dunkerton,1978), so that the calculated residual mean
circulation is quite similar to the Eulerian mean circulation
calculated in case A in which tidal effects are not included.

Fig.15 shows the calculated Lagrangian mean meridional
circulation (Andrews and McIntyre,1978b). In the present calculation,
the Stokes drift is calculated up to the second order of the tidal

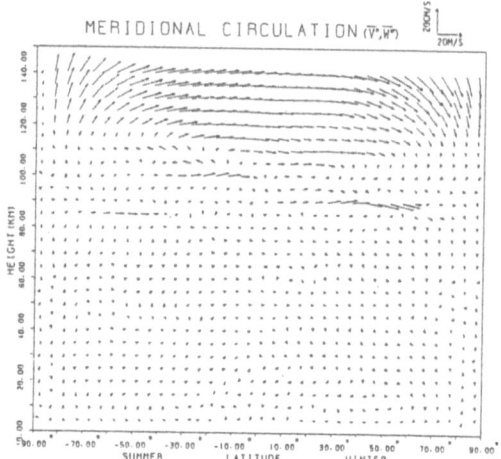

Fig.14 Meridional cross section of the calculated residual mean
meridional circulation.

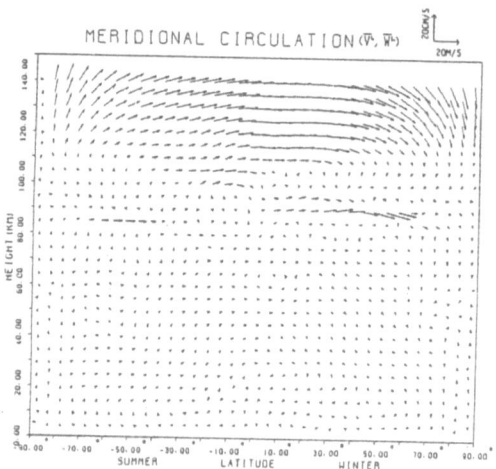

Fig.15 Meridional cross section of the calculated Lagrangian mean
meridional circulation.

wave amplitude, and the effect of the mean wind shear is not
included. The Eulerian mean meridional circulation induced by a wave
has a tendency to balance the Stokes drift, so that the calculated
Lagrangian mean circulation is also similar to the Eulerian
circulation of case A.

The difference between the Eulerian mean circulation (Fig.13c)
and the Lagrangian mean circulation (Fig.15) shows that the observed
results of the meridional circulation in the upper mesosphere and the
lower thermosphere may greatly depend on the method of observation as
in the case of stratospheric and mesospheric circulations.

4. CONCLUDING REMARKS

By introducing the parameterization of the drag force due to internal gravity waves which was proposed by Matsuno(1982) and the acceleration due to dissipating diurnal tidal waves into a two dimensional numerical model of the middle atmosphere circulation, we computed the general circulation of the middle atmosphere.

Although the model is simplified, the calculated results are quite similar to the observed results. It is shown that dissipating internal gravity waves and solar diurnal tidal waves play important roles in the dynamics of the general circulation of the mesosphere and the lower thermosphere.

However, the model used has the following limitations:

1. We assume an ad hoc profile of eddy diffusivity. However, we must keep in mind that the breakdown of internal gravity waves and tidal waves itself is one of the important causes of the turbulence (Lindzen,1981).

2. The effect of large scale waves, for instance tidal waves and planetary waves, on the propagation of internal gravity waves is neglected. The large amplitude waves may greatly affect the propagation of internal gravity waves (Walterscheid,1981).

3. The effect of mean zonal winds on the propagation of tidal waves is neglected. The effect on solar diurnal tidal waves has not yet been discussed.

4. The tidal wave is treated as a steady wave, while the observed data shows day-to-day variations in the lower thermosphere (e.g. Aso and Kato,1980). Few discussions have been conducted (Aso and Kato,1980, Walterscheid,1981), and the cause of day-to-day variations is still obscure.

5. The effects of planetary waves and equatorial waves are neglected.

These problems must be discussed to improve the present model.

ACKNOWLEDGEMENTS

The author wishes to express his thanks to Prof. M. Uryu for his discussions and criticisms. The author also thanks Prof. J. R. Holton and Dr. M. R. Schoeberl for help in clarifying the manuscript. Thanks are extended to Miss Motomura for typing. This work was financially supported in part by MAP of Japan. The computations were performed by the use of the FACOM M-200 computer at the Computer Center of Kyushu University.

REFERENCES

Andrews, D.G., and M.E. McIntyre,1976: Planetary waves in horizontal and vertical shear; The generalized Eliassen-Palm relation and mean zonal acceleration. J. Atmos. Sci.,33,2031-2048.
Andrews, D.G., and M.E. McIntyre,1978a: Generalized Eliassen-Palm and

Charney-Drazin theorem for waves on axisymmetric mean flows in compressible atmosphere. J. Atmos. Sci.,35,175-185.

Andrews, D.G., and M.E. McIntyre,1978b: An exact theory of nonlinear waves on a Lagrangian-mean flow. J. Fluid Mech.,89,609-646.

Aso, T., and S. Kato, 1980: Simulation of atmospheric tides. J. Meteorol. Soc. Japan, 58,286-291.

CIRA,1972: COSPAR International Reference Atmosphere (CIRA). Akademie-Verlag, Berlin.

Dickinson, R.E., et al., 1977: Meridional circulation in the thermosphere,II, Solstice conditions. J. Atmos. Sci.,34, 178-192.

Dunkerton, T., 1978: On the mean meridional motions of the stratosphere and mesosphere. J. Atmos. Sci.,35, 2325-2333.

Dunkerton, T., 1982: Stochastic parameterization of gravity wave stress. J. Atmos. Sci.,39,2325-2333.

Forbes, J.M., and H.B. Garrett, 1976: Solar diurnal tide in the thermosphere. J. Atmos. Sci.,33,2226-2241.

Groves, G.V., 1969: Wind models from 60 to 130km altitude for different months and latitudes. J. Brit. Interplan. Soc. 22,285-307.

Groves, G.V., 1972: Annual and semiannual zonal wind components and corresponding temperature and density variations, 60-130km. Planet. Space Sci.,20,2099-2112.

Groves, G.V., 1982: Hough components of ozone heating. J. Atmos. Terr. Phys.,44,111-121.

Hirota, I., 1978: Equatorial waves in the upper stratosphere and mesosphere in relation to the semi-annual oscillation. J. Atmos. Sci.,35,714-722.

Holton, J.R., 1982: The role of gravity wave induced drag and diffusion in the momentum budget of the mesosphere. J. Atmos. Sci.,39,791-799.

Holton, J.R. and R.S. Lindzen, 1972: An update theory for the quasi-biennial cycle of the tropical stratosphere. J. Atmos.Sci., 29,1076-1080.

Holton, J.R. and W.M. Wehrbein, 1980: A numerical model of the zonal mean circulation of the middle atmosphere. Pure. Appl. Geophy., 118,284-306.

Houghton, J.T., 1978: The stratosphere and mesosphere. Quart. J. Roy. Meteorol. Soc.,104,1-28.

Iljichev, Yu.D., and Yu.I. Portnyagin,1977: Schemes of the groval height-latitude cross sections of the wind field up to 100km. IAMAP,CMUA collection of extended summaries of contributions presented at CMUA sessions, IAGA/IAMAP Joint Assembly, 1977, Seatle,p33-1-p-33-5.

Lacis, A.A., and J.E. Hansen, 1974: A parameterization for the absorption of solar radiation in the earth's atmosphere. J. Atmos. Sci.,31,118-133.

Leovy, C.B., 1964: Simple models of thermally driven mesospheric circulation. J. Atmos. Sci.,25,327-341.

Lindzen, R.S., 1981: Turbulence and stress due to gravity wave and tidal breakdown. J.G.R.,86,9707-9714.

Matsuno, T. 1982: A quasi one-dimensional model of the middle
atmosphere circulation interacting with internal gravity waves.
J. Meteorol. Soc. Japan,60,215-226.

Miyahara, S., 1981: Zonal mean winds induced by solar diurnal tides
in the lower thermosphere. J. Meteorol. Soc. Japan,59,303-319.

Miyahara, S., 1983: Zonal mean winds induced by solar diurnal tides
in the lower thermosphere for a solstice condition. In this issue.

Plumb, R.A., and A.D. McEwan, 1978: The instability of a forced
steady wave in a viscous stratified fluid: A laboratory analogue
of the quasi-biennial oscillation. J. Atmos. Sci.,35,1827-1839.

Schoeberl, M.R., and D.F. Strobel, 1978: The zonally averaged
circulation of the middle atmosphere. J. Atmos. Sci.,35,577-591.

Walterscheid, R.L., 1981: Inertio-gravity wave induced acceleration
of mean flow having an imposed periodic component: Implications
for tidal observations in the meteor region. J.G.R.,86,9698-9706.

J. R. Holton and T. Matsuno, Dynamics of the Middle Atmosphere, 289-306.
Copyright © 1984 by Terra Scientific Publishing Company.

AN OVERVIEW OF WAVE-MEAN FLOW INTERACTIONS DURING
THE WINTER OF 1978-79 DERIVED FROM LIMS OBSERVATIONS

John C. Gille and Lawrence V. Lyjak

National Center for Atmospheric Research
Boulder, Colorado 80307 USA

ABSTRACT

Gradient winds, Eliassen-Palm (EP) fluxes and flux divergences, and
the squared refractive index for planetary waves have been calculated
from mapped data from the Limb Infrared Monitor of the Stratosphere
(LIMS) experiment on Nimbus 7. The changes in the zonal mean atmospheric state, from early winter through 3 disturbances, is described.
Convergence or divergence of the EP fluxes clearly produces changes in
the zonal mean wind. The steering of the waves by the refractive index
structure is not as clear on a daily basis.

1. INTRODUCTION

Recent work has resulted in a considerable clarification in understanding of the interactions between planetary waves and the mean zonal
flow in the stratosphere. Matsuno (1970) described the propagation of
stationary planetary waves through the atmosphere in spherical coordinates,
and defined a quantity which is analogous to the square of the refractive
index for electromagnetic waves. Karoly and Hoskins (1982), to cite
only one of the most recent papers, have shown how the refractive index,
which depends primarily on the structure of the zonal wind, guides
planetary wave propagation. Andrews and McIntyre (1976, 1978), and
Edmon et al., (1980) introduced the Eliassen-Palm (EP) fluxes to describe
the transport of wave activity, and the processes by which waves act to
change the speed of the zonal flow, and drive motions in the meridional
plane.

These mechanisms of course are very relevant to disturbances in the
winter stratosphere, since a deceleration of the polar westerlies forces
a meridional circulation with descent in the polar regions. The accompanying adiabatic heating and temperature rise restore geostrophic
balance. The temperature increase, which can be quite large, is usually
referred to as a stratospheric warming; in a major warming the westerlies
are reversed to easterlies.

With these ideas in mind, this paper makes an initial exploratory application of the data from the Limb Infrared Monitor of the Stratosphere (LIMS) experiment, which flew on Nimbus 7, to the disturbed Northern Hemisphere winter of 1978-79. The LIMS data provide almost complete coverage of the Northern Hemisphere for this period, with higher vertical resolution than other satellite instruments, as well as good accuracy and precision from 100-.1 mb. We do not propose to present a detailed quantitative study at this time, but rather to present an overview of the progression of states characterizing the winter, and the evolution of the three large disturbances which took place.

In the following sections we shall briefly describe the LIMS data, provide an overview of the course of the winter, then discuss the dynamical events in the stratosphere in four periods covering the pre-disturbance period, and three disturbances in late January, early February, and late February. The paper concludes with a discussion of the balances between terms in the transformed zonal mean heat and momentum equations, and a summary of conclusions.

2. THE LIMS DATA

The LIMS instrument scans vertically across the Earth's limb as it measures radiation emitted by the atmosphere. The measurements are made in 6 spectral channels between 6-16 μm, from which the vertical distribution of temperature, ozone, water vapor, nitric acid and nitrogen dioxide are inferred. The experiment has been described by Gille and Russell (1983), Gille et al. (1980) and Russell and Gille (1978).

The temperature data have been validated through error simulation studies, calculations of along orbit variability, and comparisons with radiosondes and rocketsondes (Gille et al., 1983b). The temperatures used in this study are from the archived version; they are accurate to 1-2K below 1 mb, and have a precision of $\sim$ 0.4K. The accuracy may be worse above 1 mb.

The temperatures were interpolated to a set of pressure levels (100, 70, 50, 30, 16, 10, 7, 5, 3, 2, 1.5, 1, 0.7, 0.5, 0.4, 0.2, and 0.1 mb), and the profiles integrated to obtain the thicknesses between them. The temperatures and the thicknesses around latitude circles spaced 4° apart were then analyzed using the Kalman filter technique described by Rodgers (1977), Kohri (1981) and Gille et al., (in preparation). This provides optimum estimates of the zonal mean value and the sine and cosine coefficients of the first 6 waves around the latitude circle. To obtain the heights, the thicknesses were built up from the 50 mb FGGE height field. These data have been evaluated by Leovy et al. (1983), who found that they are in good agreement with conventional data where the latter are available.

From the height fields, the gradient zonal wind $\bar{u}$, and geostrophic eddy zonal and meridional winds, u' and v' were computed, and used to calculate the eddy fluxes of momentum and potential temperature, $\overline{u'v'}$ and $\overline{v'\theta'}$. From these, the components of the EP flux vector $\vec{F}$ were calculated according to

$$\left\{ F_\phi, \ F_z \right\} = \rho_s \ e^{-z/H} \ a \ \cos\phi \ \left\{ - \overline{u'v'}, \ f/\bar{\theta}_z \ \overline{v'\theta'} \right\}$$

where ϕ is latitude, F_ϕ, F_z are the latitudinal and vertical components of $\vec{F}$, ρ_s is a standard density at the level above which the vertical coordinate z measures altitude, H is a constant 7 km scale height, a is the radius of the earth, f is the Coriolis parameter, θ_z is the vertical derivative of potential temperature, and overbar indicates a zonal mean.

Then as noted elsewhere (e.g. by Palmer, 1981a, b; Kanzawa, 1983) the transformed momentum and heat equations may be written

$$\frac{\partial\bar{u}}{\partial t} = f\bar{v}^* + \frac{1}{\rho_s e^{-z/H} \ a \ \cos\phi} \ \nabla \cdot \vec{F} = f\bar{v}^* + D_F \tag{1}$$

$$\frac{\partial\bar{\theta}}{\partial t} = - \bar{\theta}_z \ \bar{w}^* + \bar{Q} \tag{2}$$

where $\bar{Q}$ is the zonal mean radiative heating rate term, and $\bar{v}^*$, $\bar{w}^*$ are the residual mean meridional and vertical circulation velocities.

The square of the refractive index used here is defined by (Palmer, 1982)

$$Q_n = \left(\frac{a^2 \ \bar{q}_y}{\bar{u}} - \frac{f^2 a^2}{4H^2 N^2} - \frac{n^2}{\cos^2\phi} \right) \frac{1}{\sin^2\phi} \quad ,$$

where $\bar{q}_y = \frac{2\Omega \ \cos\phi}{a} - \frac{\partial}{\partial y} \left(\frac{1}{\cos\phi} \frac{\partial}{\partial y} \ \bar{u} \ \cos\phi \right) - \frac{f^2}{\rho} \frac{\partial}{\partial z} \left(\frac{\rho}{N^2} \frac{\partial\bar{u}}{\partial z} \right)$

is the latitudinal derivative of the quasi-geostrophic potential vorticity, n is the longitudinal wave number, N^2 is the square of the Brunt-Vaisala frequency, taken to be constant $4 \cdot 10^{-4} \ sec^{-2}$, and Ω is the rotation rate of the earth.

Smith (1983a) has discussed the significance of Q_n. Clearly, for propagation we must have $Q_n > 0$, and waves will be bent in the direction of increasing Q_n. The group velocity will be small in regions where Q_n is large, and also where $\bar{u}$ is small; at these locations the waves will be more strongly affected by dissipation.

3. SYNOPSIS OF THE WINTER

The zonal mean temperature at 10 mb, 80°N is shown in Fig. 1 as a function of time for the LIMS mission. At this location the temperature cools monotonically through the autumn and early winter, reaching a minimum of 198K in early January. There follow three major temperature pulses, in late January, early February and at the end of February. The final one reaches 246K, after which the temperature agains cools to ∿ 220K before continuing a seasonal increase. The three pulses are mirrored by decreases at 0.05 mb, 80°N and 10 mb, 20°N, and weakly mimicked by increases at 0.05 mb, 20°N (Gille et al., 1983a), showing the global nature of the disturbances and indicating meridional circulation systems like those suggested by Matsuno and Nakamura (1979).

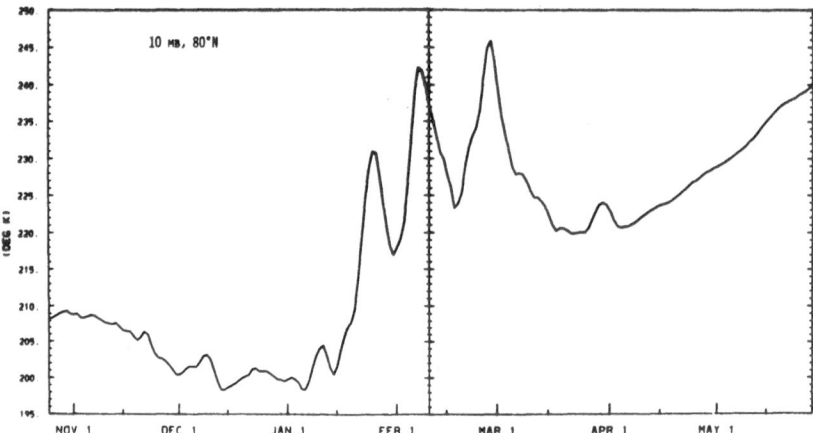

Fig. 1   Variation of zonal mean temperature at 10 mb, 80°N, from October 1978 through May 1979.

Time-altitude sections of the amplitudes of waves 1 and 2 in the height field at 60°N are shown in Fig. 2. Wave 1 shows three peaks of large amplitude ($\geq$ 2000 m), with roughly a 13 day periodicity during January, and peak altitudes becoming successively lower. The pulsating amplitude during the latter part of January appears to arise from alternating constructive and destructive interference of stationary and traveling wave 1 (Madden and Labitzke, 1981). It is interesting to note the lower altitudes of the wave peaks with each successive maximum.

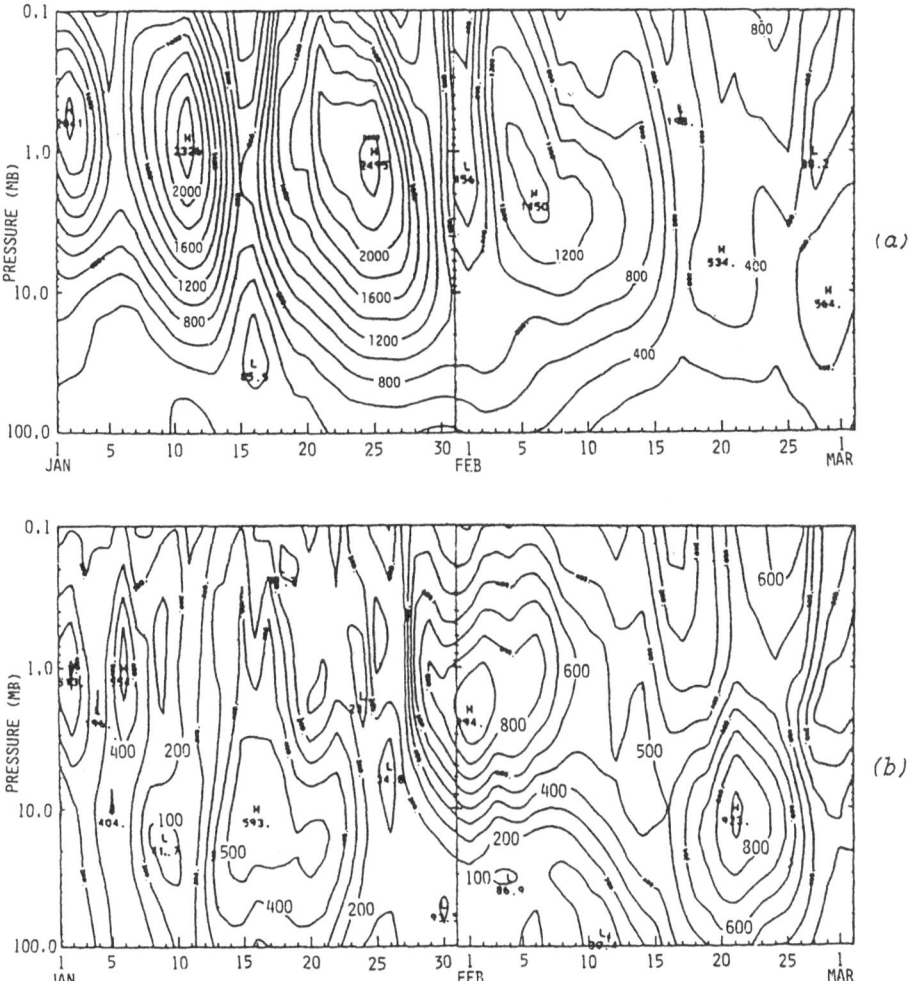

Fig. 2   *Time-altitude cross sections of the amplitude of geopotential*
             waves (meters). a) wave 1.   b) wave 2.

The peak amplitudes of wave 2 are smaller (< 1000 m) and, except in
early February, occur at times of small wave 1 and generally at lower
altitudes than wave 1. The maxima in mid-January and late February are
centered around 10 mb, but the early February event is near 2 mb. Smith
(1983b) has shown that this alternation results from wave-wave interactions. The latter two are apparently associated with the second and
third temperature peaks. In the following sections these periods will
be looked at in greater detail.

294

4. PROGRESSION OF ATMOSPHERIC STATES DURING THE NORTHERN HEMISPHERE WINTER OF 1978-79.

4.1 Pre-disturbance period (1 December-20 January)

During late autumn the radiative cooling of the winter polar atmosphere results in an increase of the zonal mean winds. With few ex-
ceptions, for the entire autumn until 13 December, maximum winds occur
in the mesosphere, near 0.4 mb. Initially the maximum is near 35°N,
with isotachs extending poleward and downward, similar to the monthly
averages presented by Quiroz (1981), Hamilton (1982), Smith (1983a) or
Geller et al. (1983). The maximum wind increases from 92 m s^{-1} on
26 November to 111 m s^{-1}, the largest value observed, on 8 December
(Fig. 3). During this period, although the refractive indices appear to
allow propagation to high altitudes, the EP fluxes do not extend above
5 mb. By 13 December the EP fluxes are still strongly refracted equatorward (Fig. 4a), but now reach altitudes above 1 mb (Fig. 4b), where

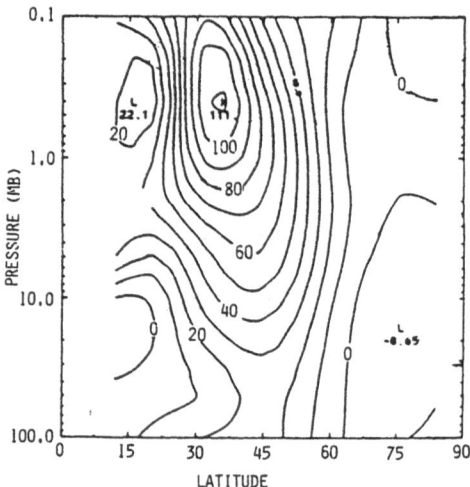

Fig. 3   Zonal mean gradient wind (m/s) on 8 December 1978.

there is a region of convergence, and the zonal wind drops to 90 m s^{-1}.
The EP flux convergence continues and the wind decreases to 67 m s^{-1}
on 15 December, after which the maximum is no longer distinct, having
been overshadowed by a new maximum located near .2 mb and 65°N which
first appeared on 16 December. This had been a region of EP flux
divergence for several days prior to the 16th. By 17 December this
polar night jet reaches a speed of 74 m s^{-1}, a value it never exceeds.
For the next month the wind maximum varies in speed and moves in latitude
as a result of wave-mean flow interactions. The net change is a decrease
in speed (to 48 m s^{-1}) on 20 January, and a broadening and shift of the
center to 1 mb, 47°N.

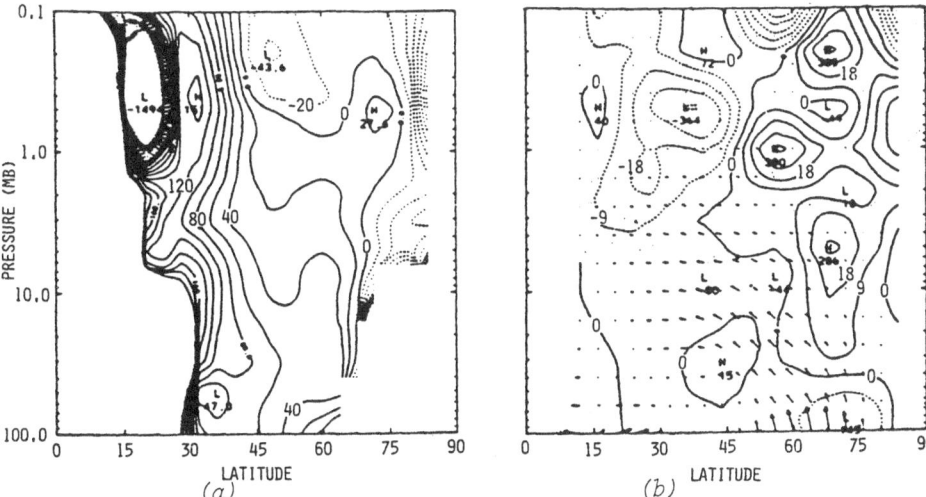

Fig. 4   (a) Cross-section of refractive index squared for Wave 1 (Q_1)
on 13 December 1978. Dashed lines indicate $Q_1 < 0$. Blank
areas indicate $|Q_1| > 200$. (b) Zonal mean cross-section of
total EP fluxes (Waves 1-6) (arrows), convergence (dashed
contours) and divergence (solid contours) for 13 December
1978. (Units are $m/sec^2 \cdot 10^6 = .086$ m s^{-1} day^{-1}.) Contour
interval, 90 units.

The wind also increases in the region of the initial wind maximum,
reaching 70 m s^{-1} at .1 mb in early January. This feature is relatively
quiescent thereafter.

In this early period the development of large winds appears to
result from radiative processes. Eliassen-Palm flux convergence may
decelerate them, but this does not always happen, even if the daily
refractive index appears to the eye to allow propagation into the
region of interest. Clearly the availability and timing of upward
propagating waves from the troposphere plays a critical role; these
factors presumably contribute strongly to interannual variability. A
general feature would appear to be the formation of a wind maximum at
$\sim$ 35°N in the mesosphere, followed by its reduction through EP flux
convergence, and the development of another, weaker maximum at higher
latitudes.

4.2 Late January (20-30 January)

This disturbance has been discussed in some detail by Smith (1982).
Beginning on 21 January, EP flux vectors are directed into the lower
polar stratosphere, leading to EP flux convergence in high latitudes,
especially at high altitudes. The refractive index and total EP flux
and its convergence for 23 January are shown in Fig. 5. Regions of

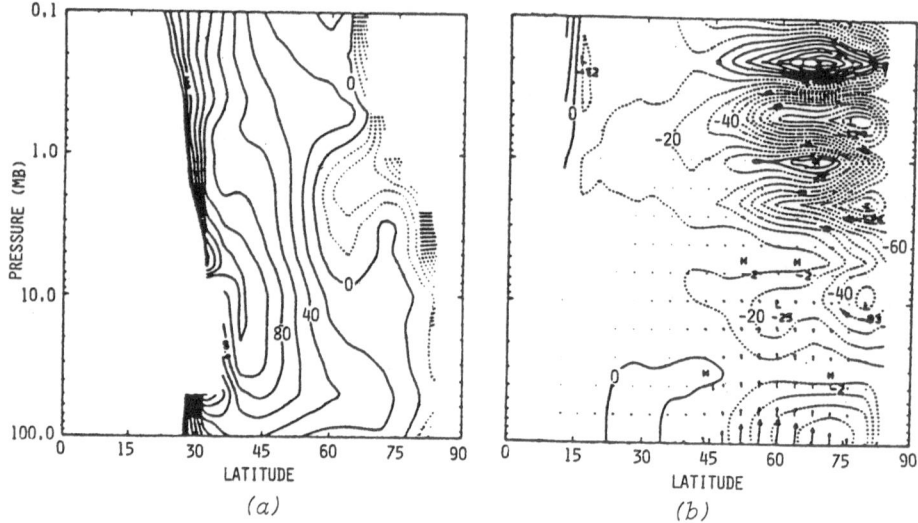

Fig. 5 (a) *As Fig. 4a, for 23 January 1979.* (b) *As Fig. 4b, for 23 January 1979, but units are m/s² · 10⁵ = .86 m s⁻¹ day⁻¹. Contour interval, 10 units.*

strong convergence occur near the horizontal $Q_1 = 0$ line and in the $Q_1 < 0$ region. The zonal wind shows large decelerations poleward of 50°N at all altitudes on 22 and 23 January, with maxima $\sim$ 17 m s⁻¹ day⁻¹ near 0.6 mb, 68°N. Easterlies first appear on the 23rd, reaching down to 1 mb at 84°N. Figure 6 displays the situation on the 25th, when maximum easterlies are - 13 m s⁻¹. Thereafter they weaken, and appear at lower altitudes, with - 11 m s⁻¹ (at 1.5 mb, 80°N) and - 3 m s⁻¹ (at 2 mb, 76°N) on the next two days.

The event is essentially short lived, however, as large EP flux divergence at high latitudes, and therefore rapid acceleration of the winds, begins on 26 January. The observed winds increase at all latitudes poleward of 55° on the 28th, so that the jet maximum reaches 50 m s⁻¹ at 2 mb, 65° on the 30th.

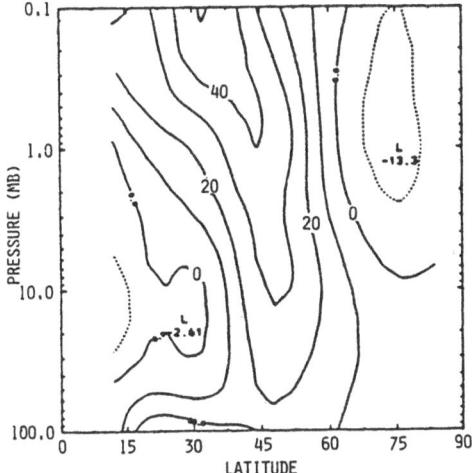

Fig. 6   As Fig. 3, for 25 January 1979.

4.3   Early February (31 January-17 February)

At the end of January and first few days of February the region of $0 < Q_1 < 200$ defines a broad upward-equatorward curving region, and thus it would seem to be a good channel for upward wave propagation. Q_1 for 2 February is shown in Fig. 7a; Q_2 is very similar to Q_1 on this day. As shown in Fig. 7b, waves do propagate through this area, with a

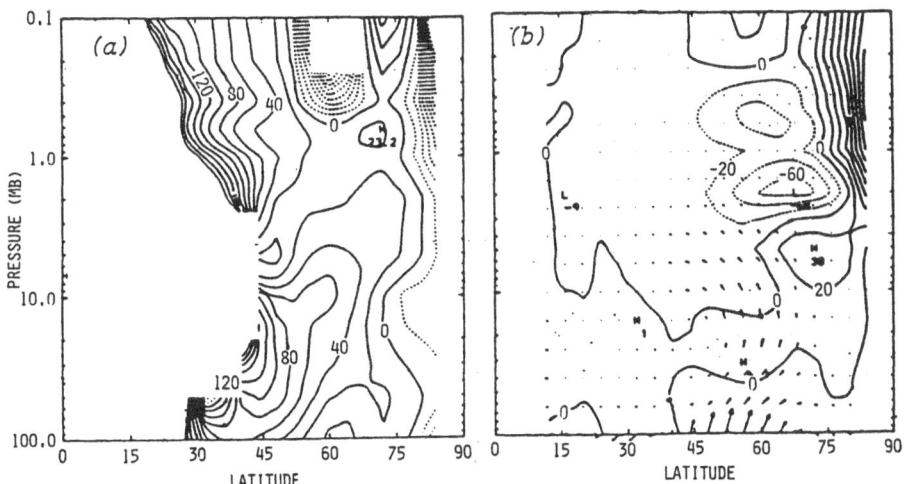

Fig. 7   (a) As Fig. 4a, for 2 February 1979.   (b) As Fig. 5b, for 2 February 1979, but contour interval is 20 units.

region of convergence ∿ 65° at 2 mb. The maximum wind, which weakened to 42 m s⁻¹ on 1 February, is centered at the same location, but decreases to 33 m s⁻¹ and is displaced to 16 mb, 68°N by 6 February. The easterlies, which first appear on 4 February below 0.2 mb, are shown in Fig. 8 to extend down to 1 mb and 60°N on 6 February.

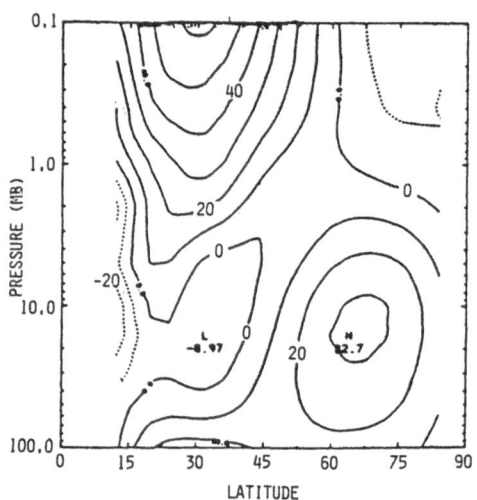

Fig. 8   As Fig. 3, for 6 February 1979.

From 6 to 10 February the region of $0 < Q_1 \leq 200$, is closed off at 1 mb, 65°, but broadens, with large regions of moderate values for the following 6 days. Initially the EP vectors still reach high altitudes, with convergence in the mesosphere and divergence in the stratosphere, but after 8 February they are concentrated below 1 mb, and after the 12th below 10 mb, somewhat at variance with the behavior of the refractive index. The polar night jet remains in the same location, gradually increasing to 49 m s⁻¹ (13 February) before dropping again to 39 m s⁻¹ on the 17th. The high altitude easterlies continue, through the 15th, but are never appreciably stronger than - 20 m s⁻¹, and that only at the highest altitudes.

4.4 Late February Warming (18 February-2 March)

The final disturbance, a dramatic major warming, began around 18 February. This has been studied earlier by Palmer (1981a, b). With a wide channel in which $0 \leq Q_2 \leq 200$ at high latitudes, EP flux vectors are steered toward the polar night jet axis and produced a strong region of convergence near 10 mb. Although the channel narrowed on subsequent days, the same pattern of flux direction and convergence continued for 4 days, resulting in easterly winds of - 33 m s^{-1} at 4 mb, 75°N. On ensuing days the region of moderately positive Q_2 narrowed further, and then was cut off until 2 March. Eliassen-Palm flux vectors now tended more southward (22-25 February), but large flux convergence still took place at 10 mb, 67°N, and the wind remained $\sim$ - 33 m s^{-1}, with easterlies down to 100 mb. On the 26th, EP flux vectors again were strongly converging into the polar cap region (Fig. 9), where, as portrayed in Fig. 10, easterly winds reached - 39 m s^{-1}.

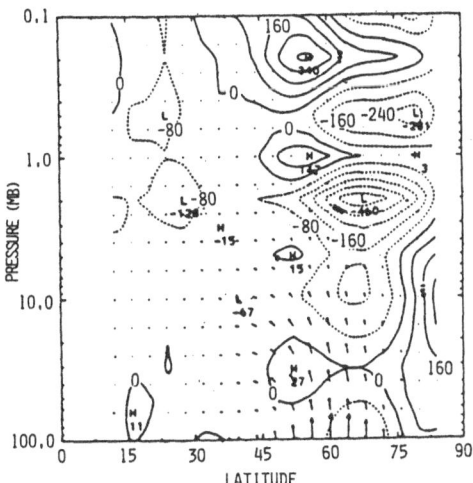

Fig. 9   As Fig. 4b, for 26 February 1979, but contour interval is 80 units.

The pattern is dramatically different on the following days, with horizontal EP vectors pointing equatorward from the polar regions, where there are extensive regions of flux divergence. The easterlies weaken to - 33, - 23, and - 10 m s^{-1} on the succeeding days. By 2 March the warming, and the active part of the winter, were over.

5.   HEAT AND MOMENTUM BUDGETS

As noted, the convergence and divergence of the EP fluxes are expected to result in deceleration and acceleration of the zonal winds. On zonal mean cross-sections, we have found that there is a rather close correspondence in the patterns of these areas, even on individual days. Figures 11a, b illustrate this for 29 January. Note however that the magnitudes of $\partial\bar{u}/\partial t$ are much less than D_F.

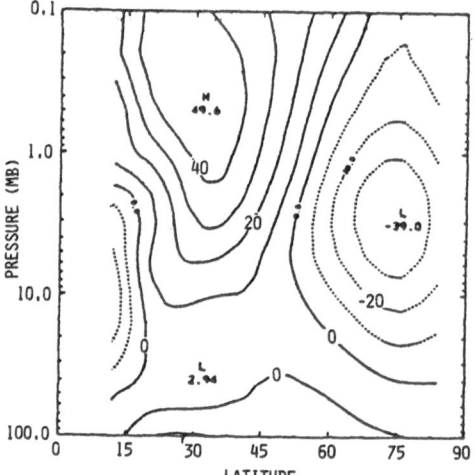

Fig. 10   As Fig. 3, for 26 February 1979.

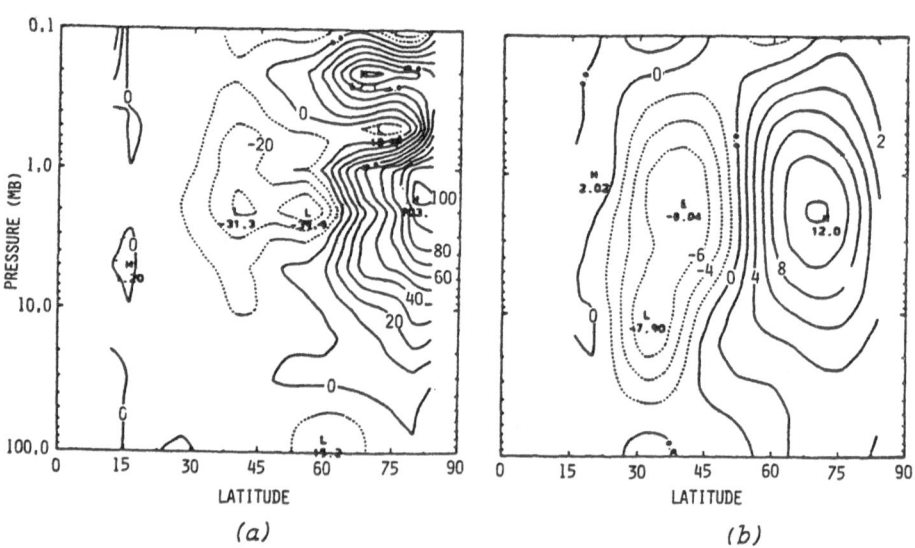

(a)   (b)

Fig. 11   (a) Wave flux divergence (m s⁻¹ day⁻¹), for 29 January 1979.
*   (b) Acceleration of zonal mean wind speed (m/s day), for*
*   29 January 1979.*

We next examine the heat budget as a function of time. Figure 12 displays the terms in Eq. (2) for 10 mb, 60°N. The radiative cooling rate was calculated from an updated version of the code of Ramanathan and Dickinson (1979), and shows little variation with time. Variations in $\partial\bar{\theta}/\partial t$ are therefore presumably due to variations in the residual mean $\bar{w}^*$ with its attendant adiabatic temperature change. The quantity $-\bar{\theta}_z\bar{w}^*$ was calculated as a residual from Eq. (2). The variations shown are typical of other latitudes and altitudes. Positive values of $-\bar{\theta}_z\bar{w}^*$ indicate $w^* < 0$, i.e., downward motion in this region, as expected. This generally downward motion varies on time scales of several days.

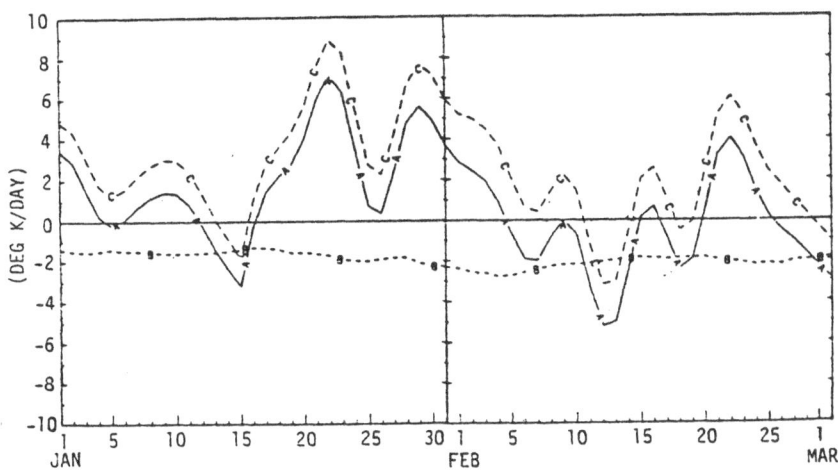

Fig. 12 *Time variation of terms in Eq. (2), for 10 mb, 60°N. Solid line (a), $\partial\bar{\theta}/\partial t$; short dashes (B), $\bar{Q}$; long dashes (c) $-\bar{\theta}_z\bar{w}^*$.*

From $\bar{w}^*$ and the continuity equation in the meridional plane, $\bar{v}^*$ may also be derived, giving the residual mean circulation based on the heat equation.

The time variation of the terms in the transformed momentum equation, Eq. (1), are displayed for 10 mb, 60°N, in Fig. 13. The $\bar{v}^*$ was calculated from the heat equation, as described above, and not derived as a residual as was done by Palmer (1981a).

These are local values which have not been averaged over the larger region in which D_F is acting, nor has D_F been smoothed in time. Both were done by, e.g., O'Neill and Youngblut (1982) or Palmer (1981a, b). Such averaging will help reduce the effects of measurement error as well as real small scale variability in the atmosphere. For the values shown, $\partial\bar{u}/\partial t$ generally shows small changes, but these can be seen to

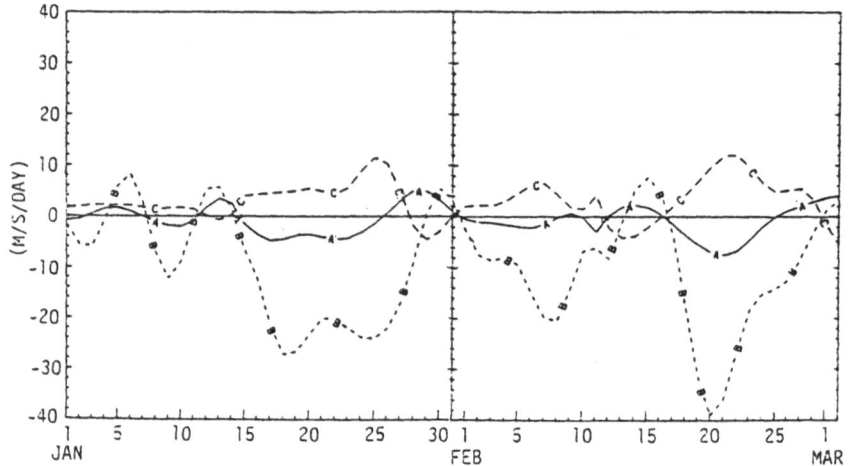

Fig. 13 *Time variation of terms in Eq. (1) for 10 mb, 60°N. Solid line (A), $\partial\bar{u}/\partial t$; short dashes (B), D_F; long dashes (C), $\overline{fv}^*$.*

vary in response to changes in D_F [as seen for the shorter data segment studied by Palmer (1981a, b)], although this local $\partial\bar{u}/\partial t$ is almost always considerably smaller and smoother than D_F.

Inspection of Fig. 13 indicates that the balance in Eq. (1) is not satisfied exactly. The residual has a positive average value, usually $< 10 \text{ m s}^{-1} \text{ day}^{-1}$. This difference may be due to measurement and calculation errors, to not averaging over a sufficiently large area and time, or in part to real physical effects.

6. CONCLUSIONS

The LIMS data are capable of providing a detailed picture of the changes and developments in the Northern Hemisphere winter of 1978-79. They are also sufficiently smooth that highly derived quantities may be calculated with some confidence. The data agree in general with earlier results; since they have more vertical resolution than other observations, they may be expected to shed additional light on the processes taking place.

From this initial look at the data, it is clear that the atmosphere is very complicated. There is considerable variability in that many processes are at work, and each disturbance has unique aspects.

The divergence of the EP fluxes do indicate the times and places of $\partial\bar{u}/\partial t$ surprisingly well, although in the local data shown here there is an indication of large forcing of $\bar{v}^*$ as well, if it is calculated as a residual in Eq. (1). Alternatively, the $\bar{w}^*$ calculated from the heat equation varies smoothly with time and space. This $\bar{w}^*$ calculated at one location appears to incorporate an intrinsic smoothing, which is not the case for $\bar{v}^*$ from Eq. (1), presumably due to the rapid variations of D_F.

The refractive index appears to provide a general indication of the direction of wave propagation, but there is often not a close correspondence on individual days, a finding similar to O'Neill and Youngblut (1982). This may be due to several factors. The theory strictly applies to small amplitude waves, and situations in which the fractional variation of Q_n is small over a wave length or wave period. All of these conditions are violated, sometimes by larger margins, during this period.

No ray tracing has been done here. This would indicate regions of low group velocity, and thus regions of high dissipation, which also has not been considered.

Wave-wave interaction also plays a large role during this winter (Smith, 1983b; Smith et al., 1983). While in theory it should not affect the total EP flux, it may modify our calculated results.

Finally, the major warming in this winter followed the development of a wind structure and refractive index field that could channel wave activity into the polar cap region. After the strong easterlies were established, there was strong absorption near and above the zero wind line.

ACKNOWLEDGMENT

We acknowledge the work of James M. Russell III of NASA Langley Research Center, Paul L. Bailey of NCAR, and Larry L. Gordley of Systems and Applied Science Corporation for many years of collaboration in achieving the present quality of the LIMS results. We thank Anne K. Smith for several stimulating conversations and helpful comments, and Donna Sanerib for patiently typing several versions of the manuscript.

This work was supported in part by the National Aeronautics and Space Administration under contracts S-70994, L-9469B, and S-10782-C.

The National Center for Atmospheric Research is sponsored by the National Science Foundation.

REFERENCES

Andrews, D. G., and M. E. McIntyre, 1976: Planetary waves in horizontal and vertical shear: The generalized Eliassen-Palm relation and the mean zonal acceleration. J. Atmos. Sci., 33, 2031-2048.

_____, and _____, 1978: Generalized Eliassen-Palm and Charney-Drazin theorems for waves on axisymmetric mean flows in compressible atmospheres. J. Atmos. Sci., 35, 175-185.

Edmon, H. J., B. J. Hoskins, and M. E. McIntyre, 1980: Eliassen-Palm cross-sections for the troposphere. J. Atmos. Sci., 37, 2600-2616.

Geller, M. A., M. F. Wu, and M. E. Gellman, 1983: Troposphere-stratosphere (surface - 55 km) monthly winter general circulation statistics for the Northern Hemisphere--four year averages. J. Atmos. Sci., 40 (in press).

Gille, J. C., P. L. Bailey, and J. M. Russell III, 1980: Temperature and composition measurements from the L.R.I.R. and L.I.M.S. experiments on Nimbus 6 and 7. Phil. Trans. R. Soc. Lond. A, 296, 205-218.

_____, P.L. Bailey, L. V. Lyjak, and J. M. Russell III, 1983a: Results from the LIMS experiment from the PMP-1 winter 1978/79. Adv. Space Res., 2, 163-167.

_____, and J. M. Russell III, 1983: The Limb Infrared Monitor of the Stratosphere (LIMS): An overview of the experiment and its results. Submitted to J. Geophys. Res.

_____, J. M. Russell III, P. L. Bailey, L. L. Gordley, E. E. Remsberg, J. H. Lienesch, W. G. Planet, F. B. House, L. V. Lyjak, and S. A. Beck, 1983b: Validation of temperature retrievals obtained by the Limb Infrared Monitor of the Stratosphere (LIMS) experiment on Nimbus 7. Submitted to J. Geophys. Res.

Hamilton, K., 1982: Some features of the climatology of the Northern Hemisphere stratosphere revealed by NMC upper atmospheric analyses. J. Atmos. Sci., 39, 2737-2749.

Kanzawa, H., 1983: Four observed sudden warmings diagnosed by the Eliassen-Palm flux and refractive index. (This volume.)

Karoly, D. J., and B. J. Hoskins, 1982: Three dimensional propagation of planetary waves. J. Meteor. Soc. Japan, 60, 109-123.

Kohri, W. J., 1981: LRIR observations of the structure and propagation of the stationary planetary waves in the Northern Hemisphere during 1975. NCAR Cooperative Ph.D. Thesis, Drexel University and National Center for Atmospheric Research, Boulder, Colorado, 312 pp.

Leovy, C. B., M. H. Hitchman, A. K. Smith, J. C. Gille, P. L. Bailey, L. V. Lyjak, and E. E. Remsberg, 1983: Properties of quasi-global fields of temperature, geopotential and wind derived from the Nimbus 7 LIMS experiment. To appear, J. Geophys. Res.

Madden, R. A., and K. Labitzke, 1981: A free Rossby wave in the tropo-
    sphere and stratosphere during January 1979. J. Geophys. Res., 86,
    1247-1254.

Matsuno, T. 1970: Vertical propagation of stationary planetary waves in
    the winter Northern Hemisphere. J. Atmos. Sci., 27, 871-883.

_____, and K. Nakamura, 1979: The Eulearian- and Lagrangian-mean
    meridional circulations in the stratosphere at the time of a
    sudden warming. J. Atmos. Sci., 36, 640-654.

O'Neill, A., and C. E. Youngblut, 1982: Stratospheric warmings diagnosed
    using the transformed Eulerian-mean equations and the effect of the
    mean state on wave propagation. J. Atmos. Sci., 39, 1370-1386.

Palmer, T. N., 1981a: Diagnostic study of a wavenumber-2 stratospheric
    sudden warming in a transformed Eulerian-mean formalism. J.
    Atmos. Sci., 38, 844-855.

_____, 1981b: Aspects of stratospheric sudden warmings studied from a
    transformed Eulerian-mean viewpoint. J. Geophys. Res., 86, 9679-9687.

_____, 1982: Properties of the Eliassen-Palm flux for planetary scale
    motions. J. Atmos. Sci., 39, 992-997.

Quiroz, R. S., 1981: The tropospheric-stratospheric mean zonal flow in
    winter. J. Geophys. Res., 86, 7378-7384.

Ramanathan, V., and R. E. Dickinson, 1979: The role of stratospheric
    ozone in the zonal and seasonal radiative energy balance of the
    earth-troposphere system. J. Atmos. Sci., 36, 1084-1104.

Rodgers, C. D., 1977: Statistical principles in inversion theory. In
    Inversion Methods in Atmospheric Remote Sounding. Academic Press,
    New York, pp. 117-138.

Russell, J. M., III, and J. C. Gille, 1978: The Limb Infrared Monitor
    of the Stratosphere (LIMS) experiment. In The Nimbus 7 Users
    Guide, C. Madrid, Ed. Goddard Space Flight Center, Greenbelt, Md.,
    pp. 71-103.

Smith, A. K., 1982: An observational study of planetary wave propagation
    in the winter stratosphere. Ph.D. Thesis, University of Washington,
    Seattle, Wash., 157 pp.

_____, 1983a: Stationary waves in the wintertime stratosphere: Seasonal
    and interannual variability. J. Atmos. Sci., 40, 245-261.

_____, 1983b: Observation of wave-wave interactions in the stratosphere.
    Accepted by J. Atmos. Sci.

_____, J. C. Gille, and L. V. Lyjak, 1983: Wave-wave interactions in the stratosphere: Observations during quiet and active wintertime periods. Submitted to J. Atmos. Sci.

FOUR OBSERVED SUDDEN WARMINGS DIAGNOSED BY THE ELIASSEN-PALM FLUX AND
REFRACTIVE INDEX

Hiroshi Kanzawa

National Institute of Polar Research

ABSTRACT

The Eliassen-Palm flux $\underline{E}$ is a measure of planetary wave propagation
in the meridional plane and the divergence of $\underline{E}$ embodies the total
forcing of the zonal mean flow by the waves, while the refractive index
Q controls the direction of planetary wave propagation. $\underline{E}$ and Q diagnostics during observed sudden stratospheric warmings were performed by
Palmer (1981a,b), O'Neill and Youngblut (1982) and Kanzawa (1982) in
order to understand the wave-zonal flow interaction dynamics of sudden
warmings. In the present paper the characteristics of sudden warmings
clarified by these observational studies are reviewed.

The characteristics common to the sudden warmings investigated in
the above mentioned studies are as follows: $\underline{E}$ due to the zonal wavenumber 1 or 2 component is focused into the polar stratosphere prior to
the circulation reversal. By this focusing there occurs an intense
convergence of $\underline{E}$ which brings about an intense deceleration of the mean
zonal wind and a poleward residual mean meridional flow in the stratosphere, leading to the sudden warmings. The maximum of Q situated in
the polar troposphere and stratosphere prior to the warming is interpreted as a main factor which determines the switching of $\underline{E}$ from equatorward to poleward. This pattern of Q comes from a mean zonal wind
profile in which a westerly maximum is situated in high latitudes so
that a tropospheric double jet structure is formed.

1. INTRODUCTION

The Eliassen-Palm (E-P) flux $\underline{E}$ is a useful and dynamically fundamental representation of wave activity (McIntyre, 1980; Edmon et al.,
1980; Sato, 1980). Although the name Eliassen-Palm flux has already
become popular, we might eaually well call the quantity the Andrews-
McIntyre flux since Andrews and McIntyre (1976) first introduced the
quantity as an important one to describe wave-zonal flow interaction.
The pattern of E-P vectors indicates the rate of transfer of the
density of "wave activity" from one height and latitude to another for
propagating planetary waves. The E-P flux divergence acts as a forcing

on the mean zonal wind. It is also known since Matsuno (1970) that planetary wave propagation depends on the "refractive index" Q which is defined as the latitudinal gradient of zonal mean quasi-geostrophic vorticity ($\overline{q}_y$) divided by the zonal mean geostrophic wind velocity ($\overline{u}_0$).

In the present study I review works which have used the E-P flux $\underline{E}$ and the refractive index Q, to diagnose sudden warmings: that is, Palmer (1981a, b), O'Neill and Youngblut (1982), and Kanzawa (1982) for the 1979 wavenumber 2 type warming, the 1980 wavenumber 1 type warming, the 1977 wavenumber 1 type warming and the 1973 wavenumber 1 type warming respectively. Numerical model studies using these diagnostics such as Dunkerton et al. (1981), Hsu (1981), Bridger and Stevens (1982) and Butchart et al. (1982) are also referred to in order to better understand the dynamics of sudden warmings. All these numerical models are constructed on the basis of the idea presented by the pioneering work of Matsuno (1971).

It was first pointed out by Kanzawa (1980) and then by a number of authors that wave-focusing by the mean zonal wind brought about by a preceding minor warming is an important factor in determining the intensity of a major warming. The nature of the profile of zonal mean wind thus preconditioned, and how planetary waves propagate in that wind profile in the real atmosphere are the main topics of the present paper. Why the preconditioning itself occurs in association with a preceding minor warming is beyond the scope of the present study but has recently been investigated by Matsuno (1983) in this volume and Palmer and Hsu (1983).

2. THEORETICAL BACKGROUND

In this section I briefly review the theoretical background for the purpose of making E-P flux and refractive index diagnoses of observed warmings. With conventional notation such as in Holton (1975), the E-P flux $\underline{E}$ for the quasi-geostrophic motion in the spherical coordinates (λ, $\overline{\theta}(=y/a)$, $z=-H\ln(p/p_s)$) is defined as

$$\underline{E} = (\ E^y,\ E^z\ ) \tag{1a}$$

where

$$E^y = -\ \rho(z)\ a\ \cos\theta\ \overline{v_0'u_0'} \tag{1b}$$

$$E^z = +\ \rho(z)\ a\ \cos\theta\ (f/N^2)\ \overline{v_0'\phi_z'} \tag{1c}$$

where ρ is basic state density: $\rho(z) = \rho_s\ e^{-z/H}$, ϕ is geopotential height, u_0 and v_0 mean eastward and northward geostrophic winds, f is the Coriolis parameter, a is the radius of the earth, N is buoyancy frequency, H is scale height, overbars and primes denote zonal mean and deviation therefrom. The transformed Eulerian-mean flow equations are

$$\partial(\rho\ a\ \cos\theta\ \overline{u}_0)/\partial t = f\ (\rho\ a\ \cos\theta\ \overline{v}*) + \underline{\nabla}\cdot\underline{E} \tag{2a}$$

$$\partial \overline{\phi}_z / \partial t = - N^2 \overline{w}* + \chi \overline{J}/H \tag{2b}$$

$$\partial (\rho \ a \ \cos\theta \ \overline{v}*)/\partial y + \partial (\rho \ a \ \cos\theta \ \overline{w}*)/\partial z = 0 \tag{2c}$$

where

$$\underline{\nabla} \cdot \underline{E} = \partial (\cos\theta \ E^y)/\cos\theta \partial y + \partial E^z/\partial z \tag{2d}$$

The residual meridional circulation ($\overline{v}*$, $\overline{w}*$) is related to the Eulerian-mean circulation ($\overline{v}$, $\overline{w}$) by the following transformations

$$\rho \ a \ \cos\theta \ \overline{v}* = \rho \ a \ \cos\theta \ \overline{v} - \partial (E^z/f)/\partial z \tag{3a}$$

$$\rho \ a \ \cos\theta \ \overline{w}* = \rho \ a \ \cos\theta \ \overline{w} + \partial (E^z/f)/\partial y \tag{3b}$$

Note that this residual mean circulation is identical to the Lagrangian-mean meridional circulation for steady linear conservative waves as shown in Matsuno and Nakamura (1979). We can interpret $\underline{E}$ as the negative angular wave momentum flux: the y-component of $\underline{E}$ (see Eq. (1b)) is the horizontal eddy flux of angular westward momentum while the z-component of $\underline{E}$ (Eq. (1c)) is proportional to the meridional heat flux. However, as Uryu (1974) first showed E^z can be interpreted as the vertical flux of angular westward momentum. We can find the following relationship between the E-P flux divergence $\underline{\nabla} \cdot \underline{E}$ and the meridional flux of potential vorticity $\overline{v_0'q'}$

$$\underline{\nabla} \cdot \underline{E} = \rho \ a \ \cos\theta \ \overline{v_0'q'} \tag{4}$$

where q' is quasi-geostrophic potential vorticity perturbation defined by

$$q' = L(\phi') \tag{5a}$$

where

$$L = \left[\partial^2/(a \ \cos\theta)^2 \partial \lambda^2 + f^2 \partial ((\cos\theta/f^2)\partial/\partial y)/\cos\theta \partial y \right.$$

$$\left. + \partial (\rho(f/N)^2 \partial/\partial z)/\rho \partial z \right]/f \tag{5b}$$

This definition of q' which takes spherical effects into account was first presented by Matsuno (1970). $\underline{\nabla} \cdot \underline{E}$ (thus $\overline{v_0'q'}$) and $\overline{v}*$ and $\overline{w}*$ are identically zero for steady, linear, conservative waves in the absence of critical levels as first shown in Andrews and McIntyre (1976). This non-acceleration theorem was originally presented by Charney and Drazin (1961) in a different manner.

The E-P flux is a measure of the rate of transfer of the density of "wave activity" from one latitude and height to another when the eddy is a propagating wave such as a Rossby wave, i.e.,

$$\underline{E} = \underline{c}_g \ A \tag{6a}$$

where $\underline{c}_g$ is the local group velocity projected onto the meridional plane, and A is the density of E-P wave activity:

$$A = \rho \ a \ \cos\theta \overline{q'^2}/2\overline{q}_y \tag{6b}$$

where $\bar{q}_y$ is the zonal mean quasi-geostrophic potential vorticity gradient:

$$\bar{q}_y = 2\Omega\cos\theta/a \; - \; \partial(\partial(\cos\theta\bar{u}_0)/\cos\theta\partial y)/\partial y$$

$$- \; \partial(\rho(f/N)^2\partial\bar{u}_0/\partial z)/\rho\partial z \tag{7}$$

We define $\underline{E}^*$ which is $\underline{E}$ multiplied by the area factor $\cos\theta$ such that

$$\underline{E}^* = \cos\theta \; \underline{E} \tag{8a}$$

then

$$\underline{\nabla}^*\cdot\underline{E}^* = \cos\theta \; \underline{\nabla}\cdot\underline{E}, \qquad \underline{\nabla}^*\cdot\underline{E}^* = \partial E^y*/\partial y + \partial E^z*/\partial z \tag{8b}$$

For representing the E-P flux graphically in the (y,z)-plane it is appropriate to draw arrows of $\underline{E}^*$ in place of $\underline{E}$, since the pattern will look nondivergent in that plane if and only if $\underline{\nabla}\cdot\underline{E}$ is zero. Also the divergence of the E-P flux will be drawn by $\underline{\nabla}^*\cdot\underline{E}^*$ in place of $\underline{\nabla}\cdot\underline{E}$. In addition, in order to see the effect of waves on the time change of mean zonal wind more clearly, the quantity D_E defined as

$$D_E = \underline{\nabla}\cdot\underline{E} \; / \; \rho(z) \; a \; \cos\theta \tag{9}$$

is useful since Eq. (2a) is rewritten as

$$\partial\bar{u}_0/\partial t = f \; \bar{v}^* \; + D_E \tag{10}$$

In his calculation of steady stationary planetary waves propagating through the winter basic zonal wind state, Matsuno (1970) introduced the refractive index squared Q_s defined as

$$Q_s = Q - s^2/(a \; \cos\theta \;)^2 - f^2/4N^2H^2 \tag{11}$$

where

$$Q = \bar{q}_y/\bar{u}_0 \tag{12}$$

and s is the zonal wavenumber of the planetary wave. We can define a refractive index for transient planetary waves if the WKB approximation holds. As the index of refraction, Karoly and Hoskins (1982) used $(\cos^2\theta \; Q_0)^{1/2}$, O'Neill and Youngblut (1982) used $Q^{1/2}$, Kanzawa (1982) used Q, and Palmer (1982) used $Q_s/\sin^2\theta$ respectively. The differences depend on the stage at which spherical effects are neglected in the WKB sense. For planetary wave propagation in high latitudes, which the present study is concerned with, the quantity Q is found to be the determining factor in all the above different forms: for the real atmospheric situation, the distribution of Q_s for s=1 and s=2 is very similar to that of Q since the second and third terms of the right hand side of Eq. (11) are smaller than Q except near the pole (see the discussion about the second term near the pole in section 3 of McIntyre (1982)).

3. PRECONDITIONED MEAN ZONAL WIND

In this section we compare the characteristics of mean zonal wind
behavior for various warmings. Figs. 1-2, 1-3, 1-4, and 1-5 show snap
shots of the latitude height cross-sections of the mean zonal wind of
the time when the mean wind was in a preconditioned state and the time
of the largest area of easterlies in the meridional plane for the 1973,
1977, 1979 and 1980 sudden warmings respectively. As a reference the
mean zonal wind profile of a normal winter is shown in Fig. 1-1. These
figures show some common features of the mean zonal wind profiles during
pre-warming periods which are different from those of normal winter;
that is, the existence of a westerly maximum in high latitudes around
70°-80°N in the stratosphere, and the extension of the westerly maximum
into the high latitude troposphere. The former fact is related to the
appearance of a westerly minimum in the middle latitude stratosphere.
The latter implies the appearance of a double jet structure in the
troposphere and has not been noticed so far except by Kanzawa (1982).

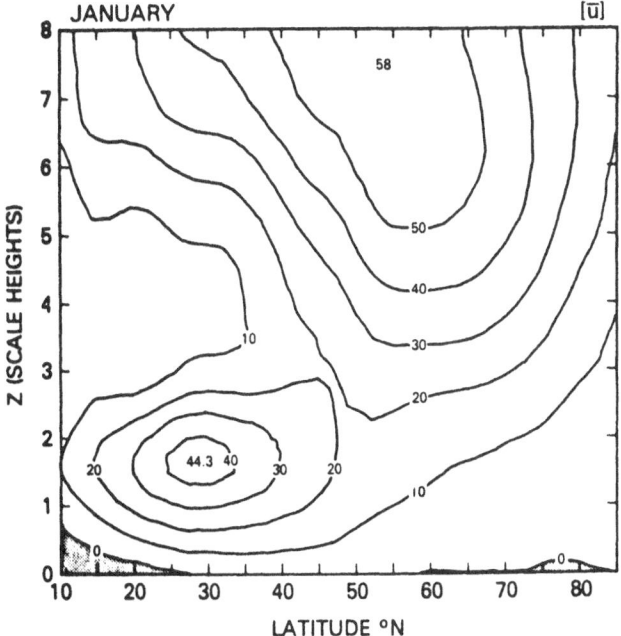

Fig. 1-1. Northern Hemisphere January mean zonal wind in m s^{-1} for
the average of the winters 1979, 1980, 1981 and 1982. Geometric
altitude is approximately the value of z (scale height) in the
ordinate multiplied by 7 km. This figure was constructed based on
the data of the NMC analysis. (after Geller et al. (1983))

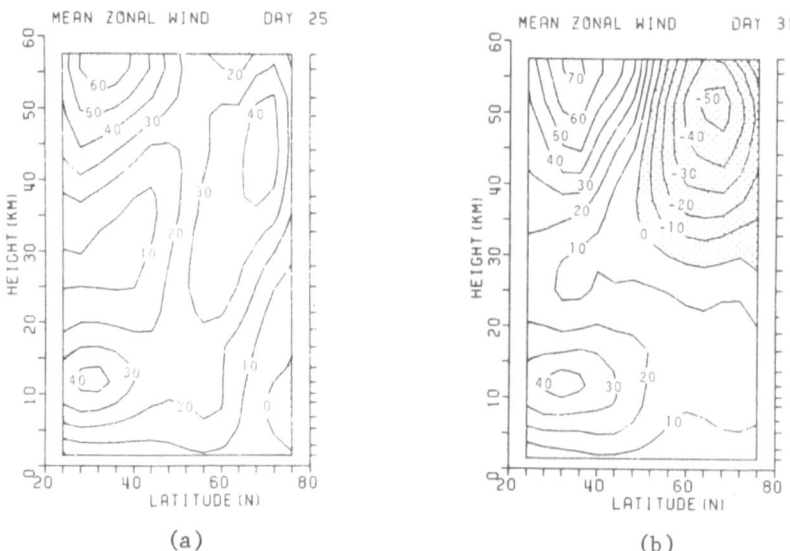

Fig. 1-2. Mean zonal wind (m s^{-1}) on 25 (a) and 31 (b) January 1973. (after Kanzawa (1980))

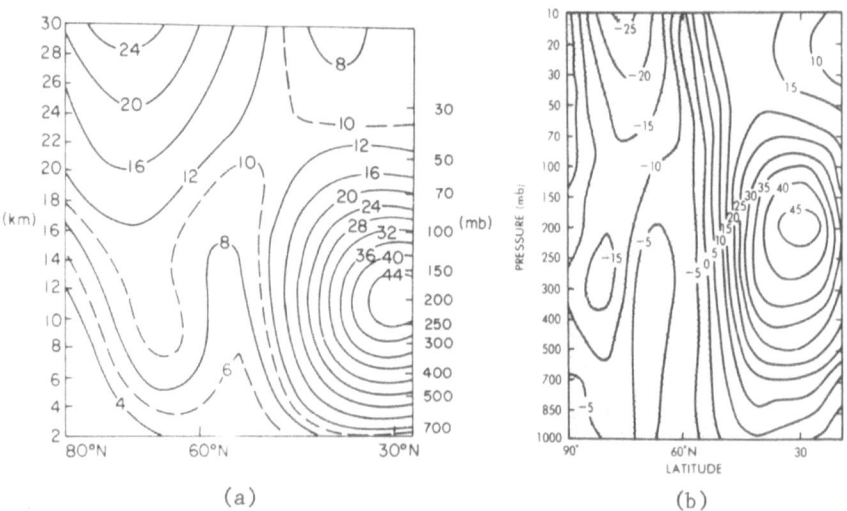

Fig. 1-3. Mean zonal wind (m s^{-1}) on 3 (a) January (averaged over the 5 days centered on this day) and 15 (b) January 1977. (after O'Neill and Youngblut (1982) for a and O'Neill and Taylor (1979) for b)

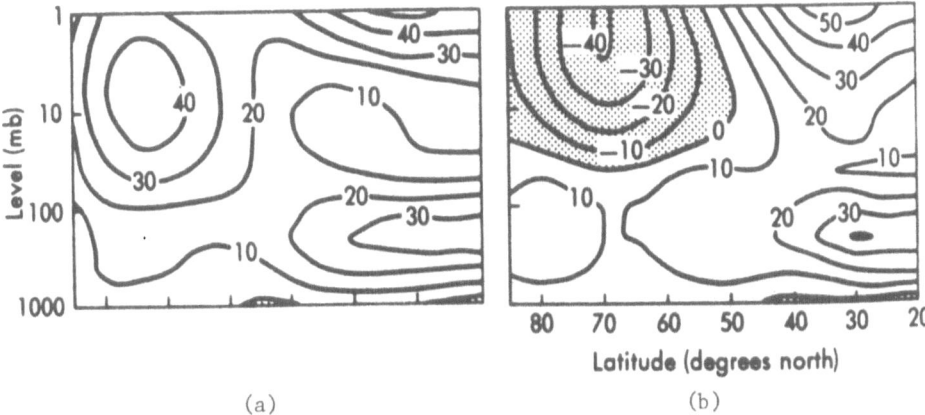

Fig. 1-4.   Mean zonal wind (m s^{-1}) on 17 (a) and 27 (b) February 1979. (after Palmer (1981a))

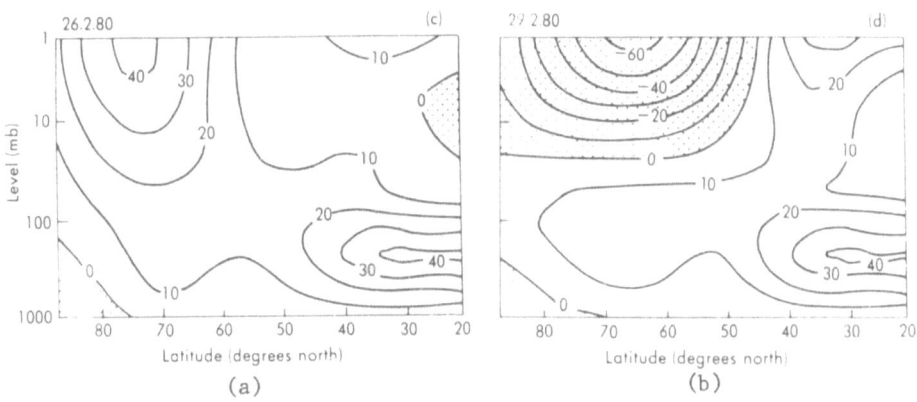

Fig. 1-5.   Mean zonal wind (m s^{-1}) on 26 (a) and 29 (b) February 1980. (after Palmer (1981b))

Next we consider the differences of the mean zonal wind behavior between the sudden warmings. It is of interest to compare the timing of the appearance of the high latitude jet. It was established 6 days (1973), 12 days (1977), 10 days (1979) and 3 days (1980) before the peak of the major warming when the easterly area became largest in the meridional plane. The mean wind was preconditioned earlier in the 1973 and the 1980 sudden warmings which were wavenumber 1 type warmings. The 1979 sudden warming was a wavenumber 2 type warming. The 1977 sudden warming was a wavenumber 1 type warming, but it had a peculiar chara-

cter, in that a circulation reversal occurred even in the polar troposphere. It seems that in wavenumber 1 type "normal" sudden warmings
(e.g., the 1973 and the 1980 sudden warmings) the circulation reversals
occur more quickly after the mean zonal flow is preconditioned than in
"abnormal" sudden warmings, although we have hardly enough cases to have
statistical significance. The other differences will be referred to
when discussing the profiles of refractive index.

4. ELIASSEN-PALM FLUX AND REFRACTIVE INDEX

4.1 Effect of the waves on the circulation reversal

Figs. 2-1, 2-2 and 2-3 show the balance of the three terms of Eq.
(10). The term $f\bar{v}*$ is estimated as a residual of the remaining terms.
All these figures show that the pattern of E-P flux divergence D_E is
similar to the pattern of mean flow deceleration and has an opposite

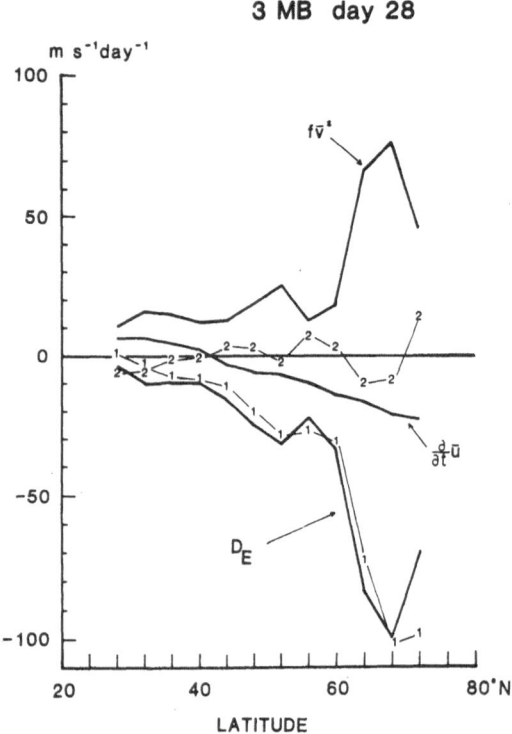

Fig. 2-1. Latitudinal profile of the three terms of the transformed
Eulerian mean equation (10) at 3 mb on day 28 with unit of 1 m s^{-1}
day^{-1}. Thin lines with figures 1 and 2 denote the contribution of
wavenumber 1 and 2 respectively. (after Kanzawa (1982))

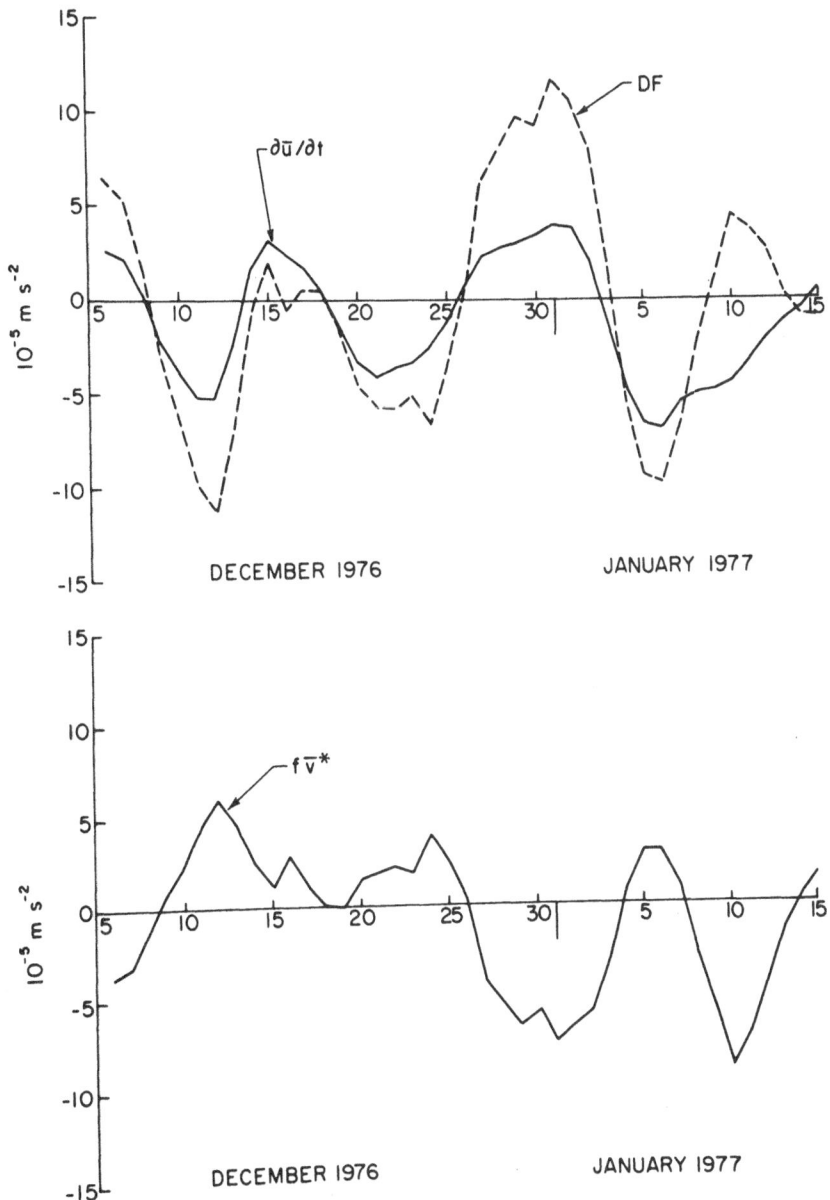

Fig. 2-2.   Terms in Eq. (10) at the 27 km level with unit of 10^{-5} m s^{-2}, averaged with respect to area in the 60°-80°N latitudinal band for the period 5 December 1976 to 15 January 1977. (after O'Neill and Youngblut (1982))

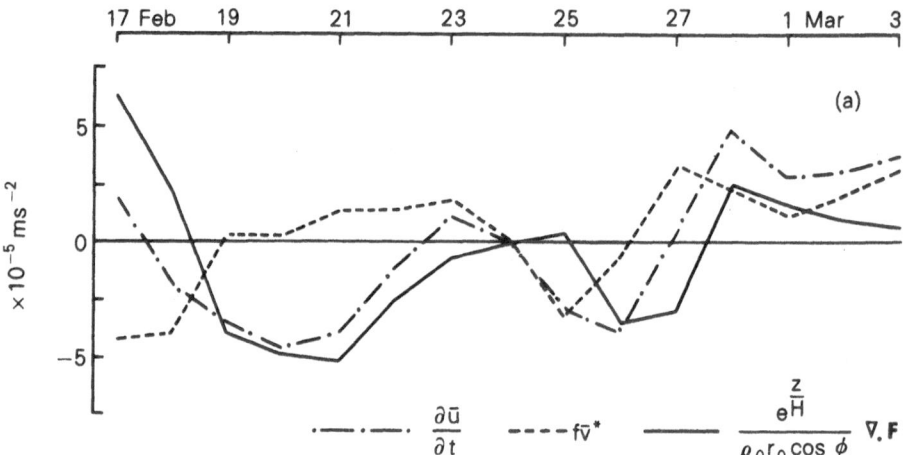

Fig. 2-3. As in Fig. 2-2 but at 1 mb for the period 17 February to 3 March 1979. (after Palmer (1981b))

sense to the pattern of the Coriolis torque acting on the residual meridional flow. The magnitude of D_E is four or five times as large as that of the mean wind deceleration at 3 mb on 28 January 1973 as shown in Fig. 2-1 while it is about the same at 1 mb in 1979 as shown in Fig. 2-3. The smallness of D_E in Fig. 2-3 is considered to come from the coarse grid intervals in the vertical direction in Palmer's analysis: this point is referred to by Gille and Lyjak (1983) in this volume. The magnitude of D_E in Fig. 2-1 is determined mainly by the wavenumber 1 component. Generally these observational results support the picture that the forcing D_E, which is negative, brings about the mean wind deceleration and at the same time a poleward residual mean meridional flow. This picture of the forcing by the waves is implicitly given theoretically by Matsuno and Nakamura (1979) and is discussed explicitly by McIntyre (1982): a zonal flow response to a given zonal force D_E includes not only a mean wind acceleration but also a meridional circulation in such a way as to preserve thermal wind balance. The speeds of $\bar{w}^*$ estimated from the mass conservation equation (2c) are about -0.02 m s^{-1} in the high latitude stratosphere; the subsidence flow brought about warming in that region by overcoming the radiative cooling (see Eq. (2b)).

The mechanism of the circulation reversal described above operates in all the warmings reproduced in the numerical models referred to in section 1. Note that the magnitude of D_E is about the same as that of mean wind deceleration in numerical models involving only the interaction between the mean flow and a single wave (see Fig. 4 of Dunkerton et al. (1981)) while about three times the deceleration in a model permitting wave-wave interaction (see Figs. 7 and 8 of Hsu (1981)) which may be nearer to the real situation.

4.2   Planetary wave propagation

Fig. 3-1 from Geller et al. (1983) shows latitude height cross-sections of the E-P flux vector direction for stationary and transient eddies of a normal winter defined as a four year average from 1979-1982. Note that the E-P vectors in this figure have constant length and so only show the wave propagation aspects. For stationary eddies we see a bifurcation of the E-P vector in the upper troposphere, that is, the

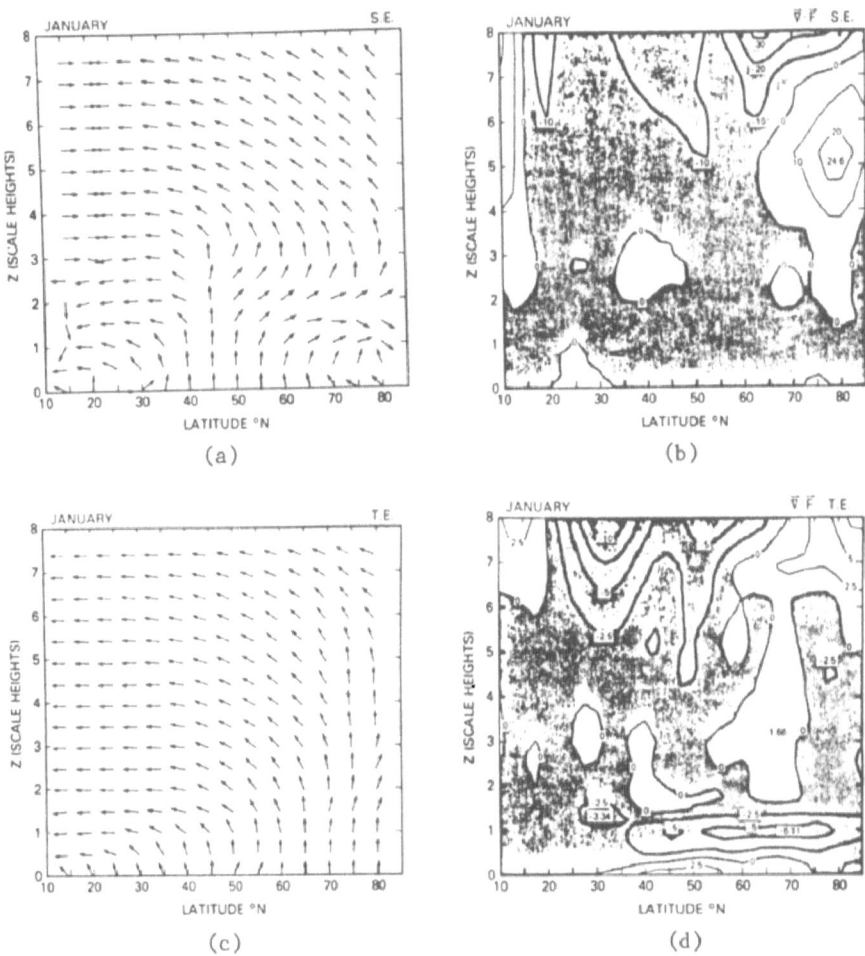

Fig. 3-1. Northern Hemisphere average January E-P flux direction and D_E (10^{-5}. m s^{-2}) from the standing eddy fluxes (a, b) and from the transient eddy fluxes (c, d).   All of the arrows are the same length.   The vertical vector component is magnified by a factor 100 with respect to the horizontal component.   The ordinate is the same as in Fig. 1-1.   (after Geller et al. (1983))

318

splitting of the E-P flux into an equatorward directed branch in low
latitudes and an upward directed branch in high latitudes. However, the
arrows turn equatorward once they enter into the stratosphere. For
transient eddies there is not such a bifurcation in the meridional
plane.

Figs. 3-2, 3-3, 3-4 and 3-5 show cross-sections of the E-P flux or
integral curves of E-P flux. Common features of these figures for the
periods prior to the peaks of the warmings are the focusing of the E-P
flux into the polar stratosphere, which has lower density and smaller
area, and the accompanying strong divergence of E-P flux in that region
(see the figures of 25 and 28 January 1973 in Fig. 3-2, 5 January 1977
in Fig. 3-3, 19 and 21 February 1979 in Fig. 3-4 and 27 February 1980 in
Fig. 3-5). Note a region with positive divergence of E-P flux in the
troposphere on 25 and 28 January 1973 in Fig. 3-2: the divergence of
the E-P flux indicates that there is a momentum source in that region.
For the 1977 and 1979 warming case displayed in Figs. 3-3 and 3-4, we
cannot find such a large region of momentum source in the troposphere.
For the 1980 warming I have no information on the E-P flux divergence.
From these facts we speculate that the 1973 warming occurred by the
action of planetary waves propagating upward from the troposphere while
the 1977 and 1979 warmings did not follow such a simple pattern.

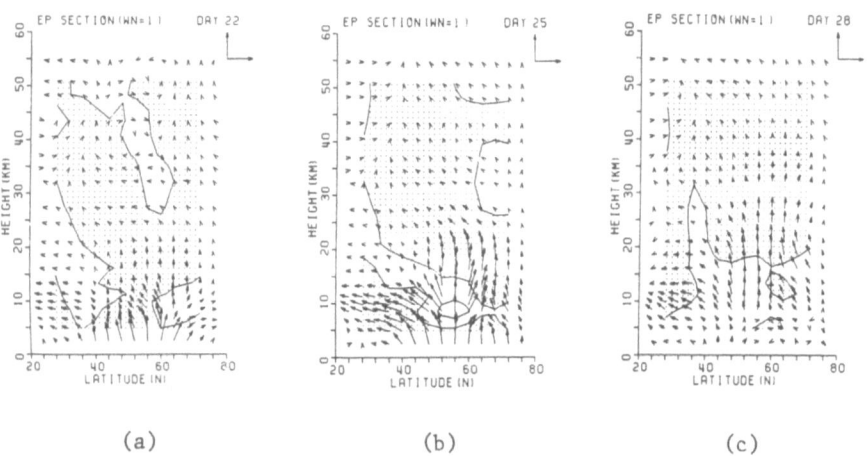

(a)   (b)   (c)

Fig. 3-2.   E-P cross-sections for wavenumber 1 on 22 (a), 25 (b) and
   28(c) January 1973. The meridional arrow at top right of cross-
   section represents a value of $E*^y$ equal to $(\rho_s a) \times (5 m^2 s^{-2})$ and
   the vertical arrow a value of $E*^z$ to $(\rho_s a) \times 5 m^2 s^{-2}$ divided by
   the factor 178. The factor 178 is selected for the same reason as
   in Fig. 3-3. $\nabla*\cdot E*$ is drawn with the interval of $(\rho_s a) \times 0.5$ m
   s^{-1} day^{-1}. Negative values are stippled. (after Kanzawa (1982))

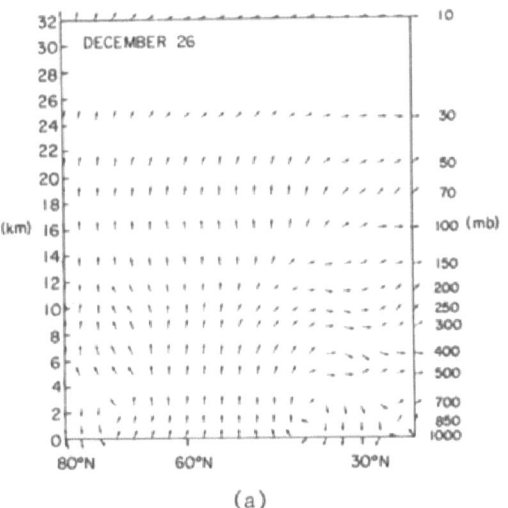

(a)

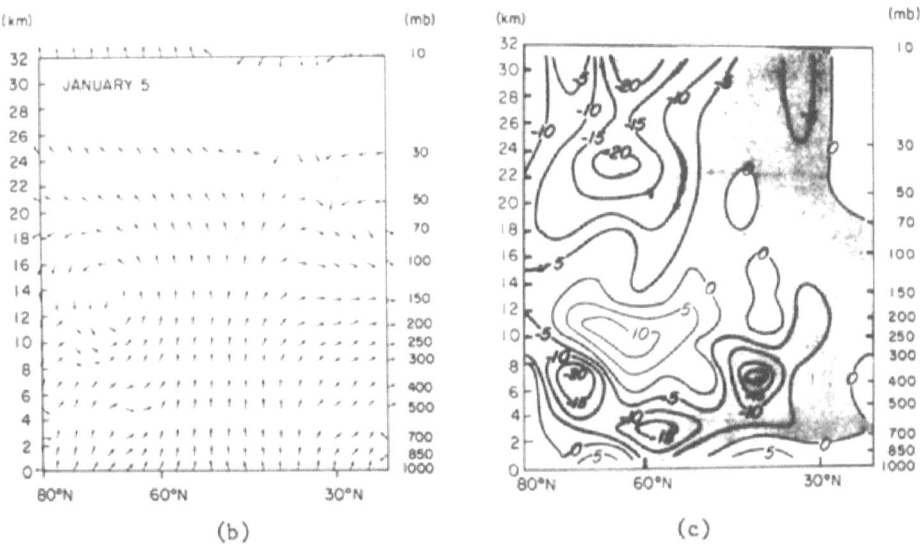

(b)   (c)

Fig. 3-3.   Directions of E-P fluxes for wavenumber 1 on 26 December
1976 (a) and 5 January 1977 (b) together with D_E (c) in units of
10^{-5} m s^{-1}.   Arrows of equal length are placed at an angle to the
horizontal which allows for stretching of the ordinate with respect
to the abscissa.   The vector is drawn at an angle A to the horizontal such that tanA = (E^Z/E^y) x ratio of distances representing 1 km
along the ordinate and abscissa.   (after O'Neill and Youngblut
(1982))

320

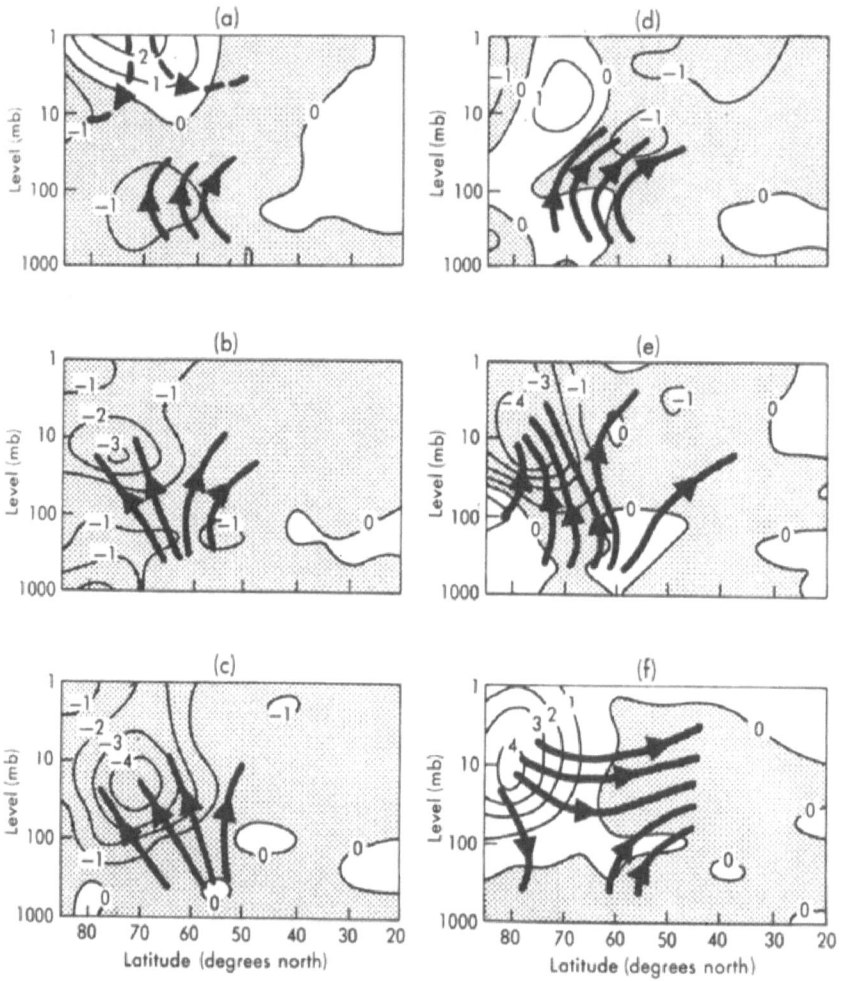

Fig. 3-4.   Contours of D_E labeled in units of 10^{-4} m s^{-2}, with some integral curves of E-P flux on 17 (a), 19 (b), 21 (c), 23 (d), 26(e) and 28(f) February 1979. Full curves are dominated by wavenumber 2 flux while dashed curves by wavenumber 1 flux. Negative values of D_E are stippled. (after Palmer (1981a))

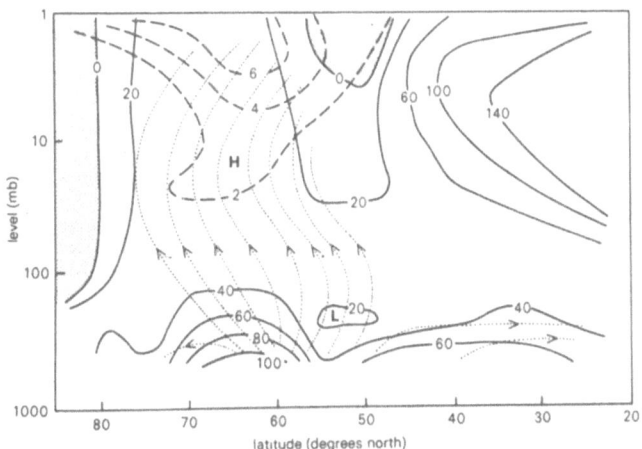

Fig. 3-5. Wavenumber 1 integral curves of E-P flux (dotted lines) and wavenumber 1 contours of D_E (dashed lines) in units of 10^{-4} m s^{-2} on 27 February 1980 together with contours of Q_1 on 26 February 1980 (solid lines). (after Palmer (1981b))

4.3 Focusing of the waves

The focusing of E-P flux into the high altitude polar cap region for the 1973 warming was attributed by Kanzawa (1980) to the existence of profile of mean zonal wind, which had a westerly maximum situated in the polar stratosphere as shown in Fig. 1-2a. A westerly maximum in the high latitude stratosphere means the existence of a high latitude stratospheric region with large quasi-geostrophic potential vorticity gradient $\bar{q}_y$ surrounded by a region with a small $\bar{q}_y$ (see Fig. 4-2a). If the WKB approximation is applicable, stationary planetary waves are refracted toward the region with larger refractive index as stated in section 2. Figs. 4-2 and 4-3 show meridional sections of the refractive index on 25 January 1973 and 3 January 1977. I have no information on the refractive index for the 1979 warming and the refractive index on 26 February 1980 is illustrated in Fig. 3-5. Note that since Q_1 minus Q is small (e.g., 3.7 (28°N), 7.5 (45°), 12.2 (60°) and 20.4 (72°) in units of Fig. 4-2b's contours) it can be said that the quantity Q determines the profile of refractive index (see the discussion given below Eq. (12)). As an example of the refractive index of normal winter we show Fig. 4-1 from Matsuno (1970) who used this wind model for his calculation of stationary planetary waves. When comparing these figures we detect the following common features in the preconditioned state. The maximum of refractive index can be traced in high latitudes from the troposphere to the stratosphere. This area is surrounded by a negative refractive index region which forbids stationary wave propagation. This

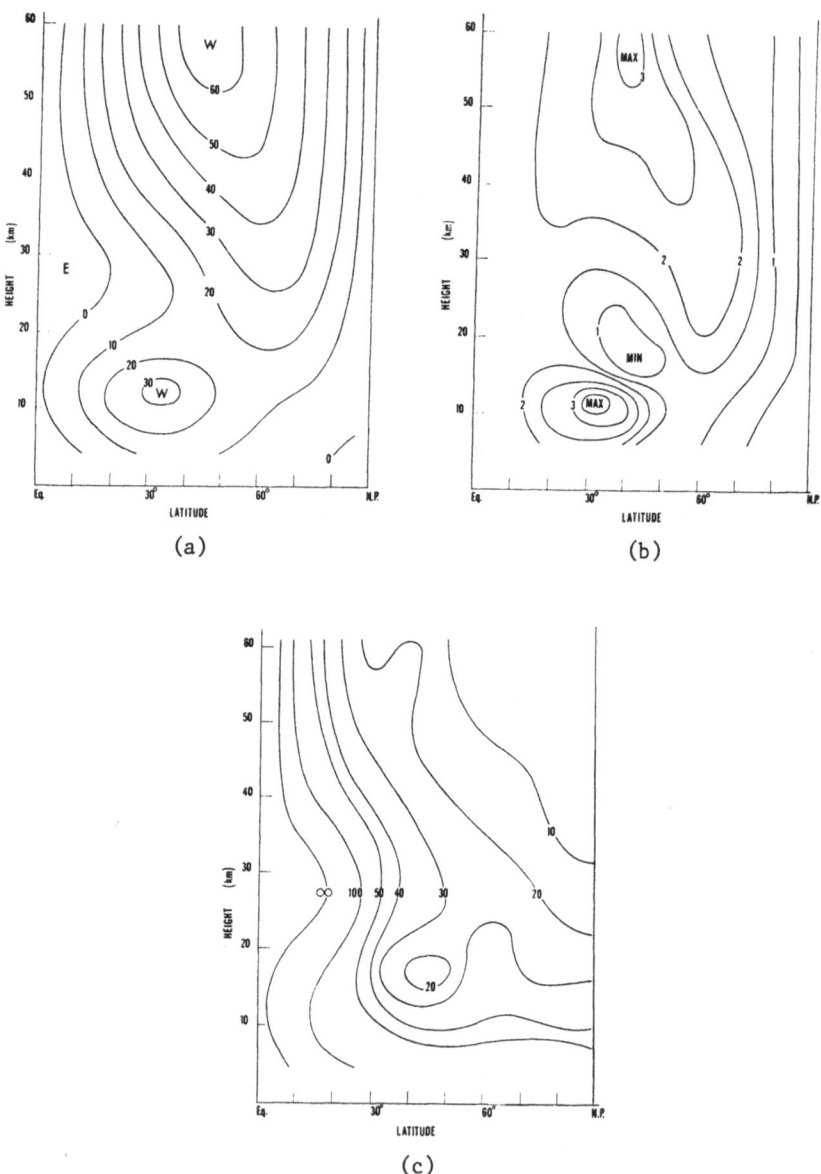

Fig. 4-1. The model basic state zonal wind distribution (m s^{-1}) in the winter Northern Hemisphere (a), the latitudinal gradient of the potential vorticity $\bar{q}_y$ (b) with unit of Ω/a and the refractive index square Q_0 (c) with unit of $1/a^2$. (after Matsuno (1970))

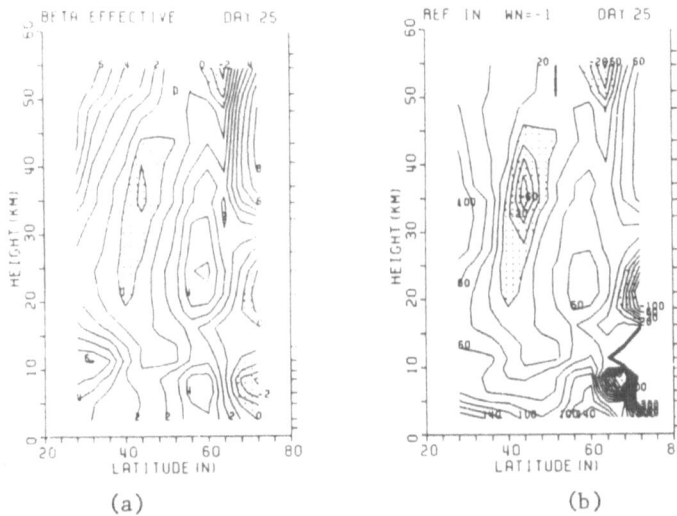

(a)   (b)

Fig. 4-2.   Latitude-height sections of $\overline{q}_y$ (a) and Q (b) on 25 January 1973.   The unit of $\overline{q}_y$ is Ω/a and that of Q is $1/a^2$. Negative values are stippled. (after Kanzawa (1982))

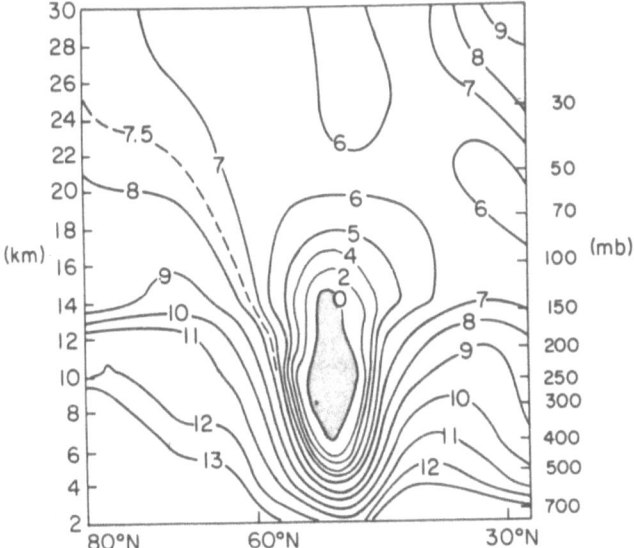

Fig. 4-3.   Refractive index $Q^{1/2}$ multiplied by the constant factor a to make it dimensionless on 3 January 1977 (computed from data averaged over the 5 days centered on this day). (after O'Neill and Youngblut (1982))

situation is in contrast to the refractive index corresponding to the normal winter wind condition which has a larger value equatorward as shown in Fig. 4-1c. For this distribution of refractive index the wave is considered to be focused into the polar stratosphere. Since the characteristics of the distribution of refractive index stated above are very similar to those of the $\bar{q}_y$'s (Fig. 4-1b), they are determined by $\bar{q}_y$ rather than by $\bar{u}_0$ itself. If we remember the definition of Q $(=\bar{q}_y/\bar{u}_0)$, we find that a maximum of $\bar{u}_0$ tends to enlarge Q by making $\bar{q}_y$ a maximum, but has also the opposing effect of weakening Q by dividing $\bar{q}_y$ by $\bar{u}_0$. We should pay attention to the maximum of refractive index in the high latitude troposphere which is associated with the double jet structure in the troposphere: Bridger and Stevens' (1982) numerical study showed that a mean wind profile producing a major warming should have relatively strong winds below the stratospheric polar night jet.

The differences between these figures of refractive index indicate the following fact: The pattern of 25 January 1973 and the pattern of 26 February 1980 are very similar, whereas the pattern of 3 January 1977 is different from these patterns: in that case the region of negative refractive index appears in the middle latitude troposphere. This pattern seems to have an intimate relation to the occurrence of the circulation reversal even in the troposphere in January 1977.

In a numerically simulated wavenumber 2 warming in a wave-zonal flow interaction model, Dunkerton et al. (1981) explained that the critical layer which formed around 40°N in the middle stratosphere reflected the wave and deflected the wave flux into the polar-cap region. However, a zero wind above 20°N in the stratosphere was not formed prior to all the real warmings reviewed in the present study (note that I have no information on mean zonal wind above the 10 mb level for the 1977 warming). In this regards, I think that an overall profile of mean zonal wind plays a more important role than a local critical layer in focusing waves during the pre-warming period. Very recently, this notion has also been presented by Butchart et al. (1982), based on the results of a series of numerical experiments.

4.4 WKB approximation

The concept of group velocity and Q introduced in section 2 comes from WKB analysis. Therefore it is necessary to discuss whether the condition for the WKB approximation to be applicable is realized or not. The condition is as follows,

$$I_t = (\partial \ln A_s/\partial t)(\partial B_s/\partial t + s\bar{\omega})^{-1} \ll 1$$

$$I_y = (\partial \ln A_s/\partial y)(\partial B_s/\partial y)^{-1} \ll 1$$

$$I_z = (\partial \ln A_s/\partial z)(\partial B_s/\partial z)^{-1} \ll 1$$

where A_s and B_s denote amplitude multiplied by $e^{-z/2H}$ and phase respectively of wavenumber=s and $\bar{\omega}$ is the angular velocity of the mean zonal wind: $\bar{\omega} = \bar{u}_0/a\cos\theta$. In the definition of I_t Kanzawa (1982) missed the

Doppler shift term $s\bar{\omega}$. Therefore the correct value of I_t is smaller than the value calculated by Kanzawa (1982): the value of I_t at 60°N and 40 km for day 25-28 is corrected from 1.7 to 1.0. According to Kanzawa's (1982) calculation of I_t, I_y and I_z for wavenumber 1 amplitude and phase prior to the 1973 warming, I_z is of the order of 0.1 while I_t and I_y often exceed 1 since the amplitude changes quickly and the constant phase line is approximately horizontal in the meridional plane (this phase structure is relevant to the fact that E-P flux is directed almost upward and indicate that quasi-normal modes were formed in the meridional direction). Palmer (1981b) stated that I_y and I_z are of the order of 1 during the build up to the 1980 warming. Thus the WKB approximation is not so good particularly in the temporal and meridional direction. However, the concept is useful for qualitative understanding of wave-mean flow interaction during sudden warmings as stated above.

4.5 After the circulation reversal

In order to examine the role of the critical line ($\bar{u}_0-c=0$ line) after the easterlies have appeared, we have Figs. 5-1, 5-2 and 5-3 which

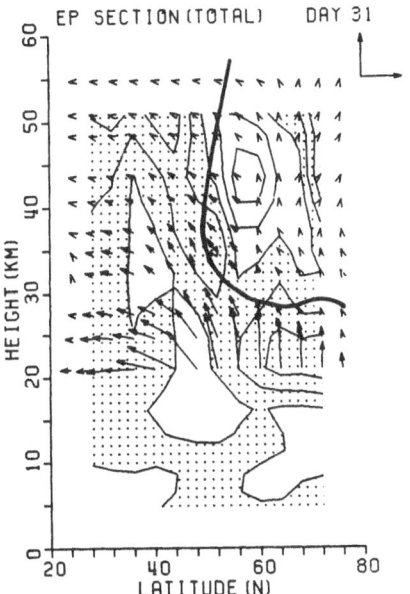

Fig. 5-1.   E-P cross-sections for wavenumber 1 on 31 January 1973, together with the zero line of mean zonal wind (thick line). The arrows are drawn only above the 20 km level with the length 2.5 times larger than that of Fig. 3-2. Contours are those of D_E drawn with the interval of 10 m s^{-1} day^{-1}. Negative values are stippled. (after Kanzawa (1982))

326

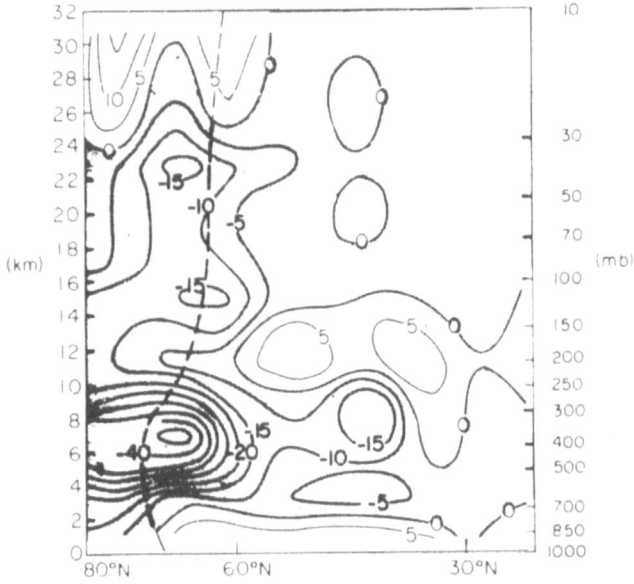

Fig. 5-2. Distribution of D$_E$ in units of 10^{-5} m s^{-2} on 10 January 1977 together with the effective critical line (dashed line) for waves retrogressing at 4 m s^{-1}, being the approximate value of the westward phase speed of waves 1 and 2 around 60°N in the stratosphere and upper troposphere. (after O'Neill and Youngblut (1982))

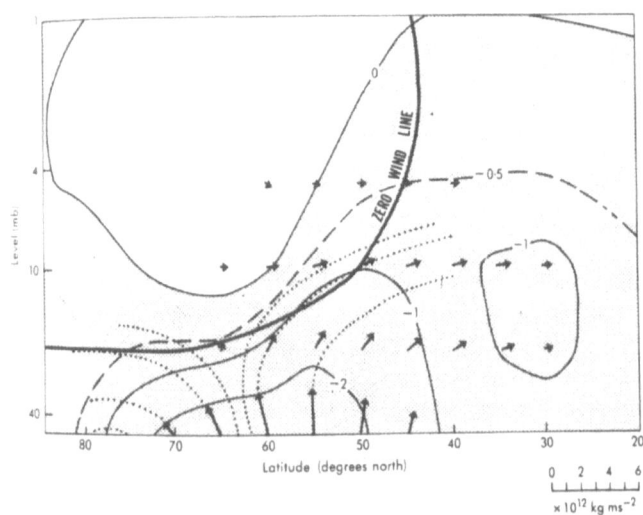

Fig. 5-3. E-P flux and divergence on 1 March 1980 together with the zero wind line. Dotted lines show some integral curves of E-P flux. (after Palmer (1981b))

show E* or D_E with the critical line in the meridional plane on day 31 January 1973, 10 January 1979 and 1 March 1980. In Figs. 5-1 and 5-3 the zero wind lines are drawn in place of the critical lines since the wave phase velocities were nearly zero. It is found that E-P vectors have small values after passing through the critical level and large negative D_E is near to the critical line. This fact suggests critical line absorption of the waves. Note that the 1979 warming was a case of a nearly vertically oriented zero-wind line at polar latitudes which is different from the other cases. The pattern of E-P section of Fig. 5-1 is found to be determined not only by wavenumber 1 component but also by wavenumber 2 and 3: the wavenumber 2 and 3 waves are speculated to be generated by nonlinear wave-wave interaction mechanism.

5. DISCUSSION AND CONCLUDING REMARKS

I have reviewed recent observational studies of sudden warmings used the E-P flux and refractive index diagnostics, which are considered to be very useful tools for diagnosing wave-mean flow interactions. The focusing of planetary waves of wavenumber 1 or 2 into the polar strato-sphere occurred prior to the sudden warmings. The focusing of the E-P flux into the polar stratosphere which has lower density and smaller area is essential to bring about the circulation reversal from the westerly to easterly winds. When the focusing occurred, the maximum of refractive index was traced in high latitudes from the troposphere to the stratosphere. We can conclude that the distribution of refractive index focused planetary waves into the polar stratosphere. The maximum of refractive index stems from that of $\bar{u}_0$ through that of $\bar{q}_y$. The maximum of mean zonal wind in high latitudes that can make a maximum of refractive index can guide planetary waves into high latitudes.

Note that the existence of the westerly maximum in the high lati-tude troposphere means that a double jet structure is formed in the troposphere. Tropospheric double jets are a rare phenomenon in the Northern Hemisphere, perhaps appearing in association with blocking and sudden warming, while we can find a double jet even in the monthly mean zonal wind cross-section for July in the Southern Hemisphere, as illus-trated in Fig. 9 of Bengtsson et al. (1982) based on FGGE data sets. It would be interesting to investigate E-P flux behavior in the Southern Hemisphere winter using the FGGE data sets.

The profile of mean zonal wind preconditioned for major sudden warmings is considered to be formed by the preceding minor warmings as first indicated by Kanzawa (1980). In the preceding minor warming of 1973, the E-P vector was not so confined to the polar area and the E-P flux convergence brought about a wind minimum in the middle latitude stratosphere. However to discuss in detail the problem of precondi-tioning by preceding minor warmings is beyond the scope of the present study.

In the present study, the mean zonal wind profile prior to major warmings is discussed from the viewpoint of whether the profile is suitable for guiding planetary waves into the polar stratosphere. However, with respect to the wave amplification problem, it would be interesting to investigate whether the wind profile has a nearly stationary free wave mode of zonal wavenumber 1 or not, by extending Tung and Lindzen's (1979) work on linear resonance theory to include meridional shear: the calculation would be a test of the resonant amplification theory. Whether Plumb's (1981) nonlinear resonance mechanism would work or not depends also on whether the zonal mean basic state satisfies the nearly resonant condition or not. Anyway, the weak or negative refractive index region surrounding the polar region is considered to play an important role in confining the wave energy to the polar region, as has already been indicated by Kanzawa (1980). On the resonant amplification hypothesis, McIntyre (1982) gave detailed discussions.

Once easterlies have appeared the mechanism of the critical layer absorption of stationary waves might occur. This speculation is based on the fact that the position of the maximum line of convergence of the E-P flux coincides with that of the critical line.

Of course, the approximations used in these theoretical concepts can not be directly applied to the real phenomenon of sudden stratospheric warming, since the spatial and temporal variations in the amplitudes of the waves are often the same order with those of the phase; the mean flow does not necessarily vary more slowly than the waves, and the effects of diabatic heating and nonlinear wave-wave interaction are not necessarily negligible. However, as shown in the present work, which reviews four sudden warmings, the theoretical concepts are very useful in order to understand the dynamical features of sudden stratospheric warmings in terms of wave-zonal flow interaction.

ACKNOWLEDGEMENTS

I wish to express my thanks to Prof. T. Matsuno of University of Tokyo for his nominating me as a member of participants in this symposium on Dynamics of the Middle Atmosphere. I am grateful to Profs. J. R. Holton and C. B. Leovy of University of Washington, the former for his comments on the original version of this paper, and the latter for his comment during the seminar on the definition of I_t in the subsection 4.4. I am also grateful to Prof. S. Kawaguchi of National Institute of Polar Research for his recommending me to attend the symposium. Thanks are also due to Drs. M. A. Geller, T. Matsuno, A. O'Neill and T. N. Palmer for permission to reproduce their figures, and Miss S. Nishikawa for typing.

REFERENCES

Andrews, D.G. and McIntyre, M.E., 1976: Planetary waves in horizontal and vertical shear: the generalized Eliassen-Palm relation and the mean zonal acceleration. J. Atmos. Sci., 33, 2031-2048.

Bengtsson, L., Kanamitsu, M., Kallberg, P. and Uppala, S., 1982: FGGE research activities at ECMWF, GARP topics No. 71. Bull. Am. Meteorol. Soc., 63, 277-303.

Bridger, A.F.C. and Stevens, D.E., 1982: Numerical model of the strato-spheric sudden warming: some sensitivity studies. J. Atmos. Sci., 39, 666-679.

Butchart, N., Clough, S.A., Palmer, T.N. and Trevelyan, P.J., 1982: Simulations of an observed stratospheric warming with quasigeostro-phic refractive index as a model diagnostic. Quart. J. Roy. Meteor. Soc., 108, 475-502.

Charney, J.G. and Drazin, P.G., 1961: Propagation of planetary-scale disturbances from the lower into the upper atmosphere. J. Geophys. Res., 66, 83-110.

Dunkerton, T., Hsu, C.-P.F. and McIntyre, M.E., 1981: Some Eulerian and Lagrangian diagnostics for a model stratospheric warming. J. Atmos. Sci., 38, 819-843.

Edmon, H.J., Hoskins, B.J. and McIntyre, M.E., 1980: Eliassen-Palm cross sections for the troposphere. J. Atmos. Sci., 37, 2600-2616.

Geller, M.A., Wu, M.F. and Gelman, M.E., 1983: Troposphere- stratosphere (surface-55km) monthly winter general circulation statistics for the northern hemisphere - four year averages. J. Atmos. Sci., 40 (in press).

Gille, J.C. and Lyjak, L.V., 1983: An overview of wave-mean flow in-teractions during the winter of 1978-79 (in this volume).

Holton, J.R., 1975: The dynamic meteorology of the stratosphere and mesosphere. Meteor. Monogr., No.37, American Meteorological Society, 218 pp.

Hsu, C.-P.F., 1981: A numerical study of the role of wave-wave interac-tions during sudden stratospheric warmings. J. Atmos. Sci., 38, 189-214.

Kanzawa, H., 1980: The behavior of mean zonal wind and planetary-scale disturbances in the troposphere and stratosphere during the 1973 sudden warming. J. Meteor. Soc. Japan, 58, 329-356.

330

_____, 1982: Eliassen-Palm flux diagnostics and the effect of the mean zonal wind on planetary wave propagation for an observed sudden stratospheric warming. J. Meteor. Soc. Japan, 60, 1063-1073.

Karoly, D.J. and Hoskins, B.J., 1982: Three dimensional propagation of planetary waves. J. Meteor. Soc. Japan., 60, 109-123.

Matsuno, T., 1970: Vertical propagation of stationary planetary waves in the winter northern hemisphere. J. Atmos. Sci., 27, 871-883.

_____, 1971: A dynamical model of the stratospheric sudden warming. J. Atmos. Sci., 28, 1479-1494.

_____, 1983: Understanding of sudden warming mechanisms by numerical experiment (in this volume).

_____ and Nakamura, K., 1979: The Eulerian- and Lagrangian- mean meridional circulations in the stratosphere at the time of a sudden warming. J. Atmos. Sci., 36, 640-654.

McIntyre, M.E., 1980: An introduction to the generalized Lagrangian-mean description of wave, mean-flow interaction. Pure Applied Geophys., 118, 152-176.

_____, 1982: How well do we understand the dynamics of stratospheric warmings? J. Meteor. Soc. Japan, 60, 37-65.

O'Neill, A. and Taylor, T.F., 1979: A study of the major stratospheric warming of 1976/77. Quart. J. Roy. Meteor. Soc., 105, 71-92.

_____ and Youngblut, C.E., 1982: Stratospheric warmings diagnosed using the transformed Eulerian-mean equations and the effect of the mean state on wave propagation. J. Atmos. Sci., 39, 1370-1386.

Palmer, T.N., 1981a: Diagnostic study of a wavenumber-2 stratospheric sudden warming in a transformed Eulerian-mean formalism. J. Atmos. Sci., 38, 844-855.

_____, 1981b: Aspects of stratospheric sudden warmings studied from a transformed Eulerian-mean viewpoint. J. Geophys. Res., 86, 9679-9687.

_____, 1982: Properties of the Eliassen-Palm flux for planetary scale motions. J. Atmos. Sci., 39, 992-997.

_____ and Hsu, C.-P.F., 1983: Stratospheric sudden coolings and the role of nonlinear wave interactions in preconditioning the circumpolar flow. J. Atmos. Sci. (to be submitted).

Plumb, R.A., 1981: Instability of the distorted polar night vortex: a theory of stratospheric sudden warmings. J. Atmos. Sci, 38, 2514-2531.

Sato, Y., 1980: Observational estimates of Eliassen and Palm flux due to quasi-stationary planetary waves. J. Meteor. Soc. Japan, 58, 430-435.

Tung, K.K. and Lindzen, R.S., 1979: A theory of stationary long waves. part II: resonant Rossby waves in the presence of realistic vertical shears. Mon. Wea. Rev., 107, 735-750.

Uryu, M., 1974: Mean zonal flows induced by a vertically propagating Rossby wave packet. J. Meteor. Soc. Japan, 52, 481-490.

J. R. Holton and T. Matsuno, Dynamics of the Middle Atmosphere, 333-351.
Copyright © 1984 by Terra Scientific Publishing Company.

DYNAMICS OF MINOR STRATOSPHERIC WARMINGS AND "PRECONDITIONING"

Taroh Matsuno

Geophysical Institute
University of Tokyo

ABSTRACT

Recent satellite observations of infrared radiances reveal that
sudden warmings in the upper stratosphere take place rather frequently,
but most of them do not spread down to the lower stratosphere; thus they
are minor warmings by definition. It is also found that after such warmings strong westerly jets are often formed at very high latitudes, which
guide succeeding planetary waves to the polar upper stratosphere to
cause major warmings (preconditioning). The present paper deals with
numerical modeling of those phenomena. A boundary condition that the
tropospheric wave forcing amplifies for a finite period and then decays
is imposed and the response of the middle atmosphere is investigated by
a numerical model which includes nonlinear interactions between wave 1
and wave 2. When a forcing with the maximum amplitude of 300 m is given,
a weak warming associated with weakening of the original westerly jet
occurs, but it is followed by the formation of a very strong westerly
jet (80 m s^{-1}) centered at 70°N, shortly after the decay of the forcing.
This result is qualitatively similar to that of Palmer and Hsu (1983)'s
work concerning the same problem. When a stronger forcing (400 m) is
given, a major warming occurs and the high latitude westerly jet does
not appear.

The evolutions of the potential vorticity distributions in these
two experiments are presented. In the minor warming case, potential
vorticity is mixed at middle latitudes whereas at high latitudes the
maximum region shifts but remains less disturbed; and it returns to the
pole after the decay of the forcing. In the case of the major warming,
the whole hemisphere undergoes irreversible mixing of potential vorticity. These evolutions are in good agreement with McIntyre's (1982)
arguments concerning the distinction between major and minor warmings
and the cause of the formation of a strong westerly jet after a minor
warming.

1. INTRODUCTION

In recent years there has been great progression in the satellite

333

334

observation of infra-red radiances and the data acquired now enable us
to visualize the flow and temperature fields of the whole stratosphere
on daily basis. A number of interesting features have been revealed
which were not known a decade ago when radiosonde observation was the
only source of the stratospheric weather maps. Among these is a minor
stratospheric warming whose true nature has only been revealed by satellite data. There seems to be no essential difference between major and
minor warmings as measured by the radiance changes during January and
February of 1979 shown in Fig. 1 (taken from Labitzke, 1981), even though
only the last peak is eligible to be called major because it spread down
to the lower stratosphere.[1]

There remains, however, a question as to whether there is no difference at all between the two types of warmings and whether the occurence
of the major warming was related to the preceeding phenomena. A number
of diagnostic studies utilizing radiance-retrieved data have been made
and it is now widely accepted that most minor warmings are followed by

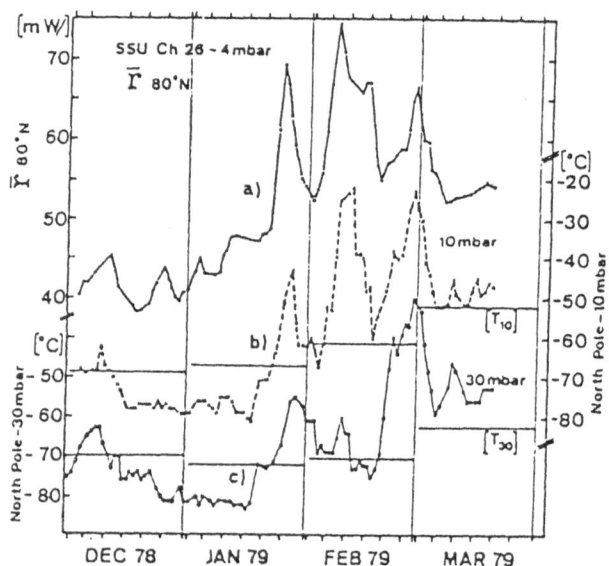

Fig. 1. Radiance and temperature over the polar region at various altitudes for the period December 1978 - March 1979. (a) radiances roughly
corresponding to the 4mb temperature. (b) temperature at the 10mb
level. (c) temperature at the 30mb level. (after Labitzke, 1981)

1) According to the WMO definiation, a stratospheric warming can be said
to be major if at 10mb or below the zonal mean temperature increases poleward from 60° latitude and an associated circulation reversal (from westerly to easterly) is observed poleward of 60° latitude.

the formation of rather strong circumpolar westerly jets at high latitudes which in turn produce favorable conditions for succeeding planetary waves to propagate deep into the polar upper stratosphere and to
easily cause major warmings (e.g. Kanzawa, 1980; Palmer, 1981a,b). This
particular chain of events is now called "preconditioning" (e.g. Labitzke,
op. cit.). The role of the strong westerly jet was also discussed on
the basis of wave propagation theory for a numerically modeled warming
(Dunkerton, Hsu and McIntyre, 1981). Butchart et al. (1982) performed
a series of numerical experiments to clarify the roles of various factors in causing a warming and demonstrated that the strong westerly jet
formed after a minor warming is indeed important for guiding the planetary waves.

An excellent review of those studies including discussions of the
above mentioned problems is given by McIntyre (1982). In that articles,
among other things, he presented a new view on the distinction between
major and minor warmings and the role of the dynamical mechanism of preconditioning, as summarized in the following. At the time of a sudden
warming, the whole stratosphere undergoes a hemisphere-wide mixing of
potential vorticity by the action of large amplitude planetary waves.
The mixing is strongest at middle latitudes where the waves attain their
maximum amplitude. Such a mixing produces a region of uniform potential
vorticity at middle latitudes which implies a distribution of the zonal
mean wind with a high latitude westerly jet as shown in Fig. 2 (thick

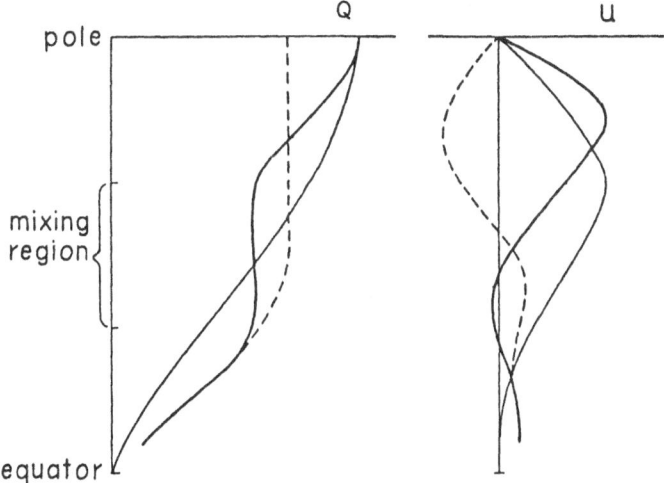

Fig. 2 Schematic latitudinal distributions of Ertel's potential vorticity Q on an isentropic surface and corresponding polar-night jet profiles u. Thin solid lines represent the initial states and thick
solid lines and dashed lines show distributions after minor and major
warmings, respectively (Adapted from McIntyre, 1982 by adding dashed
curves).

solid line). If the mixing is very strong, even the polar cap region may be involved and the result would be to produce a uniform distribution of potential vorticity including this region; an easterly may establish over the pole in this case (dashed line in Fig. 2). Thus the difference between major and minor warmings is due to the difference in the extent of potential vorticity mixing.

In the present work, the author attempts to reproduce these phenomena by use of a numerical model and to analyze the results from the view point of potential vorticity mixing. A substantial part of the results reported in this article was presented at the U.S.-Japan Seminar on the Dynamics of the Middle Atmosphere, but the analysis of potential vorticity was performed later in response to a question raised by Dr. Alan Plumb.

Recently Palmer and Hsu (1983) carried out a series of numerical experiments to identify the mechanism leading to preconditioning or "sudden cooling". They were able to reproduce a circumpolar westerly jet after the occurence of a warming in one experiment and demonstrated that inclusion of wave-wave interaction is indispensable to bringing about those results. The results of the present work are similar to theirs. However, they diagnosed the process in terms of the Eliassen-Palm fluxes and their divergences; an analysis of the potential vorticity field was not conducted.

In a recent paper McIntyre and Palmer (1983) presented the evolution of Ertel's potential vorticity on an isentropic surface for the January-February 1979 warming event. They have shown that erosion of the large potential vorticity area occurred at middle latitudes at the minor warming stage and also that a major part of the large potential vorticity area broke down into two pieces at the major warming stage.

2. MODEL

The numerical model used in the present work is basically the same as that of Matsuno (1971), which is a hemispherical geostrophic model treating mutual interactions of the zonal mean fields and one wave component, and the governing equations are discretized by use of a semi-spectral method. In the present work, the following modifications were made.

Wave - Wave interactions

The model treats two wave components, i.e., zonal wavenumbers 1 and 2 and includes nonlinear interactions between the two waves. Accordingly the equations for waves 1 and 2 are written as follows.

$$(\frac{\partial}{\partial t} + i\bar{\omega})L_1(\psi_1) + \frac{\partial \bar{q}}{\partial \theta} \frac{i}{\sin^2\theta\cos\theta}\psi_1$$

$$= - \frac{i \exp(z/2H)}{2fa^2\cos\theta}[2\psi_2\frac{\partial q_1^*}{\partial \theta} - \psi_1^*\frac{\partial q_2}{\partial \theta} + \frac{\partial \psi_2}{\partial \theta}q_1^* - 2\frac{\partial \psi_1^*}{\partial \theta}q_2] \quad (2-1)$$

$$(\frac{\partial}{\partial t} + 2i\bar{\omega})L_2(\psi_2) + \frac{\partial\bar{q}}{\partial\theta}\frac{2i}{\sin^2\theta\cos\theta}\psi_2$$

$$= -\frac{i\ \exp(z/2H)}{2fa\ \cos\theta}[\psi_1\frac{\partial q_1}{\partial\theta} - \frac{\partial\psi_1}{\partial\theta}q_1] \qquad (2-2)$$

Equation (2-1) is for wave 1 and (2-2) is for wave 2. The full notation is found in Matsuno (1971) and may be omitted here. The modification is the inclusion of the right hand side terms, which represent exchanges of q (the potential vorticities in a slightly modified form) between the two waves. The zonal mean equation remains the same as the original one except that it now includes quadratic terms associated with the two waves simultaneously. The system does not completely conserve the total potential enstrophy in the modified form as pointed out by Schoeberl (1983).

Dissipative effects

Dissipative effects i.e., a Newtonian cooling and a Rayleigh friction are incorporated. For waves the both effects are formulated so as to diminish the wave amplitudes, but for the zonal mean fields they are assumed to act to restore the fields to a prescribed equilibrium state. The rate coefficients of these effects are functions of height as shown in Fig. 3. The Newtonian cooling coefficient is approximately the same as the one proposed by Dickinson (1973). The Rayleigh friction was designed in order to absorb waves at the uppermost levels.

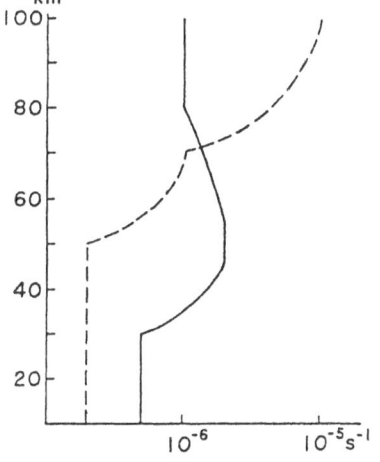

Fig. 3 Vertical distributions of the Newtonian cooling coefficient (solid line) and the Rayleigh friction coefficient (dashed line) adopted in the present experiments.

The above system is essentially the same as those due to Lodi et al. (1980) and Hsu (1981), though these authors used primitive equation models.

The integration domain is 0° - 90° in the latitudinal direction and 10 - 110 km in the vertical. The wave amplitudes and phases can be given as functions of latitude and time at the bottom boundary. In some experiments simultaneous forcing of two waves were attempted, but in the experiments reported in this article forcing is limited to zonal wavenumber 1; the amplitude is distributed from 30° to 90° with the maximum at 60° and the phase is constant (stationary waves). In these respects the experiments are similar to those of Lordi et al. (1980) and Hsu (1981). However, the time dependence of the forcing f(t) in the present study is quite different from the experiments performed by the above mentioned authors.

338

The functional form of the Boundary forcing

Among the functional forms of f(t) adopted in experiments, that which will be referred to most often is shown in Fig. 4. This form was designed so as to simulate qualitative- ly the evolution observed in the January 1973 event, according to the analysis by Kanzawa (1980). From preliminary experiments,

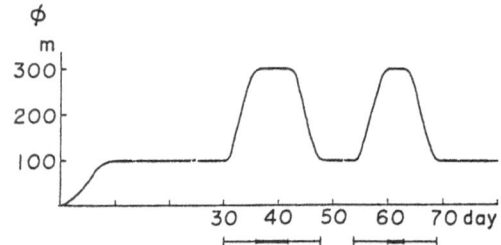

Fig. 4 Amplitude of wavenumber 1 forced at the bottom boundary as a function of time. The maximum level is 300 m for the minor warming experiment and 400 m for the major warming case.

which will be described later, it turned out that the presence of a quasi- steady wave of moderate amplitude prior to growth of the wave is impor- tant for producing an upper level warming. Thus the first 30 days are devoted to obtaining a quasi-steady state including wave 1 which is forced at the bottom with a constant amplitude (100 m). During the 39 days following day 30, the bottom forcing is amplified twice to reach a maximum amplitude of 300 m or 400 m, with a short intermission in bet- ween. Originally it was intended to simulate the 1973 event, so that the first amplification is assumed to correspond to the minor warming and the second one to the major warming. However, later it turned out that it is better to use the same functional form with different peak ampli- tudes for the purpose of investigating differences between major and minor warmings. Thus we will focus our attention on the first amplifi- cation with two different maximum levels and discard the second amplifi- cation, unless otherwise stated.

In their recent work, Palmer and Hsu (1983) performed a numerical experiment to simulate the January 1979 event. For this purpose they adopted a forcing which represent a westward moving wave with a speed of 12°/day, since Butchard et al. (1982) had found that this provided a more realistic simulation than did a stationary forcing.

3. RESULTS

3.1 Preliminary experiments

The most apparent difference between a minor warming and a major warming is that the former is limited to the upper stratosphere (see Fig. 1). In order to simulate such a warming by the numerical model, a linear version of the equations which retain only a wavenumber 1 component was integrated by imposing boundary conditions of a few different functional forms f(t). In view of the results of earlier model studies and their interpretations (e.g. Matsuno 1971; Holton, 1976), first it was naïvely supposed that a forcing of a relatively short duration might produce a warming in the upper stratosphere, because in those models warmings begin at upper levels. However, the results did not produce an upper level warming but only a weak warming at lower levels. This is because for a forcing of short duration the wave amplitude at upper levels does not grow to a sufficient magnitude, as long as the wave amplitude starts from zero. In the real atmosphere, planetary wave 1 develops in the whole stratosphere from the beginning of the autumn, and a wave of significant amplitude always exists. Warmings, whether major or minor, are produced by temporal enhancement of the tropospheric forcing.

Thus a quasi-steady wave of zonal wavenumber 1 was introduced with a constant forcing amplitude of 100 m. By day 30 both the wave and the mean field reach quasi-equilibrium states. The amplitude of the wave is 370 m at the 30 km level, which may be acceptable as representative of less disturbed winter states. Then the forcing amplitude was changed as shown in Fig. 4. The results of this experiment are reported in Matsuno (1983). Briefly, the wave amplitude developed quickly at upper levels which caused the desired upper level warming with a maximum temperature rise of about 50 K spreading from the 60 km level down to the 30 km level. Associated with this an easterly developed above about 50 km. These features are qualitatively similar to those of observed minor warmings. In this experiment the second amplification produced a low level warming at the altitude range 20 - 30 km. The whole evolution appeared to possess characteristics observed in the January 1973 event (see Kanzawa, 1980). However, as will be described later, the evolution becomes rather different when the wave-wave interactions are included. Therefore the above agreement might be partly accidental, and the results should be taken with caution.

3.2 Integration of the non-linear model with a weaker forcing

The full nonlinear system of equations was integrated with a boundary forcing for wave 1 only for a 75 day period. The functional form of the forcing is the one shown in Fig. 4, the maximum amplitude being 300 m. The results are shown in Fig. 5a - 5d, which are time-height sections of the amplitude of wave 1 at 60°N (5a, in decameter unit), wave 2 (b), the mean zonal velocity at 60°N (c), and the temperature at the pole (d). The last of these is represented by the departure from the initial value. As mentioned previously, the first 30 day period is used to obtain a quasi-steady state with a small amplitude wave 1, so that only evolutions in the period later than the day 30 are shown. Day 30 is effectively the initial time for the experiments of our main concern. However, the original time label will be retained in order to avoid confusion.

As seen in the figure, soon after the forcing at the bottom is amplified the wave amplitudes at upper levels also increase and the maximum appears with a rather short delay from the corresponding one in the bottom forcing. After the first amplification ends, the correspondence between the amplitude at upper levels and that of the bottom becomes unclear; a minimum occurs when the forcing is maximum. In the

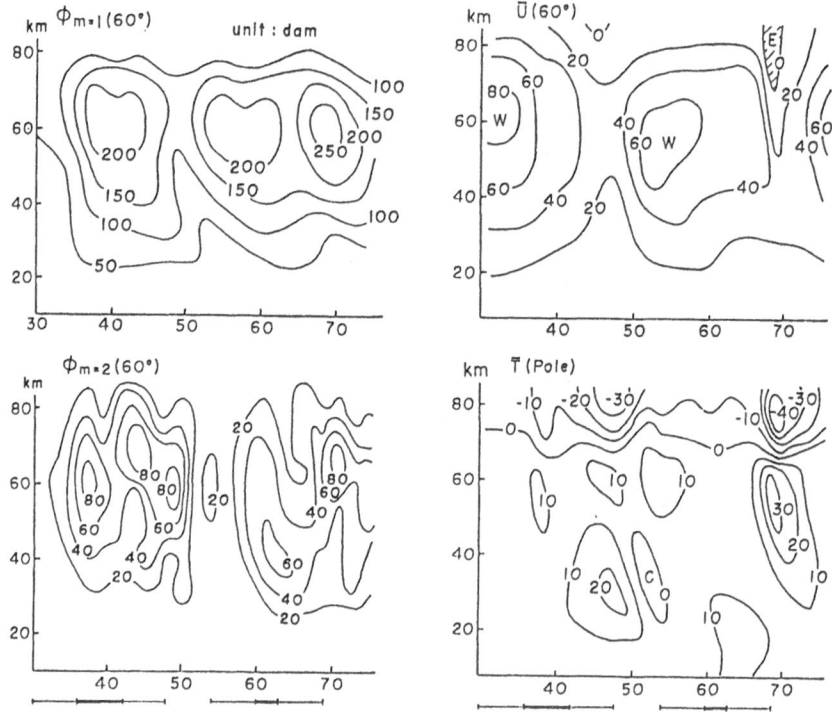

Fig. 5. Time-height sections of the amplitude of height wave 1 at 60°N
(a), that of wave 2 (b), the mean zonal wind at 60°N (c), and the temperature at the pole, in the case of the minor warming experiment.
Unit of heights in (a) and (b) is decameters.

real warming event in January 1979 a similar discordance between the wave
amplitude in the upper stratosphere and that at the tropopause level was
observed and it has been attributed to the presence of a free oscillation
mode (Madden and Labitzke, 1981). It appears that also in the model a
free oscillation is excited by the first pulse, as has been pointed out
by Schoeberl and Strobel (1980), and consequently the subsequent evolution becomes complicated. It is possible that the length of the interval
between two successive pulses might be important for causing a warming.
However, this problem is not a main concern of the present study and will
not be discussed further.

In accord with the amplification of wave 1 the mean zonal westerly
at 60°N weakens during day 40 - day 50 and simultaneously the pole temperature rises as seen Figs. 5c and 5d. Above about 65 km a cooling
occurs. These features show that a kind of sudden warming takes place.
However, this model warming is not too much like a minor warming in that
the warming layer is not centered in the upper stratosphere and the peak
temperature rise is less than 30 K. A weak easterly appeared at about

80 km but that level is too high. Compared with this event, the later one at about day 70 has greater similarity to observed minor warmings. This one was originally intended to correspond to the major warming of January 1973 and in fact the desired result was obtained in the linear experiment (Matsuno, 1983). Notwithstanding these shortcomings, we shall examine the first event as an example of a model minor warming in the present study.

An interesting feature of this event is the rather quick recovery of the mean westerlies after the warming (5c). The time constant of the Newtonian cooling is 5 days around the stratopause level (see Fig. 3), so that the westerly recovery might be due to this effect. Though this effect must be working to some degree, the westerly acceleration is much more, especially at higher latitude as shown in Fig. 6, which is the time height section of the mean zonal wind at 70°N. Very strong westerly winds exceeding 80 m s^{-1}, stronger than the quasi-equilibrium value appear after the minor warming. Recalling the thermal wind relationship, we suppose that a larger meridional temperature gradient should be produced at this latitude. Indeed the polar temperature (Fig. 5d) shows a slight cooling compared with day 0, the polar temperature at day 30 being 3 K. It is suggested that the westerly acceleration at high latitude

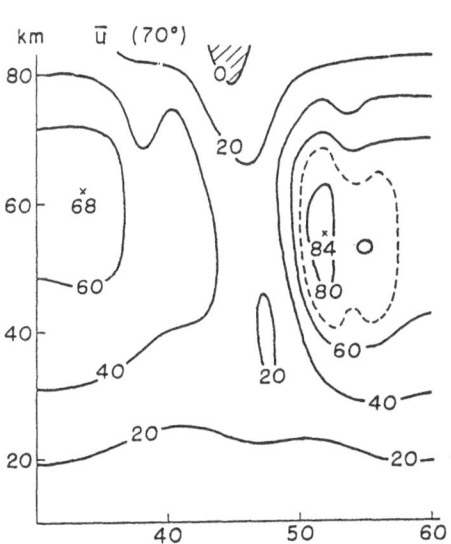

Fig. 6. Time-height section of $\bar{u}$ at 70° N.

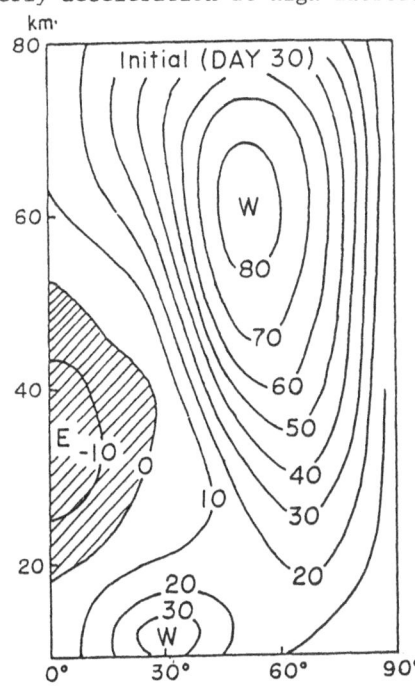

Fig. 7. Distribution of the mean zonal winds in the meridional plane at day 30 (quasi-steady state including a small amplitude wave).

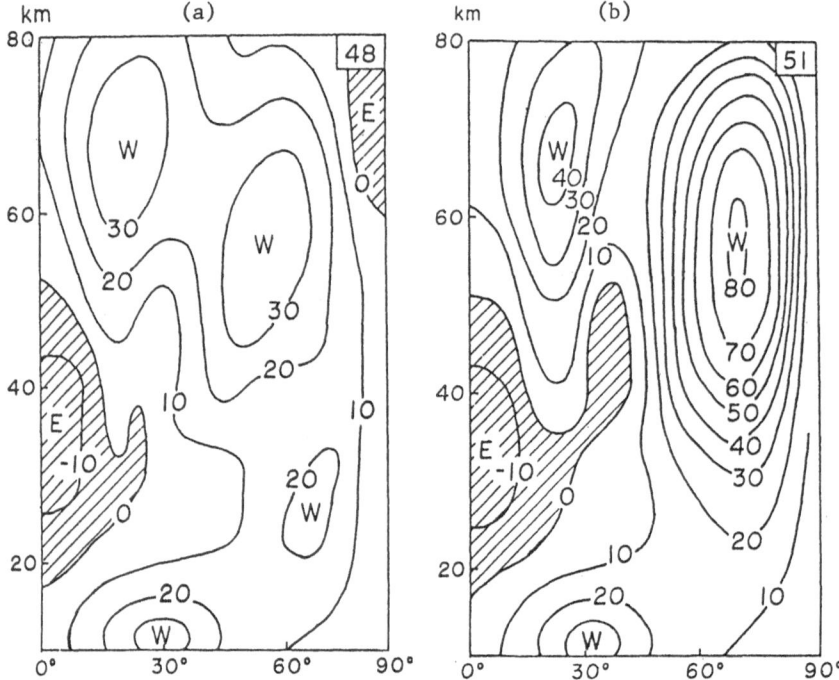

Fig. 8 Same as Fig. 7 but for day 48 (a) and day 51 (b) in the case of the minor warming experiment.

is not due to thermal relaxation but is due to dynamical effects. Such a phenomenon is known to exist in the real stratosphere as mentioned in the Introduction, and Mahlman (1967) has named it "the sudden cooling" (see Palmer and Hsu, 1983).

The above described change is best seen in the meridional sections of the mean zonal wind $\bar{u}$, as shown below. Fig. 7 is the $\bar{u}$-distribution at day 30, which is the quasi-steady state including a small amplitude wave. The equilibrium state (day 0) to which the zonal mean quantities are assumed to restore is very little different from this distribution. Fig. 8a is the meridional cross section at day 48, the time when the minor warming attains its maximum. The strato-mesospheric westerly jet is weakened at upper levels and a region of weak easterlies appears in the polar mesosphere. The westerlies at lower stratospheric levels remain less disturbed. Remarkable changes take place between this time and day 51. As seen in Fig. 8b a strong westerly jet is generated in the whole middle atmosphere with the maximum wind speed at 70°N. The change occurred only in a 3 day period and hence the Newtonian cooling effect must have only a small contribution. This point can be confirmed by a comparison with Fig. 7, the $\bar{u}$-distribution to which Newtonian cooling acts to restore the flow. Some dynamical effects must have produced the jet. Comparing Figs. 8a and 8b more closely, we see that below 60 km

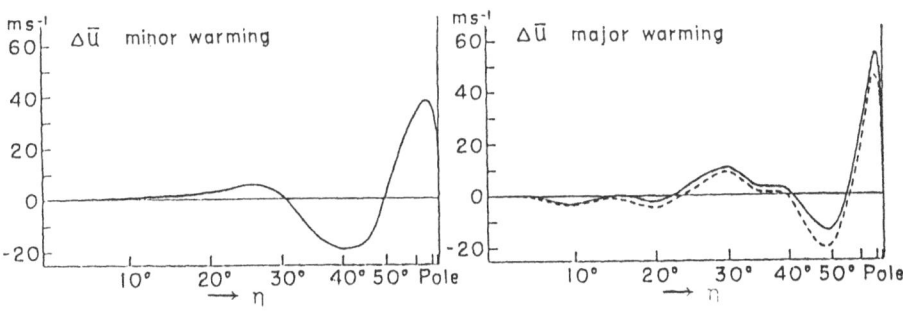

Fig. 9 Changes of the zonal mean wind, Δu = u(day 51) − u(day 48) at the
40 km level shown as functions of latitude (abscissa) whose scale is
deformed in such a way that areas enclosed by curves and the axis
give angular momentum. Left diagram: minor warming experiment.
Right: major warming experiment with and without Newtonian cooling
(solid and dashed lines).

the westerly wind speed at 30° − 40° decreases by about 10 m s⁻¹ in this
3 day period. In Fig. 9 (left panel) the change of the mean zonal wind
at the 40 km level from day 48 to day 51 is shown as a function of η
which is related with latitude θ as dη = cos²θdθ, so that areas enclosed
by the curve and the η-axis are proportional to angular momentum. As
seen from the figure the total angular momentum in this layer changed
little, but a redistribution occurred. The horizontal momentum flux or
equivalently the horizontal component of the Eliassen-Palm flux is
thought to be an agent for producing the high latitude jet.

    Palmer and Hsu (1983) carried out numerical experiments concerning
the same phenomenon   as well as analyses of the observed sudden cooling in January 1979. They demonstrated that the Eliassen-Palm flux associated with wave 2 is responsible for producing a westerly jet at high
latitudes both in the numerical experiment and the observed event. In
the present experiment diagnosis using the Eliassen-Palm flux has not
been done yet. However, from the phase distributions of the two waves
it is speculated that the horizontal momentum flux due to wave 1 is more
important to the westerly acceleration. In fact, in the work of Palmer
and Hsu, the observed Eliassen-Palm flux divergences (Fig. 7 of Palmer
and Hsu, 1983) indicate that wave 1 contributed to the westerly accelerations at very high latitudes (∿70°N) in the January 1979 event. More
quantitative analyses are needed. However potential vorticity patterns
give us a simpler picture of the mechanism of the high latitude jet formation as suggested by McIntyre (1982), so that we may leave the momentum
budget calculation as a future problem.

3.3 Integration with a stronger forcing

    A similar integration but with a stronger forcing which attains
400 m at the maximum level shown in Fig. 4 was carried out. Evolutions
of the wave and mean fields are more or less similar to those in the

344

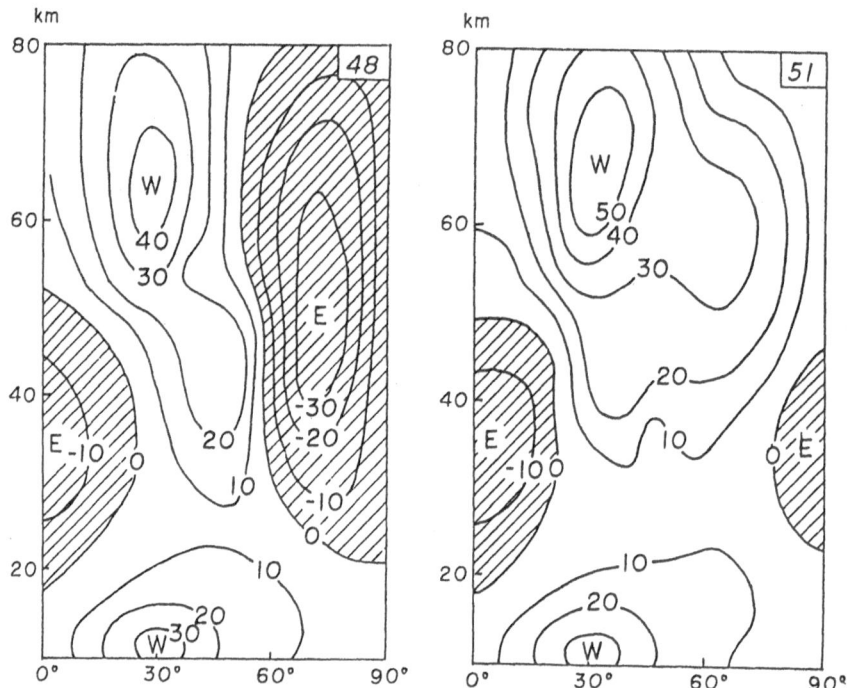

Fig. 10 Same as Fig. 8 but for the major warming experiment.

previous case. However, in the present case the mean zonal wind changed
from westerly to easterly first at upper levels and then spread downward
to reach ∿20 km. Corresponding to this a stronger warming with the maximum temperature rise of near 40 K occurred in the whole stratosphere, the
temperature rise at the 20 km level being 30 K. Thus this event can be
regarded as a major warming.

The meridional cross sections of the mean zonal wind are shown in
Figs. 10a and 10b. Remarkable differences from the corresponding figures
in the minor warming case may be apparent. The easterlies at day 48
(the time of the maximum warming) occupy a much larger domain and are
stronger in magnitude. At day 51, 3 days after the bottom forcing has
been reduced to the lowest value, easterly winds disappear in the upper
stratosphere and westerlies develop.

In this case a strong westerly jet at high latitudes as seen on
day 51 of the minor warming experiment did not appear. However, change
of the mean zonal flow from day 48 to day 51 is quite large especially
at very high latitudes. On the right of Fig. 9 the change of $\bar{u}$, i.e.,
$\Delta\bar{u} = \bar{u}(\text{day } 51) - \bar{u}(\text{day } 48)$ is shown as a function of the deformed latitude η. Apparently the distribution of $\Delta\bar{u}$ has a remarkable similarity
to that in the minor warming case. The quick change from easterly to

westerly winds at high latitude stratosphere took place at the expense
of depletion of westerlies at middle latitudes; poleward momentum transport should be responsible to the change. The change is not due to the
Newtonian cooling effect even though it contribute to westerly acceleration in this situation. The point is demonstrated by the dashed line
in the right figure which is the change of $\bar{u}$ in the same 3 day period
but for an experiment in which both Newtonian cooling and Rayleigh friction below 70 km were removed after day 48. Thus we understand that
there was a particular configuration of waves to accelerate westerlies
at very high latitudes in the recession phase of planetary wave activity
even in the major warming experiment. Comparison of this result with
observations is not yet undertaken.

3.4   Interpretation in terms of potential vorticity mixing

   As mentioned in Introduction, McIntyre (1982) suggested that the
difference between minor and major warmings and the mechanism of the high
latitude jet formation can be well understood in terms of potential vorticity mixing. Therefore, the horizontal distributions of potential vorticity at the 40 km level are examined. Fig. 11 shows rough distributions
of potential vorticity superposed upon isobaric height patterns at selected dates for the minor and major warming experiments. In the early
stages of wave amplification the potential vorticity patterns are rather
similar between the two cases. The wavenumber 1 disturbance advects low
potential vorticity from low to high latitudes in a spiral pattern. The
largest potential vorticity, which was originaly situated over the pole,
is shifted by the disturbance. There is a slight difference between the
two patterns at day 45. In the major warming case, weakening of the circumpolar westerlies is more pronounced (weak easterlies in the mean
zonal flow), so that the low potential vorticity region has penetrated
closer to the pole than in the minor warming case. This can be well
understood from kinematical considerations; the ratio of the meridional
to zonal velocities (or deflection angle measured from the zonal direction) is larger for the major warming case.

   At   day 48, the time of the maximum warming, the difference between
the two warmings become more pronounced. In the minor warming case, the
core of the maximum potential vorticity still retains a nearly circular
shape and a region of minimum (anti-cyclonic) vorticity is located off
the north pole, which has been cut off from the spiral arm shown at the
previous stage. In the major warming case, the maximum vorticity region
is no longer circular but much elongated and pushed away from the pole,
and a cut off anti-cyclonic vorticity region encroaches onto the polar
cap. The overall pattern may be regarded as a mixing (or turn-over) of
potential vorticity on the hemispheric scale. In the minor warming case,
a similar mixing occurs but is limited to the low and middle latitude
regions, precisely in accord with the arguments by McIntyre (1982).

   Consequently, the post-warming evolutions of the potential vorticity patterns are quite different between the two cases. In the minor

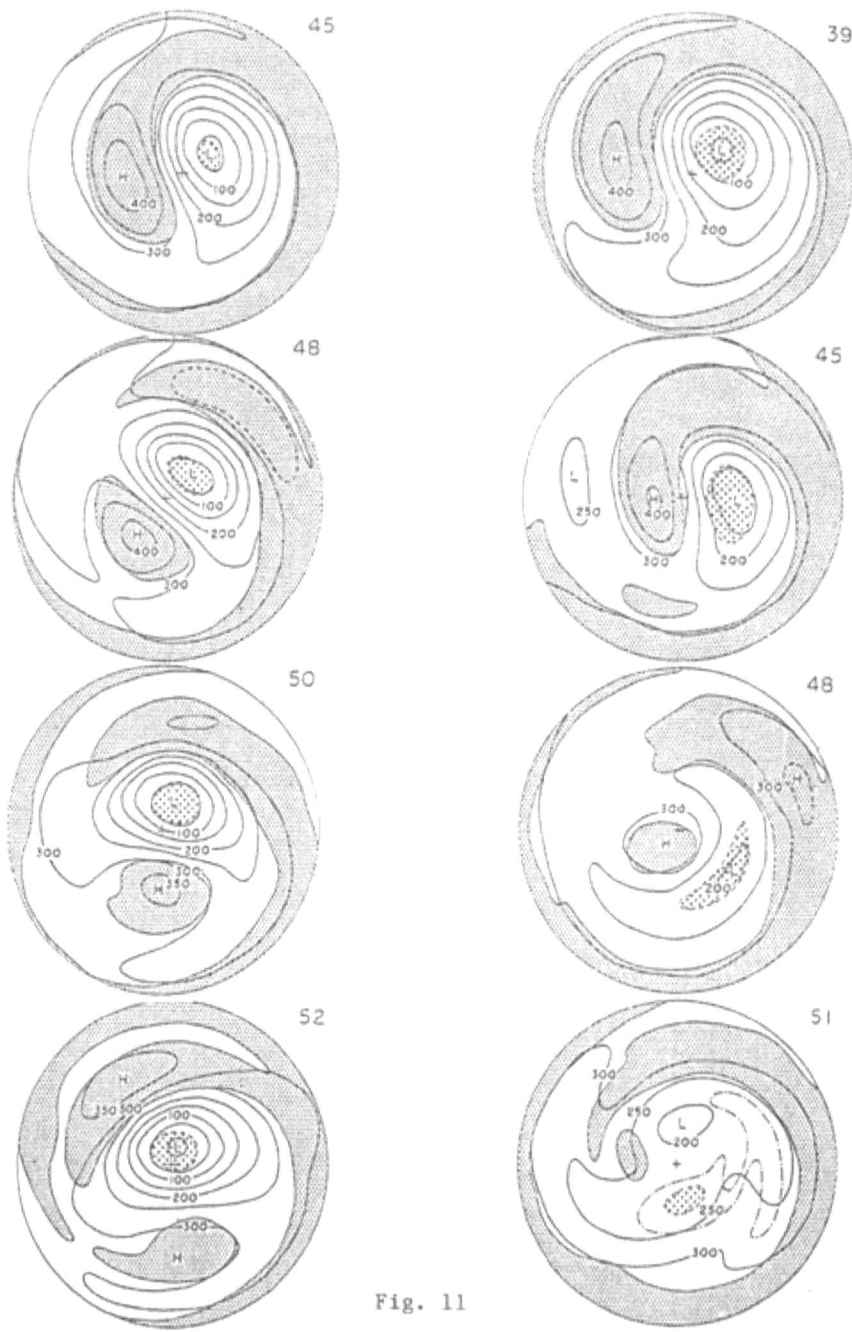

Fig. 11

Evolution of the isobaric height and potential vorticity patterns (40 km).
Left: minor warming, Right: major warming. (see bottom of next page)

warming case, the region of the maximum vorticity gradually return towards the pole and the circumpolar westerly jet is reestablished. It
should be noted that the maximum value of potential vorticity has increased slightly during the warming period. The process of the increase
has not yet been investigated in detail, but the diabatic heating could
be responsible for it, because the time constant of Newtonian cooling is
comparable with the duration of the warming. Thus when the maximum potential vorticity region comes back to the pole, it is accompanied by a
stronger westerly jet as we have seen in Fig. 8. In the major warming
case, the potential vorticity mixing is so complete that no solid core
of cyclonic vorticity has been left, as evidently seen in the pattern of
day 51. In this figure a contour line of $q = 1.5 \times 10^{-4}$ sec^{-1} is drawn in
order to show how the maximum vorticity region has been distorted. We
can thus understand rather easily the difference between the two warmings
and the origin of the high latitude westerly jet, from the viewpoint of
potential vorticity mixing as proposed by McIntyre (1982). It is also
noted that the whole evolution of potential vorticity patterns in the
major warming experiment is rather similar to that obtained by Dunkerton
et al. (1981, their Fig. 12). In this study a wave 2 forcing with a constant amplitude (except for the initial switch on) was given and the
resultant warming was "major".

Fig. 12a and 12b show the latitudinal profiles of the zonal mean
potential vorticity at the 40 km level at selected dates for the minor
and major warming experiments, respectively. These confirm the characteristics of the potential vorticity evolutions as explained previously.
Fig. 12a corresponds to McIntyre's schematic picture shown in Fig. 2.
The change from day 30 to day 51 (indicated by arrows) is similar to that
in Fig. 2, except that the maximum value at the pole increases in the
course of the warming. In the major warming case the final state (day
51) shows that a nearly uniform distribution of potential vorticity has
been established poleward of 40°N.

It should be noted that the near uniform distribution of the zonal
mean potential vorticity at high latitudes seen at day 48 of the minor
warming experiment is not the result of irreversible mixing. It is due
to the temporal juxtaposition of airmasses from different (high and low
latitude) origins caused by the temporal enhancement of the waves. When
the waves decay and undulations of material surfaces decrease, the original configuration is nearly recovered. In a numerical experiment concerning Lagrangian tracing of air parcels in a similar model with a forcing of finite duration (Kohno, 1983) it is seen that air parcels originally

---

Legend of Fig. 11
Evolution of the isobaric height and potential vorticity patterns at the
40 km level. Left diagrams (day 45, 48, 50, 52) for minor warming experiment and right diagrams (day 39, 45, 48, 51) for major warming experiment.
Solid lines are isobaric height contours with 50 dam intervals. Small
potential vorticity ($q < 2.0 \times 10^{-4}$ s^{-1}) is shaded. Stippled areas indicate
large potential vorticity ($q > 2.5 \times 10^{-4}$ s^{-1} in minor warming and $q > 2.0 \times 10^{-4}$
s^{-1} in major warming.)

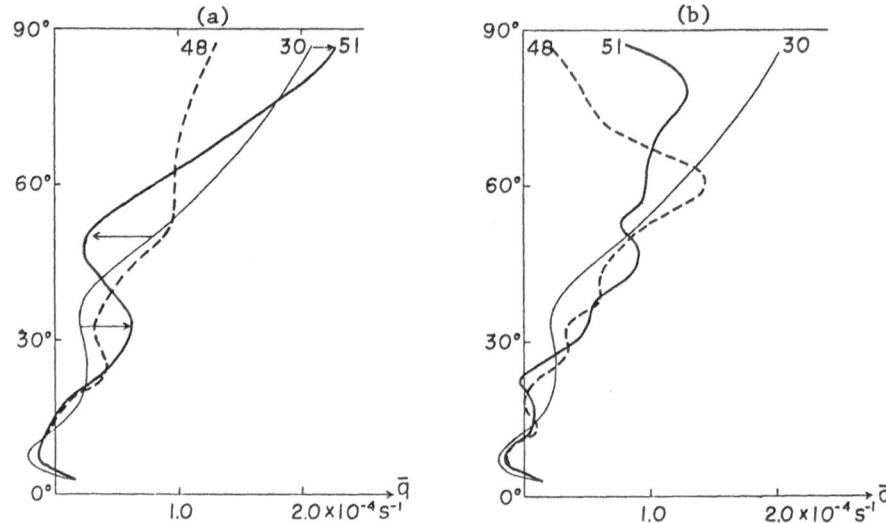

Fig. 12   Latitude profiles of the zonal mean potential vorticity at the
40 km level for selected days.   (a) for minor warming experiment,
(b) for major warming experiment.

aligned along a latitude circle are scattered over a latitude range by
the planetary waves but they return to the original latitude after the
decay of the wave, if their original positions are at high latitudes.
Apparently some effects, such as critical latitude or potential enstrophy
cascade are necessary and may be efficient for the irreversibility of
potential vorticity mixing (McIntyre, 1982). The possibility can be
inferred from the results of numerical experiments treating Lagrangian
motions of air parcels (Hsu, 1980; 1981).

4. CONCLUSION AND REMARKS

By use of a non-linear model including wave-wave interactions and
imposing a boundary forcing for wave 1 which amplifies for about two
weeks and then decays, phenomena resembling a minor warming and the subsequent generation of a strong westerly jet at very high latitude were
reproduced in the model when the maximum amplitude of the forcing was
300 m. In an experiment with a stronger (400 m) forcing, a major warming
was obtained and it was not followed by strong jet formation in accord
with observations.

The hypotheses given by McIntyre (1982) concerning the difference
between minor and major warmings and the reason for the generation of a
strong polar jet after a minor warming were successfully verified by
the examination of the evolution of the potential vorticity patterns.
From a comparison with the results of a model excluding wave-wave interaction (Matsuno, 1983), it turned out that the inclusion of the wavenumber

2 component is necessary for obtaining a strong jet after a minor warming. The wave 2 is thought to be required to cause irreversible mixing of potential vorticity (McIntyre, personal communication). From the results of similar numerical experiments and analysis of observed situations, Palmer and Hsu (1983) found that wave 2 is necessary for the generation of the high latitude jet, because the westerly acceleration forcing in the zonal mean momentum equation arises from the divergence of the Eliassen-Palm flux due to this wave component. The two arguments are made from different view point, but there may be some relation between the two processes.

Palmer and Hsu (1983) proposed that when the polar night jet is displaced away from the pole by the presence of a large amplitude wave 1 component, the geographical pole has no significance for wave dynamics but wave propagation should occur in the wave guide formed by the displaced potential vorticity pole. Though experiments directly applicable to this argument were not conducted, from the evolutions of the minor warming experiment it is speculated that wave activity (Andrews and McIntyre, 1978) can be transmitted through the polar cap region in this deformed coordinate and this transmittivity allows the easterly angular momentum associated with the wave activity to escape from the polar stratosphere, in a manner similar to the case of a small amplitude disturbance as treated by Uryu (1974a, b). This process might contribute to the quick recovery of westerlies after the end of the minor warming in the present model.

ACKNOWLEDGEMENTS

I would like to express my gratitudes to Dr. M.E. McIntyre for his many valuable suggestions during the course of this study and to Dr. J. R. Holton for his penetrating review of the first version of this paper, especially for pointing out the similarity of Δu in the minor and major warming experiments. I am also grateful to Mrs. K. Kudo and Mr. Y. Fujiki for their assistances in the preparation of the manuscript.

REFERENCES

Andrews, D.G., and M.E. McIntyre, 1978: On wave-action and its relatives, J. Fluid Mech., 89, 647-664.

Butchart, N., S.A. Clough, T.N. Palmer, and P.J. Trevelyan, 1982: Simulations of an observed stratospheric warming with quasi-geostrophic refractive index as a model diagnostic. Quart. J. Roy. Meteor. Soc., 108, 475-502.

Dickinson, R.E., 1973: Method of parameterization for infrared cooling between altitudes of 30 and 70 kms. J. Geophys. Res., 78, 4451-4457.

Dunkerton, T., C.-P.F. Hsu and M.E. McIntyre, 1981: Some Eulerian and Lagrangian diagnostics for a model stratospheric warming. J. Atmos. Sci., 38, 819-843.

Holton, J.R., 1976: A semi-spectral numerical model for wave-mean flow interactions in the stratosphere: application to sudden stratospheric warmings. J. Atmos. Sci., 33, 1639-1649.

Hsu, C.-P.F., 1980: Air parcel motions during a numerically simulated sudden stratospheric warming. J. Atmos. Sci., 37, 2768-2792.

_____, 1981: A numerical study of the role of wave-wave interactions during stratospheric warmings. J. Atmos. Sci., 38, 189-214.

Kanzawa, H., 1980: The behavior of mean zonal wind and planetary-scale disturbances in the troposphere and stratosphere during the 1973 sudden warming. J. Meteor. Soc. Japan, 58, 329-356.

Kohno, J., 1983: Stratospheric ozone transport due to transient large amplitude planetary wave. (submitted to J. Meteor. Soc. Japan)

Labitzke, K., 1981: The amplification of height wave 1 in January 1979: a characteristic precondition for the major warming in February. Month. Wea. Rev., 109, 983-989.

Lordi, N.J., A. Kasahara, and S.K. Kao, 1980: Numerical simulation of stratospheric sudden warmings with a primitive equation spectral model. J. Atmos. Sci., 37, 2746-2767.

Madden, R.A., and K. Labitzke, 1981: A free Rossby wave in the troposphere and stratosphere during January 1979. J. Geophys. Res., 86, 1247-1254.

Mahlman, J.D., 1967: Further studies on atmospheric general circulation and transport of radioactive debris. Atmos. Sci. Paper No. 103, Colorado State University. 184pp.

Matsuno, T., 1971: A dynamical model of the stratospheric sudden warming. J. Atmos. Sci., 28, 1479-1494.

_____, 1983: Circulation and waves in the middle atmosphere in winter. Space Sci. Rev., 34, 387-396.

McIntyre, M.E., 1982: How well do we understand the dynamics of stratospheric warmings? J. Meteor. Soc. Japan, 60, 37-65.

_____ and T.N. Palmer, 1983: Breaking planetary waves in the stratosphere. Nature (submitted)

Palmer, T.N., 1981a: Diagnostic study of a wavenumber-2 stratospheric sudden warming in a transformed Eulerian-mean formalism. J. Atmos. Sci., 38, 844-855.

_____, 1981b: Aspects of stratospheric sudden warmings studied from a transformed Eulerian-mean viewpoint. J. Geophys. Res., 86, 9679-9687.

_____ and C.-P.F. Hsu, 1983: Stratospheric sudden coolings and the role of nonlinear interactions in preconditioning the circumpolar flow. J. Atmos. Sci., 40, (in press).

Schoeberl, M.R., 1983: A note on the conservation properties of Matsuno's potential vorticity equations. J. Atmos. Sci. (submitted).

_____ and D.F. Strobel, 1980: Numerical simulation of sudden strato-spheric warmings. J. Atmos. Sci., 37, 214-236.

Uryu, M., 1974a: Induction and transmission of mean zonal flow by quasi-geostrophic disturbances. J. Meteor. Soc. Japan, 52, 341-364.

_____, 1974b: Mean zonal flows induced by a vertically propagating Rossby wave packet. J. Meteor. Soc. Japan, 52, 481-490.

RADIATION

J. R. Holton and T. Matsuno, Dynamics of the Middle Atmosphere, 355–366.
Copyright © 1984 by Terra Scientific Publishing Company.

INFRARED RADIATIVE EXCHANGE IN THE MIDDLE ATMOSPHERE IN THE
15 MICRON BAND OF CARBON DIOXIDE

Conway B. Leovy

Department of Atmospheric Sciences and Geophysics Program
University of Washington
Seattle, Washington 98195 U.S.A.

ABSTRACT

   The relative roles of radiation-to-space and exchanges between
atmospheric layers for 15 micron band radiation are discussed.
Although radiation-to-space is generally a good approximation in
the stratosphere, it may break down in the mesosphere where exchange
with underlying layers can be of comparable importance.

1. INTRODUCTION

   Accurate modeling of the dynamics of the middle atmosphere
requires accurate treatment of infrared radiative heating and cooling
rates. Carbon dioxide, ozone, and water vapor contribute
significantly to the infrared exchange and consequent heating and
cooling, but of these, carbon dioxide is by far the most important.
In the upper stratosphere and mesosphere, the problem is complicated
by the Voigt line shape and by departures from local thermodynamic
equilibrium (LTE). On the other hand, above about 35 km, line
overlap can be neglected for most bands so that transmission functions
can be evaluated for relatively few line groups sorted according to
strength, thereby avoiding the need for cumbersome line-by-line
calculations or the possible inaccuracies of band models.

   Plass (1956a,b) carried out an early series of calculations of
infrared heating and cooling rates due to both the 9.6 micron ozone
band and the 15 micron carbon dioxide band, and showed that cooling
rates increase upward in the stratosphere for both bands. Murgatroyd
and Goody (1958) used a matrix method developed by Curtis (1956)
together with the temperature climatology developed by Murgatroyd
(1957) to calculate heating and cooling due to ozone and molecular
oxygen in the ultraviolet as well as in the infrared. Their infrared
calculation allowed for Voigt line shape and incorporated a
simplified treatment of non-LTE effects. The results indicated that
radiative equilibrium prevails throughout most of the middle

355

atmosphere, but strong departures occur at high latitudes during solstice seasons, especially in the mesosphere. They also noted an approximately linear dependence of carbon dioxide infrared cooling on the local Planck function, a result that has served as the basis for the now widely used Newtonian cooling approximation.

The intimate connection between the net heating distribution and large scale circulation was illustrated soon afterward when Murgatroyd and Singleton (1961) published calculations of the mean meridional circulation driven by the net heating distribution at the solstices deduced by Murgatroyd and Goody. This circulation corresponds to what we now call the "residual circulation", the part of the mean meridional circulation which is not compensated by eddy heat flux divergence (Dunkerton, 1978). Although there have been a number of more recent calculations of infrared heating in the middle atmosphere (e.g., Drayson, 1967; Kuhn and London, 1969; Houghton, 1969; Dickinson, 1973; Kutepov and Shved, 1978; Fels and Schwarzkopf, 1981; Wehrbein and Leovy, 1982; Apruzese, Schoeberl, and Strobel, 1982), the early work of Curtis, Murgatroyd and Goody, and Murgatroyd and Singleton has held up remarkably well.

From the point of view of the dynamical modeler, the two most important results of middle atmosphere radiative calculations are the seasonally varying zonal mean heating distribution (the driver for the residual circulation), and the radiative damping rates corresponding to disturbances of various vertical scales. In this note, the influence of radiative exchange and departures from LTE on heating rates and radiative damping due to the 15 micron band will be discussed.

2. EXCHANGE FORMULATION FOR INFRARED HEATING RATES

For a relatively narrow band, such as the 15 micron carbon dioxide band or 9.6 micron ozone band, net heating per unit mass $Q(z)$ can be expressed in a form which explicitly displays the exchange processes (Curtis, 1956):

$$Q(z) = 2\pi c(z) S(z) c_p^{-1} \left\{ - F(z,z_+) J(z) \right.$$

$$+ \int_{F(z,z_+)}^{1} [J(z') - J(z)] \, dF(z,z')$$

$$+ \int_{F(z,z_-)}^{1} [J(z') - J(z)] \, dF(z,z')$$

$$\left. + [J(z_-) - J(z)] F(z,z_-) \right\} \quad , \tag{1}$$

where $c(z) = (\rho_a/\rho)$ = mixing ratio of the radiatively active gas
whose density is ρ_a (ρ is the air density),

c_p = constant pressure specific heat,

S = strength of the band in units of area times frequency per molecular mass,

J = source function at the band center (power per steradian per unit area and frequency); under LTE conditions, J is the Planck function B,

z = any vertical coordinate which increases monotonically upward (e.g., $z = - \ln(p/p_0)$ for log-pressure coordinates). z_- and z_+ are values corresponding to lower and upper boundaries ($z_+ \to \infty$ for height or log-pressure coordinates).

$F(z,z')$ = exchange function, to be defined more precisely below.

The four terms inside the curly brackets in eq. (1) correspond respectively to escape-to-space, exchange with layers above z ($z' > z$), exchange with layers below z ($z' < z$), and exchange with the underlying boundary (ground or cloud). The exchange function $F(z,z')$ characterizing these exchanges varies from zero for a totally opaque layer between z and z' to unity for a totally transparent layer (or for $z' = z$). The exchange function for a plane parallel atmosphere can be expressed as follows:

$$F(z,z') = S(z)^{-1} \int_0^1 d\mu \int^{\Delta\nu} d\nu k_\nu(z) exp[-\Delta\tau_\nu(z,z')/\mu] \quad ,$$

where $\Delta\nu$ is the band width, k_ν is the monochromatic absorption coefficient, and $\Delta\tau_\nu(z,z')$ is the vertical optical path between z and z':

$$\Delta\tau_\nu(z,z') = \left| \int_z^{z'} k_\nu(z'')\rho_a(z'')dz'' \right| \quad .$$

Note that the exponential is just the parallel beam transmission function for the slant path between z and z' at an angle $cos^{-1} \mu$ with the vertical.

Typical behavior of $F(z,z')$ for carbon dioxide in the stratosphere is illustrated in Fig. 1. The curves shown were derived using the Malkmus model, Lorentz line shape, the Curtis-Godson approximation, an isothermal atmosphere at 220K, and band parameters

358

from Houghton (1977, pp. 184-5). Despite these approximations, these curves illustrate several features of the exchange problem. Note the extremely sharp peaking near $z' = z$, negligible values of F as the lower boundary is approached corresponding to negligible contribution to heating from the lower boundary, and the small but significant asymptotic value of F for large values of z' corresponding to the radiation-to-space contribution.

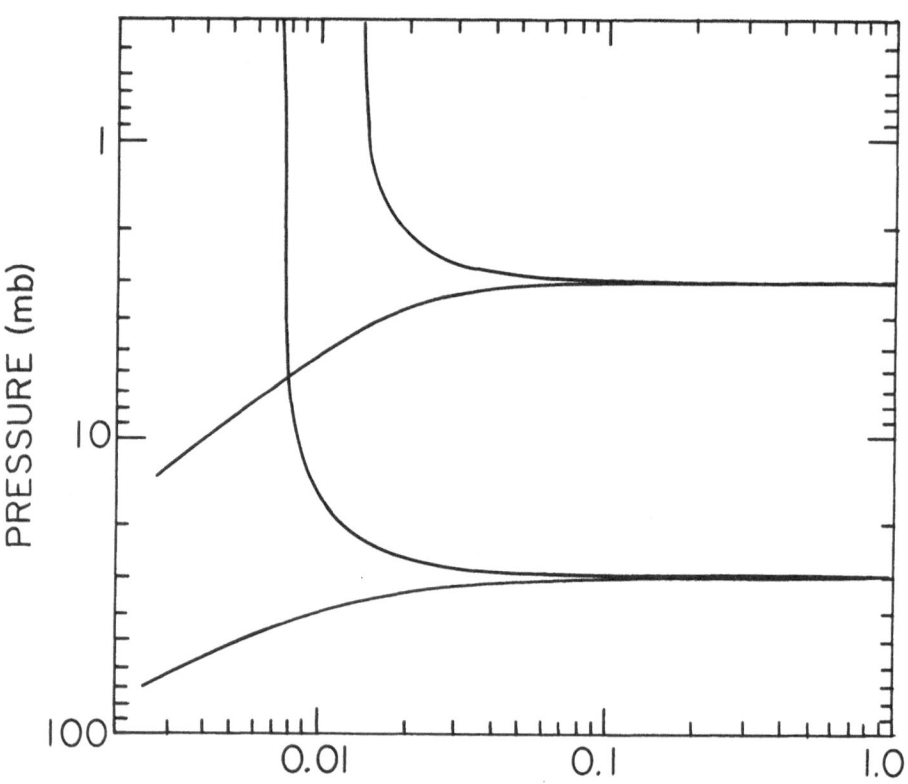

Fig. 1: Exchange functions $F(z,z')$ centered at 3 mb and 30 mb.

For typical temperature profiles in the stratosphere, radiation-to-space provides a fair approximation to the total cooling rate. This is because the variations in F are so localized that they are associated with small differences between $B(z')$ and $B(z)$, and only small contributions from the integrals in eq. (1). As a consequence, the cooling rate at stratospheric levels can be well represented by a Newtonian cooling term. Above the stratopause, however, the situation changes. The exchange function is asymmetric, and small but non-negligible variations in F occur at distances of the order of a scale height below the peak, though not at correspondingly large distances above the peak. As a consequence, in regions with positive lapse rate in which the Planck function decreases upward, exchange with the underlying atmosphere can make an important contribution. For this reason, 15 micron band cooling rates above the stratopause are not necessarily dominated by the radiation-to-space term.

Actual cooling rates usually decrease upward from the stratopause. This is due in part to the upward decrease of the Planck function, but it also reflects a more delicate balance between cooling due to radiation-to-space and warming due to exchange with underlying layers. At high latitudes during summer solstice, the latter contribution produces net warming by the 15 micron band near 70 km. On the other hand, during winter at high latitudes, the mesosphere tends to be roughly isothermal, and cooling due to the radiation-to-space term is a good approximation to the total cooling rate. Note that, according to eq. (1), radiation-to-space is the only non-zero term for an isothermal atmosphere.

A useful alternative representation of eq. (1) is obtained by quadrature approximation to the integrals. The corresponding expression for the heating rate under LTE conditions is

$$Q(z_i) = \sum_j R_{ij} B(z_j) = B(z_i)(\sum_j R_{ij}) + \sum_j R_{ij}[B(z_j) - B(z_i)] \quad (2)$$

giving the heating rate of a layer of finite thickness centered at z_i in terms of the Planck function in all other finite layers centered at levels z_j. The last expression in eq. (2) explicitly displays the radiation-to-space and exchange terms in this representation. Elements R_{ij} comprise the Curtis Matrix. A useful feature of the Curtis Matrix is this explicit display of the terms corresponding to coupling between layers. Curtis Matrices for the 15 micron band depend on line strength distributions and the corresponding temperature distributions. 15 micron band Curtis Matrices computed for three temperature distributions under various approximations are given by Wehrbein and Leovy (1982)*. In Fig. 2,

---

*Note that the 15 micron band matrix elements and radiation-to-space factors given in Wehrbein and Leovy (1982) should be multiplied by the factor 1.94 (see Wehrbein and Leovy, 1983).

the contributions of radiation-to-space and exchanges with underlying and overlying layers to net 15 micron band cooling are compared using the Curtis Matrix elements of Wehrbein and Leovy. Radiation-to-space dominates below 55 km, but above 65 km there is an approximate balance between exchange with underlying layers and radiation-to-space.

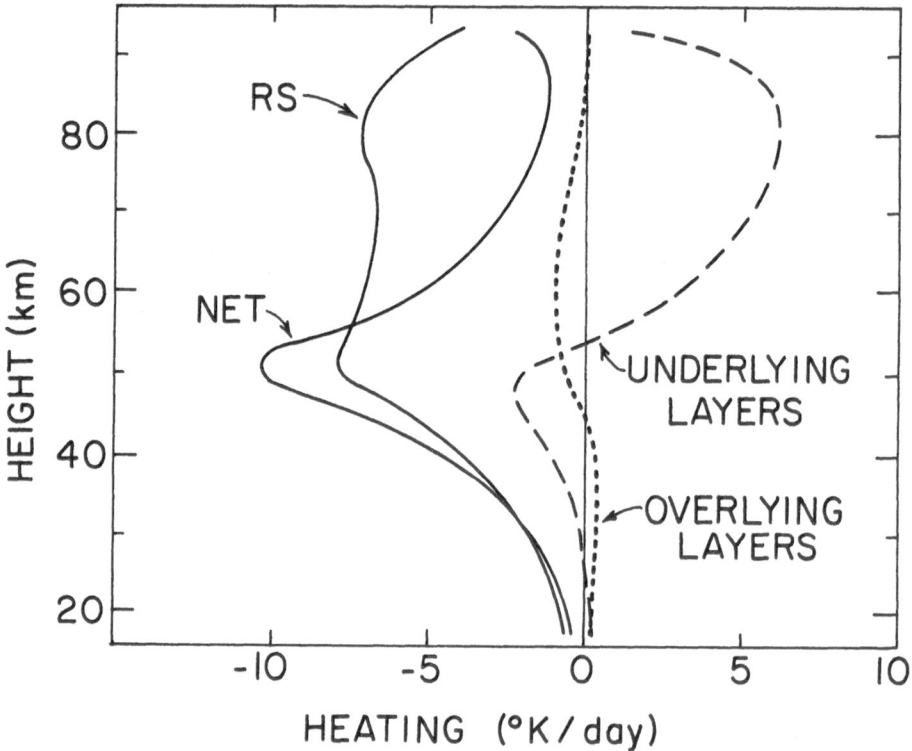

Fig. 2: Comparison of radiation-to-space (RS), underlying layer exchange, and overlying layer exchange for the standard atmosphere temperature profile and Curtis matrix elements of Wehrbein and Leovy.

Since the matrix elements involve a quadrature, the assumed variation of the Planck function between finite layer centers enters the calculation of R_{ij}. This quadrature assumption can be quite important for thick layers, especially in the summer mesosphere where exchange between layers is particularly significant (Drayson, 1967).

Wehrbein (1983) has found cooling rate differences of up to 40% in the summer mesosphere between Curtis matrices for 5 km thick iso-thermal slabs and Curtis matrices for linear variation of the Planck function between layers centered at 5 km intervals. The temperature structure implicit in the quadrature representation of the integral is one of the largest sources of inaccuracy in coarse vertical resolution mesospheric heating models.

3. BREAKDOWN OF THE PARAMETERIZED METHOD

Dickinson (1973) suggested a parameterized approach for calcu-lating net 15 micron band cooling rates based on the Newtonian cooling formulation. He proposed that, as a rough approximation, the middle atmosphere cooling rate could be separated into a contribution for the standard atmosphere plus an additional contribution propor-tional to the local departure of temperature from the standard atmosphere, and he gave values for both the cooling rate of the standard and the height dependent proportionality coefficient. The latter was determined from calculations in which the temperature profile was uniformly perturbed from the standard. This approach corresponds closely to the radiation-to-space approximation, differing only in that ΣR_{ij} in eq. (2) is replaced by $\Sigma R_{ij}(dB/dT)_j$. Each element R_{ij} in the summation is weighted by the derivative of the Planck function at the local temperature. It gives a good representation of the incremental heating or cooling due to small perturbations of very large vertical scale, but it fails in two respects. Since the lapse rate is fixed, this parameterization does not adequately represent the net 15 micron band heating rate in the mesosphere where exchanges of radiation dependent on lapse rate are important. Consequently, it underestimates the differential heating between the large lapse rate region of the summer mesosphere and the small lapse rate region of the winter mesosphere. It also underestimates the Newtonian cooling coefficient for disturbances of small vertical scale, a point to be explored more fully below.

Dickinson's treatment of the 15 micron band transmission function was carried out with great care (Dickinson, 1972, 1973). Because of this, and because of its simplicity, this parameterized approach was widely used by dynamicists for middle atmosphere calculations. Dickinson himself realized that the method produces serious errors in the mesosphere. Fig. 4 of Dickinson (1973) shows that the cooling rate of an isothermal atmosphere at 230K is underestimated by almost a factor of 3 at 70 km. In retrospect then, it should not be surprising that recent dynamical models which do not employ this parameterization yield much stronger solstice heating rates, larger radiative damping rates, and more intense residual circulations than models which used the Dickinson parameterization (Wehrbein and Leovy, 1982; Apruzese, Schoeberl, and Strobel, 1982).

4. NET HEATING NEAR THE MESOPAUSE

The mesopause region is particularly delicate for radiative calculations. For one thing, net heating rates tend to be small because exchange with underlying and overlying layers tends to balance radiation-to-space. Vibrational relaxation and the consequent departures from LTE also complicate the calculation of net infrared heating rates near the mesopause. There is net infrared warming near the cold summer mesopause and strong cooling near the warm winter mesopause, but it is not yet clear whether the mesopause region is being warmed or cooled by infrared radiative exchange in the global mean.

For a single vibrational band interacting with the ground state, the non-LTE aspect of the problem can be treated by simple manipulations involving the Curtis Matrix and the height-dependent ratio of the radiative and collisional time constants (Houghton, 1977, pp. 60-63). However the problem is complicated by the presence of hot and isotopic bands in the 15 micron region. In the mesosphere, the net heating rate associated with these bands is governed by the combination of radiation-to-space and exchange with the underlying atmosphere. Since these bands are quite transparent in the mesosphere, the downward exchange extends over a large altitude range, and the balance between the two terms is controlled essentially by the difference between the local source function and the source function near the stratopause. Lines in the more opaque fundamental band, by contrast, do not "see" so deeply into the atmosphere. If the source functions of the hot and isotopic bands are closely coupled to the source function of the fundamental, they will remain close to LTE up to about 80 km, and these bands can contribute significantly to the local heating or cooling rate. The greatest relative contribution to net heating by the hot and isotopic bands occurs between 65 and 75 km where the contrast in transparency between these bands and the fundamental has the greatest effect. However, if the source functions for the different bands are only weakly coupled, the hot and isotopic bands can depart strongly from LTE below 70 km, and these bands will have much less influence on the net heating. Depending on the degree of coupling between bands and on the steepness of the mesospheric lapse rate, the hot bands may dominate the net heating in the 15 micron region near 70 km. Wehrbein and Leovy (1982) found differences of up to 25% in net heating rate near 70 km for the standard atmosphere for fully coupled and fully decoupled models, but the dependence on coupling may be larger at high latitudes near summer solstice. Above 80 km, the problem simplifies, the fundamental band becomes transparent and the relative contribution of hot and isotopic bands is negligible, regardless of the coupling of source functions.

5. RADIATIVE DAMPING

Because of the sharply peaked character of the exchange function, the radiative damping rate is strongly dependent on the vertical scale of temperature disturbances (Goody, 1964; Goody and Belton, 1967). Fels (1982) has shown how scale dependent damping can be applied to waves in the middle atmosphere with the aid of the WKBJ approximation. The Curtis Matrix formulation provides an alternative natural description of scale dependent damping. For example, the differential cooling rate per unit of temperature increase can be readily calculated for uniform temperature perturbations of various widths. This quantity is the Newtonian damping coefficient for these "boxcar" disturbances. In Fig. 3, Newtonian cooling coefficients are shown

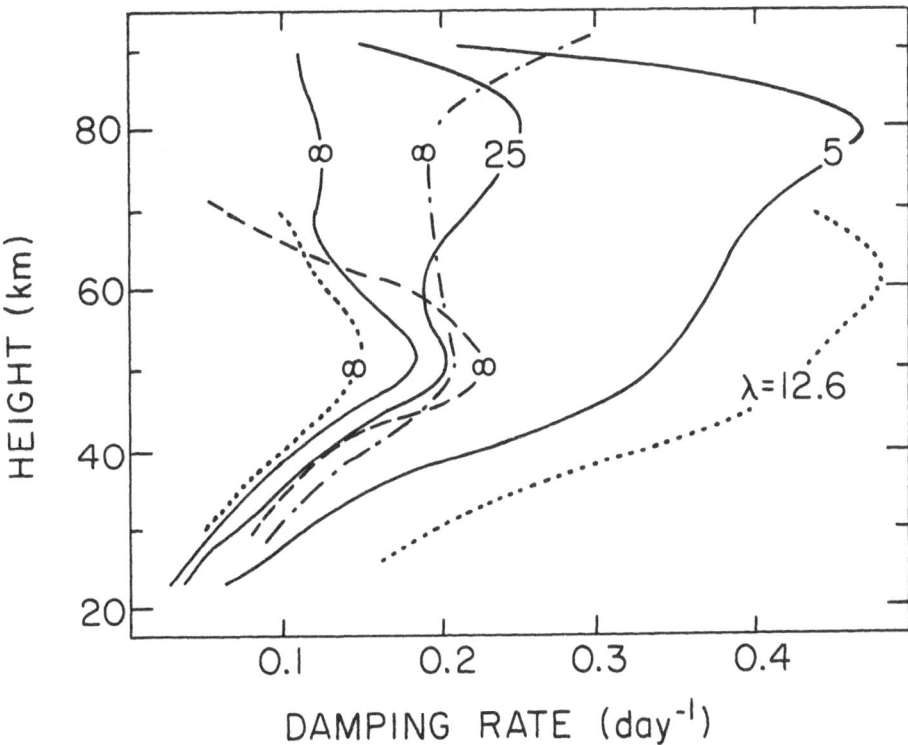

Fig. 3: Damping rates for the 15 micron band for perturbations about the standard atmosphere: Wehrbein and Leovy (1982, thin solid), Dickinson (1973, dashed), Fels (1982, dotted), Apruzese et al. (1982, dash-dot). Labels are for radiation-to-space (∞), "boxcar" (5 and 25 km), and wavelength in the WKBJ approximation ($\lambda = 12.6$ km).

for perturbations about the standard atmosphere temperature profile
of infinite vertical scale (the radiation-to-space approximation),
for various "boxcar" cases, and for a perturbation of well-defined
vertical wavelength in the WKBJ approximation. Note that different
calculations are in general agreement for the radiation-to-space
and "boxcar" cases, and that the 5 km "boxcar" case agrees well with
the 12.6 km vertical wavelength perturbation of Fels, as it should.
The WKBJ approach of Fels is most appropriate for disturbances of
small vertical scale, while Curtis Matrices for relatively thick
layers may be more applicable to disturbances of large vertical scale.
In any case, scale dependence of the damping rate is an important
property of the mesosphere.

6. CONCLUSIONS

Because net infrared heating by the 15 micron band in the mesosphere depends on both radiation-to-space and exchange with the
underlying atmosphere, mesospheric radiative calculations may be
relatively delicate, particularly during summer in high latitudes
where the lapse rate is large. For this reason, the radiation-to-
space parameterization breaks down in the calculation of seasonal
heating differences in the mesosphere, significant errors can arise
from overly crude modeling of the vertical temperature structure,
and calculations are sensitive to the extent of departures from LTE
in the hot and isotopic bands. Despite these difficulties, recent
radiative calculations are in general agreement with the pioneering
work of Murgatroyd and Goody in their assessments of the distribution
of net heating in the mesosphere at the solstices. The 15 micron
band transmission functions derived by Fels and Schwarzkopf are the
most accurate available for application to this problem.

Several recent calculations have helped to clarify the
importance of scale dependent radiative damping in the mesosphere.
This can be treated conveniently by either the use of Fourier
decomposition of the temperature structure coupled with the WKBJ
approximation, or by the use of Curtis Matrices.

ACKNWLEDGEMENT: I am indebted to W. M. Wehrbein and J. R. Holton
for helpful discussions of this problem.

REFERENCES

Apruzese, J. P., M. R. Schoeberl, and D. F. Strobel, 1982: Parameterization of IR cooling in a middle atmosphere dynamics model 1.
Effects on the zonally averaged circulation. *J. Geophys. Res.*,
87, 8951-8966.

Curtis, A. R., 1956: The computation of radiative heating rates
in the atmosphere. *Proc. Roy. Soc., London, A236*, 156-159.

Dickinson, R. E., 1972: Infrared heating and cooling in the Venusian
atmosphere. I: Global mean radiative equilibrium. *J. Atmos.
Sci., 29*, 1531-1556.

Dickinson, R. E., 1973: A method of parameterization for infrared
cooling between altitudes of 30 and 70 km. *J. Geophys. Res.*,
78, 4451-4457.

Drayson, S. R., 1967: Calculation of long-wave radiative transfer in
planetary atmospheres. Ph.D. Thesis [Rep. 07584-1-T], College
of Engineering, University of Michigan, Ann Arbor, 110 pp.

Dunkerton, T., 1978: On the mean meridional mass motions of the
stratosphere and mesosphere. *J. Atmos. Sci., 35*, 2325-2333.

Fels, S. B., 1982: A parameterization of scale-dependent radiative
damping rates in the middle atmosphere. *J. Atmos. Sci., 39*,
1141-1152.

Fels, S. B., and M. D. Schwarzkopf, 1981: An efficient, accurate
algorithm for calculating CO_2 15 micron band cooling rates.
J. Geophys. Res., 86, 1205-1232.

Goody, R. M., 1964: *Atmospheric Radiation I: Theoretical Basis.*
Clarendon Press, Oxford, 436 pp.

Goody, R. M., and M. J. S. Belton, 1967: Radiative relaxation times
for Mars: a discussion of Martian atmosphere dynamics.
Planetary Space Sci., 15, 247-256.

Houghton, J. T., 1969: Absorption and emission by carbon dioxide
in the mesosphere. *Quart. J. Roy. Meteorol. Soc., 95*, 1-20.

Houghton, J. T., 1977: *The Physics of Atmospheres.* Cambridge
University Press, 203 pp.

Kuhn, W. R., and J. London, 1969: Infrared radiative cooling in the
middle atmosphere (30-110 km). *J. Atmos. Sci., 26*, 189-204.

Kutepov, A. A., and G. M. Shved, 1978: Radiative transfer in the 15 micron CO_2 band with non-LTE in the Earth's atmosphere. *Atmos. Oceanic Phys., 14,* 28-43.

Murgatroyd, R. J., 1957: Winds and temperatures between 20 and 100 km -- a review. *Quart. J. Roy. Meteorol. Soc., 83,* 417-458.

Murgatroyd, R. J., and R. M. Goody, 1958: Sources and sinks of radiative energy from 30 to 90 km. *Quart. J. Roy. Meteorol. Soc., 87,* 225,234.

Murgatroyd, R. J., and F. Singleton, 1961: Possible meridional circulations in the stratosphere and mesosphere. *Quart. J. Roy. Meteorol. Soc., 87,* 137-158.

Plass, G. N., 1956a: The influence of the 9.6 micron ozone band on the atmospheric infrared cooling rate. *Quart. J. Roy. Meteorol. Soc., 82,* 30-44.

Plass, G. N., 1956b: The influence of the 15 micron carbon dioxide band on the atmospheric infrared cooling rate. *Quart. J. Roy. Meteorol. Soc., 82,* 310-324.

Wehrbein, W. M., 1983: Personal communication.

Wehrbein, W. M., and C. B. Leovy, 1982: An accurate radiative heating and cooling algorithm for use in a dynamical model of the middle atmosphere. *J. Atmos. Sci., 39,* 1532-1544.

Wehrbein, W. M., and C. B. Leovy, 1983: Corrections to "An accurate radiative heating and cooling algorithm for use in a dynamical model of the middle atmosphere". *J. Atmos. Sci.,* (submitted).

TRANSPORT OF TRACERS

J. R. Holton and T. Matsuno, Dynamics of the Middle Atmosphere, 369-385.
Copyright © 1984 by Terra Scientific Publishing Company.

TROPOSPHERE-STRATOSPHERE EXCHANGE OF TRACE CONSTITUENTS:
THE WATER VAPOR PUZZLE

James R. Holton

Department of Atmospheric Sciences
University of Washington
Seattle, Washington 98195 U.S.A.

ABSTRACT

Recent observations of the water vapor mixing ratio in the
stratosphere indicate that the injection of tropospheric trace
constituents into the stratosphere is a more complex process than
traditionally believed. Current evidence for spatial and temporal
variability of troposphere-stratosphere exchange is reviewed, and
possible mechanisms which might account for the observed dryness of
the stratosphere are discussed.

1. INTRODUCTION

In problems of atmospheric chemistry the stratosphere and the
troposphere are often treated as separate systems which are coupled
only through the exchange of trace constituents. This exchange plays
a crucial role, for example, in the chemistry of the ozone layer.
Because many trace constituents (including ozone) are radiatively
active molecules, tropospheric-stratospheric exchange also plays a
role in determining the state of the climate. The dynamics of
the climate in turn control the geographical and temporal distributions
of this exchange. Thus, troposphere-stratosphere exchange is a
fundamental part of the coupling among radiation, chemistry, and
dynamics in the atmosphere.

In the analysis of troposphere-stratosphere exchange of trace
species it is useful to focus on two types of tracers: i) substances
whose sources are in the troposphere and which are transported into
the stratosphere where they are destroyed by photochemical processes
(e.g., nitrous oxide, water vapor, methane, and halocarbons);
ii) substances whose sources are in the stratosphere and which are
transported into the troposphere where they are destroyed by a variety
of processes (e.g., ozone, and stratospheric aerosols). Transport
of trace substances, both horizontally and vertically, is important
to the understanding of tracer behavior in both the troposphere and

the stratosphere. However, the vertical flux of trace constituents
across the tropopause and through the lowest few kilometers of the
stratosphere is especially important because it is this region which
primarily determines the rate of transport between source and sink
regions. Indeed, the apparent slowness of transport in this region
is responsible for such features as the deduced long stratospheric
residence times for substances such as radioactive tracers
(e.g., Telegadas and List, 1969) and the long time required for
establishment of a steady state ozone perturbation for a constant
tropospheric release rate of halocarbons (WMO, 1982).

The study of troposphere-stratosphere exchange is intimately
related to the problem of tracer transport within the lower
stratosphere. Mahlman *et al.* (1983) provide an interesting historical
survey of efforts to understand tracer transport, and also outline an
interesting conceptual model of transport. A useful introduction
to the stratosphere-troposphere exchange problem is given in
Robinson (1980). In this review we will focus mainly on the constraints
imposed on exchange by the observed aridity of the stratosphere.

2. THE CONVENTIONAL MODEL

The simplest qualitatively plausible model for troposphere-
stratosphere exchange consists of bulk advection by a single mean
meridional cell in each hemisphere with uniform rising motion across
the tropical tropopause, poleward drift in the stratosphere, and by
continuity of mass, a return flow into the troposphere in the
extratropics. Such a circulation was proposed by Brewer (1949) who
argued that a scheme in which the upward moving air passed through the
"cold trap" of the high cold tropical tropopause seemed to be required
to explain the observed low water vapor mixing ratios in the stratosphere. Somewhat later Dobson (1956) pointed out that the poleward
and downward portion of this mean circulation was qualitatively
consistent with the observed high concentration of ozone in the lower
polar stratosphere, far from the region of photochemical production.
Although the Brewer-Dobson cell provides a useful partial model for
troposphere-stratosphere exchange and transport within the
stratosphere, it does not by any means represent a complete physical
description of the exchange process.

Evidence that the Brewer-Dobson cell model is an oversimplification
has come from two types of observational studies. On the one hand,
diagnostic studies of the Eulerian mean meridional circulation in
the winter stratosphere (e.g., Vincent, 1968) have revealed a two
cell pattern with rising in the tropics and polar regions and
descending motion in midlatitudes. On the other hand, observational
studies of the transport of ozone and radioactive tracers have revealed
that large scale eddy motions play an important part in tracer
transport (e.g., Reed, 1953; Feely and Spar, 1960). The comparative

roles of the eddies and the Eulerian mean are somewhat obscured by
the fact that there often tends to be almost complete cancellation
between eddy and mean flow transport. When the mechanisms of exchange
in midlatitudes are considered, however, there can be little doubt that
eddy processes are dominant. Observations reveal that the transfer
of trace constituents from the stratosphere to the troposphere is
concentrated in midlatitudes and is dominated not by the mean circulation,
but by mesoscale eddy processes associated with tropospheric cyclo-
genesis. Case studies of radioactive tracers and dynamical tracers
(potential temperature and potential vorticity) in the vicinity of
the tropospheric jetstream indicate that considerable stratospheric
air is mixed into the troposphere by intrusions which occur in
conjunction with upper level frontogenesis (Danielsen, 1968; Shapiro,
1978). These intrusions, which occur in thin layers of 100 km horizontal and 1 km vertical scale are eventually destroyed by irreversible
vertical mixing in the troposphere (Shapiro, 1980). Although some
tropospheric air no doubt is mixed into the stratosphere by slow
meridional circulations associated with the jet stream (Mahlman, 1973)
the extreme dryness of stratospheric air suggests that the primary
transport of mass from the troposphere into the stratosphere indeed
takes place in the equatorial region in accordance with the
Brewer-Dobson model.

In the past few years much progress has been made towards
resolving the eddy vs. mean cell controversy. The major contribution
to this resolution has been the work of Andrews and McIntyre (1976,
1978) who stressed the fundamental difference between Eulerian zonal
averages and Lagrangian averages. Dunkerton (1978) used the
Andrews and McIntyre framework to show that the Brewer-Dobson circulation should be interpreted as a Lagrangian mean mass circulation,
and that to a good approximation this circulation could be approximated
by the "diabatic" circulation. The latter circulation is just the
mean meridional circulation for which the adiabatic heating/cooling
due to the mean vertical motion just balances the zonal mean diabatic
heating. Since the lower stratosphere is radiatively heated at low
latitudes and radiatively cooled at high latitudes, the diabatic
circulation (shown schematically in Fig. 1) is consistent with a mass
circulation in which air flow from the troposphere into the stratosphere is limited to the tropics.

Although it is certainly true that air parcels moving from the
troposphere into the stratosphere must increase their potential
temperature through diabatic heating and air parcels moving out of
the stratosphere into the troposphere must reduce their potential
temperatures through diabatic cooling, the simple picture provided
by the diabatic circulation is not consistent with observed tracer
distributions. As Mahlman *et al.* (1983) have stressed, the combination
of upward advection in the tropics and downward advection at higher
latitudes implied by the diabatic circulation would cause the mean

isopleths of vertically stratified tracers to slope steeply downwards
from equator to pole. Although tracers such as methane and nitrous
oxide do have isopleths which slope in this sense, the actual
slopes are much less than those implied by a balance between the
advection by the diabatic circulation and photochemical destruction.
The additional process required to explain the observations is
rapid meridional mixing by quasi-isentropic eddies. This process
is demonstrated clearly in the simulation studies of Levy et $al.$
(1979). Thus, as summarized in Fig. 1, the gross characteristics
of stratosphere-troposphere exchange and transport within the lower
stratosphere can be modeled in terms of a combination of a mean
diabatic circulation and quasi-isentropic mixing by eddies.

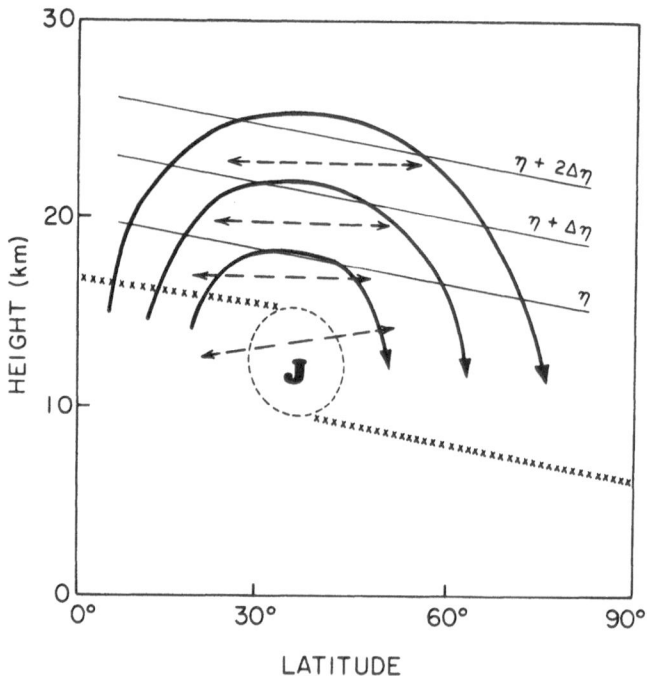

Fig. 1: *A schematic diagram of transport in the lower stratosphere. Heavy arrows show Brewer-Dobson circulation.
Dashed arrows indicate quasi-isentropic mixing by
eddies. The mean tropopause is indicated by crosses
and the J indicates the mean jetstream. Light lines
labelled with mixing ratio values (η) show the mean
slope of a long lived vertically stratified tracer.*

This conceptual model has been beautifully illustrated by
Kida (1977, 1983) in a series of experiments with a simplified
general circulation model in which a large number of "tagged"
particles were followed over along integration periods in order to
deduce the character of the mean mass circulation. Results showing
the time evolution of the projection in the meridional plane of a
set of parcels initially released in the tropical lower stratosphere
are shown in Fig. 2. The model clearly shows the tendency of parcels

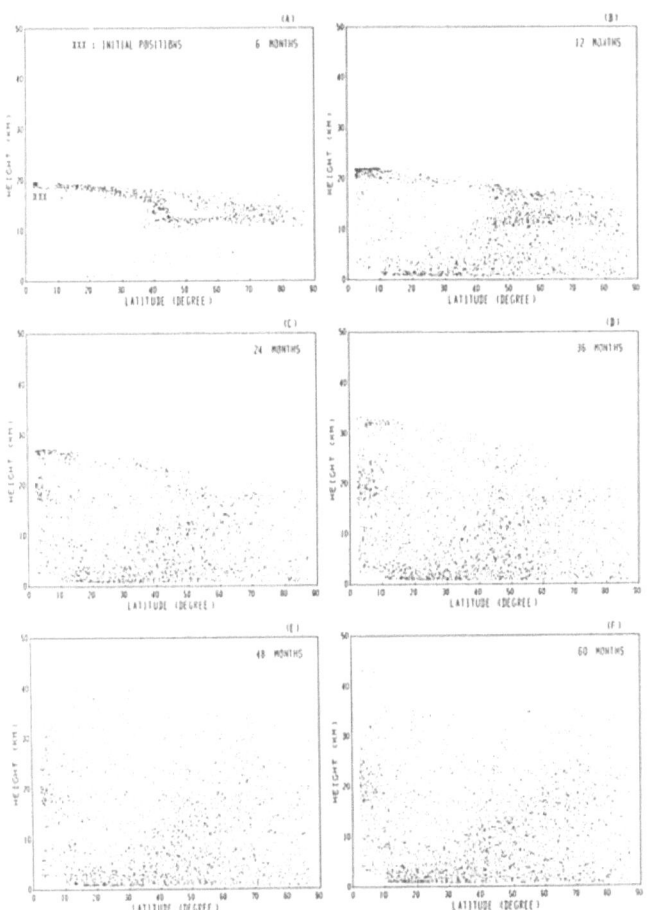

*Fig. 2: The projections on the meridonal plane at various
times for "tagged" parcels originally released
above the tropical tropopause at the positions
marked by crosses in panel A. After Kida, 1983.*

to slowly rise in the tropics and sink at high latitudes. But in addition to this slow diabatic circulation there is rapid quasi-isentropic poleward dispersion which leads to downward transfer of parcels into the troposphere at the mean latitude of the tropospheric jetstream. The results of Kida's model provide additional evidence that the upward transfer of air into the stratosphere occurs only in the tropical regions and that the rate of mass transfer is primarily controlled by diabatic heating/cooling. Thus Brewer's explanation for the observed dryness of the stratosphere appears to be qualitatively valid.

3. THE WATER VAPOR PUZZLE

Although there can be little doubt that the freeze drying of air due to upward passage through the tropical cold trap is a qualitatively reasonable explanation for the extreme dryness of the stratosphere, recent observational studies have revealed a number of problems with the Brewer-Dobson cell hypothesis. Measurement of water vapor in the stratosphere is an extremely difficult experimental problem. Attempts to summarize measurements from a variety of sources have been made by Harries (1976), Robinson (1980) and Elsaesser *at al.* (1980). The bulk of current evidence suggests a mean mixing ratio in the lower stratosphere of about 4 parts per million by volume (ppmv).

Dehydration to such a low mixing ratio would require that air entering the stratosphere pass through a cold trap with a temperature less that 191K at 100 mb, and that such air not carry ice crystals into the stratosphere, so that the total water content would be limited by the saturation mixing ratio at these conditions.

Climatological data, however, indicates that for much of the tropics, even in the active intertropical convergence zone (ITCZ), the temperatures at the tropical tropopause are insufficiently cold to allow freeze drying of air to a mixing ratio of 4 ppmv. Furthermore, the Brewer-Dobson model of a slow mean circulation rising through the tropical tropopause requires that a thick uniform cloud layer be formed as the rising air parcels are freeze dried. Such clouds are not observed. Rather, the cloudiness in the ITCZ is primarily convective in nature. As discussed in Holton (1979), for example, current evidence suggests that nearly all the upwards mass flux in the tropical troposphere is confined to the updrafts of individual cumulonimbus convective cells. These so called "hot towers" may overshoot their levels of equilibrium buoyancy and penetrate into the lower stratosphere. Thus, an understanding of the mechanism and distribution of troposphere-stratosphere exchange in the tropics requires consideration of the convective scale of motion.

Some important progress in understanding the role of convective motions in tropical exchange has occurred in the past few years as a result of aircraft experiments conducted in the Panama Canal Zone in 1977 and 1980 by NASA Ames Research Center. (Some results of these experiments are reported in Poppoff *et al.*, 1979; Danielsen, 1982a,b; Kley *et al.*, 1982; and Knollenberg, 1982). These experiments employed a U-2 aircraft equipped with a number of instruments for in-situ measurements of various meteorological parameters and trace constituents of both tropospheric and stratospheric origin. Although the 1977 experiment did not have an in-situ water vapor measurement capability, remote measurements of the water vapor overburden by a water vapor radiometer did provide evidence that cumulonimbus convective elements can penetrate into the stratosphere and mix with stratospheric air, thus providing a means for transfer of tropospheric tracers into the stratosphere.

An additional critical finding of the 1977 experiment was evidence of very large amplitude internal gravity wave motions in the tropical stratosphere, apparently excited by convective activity (Poppoff *et al.*, 1979). High temporal resolution measurements of ozone and temperature on the U-2 aircraft established that stable layers with wavelike vertical excursions of nearly 1 km were interspersed with turbulent adiabatic layers. The confirmation of such waves may have important implications for the dynamics of the equatorial middle atmosphere. However, their role in the exchange process is not clear. Since no air motion sensors were flown on the U-2, direct deductions concerning possible vertical fluxes of trace species associated with the wave motions was not possible.

The suggestion from the 1977 experiment that cumulus convective elements might be the physical entities responsible for most of the vertical transfer of trace species across the tropical tropopause, as well as the realization that an understanding of these processes was essential to elucidating the source of the extreme aridity of the stratosphere, led to a focus on the water vapor exchange problem in the 1980 Panama experiment. In that experiment the U-2 was equipped with two new in-situ water vapor instruments, a Lyman-alpha hygrometer with very fast response, and an electronic frost point hygrometer. In addition there were a pair of particle size spectrometers for measuring the aerosol and ice crystal budgets.

This experiment confirmed without reasonable doubt that the very low temperatures which exist in overshooting convective turrets can not directly account for freeze drying of the stratospheric air since air parcels in the overshooting towers are filled with ice crystals which will evaporate when the parcels are mixed with stratospheric air (Kley *et al.*, 1982). Thus, the immediate effect of convection is to hydrate the stratosphere.

An additional difficulty with the Brewer-Dobson hypothesis was revealed by the vertical profiles of the water vapor mixing ratio obtained during the 1980 experiment. As shown in Fig. 3, these profiles reveal a distinct minimum in the mixing ratio at 19 km (well above the tropopause) in agreement with some previous balloon measurements (Kley *et al.*, 1982). This minimum is significantly

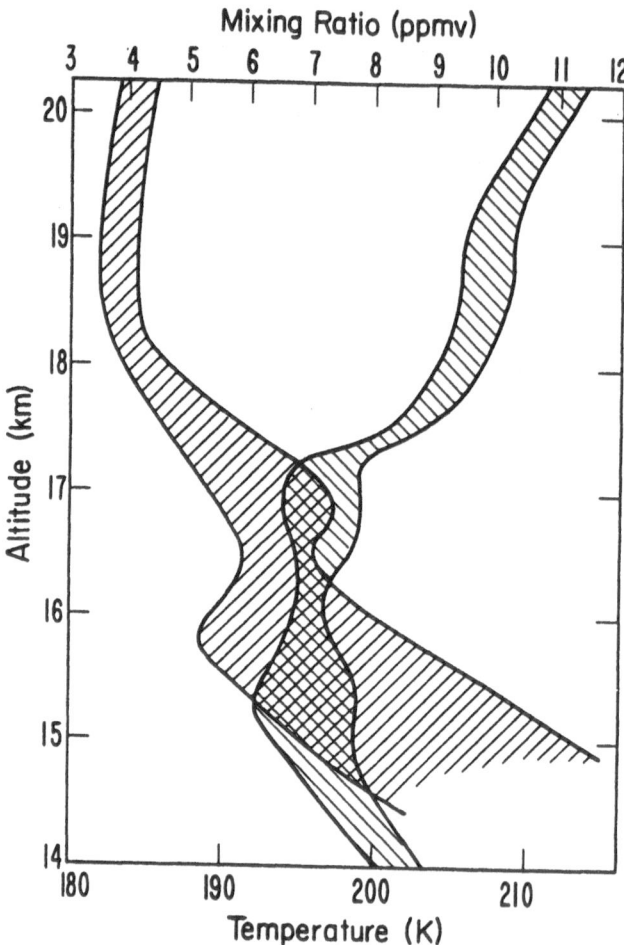

Fig. 3: *The range of temperature and water vapor mixing ratio measurements for 8 flights during the 1980 Panama experiment. Areas enclosed by cross hatching show plus and minus 1 standard deviation from the mean. After Kley et al., 1982.*

smaller than the saturation mixing ratio at the tropopause level and occurs well above the level of highest cumulus turret penetration in the Panama region. Thus it seems clear that the water vapor mixing ratio profile in the Panama region can not be explained on the basis of local vertical transport processes. Rather, long range quasi-horizonal transport must be invoked to account for the minimum at 19 km.

Newell and Gould-Steward (1981) suggested that the stratospheric water vapor observations are consistent with a model, which they called the "stratospheric fountain" in which most of the flux of mass from the troposphere into the stratosphere is concentrated in relatively limited regions of the tropics where tropopause temperatures are observed to be significantly colder than those at Panama during the summer. They found that 100 mb temperatures less than 191K are generally limited to the Indonesian "maritime continent" during Northern Hemisphere winter, and to the Indian monsoon region during Northern Hemisphere summer. Atticks and Robinson (1982) utilized the enhanced network of tropical soundings during the FGGE year of 1979 to further examine the temporal and spatial distribution of tropopause temperatures. Their results were generally in accord with those of Newell and Gould-Stewart, however they argued that during Northern Hemisphere winter sufficiently cold tropopause conditions occurred over a broader range of longitudes than suggested by Newell and Gould-Stewart. In any case, it is clear that the tropical tropopause is on the average highest and coldest in both hemispheres during the Northern Hemisphere winter. Only in the Asian monsoon area are conditions favorable for freeze drying air to the observed stratospheric mixing ratios during the Northern Hemisphere summer. This is at least qualitatively in accord with the observed annual cycle of stratospheric water vapor in midlatitudes (Ellsaesser *et al.*, 1980).

4. A POSSIBLE DEHYDRATION MECHANISM

The stratospheric fountain hypothesis of Newell and Gould-Stewart (1981) provides a basis for partial understanding of the puzzles posed by the water vapor observations over Panama, if it is assumed that the mixing ratio minimum observed at the 19 km level originated from the inflow of air into the stratosphere in the Indonesian region where the air might be freeze dried to a mixing ratio of about 2 ppmv when entering the stratosphere. Due to the slow rate of mixing in the lower tropical stratosphere, air which entered over Indonesia in January might well preserve its identity long enough to account for the 4 ppmv mixing ratio at 19 km over Panama in September.

However, the stratospheric fountain hypothesis does not itself provide a dynamical mechanism for assuring that air passing through the cold trap will have its water content reduced to the saturation mixing ratio at the tropopause. The occurrence of strong convection penetrating into the lower stratosphere in the Indonesian region during the Northern winter has been amply documented in the Winter Monex experiment (Johnson and Kriete, 1982). Thus in this region, as in Panama, overshooting cumulus turrets would tend to carry ice crystals into the stratosphere so that the immediate effect of cumulus penetration should be to hydrate rather than dehydrate the stratosphere.

In order to fully understand the dehydration of the stratosphere it appears necessary to consider the dynamical and microphysical processes which occur over the full life cycle of tropical convective systems. It is well known that the sinking due to negative buoyancy which follows the overshooting of cumulonimbus turrets in the tropics produces massive cirrus anvil clouds, far larger than produced at midlatitudes (e.g., Houze, 1982; Danielsen, 1982a, Johnson and Kreite, 1982). Such anvils may last for 5-10 hours or more. Measurements of stratospheric aerosols by Knollenberg et al. (1982) during the 1980 Panama experiment indicated that up to 40% strato-spheric air may be entrained into the collapsing turrets which form the anvil clouds. Thus, the mean potential temperature in the anvils must exceed that of the convective turrets, and their levels of neutral equilibrium will rise above those of unmixed parcels. Consequently, the cirrus anvils formed by the sinking and spreading of the air in tropical convective plumes may have their upper surfaces in the stratosphere as shown schematically in Fig. 4 taken from Danielsen (1982b).

Danielsen (1982b) proposed an intriguing hypothesis for stratospheric dehydration which depends on the radiative, dynamical, and microphysical processes which occur in anvil clouds. Danielsen pointed out that there will be rapid radiative cooling to space from the top surface of a cirrus anvil, while if there is little cloudiness between the lower surface and the ground, the lower surface should be radiatively heated. Thus, radiation will tend to produce a moist adiabatic lapse rate within the cloud as shown in Fig. 4. In this model the radiative cooling at cloud top is balanced not by subsidence of the cloud, but by turbulent transport of heat upward within the cloud. Danielsen showed that the vertical temperature flux within the cloud layer necessary to balance a 10 K/day cooling rate from the cloud top was only 5K cm/s. However, the large difference in saturation vapor pressure between the cloud bottom and cloud top implies a strong upward flux of water vapor which would lead to supersaturation and rapid ice crystal growth near cloud top with subsequent fallout of precipitation. Thus, this thermally driven in cloud circulation should produce upward heat and vapor

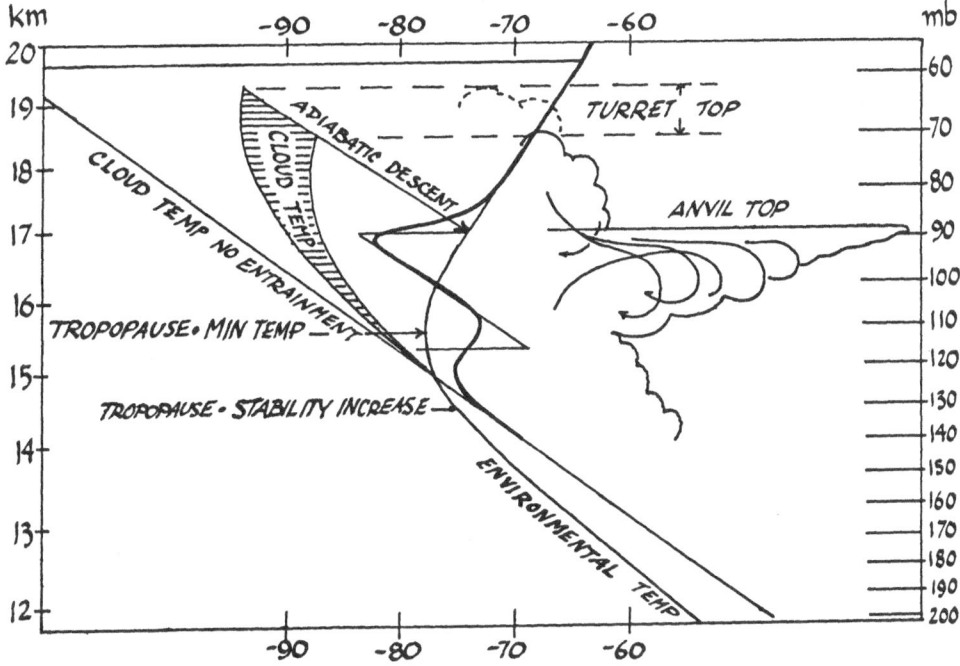

*Fig. 4: A schematic of cirrus anvil formation in the stratosphere.
The shaded area shows range of temperatures in overshooting
cumulonimbus turrets due to entrainment of stratospheric
air. Heavy line shows temperature profile which would
result from anvil formation (see text for details). After
Danielsen, 1982.*

fluxes and a downward flux of ice crystals. Danielsen (1982b)
argued that the net effect should be to dehydrate the radiatively
cooled air near cloud top. Therefore, the net effect of tropical
convection, when averaged over the entire life cycle of a tropical
convective system, should be to transfer very dry air into the
stratosphere.

Danielsen's dehydration hypothesis can only be tested by a
carefully designed program of in-situ measurements with research
aircraft. Plans for such experiments are currently being developed
at NASA Ames Research Center. In the meantime, it is possible to
at least confirm that currently available evidence is consistent
with a model in which cirrus anvil clouds play a major role.

380

Studies of convective storms during the winter MONEX program in December 1978 (Houze *et al.*, 1981) revealed that mesoscale anvils of 200 km horizontal dimension were regular features of the diurnally varying convective cycle in the vicinity of North Borneo. Johnson and Kriete (1982) have carefully analyzed rawinsonde data from the special MONEX network in order to elucidate the dynamical characteristics of this convection during the period of 9-11 December, 1978. The time series of rawinsondes taken on a Soviet research vessel during this period is shown in Fig. 5. During each of these

Fig. 5: (a) Rawinsonde time series during winter MONEX for 9-11 December 1978. Stippling denotes regions of greater than 80% relative humidity. Solid contours are temperature deviations (K) from the 6-28 December mean. Wind speeds are in m/s (full barb - 5 m/s). Dashed line marks the tropopause. Bars at top indicate fraction of ship array covered by bright IR satellite cloudiness. (b) Vertical pressure velocity in units of 100 mb/day for the ship array computed by the kinematic method. Distance scale indicated represents advective length scale for 6 m/s motion of the anvil clouds. After Johnson and Kriete, 1982.

3 days strong convective systems which developed in the morning hours gave rise to long lived mesoscale anvil clouds. The formation of the anvils was accompanied by large negative temperature anomalies just above the tropopause near 18 km. Johnson and Kriete pointed out that such thermal anomalies might result from the upward extension of the forced mesoscale ascent in the middle and upper troposphere shown on the lower panel in Fig. 5. However, the anomalies are also consistent with the Danielsen model of radiative cooling at the cloud tops accompanied by a convective overturning within the anvil. The tephigram plot of temperature soundings before and after anvil formation on 10 December shown in Fig. 6 (from Johnson and Kriete, 1982) seems completely consistent with the Danielsen model since the anvil formation clearly leads to the formation of a near adiabatic lapse rate and strong cooling in the region of the tropopause. Note that the afternoon sounding has a temperature near 185K at 100 mb which implies a saturation mixing ratio near 1 ppmv. Clearly if air is transferred from the troposphere to the stratosphere in such circumstances without at the same time transferring significant amounts of ice crystals, a dramatic dehydration would occur.

5. SUMMARY

This overview has attempted to show that the traditional Brewer-Dobson cell model provides an incomplete and oversimplified view of stratosphere-troposphere exchange in the tropical regions. The observed distribution of water vapor mixing ratios in the stratosphere provides strong constraints on the mechanisms and spatial distribution of mass exchange. In particular, the observed extreme aridity of the stratosphere appears to require that most of the upward transfer into the stratosphere occur in regions where the tropical tropopause is colder and higher than average. The Indonesian "maritime continent" is a particularly favorable region for such exchange during the Northern Hemisphere winter. However, even in favorable regions the mass transport into the stratosphere cannot occur in the form of a slow mean rising motion since such motions would carry the ice crystals formed by freeze out in the upper troposphere upward into the stratosphere where they would evaporate and hydrate the stratosphere. Thus it seems likely that the actual transport into the lowest levels of the stratosphere is associated with convective scale processes, but that the transport has a strong longitudinal and seasonal dependence controlled by the regional variability in the frequency and intensity of cumulonimbus convective disturbances.

A physical mechanism suggested by Danielsen (1982b) which is based on the combined effects of radiation, dynamics, and cloud microphysics acting on cirrus anvil clouds produced by tropical convection provides a plausible model for dehydration of air entering the stratosphere. However, this model has not yet been verified by detailed observational studies.

382

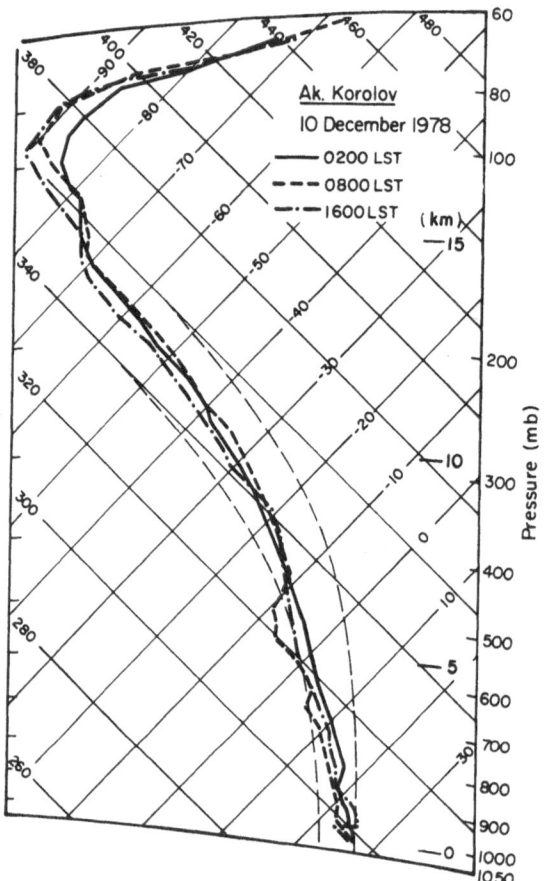

*Fig. 6: Tephigram plot of temperature soundings at the Soviet ship
Ak. Korolov prior to and during the passage of mesoscale
anvil system on 10 December, 1978. After Johnson and
Kriete, 1982.*

ACKNOWLEDGMENTS: I am indebted to Dr. Edwin Danielsen for many
illuminating discussions of troposphere-stratosphere exchange. This
work was supported by the National Science Foundation's Atmospheric
Research Section, NSF Grant ATM79-24687.

REFERENCES

Andrews, D. G., and M. E. McIntyre, 1976: Planetary waves in horizontal and vertical shear: The generalized Eliassen-Palm relation and the mean zonal acceleration. *J. Atmos. Sci.*, *33*, 2031-2048.

Andrews, D. G., and M. E. McIntyre, 1978: An exact theory of nonlinear waves on a Lagrangian mean flow. *J. Fluid. Mech.*, *89*, 609-646.

Atticks, M. G., and G. D. Robinson, 1983: Some features of the structure of the tropical tropopause. *Quart. J. Roy. Meteor. Soc.*, *109*, 295-308.

Brewer, A. M., 1949: Evidence for a world circulation provided by the measurements of helium and water vapor distribution in the stratosphere. *Quart. J. Roy. Meteor. Soc.*, *75*, 351-363.

Danielsen, E. F., 1968: Stratospheric-tropospheric exchange based on radioactivity, ozone, and potential vorticity. *J. Atmos. Sci.*, *25*, 502-518.

Danielsen, E. F., 1982a: Statistics of cold cumulonimbus anvils based on enhanced infrared photographs. *Geophys. Res. Lett.*, *9*, 601-604.

Danielsen, E. F., 1982b: A dehydration mechanism for the stratosphere. *Geophys. Res. Lett.*, *9*, 605-608.

Dobson, G. M. B., 1956: Origin and distribution of polyatomic molecules in the atmosphere. *Proc. Roy. Soc. London*, *A236*, 187-193.

Dunkerton, T., 1978: On the mean meridional mass motions of the stratosphere and mesosphere. *J. Atmos. Sci.*, *35*, 2325-2333.

Ellsaesser, H. W., J. E. Harries, D. Kley, and R. Penndorf, 1980: Stratospheric water vapor. *Planet. Space Sci.*, *28*, 827-835.

Feely, H. W., and J. Spar, 1960: Tungsten-185 from nuclear bomb tests as a tracer for stratospheric meteorology. *Nature*, *188*, 1062-1064.

Harries, J. E., 1976: The distribution of water vapor in the stratosphere. *Rev. Geophys. Space Phys.*, *14*, 565-575.

Holton, J. R., 1979: *An Introduction to Dynamic Meteorology*, Academic Press, 391 pp.

Houze, R. A., 1982: Cloud clusters and large-scale vertical motions in the tropics. *J. Meteor. Soc. Japan*, *60*, 396-409.

Houze, R. A., S. G. Geotis, F. D. Marks, and A. K. West, 1981: Winter monsoon convection in the vicinity of North Borneo. Part I: Structure and time variation of the clouds and precipitation. *Mon. Wea. Rev.*, *108*, 1595-1614.

Johnson, R. H., and D. C. Kriete, 1982: Thermodynamic and circulation characteristics of winter monsoon tropical mesoscale convection. *Mon. Wea. Rev.*, *110*, 1898-1911.

Kida, H., 1977: A numerical investigation of the atmospheric general circulation and stratospheric-tropospheric mass exchange: I. Long-term integration of a simplified general circulation model. II. Lagrangian motion of the atmosphere. *J. Meteor. Soc. Japan*, *55*, 52-88.

Kida, H., 1983: General circulation of air parcels and transport in stratosphere and troposphere derived by GCM: 1. Mean mass motion in the lower stratosphere. 2. Very long-term motions of air parcels. Submitted to *J. Meteor. Soc. Japan*.

Kley, D., A. L. Schmeltekopf, K. Kelly, R. H. Winkler, T. L. Thompson and M. McFarland, 1982: Transport of water vapor through the tropical tropopause. *Geophys. Res. Lett.*, *9*, 617-620.

Knollenberg, R. G., A. Dascher, and D. Huffman, 1982: Measurements of the aerosol and ice crystal population in tropical stratospheric cumulonimbus anvils. *Geophys. Res. Lett.*, *9*, 613-616.

Levy, H. II, J. D. Mahlman, and W. J. Moxim, 1979: A preliminary report on the numerical simulation of the three-dimensional structure and variability of atmospheric N_2O. *Geophys. Res. Lett*, *6*, 155-158.

Mahlman, J. D., 1973: On the maintenance of the polar front jet stream. *J. Atmos. Sci.*, *30*, 544-557.

Mahlman, J. D., D. G. Andrews, H. U. Dutsch, D. L. Hartmann, T. Matsuno, and R. J. Murgatroyd, 1983: Transport of trace constituents in the stratosphere. (this volume)

Newell, R. E., and S. Gould-Stewart, 1981: A stratospheric fountain? *J. Atmos. Sci.*, *38*, 2789-2796.

Poppoff, I. G., W. A. Page, and A. P. Margozzi (eds.), 1979: *The 1977 Intertropical Convergence Zone Experiment. NASA Technical Memorandum 78577.* 488 pp.

Reed, R. J., 1953: Large-scale eddy flux as a mechanism for vertical transport of ozone. *J. Meteor.*, *7*, 263-267.

Robinson, G. D., 1980: The transport of minor atmospheric constituents between troposphere and stratosphere. *Quart. J. Roy. Meteor. Soc.*, *106*, 227-253.

Shapiro, M. A., 1980: Turbulent mixing within tropopause folds as a mechanism for the exchange of chemical constituents between the stratosphere and troposphere. *Mon. Wea. Rev.*, *93*, 313-321.

Stanford, J. L., 1973: Possible sink for stratospheric water vapor at the winter Antarctic Pole. *J. Atmos. Sci.*, *30*, 1431.

Telegadas, K., and R. J. List, 1969: Are particulate radioactive tracters indicative of stratospheric motions? *J. Geophys. Res.*, *74*, 1339-1350.

Vincent, D. G., 1968: Mean meridional circulations in the Northern Hemisphere lower stratosphere during 1964 and 1965. *Quart. J. Roy. Meteor. Soc.*, *94*, 333-349.

World Meteorological Organization (WMO), 1982: *The Stratosphere 1981 Theory and Measurements. WMO Global Ozone Research and Monitoring Project Report No. 11.*

J. R. Holton and T. Matsuno, Dynamics of the Middle Atmosphere, 387–416.
Copyright © 1984 by Terra Scientific Publishing Company.

TRANSPORT OF TRACE CONSTITUENTS IN THE STRATOSPHERE

J. D. Mahlman, D. G. Andrews, D. L. Hartmann, T. Matsuno,
R. G. Murgatroyd

ABSTRACT

Our understanding of stratospheric trace constituent transport is
reviewed, beginning with the evolution of viewpoints on the subject
over the past half century. We then discuss the available observed
characteristics of trace constituents and the challenges posed by them.

To provide a framework for further developments, we present a
simplified conceptual view of processes leading to irreversible transport. Next, methods of approach to such problems are outlined, including discussions of possibilities for physically based parameterized
transport models. Finally, we pose a set of unanswered questions as
possible guides to future research on the subject.

1. INTRODUCTION AND HISTORICAL BACKGROUND

The problem of how trace constituents are transported in the
stratosphere has historically received attention mainly in relation to
other problems. The earliest case arose when it was observed that
ozone exhibited behavior contrary to expectations from photochemical
arguments (e.g., Dobson, Harrison, and Lawrence, 1926; Meetham, 1937).
A number of investigators attempted to explain the observed local
total ozone variations by horizontal and vertical advective and/or
flux divergence mechanisms in various combinations (Dobson, 1930;
Haurwitz, 1938; Nicolet, 1945; Dütsch, 1946; Craig, 1950; Reed, 1950;
Normand, 1953).

Toward the end of this period, largely because of the impact of
nuclear weapons detonations in the atmosphere, attention began to
shift toward a more global view of transport phenomena. This led to a
strong emphasis on the use of zonal averages as a means of simplifying
the problem of understanding trace constituent transport. In this
framework the problem was reduced to the one of deciding whether
material is moved across latitude-height lines by zonally symmetric
"mean" processes, or by asymmetric "eddy" processes.

The first to propose a plausible mean meridional circulation as a transport agent appears to have been Dütsch (1946), who suggested the importance of a pole-to-pole circulation with sinking in the winter hemisphere. Later Brewer (1949) hypothesized that a simple one cell circulation in each hemisphere with rising motion at and above the tropical tropopause and sinking in the polar regions is compatible with the observed dryness of the stratosphere. This view was echoed by a number of workers such as Dobson (1956), Goldie (1958), Palmer (1959), Burton and Stewart (1960), Dyer and Yeo (1960), and Murgatroyd and Singleton (1961).

The effect of eddies on average meridional and vertical transport was apparently emphasized first by Reed and Julius (1951) and by Reed (1953). Similar viewpoints on ozone transport were expressed by Ramanathan and Kulkarni (1960) and Godson (1960). These inferences received some support from the analysis of radioactive tungsten by Feely and Spar (1960) which suggested a dominant influence of eddy fluxes. These ideas were reinforced by the more quantitative approach of Newell (1961, 1963a,b, 1964, 1965). His work also showed how forced transient eddies can transport heat and tracers poleward in such manner as to be consistent with the annual mean "reversed" meridional temperature gradient in the lower stratosphere. He showed that such a countergradient eddy heat flux is possible as long as poleward moving air is subsiding and equatorward moving air is ascending. This was found to be so in lower and middle latitudes by Molla and Loisel (1962). This result was confirmed and expanded upon by Miller (1967) and Mahlman (1970a).

Thus, by the 1960's the stratospheric transport problem had evolved into a mean cell vs. eddies question. However, all at about the same time, three studies showed that the winter lower stratospheric mean meridional circulation is quite unlike that visualized by Brewer (1949) (Miyakoda, 1963; Reed, Wolfe, and Nishimoto, 1963; and Teweles, 1963). These results indicated a two cell indirect circulation with ascending motion in the cold tropics and the cold polar region and descending motion near the winter mid-latitude warm belt. These basic results were put on firmer ground by Julian and Labitzke (1965), Murakami (1965), Muench (1965), Newell and Miller (1965), Mahlman (1966, 1969), Perry (1967), and Vincent (1968).

In the southern hemisphere, investigators found that the winter stratospheric meridional circulation has a third cell with sinking over the pole (Adler, 1975; Hartmann, 1976). During disturbed conditions, however, Adler showed that the more familiar type of two-cell pattern emerges. This third cell was also obtained for the southern hemisphere in the general circulation model (GCM) study of Manabe and Mahlman (1976).

The above results made it clear that the indirect stratospheric meridional circulation is not capable of an existence independent of the zonally asymmetric disturbances or eddies. In Mahlman (1966, 1969) it was noted that the stratospheric heat balance is such that the heating effect of the meridional circulation acts in opposition to that produced by large-scale eddy heat flux convergence. This led Mahlman (1966) to hypothesize that the stratospheric dispersion of trace constituents is a rather complicated function of interactive mean cell and eddy transports. This point was made clear in a general circulation model (GCM) experiment by Hunt and Manabe (1968). Their results from an annual mean model showed the presence of a large degree of cancellation between mean cell and eddy flux convergences in the zonally averaged trace constituent balances. Similar cancellations had already been obtained in the lower stratospheric heat balances of the GCM of Smagorinsky, Manabe, and Holloway (1965).

Partially in response to the above difficulties, alternatives to the traditional zonal averaging approaches began to be considered. For example, Danielsen, Bergman, and Paulsen (1962) and Danielsen (1968) emphasized the importance of meridional gradients in diabatic heating for producing the observed poleward-downward slopes of tracer isolines. This approach arose from the recognition that adiabatic motions constrain air parcels to remain on isentropic surfaces. Those authors also pointed out that meridional gradients of diabatic heating can produce horizontal tracer gradients on isentropic surfaces which then can be mixed isentropically. A similar point was stressed by Newell (1963b).

Another alternative perspective was offered by Mahlman (1966) who examined the local vertical velocity structure responsible for the large mean cell-eddy cancellation in the heat balance. From that basis, he suggested that the zonal average was not the most physically motivated averaging approach. He showed that an average relative to the geopotential height contours (with the zero axis defined as the contour containing the strongest mean geostrophic wind speed) gives a direct "mean" circulation, opposite in sign to the mean meridional circulation (Mahlman, 1966, 1969). The use of this type of approach was also emphasized by Züllig (1973).

During this same era, a parallel theoretical development was initiated by Eliassen and Palm (1961) and by Charney and Drazin (1961). They pointed out that waves do not necessarily lead to a systematic acceleration of zonal mean flows. Those results were successively made less restrictive by Dickinson (1969), Holton (1974), Boyd (1976) and Andrews and McIntyre (1976), 1978a,b). The generalized theorem states that steady, frictionless, adiabatic waves of small amplitude, propagating in a basically zonal flow, exert no net effects on the mean flow. Rather, such waves induce meridional circulations at second order in wave amplitude which exactly cancel the eddy flux convergences due to the waves. Further results are available for finite amplitude disturbances, but must be given in terms of "Langrangian mean" quantities. Note that no general results of this kind are available for waves propagating in basic states which already include violations of the requisite conditions for "non-acceleration" to hold. A little later various authors offered arguments indicating that "non-acceleration" conditions might also imply "non-transport" for suitably distributed conservative tracers (e.g., Andrews and McIntyre, 1978b; Clark and Rogers, 1978; Mahlman and Moxim, 1978; Wallace, 1978; Plumb, 1979; Holton, 1980a; Mahlman, Levy and Moxim, 1980; Matsuno, 1980; Pyle and Rogers, 1980a).

The above developments gave a more rigorous basis for the demonstrated connection between the meridional circulation and eddy transports by showing that, under these special circumstances, the induced indirect meridional circulation is a property of the wave field itself. This helped provide an interpretation of the mutually compensating tracer transport convergences found in the numerical simulations of Hunt and Manabe (1968) as well as those found by Mahlman (1973b), Newson (1974), Cunnold, et al. (1975,1980), Mahlman and Moxim (1978), Schlesinger and Mintz (1979), and Mahlman, Levy, and Moxim (1980). The disappearance of this compensating effect when chemical sources and sinks are strong has been shown in the above studies as well as by Hunt (1969) and Clark (1970). Mechanistic interpretations of this effect are given in Hartmann and Garcia (1979) and Garcia and Hartmann (1980). In addition, Mahlman and Moxim (1978) showed that the compensation effect diminishes markedly during seasonal transitions in middle latitudes, and most of the time in lower latitudes.

Thus, even though there are many cases in which the traditional partitioning into zonal means and eddies gives straightforward results, there are many others where more enlightened approaches are required. Important advances in this regard were provided in the work of Matsuno (1972), Uryu (1974), and Andrews and McIntyre (1976, 1978b) which demonstrated the power of utilizing Lagrangian displacements (from simple undisturbed states) for gaining analytical simplicity and physical understanding. Using these ideas Matsuno (1972, 1980) emphasized the zonal mean diabatic heating (Murgatroyd and Singleton, 1961) as the most fundamental portion of the meridional circulation, somewhat analogous to that suggested earlier by Danielsen, Bergman, and Paulson (1962). The utility of this idea was demonstrated by Kida (1977) in a simplified GCM which calculated extended particle trajectories. The point was clarified by Dunkerton (1978) who argued that the motion related to the zonal mean diabatic heating may serve as a good approximation to the "Lagrangian mean circulation" (the average meridional and vertical drift of a material tube of particles (Matsuno and Nakamura, 1979)). On the other hand, the results of Mahlman, Levy, and Moxim (1980) suggest that, in the lower stratosphere at least, the effects of eddy diabatic heating are important contributors to the Lagrangian mean circulation, as well as to the dispersion of particles about that mean. In that study, the argument was based on the problem of determining the equilibrium structure of the meridional slopes of tracer isolines. The very significant effect of dispersion about the Lagrangian mean was demonstrated rather dramatically in a study of particle trajectories associated with a model sudden stratospheric warming (Hsu, 1980). As will be argued later, proper consideration of such dispersive effects is essential for obtaining a quantitatively correct calculation of trace constituent structure. A number of possibilities and problems associated with application of Lagrangian mean concepts to tracer transport problems are addressed in a perceptive summary by McIntyre (1980).

In this paper we will examine a number of aspects of the trace constituent transport problem under current emphasis. After a brief summary of the challenges offered by the various observations, we will present our effort to obtain a synthesized theoretical view of the transport problem, followed by an examination of the applicable methodologies. Then, because of their intense current interest, we will examine the status and prospects for the 1-D and 2-D parameterized transport models. Finally, we will focus on the remaining questions and what might be done in the future toward their solution.

2. OBSERVED BEHAVIOR OF TRACE CONSTITUENTS

The subject of trace constituent transport is similar to many others in geophysics in the sense that theories have arisen mainly in the effort to explain observed behavior. Because of this we will review here some of the most important aspects of trace constituent structure and climatology. For a more general review, see Reiter (1971).

The first recognition of a complex trace constituent structure arose through the early measurements of total ozone (e.g., Dobson, Harrison, and Lawrence, 1926). Once the photochemical theories of Chapman (1930) and Wulf and Deming (1936) took hold, it was realized that ozone exhibited behavior requiring transport arguments for its explanation. In particular, the total ozone network established by Dobson found that the largest ozone column amounts are in higher latitudes, opposite to that expected under photochemical equilibrium. In addition, they measured a mid-latitude spring maximum and fall minimum in ozone, as well as day-to-day variability which correlated well with the passage of surface weather systems.

It thus has been long recognized that downward transport of ozone from the middle stratosphere is required to explain the above observed features. Interestingly, since it was so difficult to make quantitative calculations of ozone behavior (it still is!), little attention was paid to the time scales of transport up until the 1950's. In fact, many investigators still assumed that diffusive separation of gases would be applicable throughout the stratosphere. However, with the advent of nuclear weapons testing, perspectives began to shift significantly. Rather than remaining in the stratosphere for long periods of time, the radioactive debris was found to return to the troposphere much faster than expected. The earliest actual measurements inferred that the residence time for such debris is on the order of 5 years (Libby, 1956; Machta and List, 1959). The early radioactivity measurements also showed that the debris returning to the troposphere is concentrated mainly in midlatitudes and in the spring (Machta, 1957). Similar behavior was later found for the cosmogenic isotope beryllium-7 (Bleichrodt and Van Abkoude, 1963).

The nuclear test moratoriums in 1958 and in 1962 allowed for more accurate assessments of the rates of removal of debris from the stratosphere. Feely et al. (1966) and Telegadas and List (1969) showed that the "residence time" (time for decay-corrected amount to reduce to $1/e$ of the original amount) for particulate debris is nearly constant in time with a value of about 15 months. On the other hand, these same authors show that the apparent excess $C^{14}O_2$ (amount above background $C^{14}O_2$) residence times after the 1962 moratorium increased from around two years during the first year to more than 4 years several years later. This difference was explained by Telegadas and List (1969) as being caused by the effect of particulate debris settling acting to keep its center of gravity in the lower stratosphere, while that of the $C^{14}O_2$ continues to rise. This effect of increasing residence time with altitude was demonstrated in the general circulation/tracer model of Mahlman and Moxim (1978). Some observationally derived residence times as a function of altitude and latitude have been presented by Reiter et al. (1975). These results provided quantitative support for the contention of Telegadas and List (1969) and Chang (1976) that the early differences between the residence times for particulate and $C^{14}O_2$ debris in 1963 must have been due to an initial deposition of the C^{14} at higher elevations because of its enhanced presence in high yield thermonuclear devices. This difference was not observed following the earlier test series -- a period when such high yield devices were not employed.

Another dominant aspect of trace constituent climatology is the tendency for pronounced meridional slopes of zonal mean isolines. This effect is evident in the ozone cross sections of Hering (1966) and the radioactivity cross sections of Feely et al. (1966) and Machta, Telegadas, and List (1970). Later compilations verify these slopes, but also show a tendency for largest winter radioactivity concentrations to appear near 60-70°N (Machta and Telegadas, 1973). They also showed that the radioactivity maximum tends to split into two regions with a minimum near 50°N, at least during the winter season. The various data sets show that the meridional tracer slopes are steeper than the corresponding slopes of the zonal mean potential temperature surfaces. In addition, Hering (1966) showed that the

zonal-mean ozone structure correlates well with the zonal-mean potential vorticity, at least in mid and high latitudes. Danielsen et al.
(1970) showed that this correlation extends as well into synoptic and
mesoscale features. The meridional tracer slopes, the winter higher
latitude bulge, and the correlation with potential vorticitiy were
simulated in the 3-D model experiment of Mahlman and Moxim (1978).

For evaluation of theories and ideas about stratospheric transport, other aspects of trace constituent behavior are most valuable.
For example, Dütsch (1969, 1974) and Pittock (1977) presented statistics of ozone temporal variations at individual stations. They
showed that ozone relative (%) standard deviations in the lower, mid-
latitude stratosphere in both hemispheres are larger than 50%. The
individual O_3 profiles show a strong layered structure which often
exhibits strong time continuity (Breiland, 1967, 1968; Attmannspacher
and Dütsch, 1970; Dobson, 1973). Using special observing periods in
the now defunct North American Ozonesonde Network and trajectory techniques, Berggren and Labitzke (1966, 1968) and Mahlman (1970b) presented daily synoptic charts of ozone mixing ratio. These charts show
large synoptic-scale variations of O_3 in the lower stratosphere, with
the highest O_3 values located in the troughs of long wave disturbances. Also, daily charts have been prepared for O_3 in the middle and
upper stratosphere using satellite based instruments (Heath, Mateer,
and Krueger, 1973; London, Frederick, and Anderson, 1977). At these
levels, the variations are larger in scale because of the dual effect
of decay of cyclone-scale disturbances with altitude and the increased
efficiency of photochemical damping.

In the southern hemisphere, there is much less ozone data available. Nevertheless, significant differences from the northern hemisphere have been identified. The total ozone compilations of Dütsch
(1969) show that the largest values are found near 50-60°S, rather
than in higher latitudes as in the northern hemisphere. The data also
shows that the largest total ozone values are found in late spring,
rather than the early spring maximum of the northern hemisphere. For
considerably more detail, see Dütsch (1978). Available satellite
data (Prabhakara et al. 1976; Hilsenrath, Heath, and Schlesinger,
1979) show a considerably higher longitudinal asymmetry in the
northern hemisphere. They show a marked exception to this in late
southern hemisphere spring, however. At that time a pronounced total
ozone maximum appears at about 65°S and centered at about 180°W. A
corresponding perturbation appears in the temperature. To date very
little attention has been given to this phenomenon.

Another aspect of trace constituent behavior providing valuable
clues about transport is the association with jet streams. For
example, Briggs and Roach (1963) showed that much larger O_3 mixing
ratios are found on the cyclonic shear side of mid-latitude jet
streams. This effect shows up strongly in the total ozone as well

(Lovill, 1972). Such structure is compatible with the direct circulation calculated to occur around the polar front jet stream by Mahlman (1973a) and as hypothesized much earlier by Namias and Clapp (1949), Nyberg (1949), Reiter (1960), Riehl (1962), Danielsen (1968) and others.

A subject which has received much observational and theoretical attention is that of the so-called sudden stratospheric warming (for reviews, see Murgatroyd (1969), McInturff (1978), Schoeberl (1978), and Holton (1980b)). It has long been known that the total ozone in high latitudes increases dramatically in association with these polar warmings (Dütsch, 1962; London, 1962). Although other interpretations are possible, it is probable that the polar ozone increases are associated with the dynamically induced poleward-downward trajectories producing the warming itself. This was shown observationally by Mahlman (1970a), interpreted theoretically by Matsuno and Nakamura (1979), and demonstrated numerically by Hsu (1980).

Another broad class of stratospheric measurements is that of vertical profiles of long-lived trace constituents such as H_2O (Brewer, 1949; Mastenbrook, 1968; Harries, 1976; Kley et al., 1979) and N_2O, CH_4, H_2, and the chlorofluoromethanes (Ehhalt, 1974; Schmeltekopf et al., 1975, 1977; Fabian et al., 1979; Goldan et al., 1980). These measurements provide a number of valuable clues into the nature of trace constituent transport. The dryness of the lower stratosphere offers indications as to the predominant (but not necessarily the only) temperatures of air entering from the troposphere. (For an alternative perspective, see Robinson (1980).) The long-lived gases provide information on the vertical transfer rates, especially in the middle and upper stratosphere. The above observations also show the types of low variability to be expected in the stratosphere for such long-lived constituents. Such observations thus offer a new set of valuable tests for quantitative simulation models (Levy, Mahlman, and Moxim, 1979).

In the future, new and improved trace constituent measurements will help answer a number of remaining questions about stratospheric transport. In addition, we may expect such measurements to offer new challenges which are unanticipated at present.

3. A CONCEPTUAL VIEW OF STRATOSPHERIC TRANSPORT

In this section, we outline some of the various considerations required to construct a usable theoretical framework for understanding stratospheric transport. As will become obvious, these various theoretical arguments have yet to be merged into a single coherent structure. Nevertheless, considerable progress has been made in the past decade or so.

Perhaps the simplest conceptual framework from which to view stratospheric transport is the so-called isentropic (potential temperature) coordinate approach. In this approach, it is first noted that air parcels cannot cross the quasi-horizontal isentropic surfaces of the stratosphere unless non-adiabatic processes are occurring. Of course adiabatic events can cause the isentropic surfaces themselves to move; indeed during sudden stratospheric warmings, this is an obvious effect. However, in the time averaged sense, the potential temperature surfaces in the stratosphere remain in essentially the same locations from year to year. This means that, for longer term systematic vertical transports, non-adiabatic processes are essential.

These considerations make it immediately obvious that the stratosphere is a region in which the vertical transport of tracers back to the troposphere must be rather inefficient. First, the potential temperatures of the stratosphere are very high, in most cases much larger than have ever been observed as sensible temperatures in the lower troposphere. Thus, stratospheric air must experience considerable cooling before it enters the troposphere. Second, the higher the static stability of the atmosphere, the smaller is the allowed vertical displacement of an air parcel relative to an isentropic surface for a given diabatic heating rate. This again points out the relatively large resistance to vertical transport in the stratosphere.

Another conceptual advantage of this approach is that some processes which appear most complex from traditional viewpoints become simpler in the isentropic view. A good example of this is the large cancellation between mean cell and eddy contributions to changes in the zonal mean mixing ratio. As pointed out in Section I, this cancellation phenomenon is produced by the effect of the waves themselves inducing an indirect Eulerian mean meridional circulation under "non-acceleration" conditions. Note that in isentropic coordinates, this type of cancellation effect would disappear under such conditions.[1] Thus, for many purposes the isentropic coordinate perspective should be simpler.

To increase our insights into the various mechanisms leading to irreversible transport, let us examine what qualitative understanding might be gained from some simple hypothetical situations. First, consider an almost frictionless stratosphere (say molecular diffusion only), which receives no upward propagating tropospheric disturbances and is subject to no zonally asymmetric instabilities. In the absence of seasonal or diurnal radiation cycles, such a system would be extremely close to radiative equilibrium. Accompanying this would be an intense zonal wind of sufficient magnitude to balance the strong

---

[1]Mahlman, J. D., and D. G. Andrews, private communication.

meridional pressure gradients necessarily present in such an atmosphere. The only transport of conservative tracer introduced into such a system would be in the zonal direction. An air parcel injected at a particular place would essentially remain "frozen" at its original latitude, height and potential temperature. The next obvious modification of this simple imaginary situation is to allow the solar heating to undergo its diurnal and annual cycles. This, of course, allows some instantaneous <u>net</u> <u>heating</u> and thus movement of air parcels relative to isentropic surfaces. However, even though this time-dependent heating would lead to transient motions in response to the resultant unbalanced pressure forces, probably relatively small <u>net</u> meridional or vertical displacements would occur over periods long relative to the radiative forcing interval.

Now consider what could happen if a significant amount of mechanical damping or friction is added to this hypothetical system. Initially, the frictional damping produces a reduction in the zonal wind, thus leading to an imbalance between the zonal wind and the meridional pressure gradient. This leads to a poleward acceleration of the meridional wind component. The resultant tendency to accumulate mass in higher latitudes leads to sinking there. This sinking, through adiabatic compression, acts to warm the polar region above its radiative equilibrium temperatures.

Once such a system finally comes into equilibrium, it is again characterized by a nearly geostrophically balanced zonal wind. However, this hypothetical atmosphere now contains a thermally direct Hadley-type meridional circulation. The poleward flow in the upper branch of the Hadley circulation acts to restore the westerlies against frictional degradation and maintain geostrophic balance. In addition, the polar sinking and equatorial rising motion act to keep the polar and equatorial temperatures above and below their respective local radiative equilibrium values. This means that the balancing radiative cooling and heating leads to downward and upward flux of air parcels through isentropic surfaces at higher and lower latitudes, respectively. Thus, systematic meridional and vertical air motions are a fundamental property of this simple system.

The next levels of complexity are far more difficult to deal with and understand in terms of their implications for stratospheric transport. Suppose we now allow various motion systems produced in the troposphere to propagate into and influence the state of the above hypothetical stratosphere. Assume a steady disturbance is "switched on" at some instant. Even if the disturbance flow is adiabatic, the stratospheric isentropic surfaces themselves will move toward their new steady-state positions compatible with the steady state forcing.[2] After this initial response, the "non-acceleration" - "non-transport"

---

[2]We must be a bit cautious here, as Holton and Mass (1976) have shown that steady forcing of sufficiently high amplitude can excite non-steady stratospheric responses.

theorems tell us that little further transport would occur, as long as the crucial assumptions are nearly valid for the basic state, as well as the disturbance.

In a superficial sense, this appears to represent the quasi-stationary state of the early winter northern hemisphere stratosphere. However, a closer look shows that this apparently equilibrated "almost non-transport" situation can lead to significant real transport in at least three ways. First, the stratospheric basic state considered here already contains departures from "non-transport" conditions because of the impact of the assumed mechanical friction. Second, the adiabatically forced displacement of the isentropic surfaces leads to actual temperature changes when viewed in physical space. This, in turn, excites increasing radiative damping, and hence more parcel motion across isentropic surfaces. Third, if such a stationary distur-bance is induced during, say, mid-winter, the eventual onset of sum-mertime radiative conditions can lead to a collapse of the effective forcing. At that time, the associated readjustment of the flow can lead to large irreversible parcel displacements (Mahlman and Moxim, 1978).

Realistic situations become even more complex in that the forcing is actually time dependent over a range of time scales. Also, distur-bances coming from the troposphere involve a spectrum of motions from smaller scale gravity waves and clear air turbulence all the way to the scale of the earth itself. The interplay of these complex motions can lead to even more radiative damping. In addition, these motion scales interact in such manner as to increase substantially the proba-bility of irreversible or "turbulent" mixing on various scales (see also Dunkerton, 1980). The effect of this process is evident in various numerical experiments (e.g., Mahlman, 1975; Hsu, 1980). These arguments imply that it is the motions themselves which determine the degree and character of dissipation, both radiative and mechanical. Once such mechanical dissipation is induced, however, increased tem-peratures at the higher latitudes are implied. This is because the increased mechanical damping leads to additional poleward accelera-tions (at least in the Lagrangian sense), and increased adiabatic heating. Thus, zonal-mean diabatic heating and cooling can be increased in response to non-zonal processes.

The above arguments emphasize the importance of various processes leading to excitation of diabatic motions in the stratosphere. Ulti-mately, however, horizontal gradients in diabatic heating will lead to horizontal gradients for any trace constituent on a given isentropic surface. This implies that essentially adiabatic processes might, under some circumstances, lead to irreversible transports on the isentropic surfaces themselves. In fact, we can argue that the equil-ibrium meridional slopes of tracers in the stratosphere result from the competition between the above two processes. If mixing along the isentropic surfaces were the only process acting, the equilibrium tracer surfaces would exhibit the same meridional slope as the isentropic surfaces. On the other hand, if the only process acting were the meridional gradient of diabatic heating, the equilibrium tracer surfaces would tend to adopt an extremely steep meridional slope, eventually becoming vertical or even inverted and convoluted.

The fact that the equilibrium tracer surfaces show a structure between two extremes strongly suggests that a balance between these two rather different processes is indeed effected in the stratosphere.

The above arguments have the advantage of simplifying the interpretation of many stratospheric transport phenomena. The price paid for this simplification can be rather high, however. Most of the various theoretical developments mentioned in Section I, as well as virtually all of the numerical models, do not view transport from such a perspective. The barrier created is more than one of investigator preference. In such more Lagrangian oriented perspectives (see also McIntyre, 1980), significant analysis difficulties can arise when realistic problems are attacked. Such considerations suggest that more difficulty will be encountered when these problems are specifically addressed than might be superficially inferred from the above arguments.

4. AVAILABLE METHODOLOGIES

In view of the fundamental difficulties in making progress in understanding transport, we review here the various approaches employed. In each case it will become clear that any approach contains notable assets and significant liabilities.

The most obvious and direct method is "measurement of trace constituents." Such activities can tell us a great deal about the amount and distribution of trace constituents. However, interpretation of such measurements requires a background theoretical/conceptual framework against which to evaluate the data. This framework must contain some information on the probable ultimate sources and sinks, intermediate transformations, and transport characteristics leading to trace constituent redistribution. This suggests that if meteorological information can accompany the trace constituent measurements, more useful information can be extracted. The most important link in such an analysis is an identification of a match or mismatch of the data with the theoretical/ conceptual model chosen for comparison. As evident in Sections I and II, there have been a number of instances in which trace constituent behavior forced a re-evaluation of our conceptual understandings of transport and chemistry.

A significant portion of our knowledge of stratospheric dynamics and transport arises from extremely simplified linear or quasi-linear calculations of wave behavior. In one sense it is surprising just how much we have learned from such studies. For example, the "non-acceleration"/"non-transport" theorems discussed in Section 1 have arisen out of quasi-linear wave theory. These simple models often capture the essence of the actual processes occurring even though they normally fail to yield a full quantitative understanding. Interestingly, though, they at times point out the need for non-linear processes and dissipative effects not readily included in such frameworks. For example "non-transport" by linear, dissipation free waves points toward the type of processes required for "real" transport to occur. Such statements are not assertions that no transport can occur.

Approaches using the so-called "mechanistic models" are most valuable in helping to close an often serious gap between the linear models and realistic dynamical/transport behavior. These models are normally designed to include transient and/or non-linear behavior in such a way as to keep other complicating factors to a minimum (e.g., see Matsuno, 1971; Holton and Lindzen, 1972; Holton, 1976; Hartmann and Garcia, 1979). As examples, the radiation and chemistry are usually highly idealized and the effects of the troposphere and the "basic state" are normally prescribed rather than computed self-consistently. Such models often yield extremely important insights, a number of which can be quantitatively compared with the actual atmosphere. Thus, at this level of complexity, comparisons with the actual atmosphere can be quite meaningful. The price paid for this advance is that simple diagnostic understanding of the model results often becomes significantly more difficult than in the linear models.

A logical extension of the mechanistic model is the "comprehensive" model or the general circulation model (GCM). In this type of approach, an attempt is made to provide a self-consistent calculation of the troposphere-stratosphere (and sometimes more) system. With such a model direct comparisons with the three-dimensional meteorological and trace constituent structure become possible. Because these models attempt to be nearly quantitative, more direct insights into deficiencies in our understanding can be achieved. In addition, results from such models can be used to develop coherent strategies for guiding atmospheric measurement programs (Mahlman, 1980). In turn, this provides a very important input for improving model capability.

Because the GCM's can generate fully self-consistent (but not necessarily correct) data sets, one can perform a far more accurate analysis of model processes than is possible for the real atmosphere using direct measurements. In the model "atmosphere" the governing equations are known (by definition) and, in principle, there are not measurement and sampling errors. However, even with these advantages, full understanding of GCM results can be elusive. This is because the

inclusion of self-consistency permits more and more interactive phenomena to exist within the model framework. Thus, in many cases, highly
imaginative analysis techniques are required before meaningful
insights can be obtained. In addition, the significant computational
and data handling requirements make this approach cumbersome and time
consuming. Because of this, use of the simpler linear and/or mechanistic models in conjunction with GCM's can be quite valuable. Nevertheless, the GCM approach is very powerful and will see increasing
application to stratospheric transport/chemistry problems in the years
ahead. In particular the mutual dependence between these models and
various observational programs will become increasingly evident as our
quantitative capabilities continue to improve.

5. PARAMETERIZED TRANSPORT MODELS

In the previous section we have argued that the most physically
correct transport model is necessarily three-dimensional (3-D) in
structure. However, in a great many problems there is considerable
advantage to be gained by employing transport models which are far
simpler than the comprehensive 3-D models.

In particular, simpler transport models tend to be much more economical computationally, often by orders of magnitude. Also, such
simplified transport models can become simpler to interpret in the
context of the problems addressed. This is particularly so when the
trace constituent source and sink chemistry is complex. In addition,
if there exist large uncertainties in the source and sink chemistry
for a given problem, often little can be gained by employing a more
realistic transport model. In fact, simple transport models often
provide a useful means for evaluating algorithms and chemical uncertainties prior to inclusion of chemistry in complex models.

Because of these considerations, we expect that simplified transport models will remain useful as long as we can forsee. Accordingly,
we review here the types of approaches being used and considered for
such models.

Perhaps the simplest approach which retains some realism is the
"2-box" model. Here the stratosphere is regarded as one well mixed
box and the troposphere another. The exchange between them is assumed
equal to the difference in mixing ratios between the two "boxes"
divided by the characteristic interbox exchange time (or residence
time). For certain trace constituent problems, this is a useful
idealization. However, the arguments in Section 3 demonstrate that
the residence time must clearly depend at least upon the altitude (or
potential temperature) of the center of mass of the tracer (see also
Hunten, 1975a).

A way to rectify the most obvious deficiency in the "box" model is to increase the number of boxes in the vertical. This is accomplished through use of a 1-D model. For such a model, the horizontal transport effects are either removed by systematically averaging in the horizontal or they are ignored. In this framework, the vertical flux of trace constituent is assumed to occur by a macro-scale analog to molecular diffusion, the "Prandtl mixing-length" hypothesis. Thus, normally, the global average vertical flux is hypothesized to be equal to the global average vertical gradient times an appropriate "eddy diffusion" coefficient (K_z) (e.g., see Hunten, 1975b). Although this approximation has been demonstrated to be far from exact (Mahlman, 1975; Tuck, 1979), it nevertheless has been found to capture the essence of the behavior of "well distributed" stratospheric trace constituents. This appears to be so, even though the above authors demonstrate that the effective K_z for a given trace constituent must depend upon its source-sink "chemistry" and its distribution in the horizontal.

In spite of these problems, the 1-D transport model will continue to see important use in the future. Thus, further attempts to refine such models appear justified, at least in the near term.

The chief disadvantage of the 1-D model representation is that effects of horizontal inhomogeneities in constituent distributions and/or chemistry are very difficult to incorporate in a reasonable manner. In principle, many of these difficulties can be avoided by averaging systematically only in the east-west direction. This is because a large fraction of the inhomogeneities of concern occur in the meridional-vertical plane. Thus, a 2-D transport model offers promise as an important intermediate step in transport modeling.

By far the most frequently employed 2-D approach is that proposed by Reed and German (1965). In this type of model the zonally asymmetric (or eddy) tracer flux is normally parameterized by a "sloping axis" eddy diffusion assumption. The parameterized eddies thus always transport tracer down their net mean gradient, but countergradient meridional fluxes are allowed because of the sloping axis.

Later it was pointed out that such an application of the Reed and German formulation may be inconsistent because it doesn't allow for the eddy flux to be self consistent with the advection due to the (prescribed) concomitant meridional circulation (Mahlman, 1975). Other investigators, however, have pointed out that the Reed and German formalism might be made more physically consistent if the eddy diffusion tensor possesses a significant antisymmetric part (Plumb, 1979; Matsuno, 1980; Pyle and Rogers, 1980a; Danielsen, 1981). In this approach the induced meridional circulation itself contributes to the antisymmetric part of the diffusion tensor. This antisymmetric part becomes advective rather than diffusive in character. A first attempt invoking such insights has been made recently by Pyle and Rogers (1980b).

At the present time, such an "improved" 2-D transport model does not yet exist, and it is not clear just what approach may eventually prove to be both mechanistically correct and empirically useful. In view of the developmental nature of this problem, we discuss other possible approaches which may lead to progress.

One scheme which already has achieved some success is that developed by Vupputuri (1973) and later by Harwood and Pyle (1977). In this method eddy heat and momentum fluxes, as well as the eddy tracer fluxes are parameterized. However, the zonal-mean heat, momentum, and mass continuity equations themselves are solved as well, so that a meridional circulation consistent with the chosen eddy flux parameterization is generated. This method thus might avoid the inconsistency arising in the essentially symmetric diffusion tensor models.

Another framework which may prove useful is an "isentropic coordinate" approach. Here, essentially the insights into transport offered in Section 3 are more or less applied directly in a zonally averaged model using isentropic coordinates (or a regular coordinate with appropriate transformations). The zonal-mean diabatic circulation becomes the only mean meridional circulation (e.g., see Dunkerton, 1978). The only other fluxes through the isentropic surface would result from eddy diabatic and mechanical mixing effects. In addition, the dispersion and irreversible mixing along zonal-mean isentropic surfaces must be parameterized. In all cases, the parameterizations must depend upon the manner in which radiative and mechanical forcing influences the stratosphere and its induced transports. Thus, as in the above approaches, further theoretical progress is required to guide these parameterizations.

A suggestion that could prove successful is to employ a second-order closure approach.[3] Here, the trace constituent and mass continuity equations, as well as the thermodynamic and momentum equations are used to develop explicit additional equations for the trace constituent eddy fluxes, rather than invoking a flux-gradient hypothesis as used in the above techniques. This approach needs input zonal mean statistics of the meteorological structure (heat and momentum fluxes, variances, etc.) in addition to requiring the meridional circulation. In this formulation, closure hypotheses must be made at the triple product level to allow solution of the time dependent equations for the tracer fluxes. Again the viability of such a system remains to be demonstrated.

Finally, it may prove more reasonable to abandon the 2-D approach altogether. A relatively economical way to achieve this appears to be

---

[3]Mahlman, J. D. and Y.-H. Lee, 1978: unpublished manuscript.

provided by severely truncated 3-D models (e.g., Clark, 1970; Cunnold, et al., 1975). Here, the nonzonal effects or "eddies" are calculated explicitly, but with a minimum of resolution or degrees of freedom. Although significant processes may be absent because of the coarse resolution (Mahlman, 1975), at least the 3-D structure is included self consistently. Another advantage of such models is that they can be compared more readily to the comprehensive 3-D models or even higher resolution versions of the same model.

It should be clear from this discussion that much additional theoretical and developmental work needs to be done to construct more physically based parameterized transport models. We believe that such parameterized models will remain useful for a number of years to come.

6. UNANSWERED QUESTIONS AND FUTURE REQUIREMENTS

The aspects of stratospheric transport reviewed here show that significant progress has been achieved, particularly in the last decade or so. For a more complete understanding, however, much remains to be done. It now seems clear that optimum progress will be achieved through careful combination of the various observational, theoretical, and modeling tools available to us. In addition, more imaginative analysis of observational and numerical trace constituent information will continue to be required. This becomes obvious by examining the selection of questions needing further clarification listed below (Note that items a - f involve processes by which the "non-transport" constraints are broken in the actual atmosphere.):

a. The role of gravity waves and small-scale turbulence in irreversible tracer mixing.

b. Actual zonal-mean and eddy net diabatic heating rates in the middle atmosphere and how motions produce them.

c. Role of chemical sources and sinks in producing net transport of trace constituents. Improved understanding of the chemistry itself.

d. Means by which standing and transient disturbances excite mechanical dissipation.

e. Role of non-linearity for increasing the probability of irreversible mixing.

f. Influence of seasonal radiative forcing on stratospheric transport.

g. Influence of seasonal and transient variations in propagation of tropospheric disturbances into the stratosphere.

h. Nature of the Lagrangian-mean circulation and how it is forced. Its relevance in increased diagnostic understanding of transport, as well as its possible use in parameterized transport models.

i. Nature of the dispersion about this Lagrangian-mean circulation. Improved theoretical understanding of the causes of this dispersion.

j. Mechanisms allowing interhemispheric exchange of air. Clarification of the possible "barrier" caused by the opposite signs of potential vorticity in each hemisphere.

k. Mechanisms by which tropospheric air is exchanged with the stratosphere in low latitudes (convection, forced equatorial waves, or Hadley circulation?).

l. Physical basis for simpler trace constituent transport models.

m. Role of motions, chemistry, and dissipation in determination of trace constituent variance budgets and spatial-temporal variations.

n. Effect of stratospheric climate changes on transport.

Before real progress can be made in understanding many of the problems raised in this list, as well as in the formulation of other relevant questions, it is necessary that significant improvements be made in the accuracy and amount of middle atmosphere trace constituent data. Currently, we know of no trace constituents whose climatology can be said to be well documented. Such information will prove to be necessary for addressing a number of well focussed questions about transport. In addition, we note that solutions to most of the problems outlined here depend upon accurate meteorological data in addition to the trace constituent data.

It must be pointed out, however, that such data will seldom answer directly the various questions posed here. Acceptable answers to these difficult questions will almost inevitably come from careful analysis of theoretical/numerical models of such processes. Without the data base, however, the credibility of the models used to address such questions will remain subject to doubt.

Finally, note that most of the questions listed here are closely related to the more general requirement that we gain an improved understanding of the joint radiative-photochemical-dynamical stratospheric system and its coupling to the troposphere and mesosphere. Thus, "transport" as a field of study must ultimately be viewed as a necessary and indeed crucial, component of the complete stratospheric physical system.

ACKNOWLEDGMENTS

The authors are grateful to our various institutions for their support of the resources required to prepare this review. We are indebted to Drs. M. E. McIntyre, C.-P. Hsu, I. M. Held, Y. Hayashi and S. B. Fels for stimulating discussions on aspects of this work. The material presented here is current only up to autumn 1981. Accordingly, some important recent contributions do not appear in this review.

An earlier version of this paper was published in "Handbook for MAP," Vol. 3, C. F. Sechrist, Jr. (ed), 14-43. In that form it appeared as the final report of MAP Study Group 2.

REFERENCES

Adler, R. F., 1975: Mean meridional circulation in the southern hemisphere based on satellite information. J. Atmos. Sci., 32, 893-898.

Andrews, D. G., and M. E. McIntyre, 1976: Planetary waves in horizontal and vertical shear: The generalized Eliassen-Palm relation and the mean zonal acceleration. J. Atmos. Sci., 33, 2031-2048.

_____ and _____, 1978a: Generalized Eliassen-Palm and Charney-Drazin theorems for waves in axisymmetric mean flows in compressible atmospheres. J. Atmos. Sci., 35, 175-185.

_____ and _____, 1978b: An exact theory of nonlinear waves on a Lagrangian mean flow. J. Fluid Mech., 89, 609-646.

Attmannspacher, W., and H.U. Dütsch, 1970: International ozonesonde intercomparison at the Observatory Hohenpeissenberg. Berichte des Deutschen Wetterdienstes, 16, No. 120.

Berggren, R., and K. Labitzke, 1966: A detailed study of the horizontal and vertical distribution of ozone. Tellus, 18, 761-772.

_____ and _____, 1968: The distribution of ozone on pressure surfaces. Tellus, 20, 88-97.

Bleichrodt, J. J., and E. R. Van Abkoude, 1963: On the deposition of cosmic ray produced beryllium-7. J. Geophys. Res., 68, 5283-5288.

Boyd, J. P., 1976: The noninteraction of waves with the zonally averaged flow on a spherical earth and the interrelationships of eddy fluxes of energy, heat and momentum. J. Atmos. Sci., 33, 2285-2291.

Breiland, J. G., 1967: Comparison of the vertical distribution of thermal stability in the lower stratosphere with the vertical distribution of atmospheric ozone. J. Atmos. Sci., 24, 569-576.

_____, 1968: Some large-scale features of the vertical distribution of atmospheric ozone associated with the thermal structure of the atmosphere. J. Geophys. Res., 73, 5021-5028.

Brewer, A. W., 1949: Evidence for a world circulation provided by the measurements of helium and water vapor distribution in the stratosphere. Quart. J. Roy. Meteor. Soc., 75, 351-363.

Briggs, J. and W. T. Roach, 1963: Aircraft observations near jet streams. Quart. J. Roy. Meteor., Soc., 89, 225-247.

Burton, W. M., and N. G. Stewart, 1960: Use of long-lived natural radio-activity as an atmospheric tracer. Nature, 186, 584-589.

Chang, J., 1976: Uncertainties in the validation of parameterized transport in 1-D models of the stratosphere. Proc. Fourth Conf. Climatic Impact Assessment Program, T. M. Hard and A. J. Broderick, Eds., DOT-TSC-75-38, 175-182 [NTIS AD-A068982].

Chapman, S., 1930: A theory of upper atmospheric ozone. Mem. Roy. Meteor. Soc., 3, 103-125.

Charney, J. G. and P. G. Drazin, 1961: Propagation of planetary-scale disturbances from the lower into the upper atmosphere. J. Geophys. Res., 66, 83-109.

Clark, J. H. E., 1970: A quasi-geostrophic model of the winter stratospheric circulation. Mon. Wea. Rev., 98, 443-461.

_____and T. G. Rogers, 1978: The transport of conservative trace gases by planetary waves. J. Atmos. Sci., 35, 2232-2235.

Craig, R. A., 1950: The observations and photochemistry of atmospheric ozone and their meteorological significance. Meteorological Monographs, 1(2), 50 pp.

Cunnold, D., F. Alyea, N. Phillips, and R. Prinn, 1975: A three-dimensional dynamical-chemical model of atmospheric ozone. J. Atmos. Sci., 32, 170-194.

_____, _____, and R. G. Prinn, 1980: Preliminary calculations concerning the maintenance of the zonal mean ozone distribution in the Northern Hemisphere. Pure. Appl. Geophys., 118, 329-354.

Danielsen, E. F., 1968: Stratospheric-tropospheric exchange based on radioactivity, ozone, and potential vorticity. J. Atmos. Sci., 25, 502-518.

_____, 1981: An objective method for determining the generalized-transport tensor for two dimensional Eulerian models. J. Atmos. Sci., 38, 1319-1339.

_____, K. H. Bergman, and C. A. Paulsen, 1962: Radioisotopes, potential temperature, and potential vorticity. A study of stratospheric-tropospheric exchange processes. Department of Meteorology and Climatology, University of Washington, 54 pp.

_____, R. Bleck, J. Shedlovsky, A. Wartburg, P. Haagenson, and W. Pollock, 1970: Observed distribution of radioactivity, ozone, and potential vorticity associated with tropopause folding. J. Geophys. Res., 75, 2353-2361.

Dickinson, R. E., 1969: Theory of planetary wave-zonal flow interaction. J. Atmos. Sci., 26, 73-81.

Dobson, G. M. B., 1930: Observations of the amount of ozone in the earth's atmosphere and its relation to other geophysical conditions - Part IV. Proc. Roy. Soc. London, A129, 411-433.

_____, D. N. Harrison, and J. Lawrence, 1929: Measurements of the amount of ozone in the earth's atmosphere and its relation to other geophysical conditions - Part III. Proc. Roy. Soc. London, A122, 456-486.

_____, 1956: Origin and distribution of polyatomic molecules in the atmosphere. Proc. Roy. Soc. London, A236, 187-193.

_____, 1973: The laminated structure of the ozone in the atmosphere. Quart. J. Roy. Meteor. Soc., 99, 599-607.

Dunkerton, T., 1978: On the mean meridional mass motions of the stratosphere and mesosphere. J. Atmos. Sci., 35, 2325-2333.

_____, 1980: A Lagrangian mean theory of wave, mean-flow interaction with applications to nonacceleration and its breakdown. Rev. Geophys. Space Phys., 18, 387-400.

Dütsch, H. U., 1946: Photochemische theorie des atmosphärischen ozons unter Berucksichtigung von Nichtgleichgewichtszustanden und Luftbewegungen. Zurich, Doctoral Dissertation.

_____, 1962: Ozone distribution and stratospheric temperature field over Europe during the sudden stratospheric warming in January/February 1958. Beitr. Physik Atmosphäre, 35, 87-107.

_____, 1969: Atmospheric ozone and ultraviolet radiation. World Survey of Climatology, Vol. 4, Climate of the free Atmosphere, D. F. Rex, Ed., Elsevier, 383-432.

408

_____, 1974: Regular ozone soundings at the aerological station of the Swiss Meteorological Office at Payerne, Switzerland, 1968-1972. Lapeth-10, Laboratorium für Atmosphärenphysik ETH, Zürich, 337 pp. [NTIS N75-2185415GA].

_____, 1978: Vertical ozone distribution on a global scale. Pure. Appl. Geophys., 16, 511-529.

Dyer, A. J., and S.-A. Yeo, 1960: A radioactive fallout study at Melbourne, Australia. Tellus, 12, 195-199.

Ehhalt, D. H., 1974: Sampling of stratospheric trace constituents. Can. J. Chem., 52, 1510-1518.

Eliassen, A., and E. Palm, 1961: On the transfer of energy in stationary mountain waves. Geofys. Publ., 22, 1-23.

Fabian, P., R. Borchers, K. H. Weiler, U. Schmidt, A. Volz, D. H. Ehhalt, W. Seiler, and F. Muller, 1979: Simultaneously measured vertical profiles of H_2, CH_4, CO, N_2O, $CFCl_3$, and CF_2Cl_2 in the mid-latitude stratosphere and troposphere. J. Geophys. Res., 84, 3149-3154

Feely, H. W., and J. Spar, 1960: Tungsten-185 from nuclear bomb tests as a tracer for stratospheric meteorology. Nature, 188, 1062-1064.

_____, H. Seitz, R. J. Lagomarsino, and P. E. Biscayne, 1966: Transport and fallout of stratospheric radioactive debris. Tellus, 18, 316-328.

Garcia, R. R., and D. L. Hartmann, 1980: The role of planetary waves in the maintenance of the zonally averaged ozone distribution of the stratosphere. J. Atmos. Sci., 37, 2248-2264.

Godson, W. L., 1960: Total ozone and the middle stratosphere over arctic and sub-arctic areas in winter and spring. Quart. J. Roy. Meteor. Soc., 86, 301-317.

Goldan, P. D., W. C. Kuster, D. L. Albritton, and A. L. Schmeltekopf, 1980: Stratospheric $CFCl_3$, CF_2Cl_2, and N_2O height profile measurements at several latitudes. J. Geophys. Res., 85(C1), 413-423.

Goldie, A. H. R., 1950: The average planetary circulation in vertical meridian planes. Cent. Proc. Roy. Meteor. Soc., 175-180.

Harries, J. E., 1976: The distribution of water vapor in the stratosphere. Rev. Geophys. Space. Phys., 14, 565-575.

Hartmann, D. L., 1976: The dynamical climatology of the stratosphere in the southern hemisphere during late winter 1972. J. Atmos. Sci., 33, 1789-1802.

_____, and R. R. Garcia, 1979: A mechanistic model of ozone transport by planetary waves in the stratosphere. J. Atmos. Sci., 36, 350-364.

Harwood, R. S., and J. A. Pyle, 1977: Studies of the ozone budget using a zonal mean circulation model and linearized photochemistry. Quart. J. Roy. Met Soc., 103, 319-343.

Haurwitz, B., 1938: Atmospheric ozone as a constituent of the atmosphere. Bull. Am. Meteor. Soc., 19, 417-424.

Heath, D. F., C. L. Mateer, and A. J. Krueger, 1973: The Nimbus-4 backscatter ultraviolet (BUV) atmospheric ozone experiment - two years operation. Goddard Space Flight Center, X-651-73-64, Greenbelt, Maryland, 24 pp.

Hering, W. S., 1966: Ozone and atmospheric transport processes. Tellus, 18, 329-336.

Hilsenrath, E., D. F. Heath, and B. M. Schlesinger, 1979: Seasonal and interannual variations in total ozone revealed by the Nimbus 4 backscattered ultraviolet experiment. J. Geophys. Res., 84, 6969-6979.

Holton, J. R., 1974: Forcing of mean flows by stationary waves. J. Atmos. Sci., 31, 942-945.

_____, 1976: A semi-spectral model for wave-mean flow interactions in the stratosphere: Application to sudden stratospheric warmings. J. Atmos. Sci., 33, 1639-1649.

_____, 1980a: Wave propagation and transport in the middle atmosphere. Phil. Tran. Roy. Soc. London, A296, 73-85.

_____, 1980b: The dynamics of sudden stratospheric warmings. Ann. Rev. Earth Planet Sci., 8, 169-190.

_____, and R. S. Lindzen, 1972: An updated theory for the quasi-biennial cycle of the equatorial stratosphere. J. Atmos. Sci., 29, 1076-1080.

_____, and C. Mass, 1976: Stratospheric vacillation cycles. J. Atmos. Sci., 33, 2218-2225.

Hsu, C.-P., 1980: Air parcel motions during a numerically simulated sudden stratospheric warming. J. Atmos. Sci., 37, 2768-2792.

Hunt, B. G., 1969: Experiments with a stratospheric general circulation model, III. Large-scale diffusion of ozone including photochemistry. Mon. Wea. Rev., 97, 287-306.

_____, and S. Manabe, 1968: Experiments with a stratospheric general circulation model, II. Large-scale diffusion of tracers in the stratosphere. Mon. Wea. Rev., 96, 503-539.

Hunten, D. M., 1975a: Residence times of aerosols and gases in the stratosphere. Geophys. Res. Lett., 2, 26-28.

_____, 1975b: The philosophy of one-dimensional modeling. Proc. Fourth Conf. Climatic Impact Assessment Program, T. M. Hard and A. J. Broderick, Eds., DOT-TSC-OST-75-38, 147-155. [NTIS AD-A068982].

Julian, P.R., and K. B. Labitzke, 1965: A study of atmospheric energetics during the January-February 1963 stratospheric warming. J. Atmos. Sci., 22, 597-610.

Kida, H., 1977: A numerical investigation of the atmospheric general circulation and stratospheric-tropospheric mass exchange: II. Lagrangian motion of the atmosphere. J. Meteor. Soc. Japan, 55, 71-88.

Kley, D., E. J. Stone, W. R. Henderson, J. W. Drummond, W. J. Harrop, A. L. Schmeltekopf, T. L. Thompson, and R. H. Winkler, 1979: In-situ measurements of the mixing ratio of water vapor in the stratosphere. J. Atmos. Sci., 36, 2513-2524.

Levy, H. II, J. D. Mahlman, and W. J. Moxim, 1979: A preliminary report on the numerical simulation of the three-dimensional structure and variability of atmospheric N_2O. Geophys. Res. Lett.,6, 155-158.

Libby, W., 1956: Radioactive strontium fallout. Proc. Natl. Acad. Sci., 42, 365-390.

London, J., 1962: Ozone variations and their relation to stratospheric warmings. Proc. Intl. Symp. Stratospheric and Mesospheric Circulation, Berlin, 299-310.

_____, J. F. Frederick, and G. P. Anderson, 1977: Satellite observations of the global distribution of stratospheric ozone. J. Geophys. Res., 82, 2543-2556.

Lovill, J. E., 1972: Characteristics of the general circulation of the atmosphere and the global distribution of total ozone as determined by the Nimbus III satellite infrared interferometer spectrometer. Atmospheric Science Paper No. 180, Colorado State University, Fort Collins, Colo., 72 pp.

Machta, L., 1957: Discussion of meteorological factors and fallout distribution. U.S. Dept. Commerce, Weather Bureau, 11 pp.

_____, and R. J. List, 1959: Analysis of stratospheric Sr-90 measurements. J. Geophys. Res., 64, 1267-1276.

_____, K. Telegadas, and R. J. List, 1970: The slopes of surfaces of maximum tracer concentration in the lower stratosphere. J. Geophys. Res., 75, 2279-2288.

_____ and _____, 1973: Examples of stratospheric transport. Proc. Second Conf. Climatic Impact Assessment Program, A. J. Broderick, Ed., DOT-TSC-OST-73-4, 47-56 [NTIS PB-221 16612].

Mahlman, J. D., 1966: Atmospheric general circulation and transport of radioactive debris. Atmospheric Science Paper No. 103, Department of Atmospheric Science, Colorado State University, 184 pp.

_____, 1969: Heat balance and mean meridional circulations during the sudden warming of January 1958. Mon. Wea. Rev., 97, 534-540.

_____, 1970a: Eddy transfer processes in the stratosphere during major and "minor" breakdowns of the polar night vortex. J. Geophys. Res., 75, 1701-1705.

_____, 1970b: Dynamical mechanisms producing large-scale transport of trace substances. U.S. Naval Postgraduate School, NPS-51MZ 70101A, 184 pp.

_____, 1973a: On the maintenance of the polar front jet stream. J. Atmos. Sci., 30, 544-557.

_____, 1973b: Preliminary results from a three-dimensional general circulation/tracer model. Proc. Second Conf. Climatic Impact Assessment Program, A. J. Broderick, Ed., DOT-TSC-OST-73-4, 321-337 [NTIS PB-221 16612].

_____, 1975: Some fundamental limitations of simplified-transport models as implied by results from a three-dimensional general-circulation/tracer model. Proc. Fourth Conf. Climatic Impact Assessment Program. T. M. Hard and A. J. Broderick, Eds., DOT-TSC-OST-75-38, 132-146 [NTIS AD-A068982].

_____, 1980: Coupling of atmospheric observations with comprehensive numerical models. Collection of extended abstracts presented at ICMUA Sessions and IUGG Symposium 18, IUGG XVII General Assembly, Canberra, Australia, December 1979, 253-259

_____, and W. J. Moxim, 1978: Tracer simulation using a global general circulation model: Results from a midlatitude instantaneous source experiment. J. Atmos. Sci., 35, 1340-1374.

_____, H. Levy II, and W. J. Moxim, 1980: Three-dimensional tracer structure and behavior as simulated in two ozone precursor experiments. J. Atmos. Sci., 37, 655-685.

Manabe, S., and J. D. Mahlman, 1976: Simulation of seasonal and interhemispheric variations in the stratospheric circulation. J. Atmos. Sci., 33, 2185-2217.

Mastenbrook, H. J., 1968: Water vapor distribution in the stratosphere and high troposphere. J. Atmos. Sci., 25, 299-311.

_____, 1974: Water vapor measurements in the lower stratosphere. Can. J. Chem., 52, 1527-1531.

Matsuno, T., 1971: A dynamical model of the stratospheric sudden warming. J. Atmos. Sci., 28, 1479-1494.

_____, 1972: Paper presented at CIAP Workshop on Computational Modeling of the Atmosphere. Pacific Grove, Calif.

_____, 1980: Lagrangian motion of air parcels in the stratosphere in the presence of planetary waves. Pure Appl. Geophys., 118, 189-216.

_____ and K. Nakamura, 1979: The Eulerian- and Lagrangian-mean meridional circulations in the stratosphere at the time of a sudden warming. J. Atmos. Sci., 36, 640-654.

McInturff, R. M. (editor), 1978: Stratospheric warmings: Synoptic, dynamic, and general-circulation aspects. NASA Ref. Publ. 1017, Science and Technical Information Office, 166 pp.

McIntyre, M. E., 1980: Towards a Lagrangian-mean description of stratospheric circulations and chemical transports. Phil. Trans. Roy. Soc. London, A296, 129-148.

Meetham, A. R., 1937: The correlation of the amount of ozone with other characteristics of the atmosphere. Quart. J. Roy. Meteor. Soc., 63, 289-307.

Miller, A. J., 1967: Note on vertical motion in the lower stratosphere. Beitr. Phys. Atmos., 40, 29-48.

Miyakoda, K., 1963: Some characteristic features of the winter circulation in the troposphere and lower stratosphere. Technical Report No. 14 to National Science Foundation (Grant NSF-GP-471), University of Chicago, 93 pp. [NTIS PB 174 308].

Molla, A. C., and C. J. Loisel, 1962: On the hemispheric correlations of vertical and meridional wind components. Geofisica Pura e Appl., 51, 166-170.

Muench, H. S., 1965: On the dynamics of the wintertime stratospheric circulation. J. Atmos. Sci., 22, 349-360.

Murakami, J., 1965: Energy cycle of the stratospheric warming in early 1958. J. Meteor. Soc. Japan, 43, 262-283.

Murgatroyd, R. J., 1969: The structure and dynamics of the stratosphere. The Global Circulation of the Atmosphere. G. A. Corby (Ed.), Royal Meteorological Society, 159-195.

_____, and J. Singleton, 1961: Possible meridional circulations in the stratosphere and mesosphere. Quart. J. Roy. Meteor. Soc., 87, 125-135.

Namias, J., and P. H. Clapp, 1949: Confluence theory of the high tropospheric jet stream. J. Meteor., 6, 330-336.

Newell, R. E., 1961: The transport of trace substances in the atmosphere and their implications for the general circulation of the stratosphere. Geofisica Pura e Appl., 49, 137-158.

_____, 1963a: The general circulation of the atmosphere and its effects on the movement of trace substances. J. Geophys. Res., 68, 3949-3962.

_____, 1963b: Transfer through the tropopause and within the stratosphere. Quart. J. Roy. Meteor. Soc., 89, 167-204.

_____, 1964: Further ozone transport calculations and the spring maximum in ozone amount. Pure Appl. Geophys., 59, 191-206.

_____, J. M. Wallace, and J. R. Mahoney, 1965: The general circulation of the atmosphere and its effects on the movement of trace substances. Part 2: Tellus, 18, 363-380.

_____ and A. J. Miller, 1965: Some aspects of the general circulation of the lower stratosphere. Radioactive Fallout from Nuclear Weapons Tests, Division of Technical Information, U.S. Atomic Energy Commission, Washington, D.C., 392-404.

Newson, R. L., 1974: An experiment with a tropospheric and stratospheric three-dimensional general circulation model. Proc. Third Conf. Climatic Impact Assessment Program, A. J. Broderick and T. M. Hard, Eds., DOT-TSC-OST-74-15, 461-473 [NTIS ADA 003 846].

Nicolet, M., 1945: L'ozone et ses relations avec la situation atmospherique. Inst. Roy. Meteor. de Belgique, Misc., Fasc. 19, 36 pp.

414

Normand, C., 1953: Atmospheric ozone and the upper air conditions. Quart. J. Roy Meteor. Soc., 79, 39-50.

Nyberg, A., 1949: An aerological study of large-scale atmospheric disturbances. Tellus, 1, 44-53.

Palmer, C. E., 1959: The stratospheric polar vortex in winter. J. Geophys. Res., 64, 749-764.

Perry, J. S., 1967: Long-wave energy processes in the 1963 sudden stratospheric warming. J. Atmos. Sci., 24, 539-550.

Pittock, A. B., 1977: Climatology of the vertical distribution of ozone over Aspendale (38°S 145°E). Quart. J. Roy. Meteor. Soc., 103, 575-584.

Plumb, R. A., 1979: Eddy fluxes of conserved quantities by small-amplitude waves. J. Atmos. Sci., 36, 1699-1704.

Prabhakara, C. E., E. B. Rodgers, B. J. Conrath, R. A. Hanel, and V. G. Kunde, 1976: The Nimbus 4 infrared spectroscopy experiment: 3. Observations of the lower stratosphere thermal structure and total ozone. J. Geophys. Res., 81, 6391-6399.

Pyle, J. A., and C. F. Rogers, 1980a: stratospheric transport by stationary planetary waves - the importance of chemical processes. Quart. J. Roy. Meteor. Soc., 106, 421-446.

_____ and _____, 1980b: A modified diabatic circulation model for stratospheric tracer transport. Nature, 287, 711-714.

Ramanathan, K. R., and R. N. Kulkarni, 1960: Mean meridional distributions of ozone in different seasons calculated from Umkehr observations and probable vertical transport mechanisms. Quart. J. Roy. Meteor. Soc., 86, 144-155.

Reed, R. J., 1950: Role of vertical motions in ozone-weather relationships. J. Meteor., 7, 263-267.

_____, 1953: Large-scale eddy flux as a mechanism for vertical transport of ozone. J. Meteor., 10, 296-297.

_____, and A. L. Julius, 1951: A quantitative analysis of two proposed mechanisms for vertical ozone transport in the lower stratosphere. J. Meteor., 8, 321-325.

_____, J. L. Wolfe, and H. Nishimoto, 1963: A spectral analysis of the stratospheric sudden warming of early 1957. J. Atmos. Sci., 20, 256-275.

_____, and K. E. German, 1965: A contribution to the problem of stratospheric diffusion by large-scale mixing. Mon. Wea. Rev., 93, 313-321.

Reiter, E. R., 1960: The detailed structure of the atmosphere near jet streams. Geofisica Pura e Appl., 46, 193-200.

_____, 1971: Atmospheric Transport Processes, Part 2: Chemical Tracers. A.E.C. Critical Review Series, U.S. Atomic Energy Commission, Division of Technical Information, 382 pp. [NTIS TID-25314].

_____, W. Carnuth, H.-J. Kantor, K. Potzl, R. Reiter, and R. Sladkovic, 1975: Measurements of stratospheric residence times. Arch. Met. Geoph. Biokl., 24, 41-51.

Riehl, H., 1962: Jet streams of the atmosphere. Technical Report No. 32, Department of Atmospheric Science, Colorado State University, Fort Collins, Colo., 117 pp.

Robinson, G. D., 1980: The transport of minor atmospheric constituents between troposphere and stratosphere. Quart. J. Roy. Meteor. Soc., 106, 227-253.

Schlesinger, M. E., and Y. Mintz, 1979: Numerical simulation of ozone production, transport, and distribution with a global atmospheric general circulation model. J. Atmos. Sci., 36, 1325-1361.

Schmeltekopf, A. L., P. D. Goldan, W. R. Henderson, W. J. Harrop, T. L. Thompson, F. C. Fehsenfeld, H. I. Schiff, P. J. Crutzen, I.S.A. Isaksen, and E. E. Fergeson, 1975: Measurements of stratospheric $CFCl_3$, CF_2Cl_2, and N_2O. Geophys. Res. Lett., 2, 393-396.

_____, D. L. Albritton, P. J. Crutzen, P. D. Goldan, W. J. Harrop, W. R. Henderson, J. R. McAfee, M. McFarland, H. I. Schiff, T. L. Thompson, D. J. Hofmann, and N. T. Kjome, 1977: Stratospheric nitrous oxide altitude profiles at various latitudes. J. Atmos. Sci., 34, 729-736.

Schoeberl, M. R., 1978: Stratospheric warmings: Observations and theory. Rev. Geophys. Space Phys., 16, 521-538.

Smagorinsky, J., S. Manabe, and J. L. Holloway, Jr., 1965: Numerical results from a nine-level general circulation model of the atmosphere. Mon. Wea. Rev., 93, 727-768.

Telegadas, K., and R. J. List, 1969: Are particulate radioactive tracers indicative of stratospheric motions? J. Geophys. Res., 74, 1339-1350.

Teweles, S., 1963: Spectral aspects of the stratospheric circulation during the IGY. Planetary circulations Project, Report No. 8, Massachusetts Institute of Technology, Cambridge, Mass., 191 pp.

Tuck, A. F., 1979: A comparison of one, two and three-dimensional model representations of stratospheric gases. Phil. Trans. Roy. Soc. London, A290, 477-498.

Uryu, M., 1974: Mean zonal flows induced by a vertically propagating Rossby wave packet. J. Meteor. Soc. Japan, 52, 481-490.

Vincent, D. G., 1968: Mean meridional circulations in the Northern Hemisphere lower stratosphere during 1964 and 1965. Quart. J. Roy. Meteor. Soc., 94, 333-349.

Vupputuri, R. K., 1973: Numerical experiments on the steady state meridional structure and ozone distribution in the stratosphere. Mon. Wea. Rev., 101, 510-527.

Wallace, J. M., 1978: Trajectory slopes, countergradient heat fluxes and mixing by lower stratospheric waves. J. Atmos. Sci., 35, 554-558.

Wulf, O. R., and L. S. Deming, 1936: The theoretical calculation of the distribution of photochemically-formed ozone in the atmosphere. Terr. Magn. Atmos. Elect., 41, 299-310.

Züllig, W., 1973: Relation between the intensity of the stratospheric circumpolar vortex and the accumulation of ozone in the winter hemisphere. Pure Appl. Geophys., 106-108, 1544-1552.

J. R. Holton and T. Matsuno, Dynamics of the Middle Atmosphere, 417-444.

MODELING OF TRACER TRANSPORT IN THE MIDDLE ATMOSPHERE

Ka Kit Tung

Department of Mathematics
Massachusetts Institute of Technology

ABSTRACT

The discussion is concerned with zonally averaged (or the so-
called 2-D) models of transport of trace gases in the middle atmosphere.
A brief review of processes affecting transport and an order of magnitude estimate of various eddy transport terms are given.

1. INTRODUCTION

For various reasons (mostly economic ones) it is sometimes desirable to calculate the zonally averaged distribution of various minor
constituents in the atmosphere directly from the zonally averaged
equations of transport. Since motion in our atmosphere is not zonally
symmetric, there arise in the averaged equations terms that represent
transports by nonsymmetric motion fields (the "eddies") which cannot
be determined consistently within the framework of the 2-D models.
These eddy transport terms have to be parameterized in terms of the
zonally averaged fields. The situation here is similar to the
closure problem in turbulence theory. It is by now known, however,
that the large-scale atmospheric waves in the middle atmosphere are
largely organized and their transports do not resemble those of
turbulence.

Given the fact that some form of parameterization of the eddy
transports is unavoidable in 2-D Eulerian models, it is desirable to
have a formulation in which the role played by the eddies can be made
as small as possible, so that the degree with which our model depends
on our ability to accurately parameterize the eddy transports can be
reduced. That this is at least conceptually feasible is demonstrated
by the generalized Lagrange mean formulation of Andrews and McIntyre
(1978). In this formulation, the transport of a tracer is accomplished
by advection of the zonally averaged flow only (with the average
taken with respect to the "displaced" position), and no eddy transport
terms appear explicitly. Practical problems encountered in the application of the theory of the Lagrangian mean to the tracer transport
problem (as pointed out by e.g. McIntyre (1980)) prompted the development of alternate Eulerian mean models which can retain some of the

417

positive attributes of the Lagrangian mean theory.

Two Eulerian formulations, the residual mean circulation
(Andrews and McIntyre, 1976; Boyd, 1976; Dunkerton, 1978; Holton,
1980; Matsuno, 1980; Holton, 1981) and the formulation in isentropic
coordinates (Mahlman et al, 1981; Tung, 1982) both have the property
that for steady adiabatic small amplitude waves, the mean circulation
reduces to the Lagrangian mean circulation. However, the manner in
which this is accomplished is very different. In the residual mean
formulation, the advective transport by the steady adiabatic waves
is subtracted from the mean velocities in the definition of a
residual circulation. In isentropic coordinates, the reduction comes
from a simplification of air trajectories as compared to those in
height or pressure coordinates.

In section 2, a brief review of the eddy transport processes is
given from the viewpoint of air trajectories. This is followed by a
parameterization of nonadiabatic and nonsteady waves in the stratosphere. This parameterization allows us to give an order of magnitude
estimate of various eddy transports terms. Results are summarized in
section 6.

2. AIR PARCEL TRAJECTORY AND EDDY TRANSPORT

The type of transport produced by the eddy field is intimately related to the nature of air displacement trajectory. If eddy displacements occur horizontally, only horizontal diffusion (or dispersion) of
tracers can result. In this case the eddy transport tensor, $\mathbb{K}$, can
have only one nonzero component — K_{yy}. If the eddy field involves
both vertical and horizontal air displacements in the meridional plane,
but the motion in the vertical direction is uncorrelated with that in
the horizontal direction, the tensor $\mathbb{K}$ then has two nonzero components — K_{yy} and K_{zz}, and no off-diagonal elements. This diagonal
transport tensor is relevant to random turbulence-like eddy
processes. The resulting transport is diffusive, and as such
the direction of transport is down-gradient, that is, from a
region of high zonal mean tracer concentration to a region of low mean
concentration.

A subtle generalization of the above-mentioned picture of Fickian
diffusion is the parameterization proposed by Reed and German (1965).
While still retaining the interpretation that the eddy processes in the
stratosphere is turbulence-like, Reed and German suggested that the
diagonal transport tensor $\mathbb{K}$ for Fickian diffusion can be generalized
to include the off-diagonal components, K_{yz} and K_{zy}, if a sloping
principal axis system for eddy displacements is adopted. The eddy
motion along one axis is still assumed (albeit implicitly in their
paper) to be uncorrelated with that in the other axis. However, when
projected into the regular y-z axes, there would now be some correlation
between the displacement in y-direction and that in the z-direction,
hence the presence of the off-diagonal components. Consistent with this

interpretation is the assumption that the tensor $\mathbb{K}$ be symmetric, i.e. $K_{yz} = K_{zy}$. (This guarantees that the tensor can be diagonalized by a real orthogonal transformation. In other words, it is assumed that there exists a (sloped) principal axis system in which the displacements along different directions are uncorrelated.) This generalization allows the possibility that diffusion along a coordinate axis (e.g. the y-axis) be countergradient. Countergradient northward eddy transports have been observed in the lower stratosphere during winter. This evidence, in fact, was the primary motivation for Reed and German to generalize the Fickian diffusion tensor.

There is no reason to believe that large-scale eddy displacements in the atmosphere should be uncorrelated in any two directions. There has been increasing evidence to suggest that large-scale atmospheric waves possess coherent structures and do not behave at all like turbulence. Matsuno (1980) showed that fluid particle trajectories induced by forced stationary planetary waves are elliptical when projected onto the meridional plane. The transport tensor $\mathbb{K}$ for such an eddy field cannot be symmetric, because the displacements that make up the elliptical trajectories are strongly correlated in any two (fixed) directions. For the case of adiabatic steady planetary waves studied by Clark and Rogers (1978), Plumb (1979) and Matsuno (1980), the transport tensor turns out to be antisymmetric (giving an advective transport), just the opposite from what one would expect for a diffusive process. Although the tensor becomes full when transient waves are taken into consideration, the above-mentioned results nevertheless serves to point out that a substantial component of the eddy field in the stratosphere produces transports that are mainly nondiffusive in nature.

There is an interesting modification to what is said in the preceding paragraph. We have noted that the eddy transport in the stratosphere is more complicated than what one would infer from a turbulent diffusion model. This is a result of the strong coherence in the horizontal and vertical displacements associated with the air trajectories, which are nonrectilinear. This fact cannot be altered by a simple coordinate transformation (and so $\mathbb{K}$ cannot be diagonalized in general). It will be a different story if dynamical coordinate transformations are allowed. In particular, if the isentropic coordinate system, with potential temperature (a dynamical variable) as the vertical coordinate, is adopted, then all adiabatic eddy displacements, including the elliptical trajectories deduced by Matsuno (1980), become rectilinear — in the horizontal direction (i.e. along isentropes) only. Of the four components in the eddy transport tensor $\mathbb{K}$, only one, K_{yy}, is nonzero. The resulting transport resembles the classical Fickian diffusion in one dimension. Air mixes along isentropic surfaces, serving to smooth out gradients of mean tracer concentration along the isentropes.

If one further assumes that the eddy field is steady, even this last component of the transport tensor vanishes. No eddy transport is accomplished by an adiabatic, steady eddy field. For this idealized case, tracers are transported solely by the advection of mean flow in isentropic coordinates (see Tung (1982)).

The eddy field in the atmosphere is in general nonadiabatic and often nonsteady. Such nonconservative processes are difficult to parameterize, but it appears valid in assuming that in the stratosphere (below the breaking height for gravity waves), a large component of the atmospheric eddy field contributing to tracer transport is not turbulence-like, but coherent waves forced in the lower atmosphere.

3. DEFINITION OF THE EDDY TRANSPORT TENSOR

Surprisingly, there does not seem to be an agreed upon definition for the eddy transport tensor $\mathbb{K}$. Consequently, there has been some confusion, in particular concerning its dependence on the chemical species being transported.

The original definition of Reed and German (1965), which is also adopted here[1], is that the four components of the $\mathbb{K}$ tensor are given by

$$K_{yy} \equiv \overline{\eta'v'}, \quad K_{zy} \equiv \overline{\eta'w'}, \quad K_{yz} \equiv \overline{\zeta'v'}, \text{ and } \quad K_{zz} \equiv \overline{\zeta'w'} \quad . \tag{3.1}$$

Here η' and ζ' are the horizontal and vertical eddy air displacement fields, respectively, and v' is the perturbation horizontal air velocity and w' is the perturbation vertical air velocity. (The same definition will be retained in isentropic coordinates, although the meaning of vertical "velocity" has to be reinterpreted.) It is important to note that the eddy field entering into the definition in (3.1) is independent of the particular minor species being transported.

For the conservative tracers that Reed and German were primarily concerned with, the eddy fluxes of a species with mass concentration χ is given by

$$\left(\begin{array}{c} \overline{v'\chi'} \\ \overline{w'\chi'} \end{array} \right) = - \mathbb{K} \cdot \vec{\nabla}\chi \quad . \tag{3.2}$$

---

[1] Even though the definition in (3.1) is the same as in Reed and German (1965), the interpretation of the eddy displacement fields η' and ζ' is not the same. Reed and German assumed η' and ζ' as mixing lengths. This permits the further assumption that $w'/v' = \zeta'/\eta'$, which then implies $K_{yz} = K_{zy}$. Here, however, the displacements are related to the velocities according to $(\partial/\partial t + \bar{u}\, \partial/\partial x)\eta' = v'$, and $(\partial/\partial t + \bar{u}\, \partial/\partial x)\zeta' = w'$. In general the tensor is neither symmetric nor antisymmetric.

When the tracers under consideration are not inert, the eddy fluxes can no longer be expressed in the form of (3.2) with the transport tensor $\mathbb{K}$ given by (3.1). This is due to the possible presence of an eddy source term, S', in the equation for χ', the eddy species concentration (see e.g., Tung (1982)). According to perturbation theory, the appropriate modification to (3.2) should be the <u>addition</u> of an extra source term on the right hand side of (3.2), as

$$\left(\begin{array}{c} \overline{v'\chi'} \\ \hline \overline{w'\chi'} \end{array} \right) = - \mathbb{K} \cdot \vec{\nabla}\overline{\chi} + \left(\begin{array}{c} \overline{v'\sigma'} \\ \hline \overline{w'\sigma'} \end{array} \right) \tag{3.3}$$

where σ' is defined through

$$(\frac{\partial}{\partial t} + \overline{u} \frac{\partial}{\partial x})\sigma' = S' \quad . \tag{3.4}$$

The definition for $\mathbb{K}$ should remain unchanged. In particular, it should <u>not</u> depend on the chemistry or concentration of the minor species being transported.

Some authors perfer to absorb the new term (the last term in Eq. (3.3)) into a generalized $\mathbb{K}$ so that the same expression (3.2) can be used for both conservative and reactive species. However, as pointed out by Strobel (1981), the eddy fluxes of some important chemical species, such as ozone, have terms in their eddy source fluxes that are proportional to the mean concentration $\overline{\chi}$, and consequently cannot be conveniently expressed in the form of (3.2), since (3.2) is a statement of the proportionality between the eddy fluxes and the <u>gradient</u> of the mean concentration.

The parameterization of the eddy source fluxes is difficult to do in general. We will postpone its discussion until section 5. In the next section, the components of the tensor $\mathbb{K}$ as defined in (3.1) (and therefore as independent of chemistry) are parameterized and their magnitudes quantitatively estimated.

4. QUANTITATIVE ESTIMATES OF $\mathbb{K}$

In the formulation of tracer transport in isentropic coordinates, or in the residual mean formulation in pressure coordinates, it is the deviation of the atmospheric eddy field from being strictly steady and adiabatic which contributes to the diffusive and advective transports of the tracers; the eddy transport tensor vanishes for adiabatic and steady eddy fields. Order of magnitude estimates of the degree of such deviations for typical eddy fields in the middle atmosphere and their effects on the diffusive and advective transports will be given in this section. The calculations are somewhat more direct in the isentropic coordinate formulation than in the residual mean formulation, and therefore the former formulation will be used first. Corresponding estimates in pressure coordinates will also be given for the purpose of comparing the two formulations.

Following a common practice, I will write the transport tensor $\mathbb{K}$ as a sum of a symmetric tensor $\mathbb{D}$ and an antisymmetric tensor ψ :

$$\mathbb{K} \equiv \mathbb{D} + \psi , \qquad (4.1)$$

where

$$\mathbb{D} = \begin{pmatrix} D_{yy} & D_{y\theta} \\ D_{y\theta} & D_{\theta\theta} \end{pmatrix} , \quad \text{and} \quad \psi \begin{pmatrix} 0 & -\psi \\ \psi & 0 \end{pmatrix} , \qquad (4.2)$$

Using the definition in (3.1), but replacing z now by the new vertical coordinate, θ, and w' by the "vertical velocity" $\dot{\theta}'$ in isentropic coordinates, one can show that (see Tung (1983)[2]):

$$D_{yy} = \frac{\partial}{\partial t} \frac{1}{2} \overline{\eta'\eta'} , \quad D_{\theta\theta} = \frac{\partial}{\partial t} \frac{1}{2} \overline{\zeta'\zeta'} , \quad D_{y\theta} = \frac{\partial}{\partial t} \frac{1}{2} \overline{\eta'\zeta'} \qquad (4.3)$$

and

$$\psi = - \frac{\partial}{\partial t} \frac{1}{2} \overline{\eta'\zeta'} + \overline{\eta'\dot{\theta}'} . \qquad (4.4)$$

Here ζ' is understood to be the "vertical displacement" in isentropic coordinates, and is given by

$$(\frac{\partial}{\partial t} + \overline{u} \frac{\partial}{\partial x})\zeta' = \dot{\theta}' . \qquad (4.5)$$

4.1 The Diffusive Transport

The transport by the symmetric $\mathbb{D}$ tensor is diffusive and down-gradient in nature provided that $D_{yy}, D_{\theta\theta} \geq 0$ and $D_{yy} D_{\theta\theta} \geq D_{y\theta}^2$ (Matsuno, 1980). In Tung (1982) it was mentioned that if the eddy field can be assumed to be adiabatic, then this $\mathbb{D}$ tensor becomes diagonal. In fact, only the first component D_{yy} remained. It will be shown here that D_{yy} remains to be the dominant term for non-adiabatic eddies under typical conditions in the stratosphere in the presence of Newtonian cooling.

Due to the difference in the scales of vertical and horizontal

---

[2]There are actually some density perturbations in the definition of these quantities if the isentropic coordinates are adopted, but these contribution were found to be small in Tung (1982) (see (D.11) and (D.12) there).

motions in the atmosphere, a simple ratio such as $|D_{\theta\theta}/D_{yy}|$ is not indicative of the ratio of vertical vs. horizontal eddy transports. A more appropriate ratio is obtained when the vertical displacement is normalized by a vertical scale "height", L_v, and horizontal displacement by a horizontal scale L_h. The more appropriate ratios that we need to consider are

$$r_1 \equiv \left| \frac{L_h^2 D_{\theta\theta}}{L_v^2 D_{yy}} \right| \quad \text{and} \quad r_2 \equiv \left| \frac{L_h D_{y\theta}}{L_v D_{yy}} \right| . \tag{4.6}$$

The horizontal scale L_h can be chosen, for our purpose of studying large-scale transports, to be the radius of the earth, a. The vertical scale L_v is more difficult to fix, as it may depend on the scale of the mean stratification of the particular species under consideration. Nevertheless, since only an order of magnitude is needed here, we will take it to be $L_v \sim 0.3\theta$ in potential temperature, approximately equivalent to a density scale height H in height coordinates.

As a reference, we note that in the Reed and German formulation, typical values of the ratios are

$$r_1 \equiv \left| \frac{a^2 K_{zz}}{H^2 K_{yy}} \right| \sim 1 \tag{4.7}$$

and

$$r_2 \equiv \left| \frac{a}{H} \frac{K_{yz}}{K_{yy}} \right| \sim 1 \tag{4.8}$$

using typical orders of magnitude from Reed and German (1965) or Luther (1973):

$$K_{yy} \sim 10^{10} \text{ cm}^2/\text{s}, \; K_{zz} \sim 10^4 \text{ cm}^2/\text{s}, \; K_{yz} \sim 10^7 \text{ cm}^2/\text{s} .$$

Thus in the diffusion type parameterization of Reed and German that is in common use, the eddy vertical and horizontal transports are roughly comparable. The situation appears to be different in isentropic coordinates using a more rational approach to eddy dynamics.

4.1.1 Newtonian Damping

In isentropic coordinates, nonadiabatic eddies have a non-vanishing perturbation "vertical velocity", $\dot\theta'$, which can be calculated directly from the thermodynamic equation

$$\dot\theta = \theta Q/T , \tag{4.9}$$

where Q is the diabatic heating range per unit mass divided by the specific heat c_p, and T is the temperature. The perturbation form of (4.9) is

$$\dot{\theta}' = \frac{\theta}{\bar{T}} [Q' - \frac{\bar{Q}}{\bar{T}} T'].$$

(4.10)

[Note that θ is not perturbed in (4.10) because it is an independent variable.]

Following Dickinson (1973), we adopt here the Newtonian damping parameterization, which gives

$$Q' = -\alpha_N T',$$

where α_N is the coefficient of Newtonian damping. Thus (4.10) gives

$$\dot{\theta}' = - \frac{\alpha\theta}{\bar{T}} \cdot T',$$

(4.11)

where

$$\alpha \equiv \alpha_N + \frac{\bar{Q}}{\bar{T}}$$

is the total Newtonian damping coefficient in isentropic coordinates. The extra term, $\bar{Q}/\bar{T}$, is typically one to two orders of magnitude smaller than α_N. A commonly used value for the Newtonian damping coefficient in the lower stratosphere is about 1/10 days. Its value is likely to be larger with increasing altitude, although there is no general agreement on the exact value. Blake and Lindzen (1973) gave a thermal relaxation time of about 3-7 days at 35 km and 1.2-2.5 days at 50 km. For our order of magnitude estimate I will adopt here a damping time scale of ∿5 days for the 30-40 km region and ∿10 days in the lower stratosphere.

4.1.2 The Ratio r

Using

$$ik(\bar{u}-c)\eta' = v'$$

and

$$ik(\bar{u}-c)\zeta' = \dot{\theta}'$$

where $\bar{u}$ is the speed of the mean flow and c is the phase speed of the wave, one finds

$$r \equiv \left| \frac{(\zeta'/L_v)}{(\eta'/L_h)} \right| \sim \frac{1}{(L_v/\theta)} \left| \frac{T'/\bar{T}}{v'/(a\alpha)} \right| . \tag{4.12}$$

In (4.12), the parameterization (4.11) is used to express the eddy vertical velocity in terms of temperature fluctuations. van Loon et al. (1973) and van Loon et al (1975) found largest stratospheric temperature oscillation to occur in wavenumber 1 during winter with an amplitude of about 10-12°C. A value of $v' \sim 13$ m/s is found for the same wave using

$$f_o v' = ik\Phi' ,$$

and adopting the value $|\Phi'/g| \sim 600$ m, also from van Loon et al. Thus

$$\begin{array}{ll} r \sim 0.07 \text{ for } \alpha = 1/10 \text{ days} \\ 0.14 \text{ for } \alpha = 1/5 \text{ days} \end{array} \tag{4.13}$$

4.1.3 An Alternative Estimate

In arriving at (4.13), I have used the observed values for T' and Φ' in __pressure coordinates__, since the corresponding values in isentropic coordinates are not readily available. Although such a procedure appears to be sufficient for our purpose of obtaining an order-of-magnitude estimate, it would be reassuring if an alternative estimate can be given that does not rely on our knowing these quantities directly.

To give a more deductive estimate for the ratio r, I use the equation for hydrostatic equilibrium

$$T' = \frac{\theta}{c_p} \frac{\partial}{\partial \theta} \Phi' . \tag{4.14}$$

The vertical displacement is then estimated using (4.14) together with (4.11) and (4.5) as

$$\zeta' \sim - \frac{\alpha \theta^2}{c_p \bar{T} ik(\bar{u}-c)} \frac{\partial}{\partial \theta} \Phi' , \tag{4.15}$$

while the horizontal displacement is, for geostropic waves,

$$\eta \sim \frac{\Phi'}{(\bar{u}-c)f_0} . \tag{4.16}$$

Thus the normalized ratio of vertical to horizontal displacements is

$$r \sim \frac{\alpha \, f_0 \, L_h}{(\Delta\theta/\theta)(L_v/\theta)k \, c_p \overline{T}} \tag{4.17}$$

In (4.17), we have written $|\partial/\partial\theta \, \Phi'| \sim |\Phi'/\Delta\theta|$, where $\Delta\theta$ is the vertical scale of the wave in isentropic coordinates. Note that the wave amplitudes cancel and therefore do not appear in (4.17). The same is true also for the factor $(\overline{u}-c)$ in (4.15) and (4.16), although $\Delta\theta$ does depend on $(\overline{u}-c)$ through the dispersion relation.

A WKB solution of the vertical structure equation for quasi-geostrophic waves gives

$$\Phi'(\theta)/\Phi'(\theta_0) = (\frac{\theta}{\theta_0})^\lambda \quad ,$$

where

$$\lambda = \frac{1}{\kappa} \{\frac{1}{2} + i[\frac{\hat\beta \, RH\Gamma^{(0)}}{f_0^2} (\frac{1}{\overline{u}-c} - \frac{1}{U_T})]^{1/2}\} \quad , \tag{4.18}$$

$\kappa = R/c_p = 2/7$

$\Gamma^{(0)} \simeq \partial T_e/\partial z + g/c_p$, the static ability parameter

and $U_T \simeq \hat\beta/[(k^2+\ell^2) + f_0^2/4RH\Gamma^{(0)}]$ is the Doppler-shifted speed at which the wave ceases to propagate vertically ($\hat\beta$ is the mean potential vorticity gradient).

For stationary ultralong waves, whose vertical wavelengths are typically larger than two scale heights, the square root in λ is typically much smaller than 1/2. Therefore

$$|\lambda| \sim \frac{1}{2\kappa} \sim 7/4 \quad ,$$

and

$$|\frac{\Delta\theta}{\theta}| \sim |\Phi'/(\theta \frac{\partial}{\partial\theta} \Phi')| \sim \frac{1}{|\lambda|} \sim 4/7 \quad ,$$

which gives

$$\begin{array}{ll} r \sim 0.07 & \text{for } \alpha = 1/10 \text{ days} \\ r \sim 0.14 & \text{for } \alpha = 1/5 \text{ days} \end{array} \quad . \tag{4.19}$$

These figures are remarkably close to the ones estimated earlier in (4.13).

The estimates in (4.19) are for wavenumber 1. The corresponding figures for wavenumber 2 should be reduced by approximately a factor of 1/2.

4.1.4 Critical Levels

The above estimates are based on the assumption that the wave encounters no critical level in its vertical propagation (in fact I have used in the estimates, $\bar{u}-c = \bar{u}{\sim}30$ m/s.) It would appear from (4.18) that if there exists a region in which u-c approaches zero, then the vertical scale of the wave should vanish, yielding an infinite r in (4.17). This, of course, does not happen when damping is present, and u-c in (4.18) should be replaced by $\bar{u}-c + \alpha/ik$ when u-c is not much larger than α/k. The quantity α/k is about 5 m/s for α=1/10 days and wavenumber 1. For u-c small, the limit for $|\lambda|$ is finite and is approximately

$$|\lambda| \sim \frac{1}{\kappa} \left[\frac{\hat{\beta}RH\Gamma^{(0)}}{f_0^2 \, \alpha/k} \right]^{1/2} \sim 9.$$

This yields

$$\begin{array}{ll} r \sim 35\% \text{ for } \alpha = 1/10 \text{ days} & \\ 50\% \text{ for } \alpha = 1/5 \text{ days} & (4.20) \end{array}$$

It appears that near critical levels, the effect of vertical displacement is about 4 to 5 times larger than the case away from critical levels, although it is still smaller than the effect of horizontal displacement since r is still less than one.

For forced stationary waves that propagate to the stratosphere, critical levels (i.e. zero wind lines) do not seem to occur frequently, away from the tropics. Exception occurs during a major sudden warming, when easterlies descend into the stratosphere. To the extent that major warmings are rare, our estimate of

$$r \sim 10^{-1} \tag{4.21}$$

in (4.19) appears to hold most of the time in mid to high latitude regions.

4.1.5 Gravity Waves

We now repeat the estimate of r for gravity waves. To estimate the horizontal displacement for gravity waves, I use

$$\left(\frac{\partial}{\partial t} + \bar{u} \frac{\partial}{\partial x} \right) v' = - \frac{\partial}{\partial y} \Phi'.$$

This gives

$$\eta' \sim i\ell\phi'/\omega^2,$$

where ℓ is the meridional wavenumber of the gravity wave, $\omega = k(\bar{u}-c)$, its Doppler-shifted frequency. Under hydrostaticity, the vertical displacement for gravity waves is also given by (4.15). Thus the normalized ratio between vertical and horizontal displacements is

$$r \sim \left| \alpha \; \frac{(\omega/\ell) \; L_h}{(\Delta\theta/\theta)(L_v/\theta) \; c_p\bar{T}} \right| . \tag{4.22}$$

The vertical wave scale $|\Delta\theta/\theta|$ can be estimated from the dispersion relation, giving

$$|\theta/\Delta\theta| \sim \frac{1}{\kappa} \left| \frac{1}{2} + i \; [\frac{R H\Gamma^{(0)}(1+\ell^2/k^2)}{(\bar{u}-c)^2}]^{1/2} \right| . \tag{4.23}$$

$$\sim \frac{[R H\Gamma^{(0)}(1+\ell^2/k^2)]^{1/2}}{\kappa|\bar{u}-c|} ,$$

since the square root in (4.23) is typically much larger than 1/2. [This is not true for the fast periodic tides, which we are not concerned with, since they seem to contribute little to transient eddy dispersion.] Thus (4.22) becomes

$$r \sim \frac{a\alpha|k/\ell|(1+\ell^2/k^2)^{1/2}}{(L_v/\theta)(c_p\bar{T})^{1/2}} . \tag{4.24}$$

Note that the dependence on $\bar{u}-c$ cancels in (4.24). This is fortunate, as this quantity can vary over a range of values for gravity waves. Furthermore, (4.24) should hold even near critical levels.

For $|\ell/k| \sim 1$, (a "guesstimate" from Lindzen (1981)), (4.24) gives

$$\begin{aligned} r &\sim 0.07 \quad \text{for } \alpha = 1/10 \text{ days} \\ r &\sim 0.14 \quad \text{for } \alpha = 1/5 \text{ days} , \end{aligned} \tag{4.25}$$

indicating also for gravity waves that the effect of vertical displacement is small compared with that of horizontal displacement.

4.1.6 Pressure Coordinates

As mentioned earlier, there are some indications that, unlike the case in isentropic coordinates, the normalized ratio r is of order unity in pressure coordinates. To illustrate the difference in the two systems, I now redo the calculation for r in pressure coordinates.

The vertical velocity w' in pressure coordinates is given by

$$(\frac{\partial}{\partial t} + \bar{u}\frac{\partial}{\partial x})T' + \Gamma^{(0)}w' = Q' \quad . \tag{4.26}$$

Note that in isentropic coordinates the temperature advection term does not appear as it does in (4.26).

Using as before the Newtonian damping parameterization:

$$Q' = -\alpha_N T'$$

and the definition for vertical displacement ζ':

$$(\frac{\partial}{\partial t} + \bar{u}\frac{\partial}{\partial x})\zeta' = w' \quad ,$$

one finds

$$\zeta' = -\frac{(\bar{u}-c + \frac{\alpha_N}{ik})T'/\Gamma^{(0)}}{(\bar{u}-c)} \tag{4.27}$$

The horizontal displacement is

$$\eta' = \frac{v'}{ik(\bar{u}-c)} \quad .$$

The normalized ratio in pressure coordinates is

$$(r_p) \equiv |\frac{\zeta'/H}{\eta'/a}| = \frac{\bar{T}}{H\Gamma^{(0)}} |(1+\frac{ik(\bar{u}-c)}{\alpha_N})| \ |\frac{T'/\bar{T}}{v'/(a\,\alpha_N)}| \quad . \tag{4.28}$$

Using the same value for T' and v' as in section 4.1.2, one finds that the ratio of the quantity r in pressure coordinates and that in isentropic coordinates is essentially given by ($H\Gamma^{(0)}/\bar{T}$ is about the same as L_v/θ):

$$\frac{(r)_p}{(r)_\theta} \sim |1 + \frac{ik(\bar{u}-c)}{\alpha_N}| \quad . \tag{4.29}$$

This ratio is always greater than one, indicating that the effect of vertical displacement is more significant in pressure coordinates.

The quantity, $(\bar{u}-c)/(\alpha_N/k)$, in (4.29) measures the relative

importance of the horizontal temperature advection vs. the diabatic
heating term (viz. the first vs. the third term in Eq. (4.26)). For a
typical value of $\bar{u}-c = u \sim 30$ m/s. This ratio is

$$(\bar{u}-c)/(\alpha_N/k) \sim \begin{array}{l} 6 \text{ for wavenumber 1} \\ 12 \text{ for wavenumber 2} \end{array}$$

in the lower stratosphere, where $\alpha = 1/10$ days is used. This then
suggests that r in pressure coordinates is about one order of
magnitude larger than the corresponding quantity in isentropic coordinates, implying that vertical and horizontal diffusion are about
comparable in pressure coordinates.

4.1.7 Ratio of Vertical to Horizontal Transports

If it can be assumed that the time dependence in the vertical displacement field is the same as that in the horizontal displacement
field, in particular, that they have the same time scales (this
assumption appears to be valid for the organized wave motion under consideration), then the smallness of the ratio of normalized vertical and
horizontal displacements implies directly the smallness of vertical
transport as compared to the horizontal transport. Specifically, the
normalized ratios of vertical vs. horizontal transports defined in
(4.6) are estimated from

$$r_1 \sim r^2 \text{ and } r_2 \sim r . \tag{4.30}$$

This gives

$$r_1 \sim 0(10^{-2}), \quad r \sim 0(10^{-1}) \tag{4.31}$$

implying that $D_{\theta\theta}$ can be neglected when compared with D_{yy}, and $D_{y\theta}$
can be approximately neglected also when compared to D_{yy}. In other
words, our estimates suggest that the dominant diffusive transport is
horizontal diffusion along horizontal gradients of mean tracer concentration. Therefore it appears justified in approximating the
diffusion tensor by a scalar:

$$\mathbb{D} \simeq \begin{bmatrix} D_{yy} & 0 \\ 0 & 0 \end{bmatrix} . \tag{4.32}$$

We will next estimate the magnitude of D_{yy}.

4.1.5 Maximum Magnitude of D_{yy}

Since the diffusion coefficient is given in terms of a time
derivative, i.e., $D_{yy} = \partial/\partial t \, 1/2 \, \overline{n'n'}$, it is clear that periodic wave
motions do not give rise to net transport. Net transport is produced

by irreversible processes, either caused by direct dissipation of the
wave or through cascade of wave energy to smaller scales and ultimate
dissipation of the small scale waves. Since wave amplification is
necessary in order to give positive diffusion, the wave must obtain its
energy from somewhere, either through forcing, wave-mean flow interaction, or wave-wave interaction.

In terms of kinetic energy, a large component of the eddy field in
the middle atmosphere is due to the stationary planetary waves forced
by topography and differential heating in the lower atmosphere. These
waves occasionally amplify in a time scale of about four or five days,
reaching an equilibrated maximum amplitude when forcing balances damping. A dramatic example of such events is the so-called sudden warming
phenomenon, when wave amplification of $\sim$ 1000 meters in geopotential
height is observed to occur in less than a week in the stratosphere
(and possibly larger values higher up). The time dependence of such an
amplification episode can be modelled approximately by

$$\eta' \simeq \eta_0' \cdot (1-e^{-(t/\tau)}) \quad . \tag{4.33}$$

The time behavior of (4.33) is depicted in Fig. 1a. Essentially it
consists of an approximately linear growth initially (for t/τ small),
reaching an <u>equilibrated maximum amplitude</u> η_0' when t is larger than
the <u>amplification time scale</u> τ. (This time behavior was deduced by Tung
and Lindzen (1978a,b) for the resonant amplification of forced
stationary wave in the presence of damping with a time scale τ. The
diffusion coefficient D_{yy} induced by such an eddy field is found to be

$$D_{yy} = \overline{\eta_0'^2} \; \frac{1}{\tau} \; e^{-(t/\tau)}(1-e^{-(t/\tau)}); \tag{4.34}$$

its behavior with time is plotted in Fig. 1b. Note that the maximum

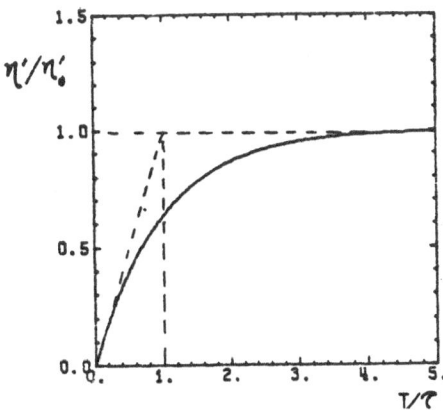 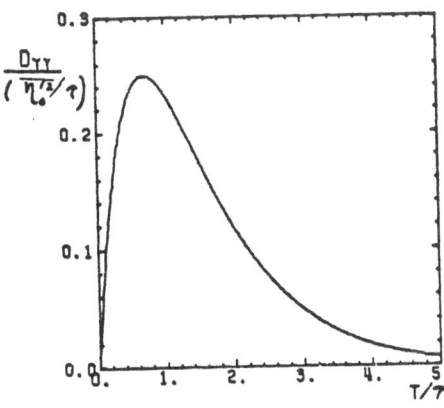

Fig. 1a Transient amplification of Fig. 1b Contribution to the eddy
    a forced wave in the   diffusion coefficient from a
    presence of damping   single episode of wave
                        amplification

value achieved for D_{yy} for the episode is

$$(D_{yy})_{max} = \frac{\overline{n_0'^2}}{4\tau} \quad . \tag{4.35}$$

To estimate this value, the eddy horizontal displacement in isentropic coordinates is needed but unfortunately it is not readily available. I will instead use the values in pressure coordinates, hoping that there is not much difference in the horizontal quantities. For the 1973 sudden warming, Kanzawa (1980) gave a peak value for the geopotential height of wave number 1 to be 1,100 meters in the stratosphere (although for this particular warming event there later developed a second peak in the mesosphere with a peak of 2,200 meters. This larger value is not used here because it occurs outside the photochemical region of primary interest here). Thus roughly n_0, $\sim \phi/f_0\overline{u} \sim 3,000$ km, achieved in approximately five days, and

$$(D_{yy})_{max} \sim 5 \times 10^{10} \ cm^2/s \quad . \tag{4.36}$$

This value is comparable to that of K_{yy} in the mixing-length formulation of Reed and German (1965). Note, however that our estimate in (4.36) is likely to be an upperbound for the typical values of D_{yy}, because major sudden warmings are rare over a period of a year. Furthermore, the peak value quoted in (4.36) is attained only for a short period (see Fig. 1b). Over longer periods of time, it is the less dramatic, but more frequent, events that contribute more to the diffusive process.

4.1.6 A Parameterization of D_{yy}

We envisage a series of amplification episodes each of the form (see Fig. 2):

$$n' \simeq \Delta n' \cdot (1 - e^{-(t-t_0)/\tau}) + n_m', \ t \geq t_0, \tag{4.37}$$

where n_m' is the time mean eddy field, $\Delta n'$ the (maximum) amplification amplitude from the time mean value and t_0, the starting time of amplification. Each such episode contributes to D_{yy} terms:

$$d_{yy} \equiv \overline{\Delta n'^2} \frac{1}{\tau} e^{-(t-t_0)/\tau} \left[1 - e^{-(t-t_0)/\tau}\right] + \overline{n_m'\Delta n'} \frac{1}{\tau} e^{-(t-t_0)\tau} \tag{4.38}$$

The superposition of contributions from all episodes gives

$$D_{yy} \simeq \frac{1}{\tau} \int_\infty d_{yy} \ dt_0 = \frac{1}{\tau} \left[\overline{n_m'\Delta n'} + \frac{1}{2} \overline{\Delta n'^2}\right] \quad , \tag{4.39}$$

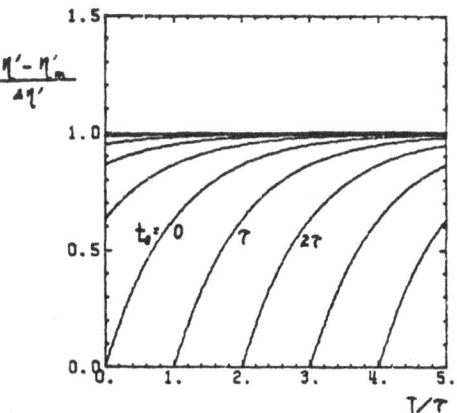

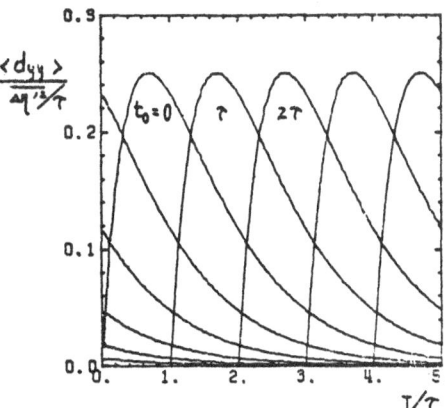

Fig. 2a A series of wave
         amplification episodes
         each starting at a
         different time t_0.

Fig. 2b Contribution to the eddy
         diffusion coefficient
         from a series of wave
         amplification episodes

assuming the episodes are uncorrelated with each other. [By assuming
that t_0 is continuous, we admittedly are overestimating the value of
D_{yy}]. Note that the expression obtained is independent of the fast
amplification time (t/τ), as desired, but may vary slowly on a
seasonal time scale as the amplitudes change. Note also that the first
term, $\overline{\eta_m' \Delta \eta'}$, may be of either sign depending on the correlation of the
phases of $\Delta \eta'$ and η_m'. However, this term disappears when (4.39) is
time averaged over periods longer than τ (as denoted by $< >$) (since
by definition $<\Delta \eta'> = 0$):

$$<D_{yy}> \simeq \frac{1}{\tau} <\overline{\Delta \eta'^2}> \quad . \tag{4.40}$$

If $<\overline{\Delta \eta'^2}>$, the distribution of the variance of the transient eddies[3],
can be obtained from observed data, one then has the required estimate
of the diffusion coefficient D_{yy}.

Lau and Oort (1982) obtained transient eddy statistics using GFDL
and NMC analyses of data below 100 mb for six winter and summer
seasons. Latitude-pressure distribution of the following quantity is
displayed:

$$(<\overline{(\Delta \phi/g^2)}>)^{1/2}$$

---

[3]Periodic travelling waves should first be removed from data.

A maximum of ~ 200 m is found near the 300 mb level during the winter
season. This is not exactly what we wanted, as the zonal mean geopotential height was not removed from $\Delta\phi/g$. I will use the figure,
200 m, as the upperbound for

$$(\overline{<(\Delta\phi'g)^2>})^{1/2} \quad .$$

So we have

$$(\overline{<\Delta\eta'^2>})^{1/2} \, \underset{\sim}{<} \, 600 \text{ km}$$

and

$$<D_{yy}> \, \underset{\sim}{<} \, 4 \times 10^9 \text{ cm}^2/\text{s} \quad . \tag{4.41}$$

This value is consistent with that estimated by Kida (1983) from his
GCM. He found

$$K_{yy} \sim 3 \times 10^9 \text{ cm}^2/\text{s} \quad .$$

At the value given by (4.41), the diffusion process does not appear to
be competitive with mean advection in transporting tracers over global
scales. A time scale for global diffusion, a^2/D_{yy}, is about three
years, while the advection time scale, $a/\bar{v}$, is about four months using
$\bar{v} \sim 1/2$ m/s. Nevertheless, the diffusion process may be more
effective over smaller scales and in high gradient regions.

Our result concerning large-scale diffusion appears to be at
variance with the conclusion of Pyle and Rogers (1980a), who found insufficient transport of columnar ozone into the high latitudes when
all the eddy diffusion terms were dropped (but with the mean circulation replaced by the approximate residual mean circulation in the
Oxford 2-D model of Harwood and Pyle (1975)). It is possible that the
numerical scheme used in the original Oxford 2-D model becomes unstable when the diffusion term is removed and numerical instability
may cause the breakdown of the large scale circulation into small
cells, which are ineffective in transporting over large distances.

4.2 The Advective Transport

The advective eddy transport is caused by the antisymmetric part
of the transport tensor. The eddy induced advective velocities are
given by

$$\bar{v}_E \simeq -\frac{1}{\bar{\rho}_\theta} \frac{\partial}{\partial\theta} \bar{\rho}_\theta \psi, \quad \bar{\dot{\theta}}_E \simeq \frac{1}{\bar{\rho}_\theta} \frac{\partial}{\partial y} \bar{\rho}_\theta \psi, \tag{4.42}$$

where

$$\psi = \overline{\eta'\dot{\theta}'} - \frac{\partial}{\partial t}\frac{1}{2}\overline{\eta'\zeta'} \ .$$

These eddy advective velocities are to be compared with the advection by the mean velocities $\overline{v}$ and $\overline{\dot{\theta}}$. In isentropic coordinates, the mean vertical velocity $\overline{\dot{\theta}}$ can be directly estimated from the thermodynamic equation (4.9) as

$$\overline{\dot{\theta}} \simeq \theta\,\frac{\overline{Q}}{\overline{T}} \ . \tag{4.43}$$

Since the eddy advective transport adds to the mean transport, the combined transport can be viewed as given by a combined diabatic heating rate $\overline{Q}*$ (Mahlman, personal communication):

$$\overline{\dot{\theta}} + \overline{\dot{\theta}}_E = \frac{\theta}{\overline{T}}\,\overline{Q}* \ , \tag{4.44}$$

where

$$\overline{Q}* \equiv \overline{Q} + \overline{Q}_E$$

$$\overline{Q}_E \equiv \frac{\overline{T}}{\theta}\,\bar{\rho}_\theta^{-1}\frac{\partial}{\partial y}\,\bar{\rho}_\theta\psi \simeq \frac{\overline{T}}{\theta}\frac{\partial}{\partial y}\,\psi \ . \tag{4.45}$$

Since the transient part in ψ: $-\partial/\partial t\ 1/2\ \overline{\eta'\zeta'}$, can be shown to be smaller than $\overline{\eta'\dot{\theta}'}$ by one order of magnitude, we use

$$\psi \sim \overline{\eta'\dot{\theta}'} = -\alpha\frac{\theta}{\overline{T}}\overline{\eta'T'}$$

and so

$$|\overline{Q}_E| \sim |\frac{1}{a}\alpha\,\overline{\eta'T'}| \sim \begin{array}{l} 0.2°\text{C/day for } \alpha=1/10 \text{ days} \\ 0.5°\text{C/day for } \alpha=1/5 \text{ days} \end{array} \tag{4.46}$$

using $T' \sim 10°C$ (van Loon et al., 1973) and $\eta' \sim 1500$ km. These values of $\overline{Q}_E$ are not negligible compared to $\overline{Q}$, which is about $\sim 1°C/day$ according to Murgatroyd and Singleton (1962), or $\sim 0.4°C/day$ according to Dopplick (1979). Nevertheless, the contribution from Q_E is within the range of uncertainty in our ability to calculate $\overline{Q}$ at the present time, and so an accurate calculation of $\overline{Q}_E$ does not seem warranted, at least for the time being.

The eddy advective transport appears to be also important in pressure coordinates. Rood and Schoeberl (1983) have found that in the presence of Newtonian damping the residual mean circulation calculated in their model tends to overestimate the "Lagrangian mean" circulation by as much as 30% with $\alpha=1/10$ days implying that the

advective eddy transport by nonadiabatic stationary planetary waves
accounts for about $\sim$ 30% of the transport in pressure coordinates.

5. CHEMICAL EDDY TERMS

Eddy fluxes in 2-D models in the chemical source term have
generally been ignored in the past even for nonconservative species,
until recently, when studies, first by Pyle and Rogers (1980b) and
later by Strobel (1981), Garcia and Solomon (1983), and Rood and
Schoeberl (1983), demonstrated the importance of these chemical eddy
terms.

As mentioned previously in section 3, the presence of a chemical
source term S in the species transport equation

$$\frac{d}{dx}\chi = S \qquad\qquad (5.1)$$

invariably gives rise to a chemical eddy source, S', when the air
motion is zonally asymmetric. The perturbation equation:

$$D_0\chi' + v'\frac{\partial}{\partial y}\bar{\chi} + w'\frac{\partial}{\partial z}\bar{\chi} = S' \qquad\qquad (5.2)$$

can be "solved" to yield

$$\chi' = -\eta'\frac{\partial}{\partial y}\bar{\chi} - \zeta'\frac{\partial}{\partial z}\bar{\chi} + \sigma' \quad, \qquad\qquad (5.3)$$

where

$$D_0 \equiv \frac{\partial}{\partial t} + \bar{u}\frac{\partial}{\partial x} \quad, \quad D_0\cdot\sigma' = S' \quad.$$

Strictly speaking, (5.3) is not a solution for χ', because the
chemical eddy term σ' may depend on χ' (and even the concentration
of other species partaking in the reaction). In general no closed form
solution can be found for χ', and hence also for the eddy fluxes of
χ (see (3.3)).

To fix ideas, let us consider the following simple example of a
conservative tracer (see also Matsuno (1980)):

$$\frac{d}{dt}\chi = -\frac{1}{\tau_0}(\chi-\bar{\chi}_0) \quad. \qquad\qquad (5.4)$$

The "source" term on the right-hand side of (5.4) describes a relaxation process about a mean equilibrium distribution $\bar{\chi}_0$, involving
a relaxation time τ_0. τ_0 is here treated as a mean quantity
dependent on the mean photochemistry involved. The perturbation source

term is thus

$$S' = - \frac{1}{\tau_0} \chi' \quad . \tag{5.5}$$

Even for this simple eddy source term, (5.2) cannot be solved explicity. We shall instead attempt to solve it asymptotically assuming that the chemical relaxation time, τ_0, is much longer than the dynamical advection time scale: $a/\overline{u}$ (about three days). Therefore,

$$\chi' = - \eta' \frac{\partial}{\partial y} \overline{\chi} - \zeta' \frac{\partial}{\partial z} \overline{\chi} + D_0^{-1} \left(- \frac{1}{\tau_0} \chi'\right) \tag{5.6}$$

$$\simeq - \eta' \frac{\partial}{\partial y} \overline{\chi} - \zeta' \frac{\partial}{\partial z} \overline{\chi} + \frac{1}{\tau_0} [D_0^{-1} \eta' \frac{\partial}{\partial y} \overline{\chi} + D_0^{-1} \zeta' \frac{\partial}{\partial z} \overline{\chi}] + 0((\frac{a}{\overline{u}\tau_0})^2)$$

The eddy fluxes of χ' are then found to be

$$\overline{v'\chi'} \simeq - K_{yy} \frac{\partial}{\partial y} \overline{\chi} - K_{yz} \frac{\partial}{\partial z} \overline{\chi} - \frac{1}{\tau_0} [K_{yy}^{(1)} \frac{\partial}{\partial y} \overline{\chi} + K_{yz}^{(1)} \frac{\partial}{\partial z} \overline{\chi}]$$

$$\overline{w'\chi'} \simeq - K_{zy} \frac{\partial}{\partial y} \overline{\chi} - K_{zz} \frac{\partial}{\partial z} \overline{\chi} - \frac{1}{\tau_0} [K_{zy}^{(1)} \frac{\partial}{\partial y} \overline{\chi} + K_{zz}^{(1)} \frac{\partial}{\partial z} \overline{\chi}] \quad , \tag{5.7}$$

where the K's are the components of the eddy transport tensor defined in section 3 and the $K^{(1)}$'s are defined here as

$$K_{yy}^{(1)} \equiv - \overline{v'D_0^{-1}\eta'} \quad , \qquad K_{yz}^{(1)} \equiv - \overline{v'D_0^{-1}\zeta'}$$

$$K_{zy}^{(1)} \equiv - \overline{w'D_0^{-1}\eta'} \quad , \qquad K_{zz}^{(1)} \equiv - \overline{w'D_0^{-1}\zeta'} \quad . \tag{5.8}$$

Note that, like the K's, the $K^{(1)}$'s are also independent of chemistry or species concentration. They depend only on the air displacements and have to be parametrized. The dependence on the mean chemistry, which is in principle determinable within the framework of 2-D theory, has been separated out in (5.7) as a factor, $1/\tau_0$, multiplying the empirical parameters, $K^{(1)}$'s, which cannot be determined within the 2-D theory. It has been shown in Tung (1982) that such a separation can in general be done.

Unlike the $\mathbb{K}$ matrix, whose symmetric components are caused entirely by transient eddies, the symmetric part of the $\mathbb{K}^{(1)}$ matrix has a significant stationary component. This can be seen more clearly if we reexpress the $K^{(1)}$'s using the identities

$$D_0 \overline{\alpha' \cdot \beta'} = \frac{\partial}{\partial t} \overline{\alpha' \beta'} - \overline{\alpha' D_0 \beta'}, \text{ and } D_0 D_0^{-1} = 1,$$

so that

$$K_{yy}^{(1)} = \overline{n'n'} - \frac{\partial}{\partial t} \overline{n' D_0^{-1} n'} = \frac{1}{2} \overline{n'n'}$$

and similarly

$$K_{zz}^{(1)} = \frac{1}{2} \overline{\zeta'\zeta'}$$

$$K_{yz}^{(1)} = \overline{n'\zeta'} - \frac{\partial}{\partial t} \overline{n' D_0^{-1} \zeta'}, \quad K_{zy}^{(1)} = \overline{n'\zeta'} - \frac{\partial}{\partial t} \overline{\zeta' D_0^{-1} n'}.$$

The $\mathbb{K}^{(1)}$ matrix is actually symmetric for steady wave fields. Comparing the magnitudes of the components of $1/\tau_0 \, \mathbb{K}^{(1)}$ with the corresponding components of $\mathbb{D}$, the symmetric part of $\mathbb{K}$, we see that the ratios are principally determined by the strength of the stationary eddy field vs. that of the transient eddy field in the atmosphere. For example, using the parameterization of transient eddies (4.30), we find

$$\left| \frac{1}{\tau_0} K_{yy}^{(1)} / D_{yy} \right| \sim \frac{\tau}{\tau_0} \frac{<n'n'>}{<\Delta n' \Delta n'>} . \tag{5.9}$$

This ratio is in general not small despite our assumption that τ_0 be much greater than $a/\bar{u} \sim$ 2-3 days.

Nevertheless, the $\mathbb{K}^{(1)}$ matrix itself can be simplified if we adopt the isentropic coordinate system. Using our earlier result that the normalized ratio between the vertical and horizontal displacements in isentropic coordinates is small (i.e. $r \ll 1$ according to (4.13)), the $\mathbb{K}^{(1)}$ matrix can be approximated by

$$\mathbb{K}^{(1)} \approx \begin{bmatrix} D_{yy}^{(1)} & 0 \\ 0 & 0 \end{bmatrix}, \quad D_{yy}^{(1)} \equiv \frac{1}{2} \overline{n'n'} \tag{5.10}$$

This approximation is the same as that used in Tung (1982).

Using data from van Loon et al (1973), who gave a value of $\sim$600 m for the geopotential height for zonal harmonic standing wave 1 in January around 10 mb level, we estimate the value for $D_{yy}^{(1)}$ to be

$$D_{yy}^{(1)} \sim 10^{16} \text{ cm}^2.$$
(5.11)

At this value, the chemical eddy team will become important compared with the dynamical diffusion term when the chemical relaxation time is less than 30 days. That is

$$\frac{1}{\tau_0} D_{yy}^{(1)} \gtrsim D_{yy} \quad \text{if} \quad \tau_0 \lesssim 30 \text{ days.}$$

There are a few more points that are worth making concerning this simple example. First, that the eddy fluxes in (5.7) can be expressed in terms of the mean species gradients is entirely a consequence of our assumption that τ_0 contains no eddy term. In more general systems, the photochemical relaxation time is rather temperature-dependent and perturbations in the temperature field should cause perturbations in τ_0, which then leads to an additional term in (5.7) proportional to $\bar{\chi}$. Secondly, our asymptotic solution procedure requires the assumption that the photochemical time scale is longer than the dynamical time scale, a/u. This assumption is probably correct in the lower stratosphere but becomes increasingly invalid higher up. In regions where the relaxation time is very short, a different assumption -- the photochemical equilibrium assumption -- can be used, namely, the dynamical transports can be neglected and (5.4) approximated by

$$0 \simeq - \frac{1}{\tau_0} (\chi - \bar{\chi}_0),$$
(5.12)

yielding the equilibrium solution $\chi \simeq \bar{\chi}_0$. The troublesome region is the so-called transition region where the photochemical and dynamical time scales are comparable. The method presented here cannot handle this problem and further study is needed.

6. SUMMARY

A main problem in the formulation of zonally averaged models of tracer transport is the treatment of eddy fluxes. As these eddy terms cannot be determined consistently within zonally averaged models, their magnitudes need to be calculated from a knowledge of the wave field in the atmosphere. I have attempted in this paper to give an order of magnitude estimate of various eddy transport teams, assuming that the eddy field in the stratosphere consists predominantly of organized (but nonadiabatic and possibly nonsteady) planetary and nonbreaking gravity waves forced in the lower atmosphere (gravity wave breaking level appears to occur higher up in

the mesophere, above the photochemical region of interest here
(see Lindzen (1981)). Though the estimates obtained depend somewhat
on the parameters chosen, the following features appear to be
typical:

(1) In isentropic coordinates, the effect of vertical dis-
     placement is found to be about one order of magnitude
     smaller than that of horizontal displacement (even after
     the difference in vertical vs. horizontal scales is
     taken into account). The eddy displacements are, there-
     fore, approximately rectilinear in isentropic coordinates.

(2) As a consequence, the dominant eddy diffusion is horizontal
     (i.e. along isentropes) across horizontal gradient of mean
     tracer concentration (the K_{yy} term in the transport
     equation). Contrast this with the situation in pressure
     coordinates where the normalized ratio of the vertical and
     horizontal diffusion coefficients is about order unity (see
     (4.7) using Reed and German's K_{yy} and K_{zz}). This ratio
     is found also by Kida (1983) to be of order one in his GCM,
     although both of his calculated K_{yy} and K_{zz} are about
     one order smaller than the values used by Reed and German.

(3) Assuming that the transient eddies are due to damped forced
     waves, I have given a parameterization of the coefficient
     K_{yy} (or called D_{yy} here), and estimated its climatological
     magnitude to be

     $$K_{yy} < 4 \times 10^9 \ cm^2/s.$$

     At this magnitude, the diffusive transport appears to be one
     order of magnitude smaller than the mean advective transport
     over scales comparable to the radius of the earth, but may
     be more effective over scales less than a thousand kilome -
     ters.

(4) Occasionally during major sudden warmings, the diffusion
     coefficient can achieve a maximum value of

     $$K_{yy} \sim 5 \times 10^{10} \ cm^2/s,$$

     for a short period of time (a few days). Though at this
     value the diffusive process is competitive with the mean
     advection in transport, the short duration at which this
     larger value is achieved and the infrequency of major
     sudden warmings tend to make the former a less effective
     means of systematic transport than the latter over
     seasonal time scales.

(5) The eddy advective transport is found to be dominated by
     the stationary planetary wave component in the presence of

eddy diabatic heating Q'. Assuming Q' is about 1-2°C/day, I have estimated the magnitude of this transport to be equivalent to that arising from a mean diabatic heating rate of about a fraction of a degree per day. Though not clearly negligible compared to the mean diabatic circulation, the eddy advective contribution is within the range of uncertainty in our ability to deduce the diabatic heating rate from observed data.

(6) For nonconservative tracers, chemical eddy terms may also be present. Since perturbations in the minor chemical species arise ultimately from perturbations in air, the chemical eddy terms should in principle be expressible in terms of air displacement correlations, which can be parameterized; the parameterization would then be independent of the minor species within the same atmosphere. We have shown here using a simple example how this can be done for the case when the photochemical time scale is longer than the advection time scale, a/u, which is about 2-3 days. It is found that the chemical eddy term in our example is important when the photochemical relaxation time is less than about 30 days.

As far as the dynamical transports are concerned, it now appears that the dominant mechanism for systematic tracer transport is advective in nature, with horizontal diffusion playing a secondary role and vertical diffusions even less significant. The approximate transport equation is then:

$$\frac{\partial}{\partial t}\,\bar{\chi} + \frac{\bar{V}^*}{\rho_\theta}\,\frac{\partial}{\partial y}\,\bar{\chi} + \frac{\bar{W}^*}{\rho_\theta}\,\frac{\partial}{\partial \theta}\,\bar{\chi} - \frac{\partial}{\partial y}\,(D_{yy}\,\frac{\partial}{\partial y}\,\bar{\chi}) = \bar{P}\ , \tag{6.1}$$

where $\bar{P}$ involves various chemical source terms. In (6.1) the advective velocities are found from (see Tung (1982) with $\bar{W}^* \equiv \bar{W} + \bar{W}_E$):

$$\bar{W}^* \simeq \bar{q}^*/\Gamma^{(0)} \tag{6.2}$$

$$\frac{\partial}{\partial y}\,\bar{V}^* + \frac{\partial}{\partial \theta}\,\bar{W}^* \simeq 0 \tag{6.3}$$

with

$$\bar{q}^* \equiv \bar{q} + \Gamma^{(0)}\,\frac{\partial}{\partial y}\,\bar{\rho}_\theta\,\psi$$

$\bar{q}$: the mean diabatic heating rate per unit volume.

The mean density $\bar{\rho}_\theta$ appearing in (6.1) can be approximated by the basic state density $\rho_\theta^{(0)}$, which is a function of vertical coordinates only:

442

$$\bar{\rho}_\theta^{(0)}(\theta) \simeq \bar{\rho}_\theta^{(0)}(\theta_0) \cdot \left(\frac{\theta}{\theta_0}\right)^{-9/2} \tag{6.4}$$

Thus from a diagnostic point of view, $\bar{q}*$ is the only quantity that needs to be specified in determining the advective transports. Because of the uncertainties in $\bar{q}$, one can replace $\bar{q}*$ by $\bar{q}$ at the present stage of model development.

ACKNOWLEDGEMENT:

The research is supported by the National Science Foundation under Grant ATM-8217616. I would like to thank Drs. F. Hasebe and M. Takahashi for reviewing the draft manuscript and offering helpful comments. I have also benefited from conversations with Drs. M. Ko and N-D. Sze during the course of the work.

REFERENCES

Andrews, D. G., and M. E. McIntyre, 1976: Planetary waves in horizontal and vertical shear: The generalized Eliassen-Palm relation and the mean zonal acceleration. J. Atmos. Sci. 33, 2031-2048.

-------, and -------, 1978: An exact theory of nonlinear waves on a Lagrangian-mean flow. J. Fluid Mech. 89, 609-646.

Blake, D. and R. S. Lindzen, 1973: The effect of photochemical models on calculated equilibria and cooling rates in the stratosphere. Mon. Wea. Rev. 101, 783-802.

Boyd, J. P., 1976: The noninteraction of waves with zonally-averaged flow on a spherical earth and the interrelationships of eddy fluxes of energy, heat and momentum. J. Atmos. Sci. 33, 2285-2291.

Clark, J. H. E., and T. G. Rogers, 1978: The transport of trace gases by planetary waves. J. Atmos. Sci. 35, 2232-2235.

Dickinson, R. E., 1973: Method of parameterization for infrared cooling between latitudes of 30 and 70 kilometers. J. Geophys. Res., 78, 4451-4457.

Dopplick, T. G., 1979: Radioactive heating of the global atmosphere. Corrigendum. J. Atmos. Sci., 36, 1812-1817.

Dunkerton, T., 1978: On the mean meridional mass motions of the stratosphere and mesosphere. J. Atmos. Sci. 35, 2325-2333.

Garcia, R. R., and S. Solomon, 1983: A numerical model of zonally averaged dynamical and chemical structure of the middle atmosphere. J. Geophys. Res., 88, 1379-1480.

Harwood, R. S., and J. A. Pyle, 1975: A two-dimensional mean circulation model for the atmosphere below 80 km., Quart. J. R. Met. Soc., 10, 723-747.

Holton, J. R., 1980: Wave propagation and transport in the middle atmosphere. Phil. Trans. Roy. Soc. London A 296, 73-85.

-----, 1981: An advective model for two-dimensional transport of stratospheric trace species. J. Geophys. Res. 86, 11989-11994.

Kanzawa, H., 1980: The behavior of mean zonal wind and planetary-scale disturbances in the troposphere and stratosphere during the 1973 sudden warming. J. Met. Soc. Japan, 58, 329-356.

Kida, H., 1983: General circulation of air parcels and transport characteristics derived from a hemispheric GCM. Part 1. A determination of advective mass flow in the lower stratosphere. J. Met. Soc. Japan 61, 171-187.

Lau, N-C, and A. H. Oort, 1982: A comparative study of observed Northern hemispheric circulation statistics based on GFDL and NMC analysis. Part II: Transient eddy statistics and energy cycle. Mon. Wea. Rev. 10, 889-906.

Lindzen, R. S., 1981: Turbulence and stress due to gravity wave and tidal breakdown. J. Geophys. Res. 86, 9707-9714.

Luther, F. M., 1973: Monthly mean values of eddy diffusion coefficients in the lower stratosphere, Lawrence Livermore Lab. Rep. UCRL-74616 Preprint.

Mahlman, J. D., D. G. Andrews, H. U. Dütsch, D. L. Hartmann, T. Matsuno and R. J. Murgatroyd, 1981: Transport of trace constituents in the stratosphere. Report Study Group 2, Middle Atmosphere Program Handbook for MAP, vol. 3.

Matsuno, T., 1980: Lagrangian motion of air parcels in the stratosphere in the presence of planetary waves. Pure Appl. Geophys. 118, 189-216.

McIntyre, M. E., 1980: Towards a Lagrangian-mean description of stratospheric circulations and chemical transports. Phil. Trans. Roy. Soc. London A 296, 129-148.

Murgatroyd, R. J. and F. Singleton, 1961: Possible meridional circulations in the stratosphere and mesosphere. Quart. J. Roy. Meteor. Soc. 87, 125-135.

Plumb, R. A., 1979: Eddy fluxes of conserved quantities by small amplitude waves. J. Atmos. Sci. 36, 1699-1704.

Pyle, J. A., and C. F. Rogers, 1980a: A modified diabatic circulation model for stratospheric tracer transport. Nature, 287, 711-714.

Pyle, J. A., and C. F. Rogers, 1980b: Stratospheric transport by stationary planetary waves -- the importance of chemical processes. Quart. J. Roy. Meteor. Soc. 106, 421-446.

Reed, R. J., and K. E. German, 1965: A contribution to the problem of stratospheric diffusion by large-scale mixing. Mon. Wea. Rev. 93, 313-321.

Rood, R. B., and M. R. Schoeberl, 1983: A mechanistic model of Eulerian, Lagrangian-mean, and Lagrangian ozone transport by steady planetary waves. J. Geophys. Res., 88, 5208-5218.

Strobel, D. F., 1981: Parameterization of linear wave chemical transport in planetary atmospheres by eddy diffusion. J. Geophys. Res. 86, 9806-9810.

Tung, K. K., 1982: On the two-dimensional transport of stratospheric trace gases in isentropic coordinates. J. Atmos. Sci. 39, 2230-2355.

Tung, K. K., and R. S. Lindzen, 1978a: A theory of stationary long waves. Part I: A simple theory of blocking, Mon. Wea. Rev. 107, 714-734.

-----, and -----, 1978b: A theory of stationary long waves. Part II: Resonant Rossby waves in the presence of realistic vertical shears. Mon. Wea. Rev. 107, 735-750.

van Loon, H., R. L. Jenne, K. Labitzke, 1973: Zonal harmonic standing waves. J. Geophys. Res. 78, 4463-4471.

van Loon, H., R. A. Madden, and R. L. Jenne: Oscillations in the winter stratosphere: Part I. Description. Mon. Wea. Rev. 103, 154-162.

J. R. Holton and T. Matsuno, Dynamics of the Middle Atmosphere, 445-464.

THE GLOBAL STRUCTURE OF THE TOTAL OZONE FLUCTUATIONS
OBSERVED ON THE TIME SCALES OF TWO TO SEVERAL YEARS

Fumio Hasebe

Geophysical Institute/Laboratory for Climatic Change Research
Kyoto University

ABSTRACT

Total ozone variations (1970-1977) are derived on a global scale
from Nimbus 4 BUV and ground-based observations. The fluctuations with
quasi-biennial and four-year periodicities are investigated.

The quasi-biennial component of the total ozone fluctuations shows
a roughly out-of-phase relation between the northern and southern
hemispheres with the phase reversal around 15°S. In the equatorial region, the phase of maxima and minima exhibit cross-equatorial northward
propagation with the maxima around the westerly phase of the quasi-
biennial zonal wind oscillation at 50 mb. An out-of-phase relation between middle and high latitudes is observed in the northern hemisphere.
The oscillation with period of about four years is almost symmetric with
respect to the equator showing an out-of-phase relation between tropics
and extratropics. Systematic phase progression in the western portion of
the southern hemisphere is also observed. For these oscillations,
planetary-scale total ozone waves of zonal wavenumber 1 in the winter
hemisphere seem to penetrate into the summer hemisphere.

The zonal mean characteristics of the total ozone fluctuations are
discussed in connection with the lower stratospheric temperature field.
Good correspondence between the ozone and temperature is obtained within
the availability of the data at the present. The quantitative reproduc-
tion of the four-year component is, however, left as a future problem.

1. INTRODUCTION

In the upper stratosphere, oxygen molecules are photodissociated by
solar ultraviolet radiation to form atomic oxygen. By the combination
reaction between atomic and molecular oxygen under the existence of an
ambient molecule, stratospheric ozone is produced. The ozone thus created is destroyed directly by recombination reactions with atomic oxygen
and indirectly through catalytic reactions with nitrogen and hydrogen
oxides (e.g., Chapman, 1930; Hunt, 1966; Crutzen, 1970). In the lower
stratosphere, on the other hand, not enough solar ultraviolet radiation

penetrates to photodissociate the oxygen molecule, and owing to the low temperature the chemical reactions are very slow as well. The photochemical life time of ozone is, therefore, so long that the ozone concentration could be regarded as a conservative tracer following the atmospheric motion. Thus the total ozone values with which we are concerned here are strongly affected by the stratospheric general circulation.

As for the interannual variations of global total ozone, previously reported are the quasi-biennial oscillation (QBO), the four-year oscillation (FYO) and those related to the solar cycle (e.g., Angell and Korshover, 1964; Wilcox et al., 1977; Hasebe, 1980). It has also been pointed out that the growth of human activities can possibly deplete the stratospheric ozone content (e.g., Johnston, 1971; Molina and Rowland, 1974; Cicerone et al., 1974; Crutzen, 1976). In the present article, we intend to show the detailed global structures of the QBO and the FYO as an extension of Hasebe (1983; hereafter referred to as HA). Since the total ozone variations are sensitive to the lower stratospheric circulation changes, such an investigation would be helpful for our understanding of some dynamical properties of the stratosphere on the time scales of two to several years.

The analysis is based on $10°$ latitude X $20°$ longitude grid point values; these are obtained by an objective analysis, a modified scheme of optimum interpolation, using simultaneously the Nimbus 4 BUV and ground-based observational data (see HA for the details). The period of the investigation is from April 1970 to May 1977, the full time span of the Nimbus 4 BUV experiment (Heath et al., 1973; Fleig et al., 1981). Missing observations of the BUV data especially after 1972 (e.g., Hilsenrath and Schlesinger, 1981) are fortunately recovered to a large extent by the ground-based observations. The results are presented separately for the QBO (section 2) and the FYO (section 3) by applying

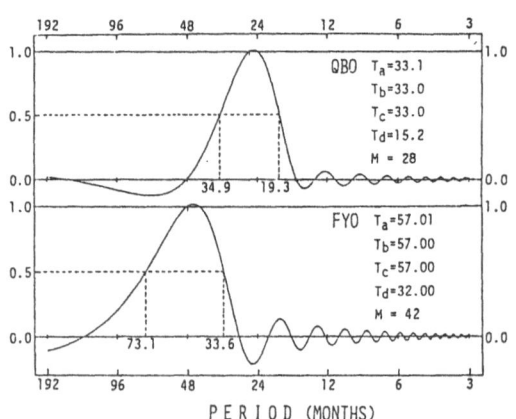

PERIOD (MONTHS)

Fig. 1. The frequency responses of the filters applied in the present analysis. The gain is normalized to unity at 26 and 48 month periods for the quasi-biennial oscillation (QBO) and the four-year oscillation (FYO), respectively. (After Hasebe, 1983.)

numerical filters, the frequency responses of which are shown in Fig. 1. The details for these filters appears in Hasebe (1980). The correspondence of such total ozone fluctuations to the lower stratospheric general circulation will be discussed in section 4.

2. QUASI-BIENNIAL OSCILLATION IN·TOTAL OZONE

Since the discovery of the quasi-biennial oscillation in total ozone (QBO) by Funk and Garnham (1962), many observational studies have been made in order to reveal its features on a global scale (e.g., Angell and Korshover, 1964, 1973; Wilcox et al., 1977; Hasebe, 1980; Tolson, 1981; Oltmans and London, 1982). Nowadays when the ozone observations from satellites are continuously made, reliable and detailed description of the QBO has become possible. Fig. 2 illustrates the quasi-biennial component of the total ozone fluctuations in the mean values of northern hemisphere (NH), southern hemisphere (SH), and globe (GL), and

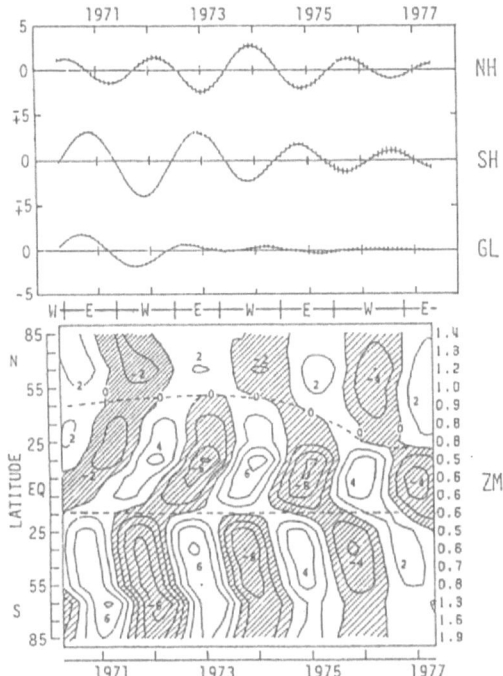

Fig. 2. Quasi-biennial oscillation of total ozone (matm-cm) in the mean values of (NH) northern hemisphere, (SH) southern hemisphere, and (GL) globe, and (ZM) zonal mean values. The isopleths in ZM are drawn with the interval of 2 matm-cm, and the shaded areas correspond to negative deviations. The tick mark is January of the given year. See the text for the letters E and W situated between GL and ZM. (After Hasebe, 1983.)

zonal mean values (ZM). The estimates of errors given by confidence limits of about 70 % (one S.D.) are indicated by vertical bars in NH, SH and GL, and by the numerals on the right column of ZM. Letters E and W situated between GL and ZM respectively denote the easterly and westerly phase of the equatorial zonal wind at 50 mb as given by Coy (1979). Exhibiting large variabilities in time and space, the structure of zonal mean QBO can be summarized as follows.

In the tropics (20°N–15°S), the positive deviations in total ozone nearly correspond to the westerly phase of the equatorial quasi-biennial zonal wind oscillation at 50 mb. However, the ozone QBO is asymmetric with respect to the equator in marked contrast with the zonal wind oscillation (e.g., Wallace, 1973). Separated by a major node at 15°S, the QBO propagates northward and southward with the phase being opposite. The northward propagation crosses the equator and continues to northern midlatitudes; the meridional phase velocity is relatively low in spring-summer and high in autumn-winter. In the northern high latitudes, the phase of the QBO is observed often to be opposite to that in the middle latitudes. The QBO in the southern midlatitudes shows a phase delay of about a half year to that in the northern midlatitudes. The northern and the southern hemispheric mean QBO are almost out-of-phase between each other in later years, and the global mean QBO exhibits no appreciable amplitude then.

In order to see the out-of-phase relation between the middle and high latitude total ozone in more detail, typical examples of the winter time spatial distributions of the QBO are presented in Fig. 3; (a) the northern hemisphere in February 1973 and (b) the southern hemisphere in August 1973. In both hemispheres, negative deviations in midlatitudes and positive ones in high latitudes are seen separated by 50–60° latitude; we can see that the zonal wavenumber 1 distributions in high latitudes are also remarkable. Considering the great importance of

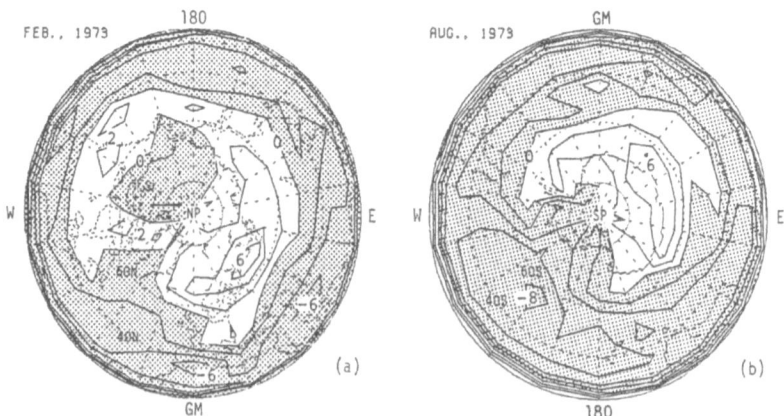

Fig. 3. Hemispheric patterns of the total ozone QBO for (a) February 1973 and (b) August 1973. The intervals of the isopleths is 2 matm-cm with the shade on the negative deviations.

stratospheric transport processes in the total ozone fluctuations, it is suggested that some dynamical properties must have changed to cause these fluctuations. The seesaw pattern revealed in the lower stratospheric geopotential height field (Holton and Tan, 1980, 1982) is evidence for such circulation changes.

The fact that zonal wavenumber 1 is dominant in high latitudes suggests that the high latitude QBO is strongly associated with the modulation of planetary waves of wavenumber 1. Actually the composite analysis of Holton and Tan (1980) revealed that such a wave at 50 mb 60-70°N in November and December is 40 % stronger in the easterly phase than in the westerly phase of equatorial zonal wind. As the total ozone waves have good negative correlation with the geopotential height waves in midlatitudes (Miller et al., 1979), investigation of the ozone waves of the QBO is also interesting. The time series of the amplitude (continuous lines) and phase (dots) of the total ozone QBO are shown together with those of the zonal mean values in Fig. 4. In each diagram, the amplitude is normalized by the maximum deviation of the zonal mean values. Considering the reduced satellite coverage in later years, we will focus on the early 1970's. Then the following characteristics may be noted.

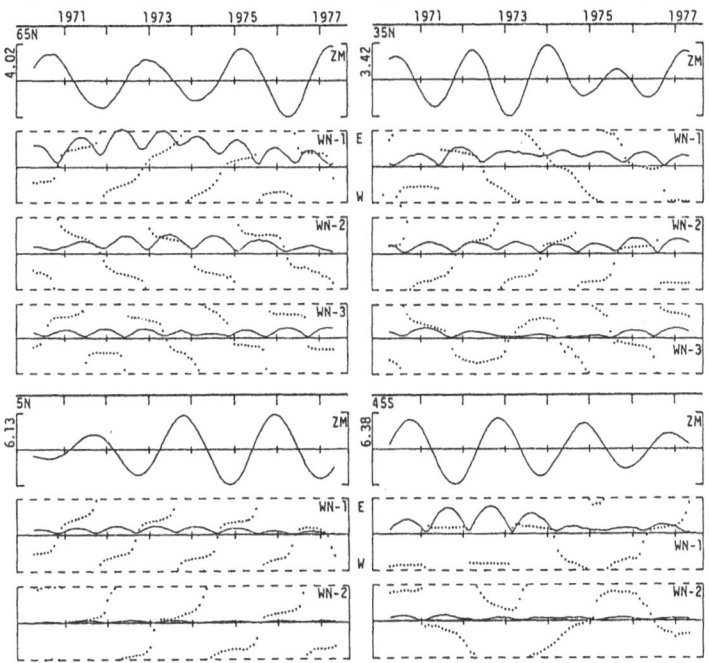

Fig. 4. The time series of zonal wave component of the QBO represented by amplitude (continuous line) and phase (dot) together with those of zonal mean values (ZM). Each diagram is scaled by the maximum deviation of ZM as indicated on the left end. The phase ranges from 0 to π for eastward (+) and westward (-) on the same scale.

The oscillations are quasi-standing especially for wavenumber 1. At the northern high latitude (65°N), it is not autumn-winter but summer (June-July) that the amplitudes of wavenumbers 1 and 2 have maxima. Following the intensification of the ozone ridge (trough) of wavenumber 1 around 90°W, zonal mean values take maxima (minima) in November-December. Thus, the deviations around 90°W associated with zonal wavenumber 1 agree qualitatively with the time derivative of zonal mean values. At the midlatitude (35°N), the maximum amplitude of wavenumber 1 appears in November-December. The maxima and minima of zonal mean values take place a few months later. The equatorial region (5°N) has very small amplitudes of wave components especially for wavenumbers greater than 1. In the southern hemisphere, wavenumber 1 is the dominant wave component. Its amplitude is maximum in August; maxima and minima of the zonal mean values are observed in October-November. Thus, the maxima of zonal mean values are attained soon after the maximum positive deviation around 140°W.

The ozone fluctuations thus described must be reflections of some modifications of the dynamical processes in the stratosphere. A brief speculation for such processes will be attempted in section 4.

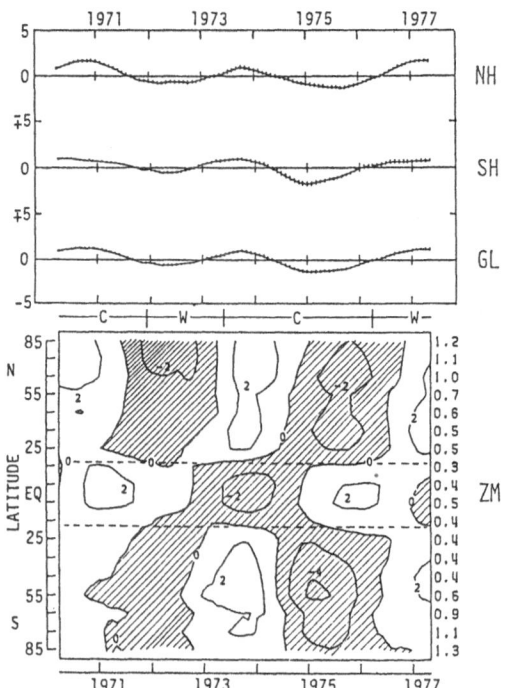

Fig. 5. The same as Fig. 2 but for four-year oscillation. See the text for the letters C and W situated between GL and ZM. (After Hasebe, 1983.)

3. FOUR-YEAR OSCILLATION IN TOTAL OZONE

Among the various time scales of interannual variations of total ozone, a periodicity of four years was noticed from the ground-based network data of fifteen years (Hasebe, 1980). This oscillation was observed in the northern mid- and high latitudes with the phase of maxima in early 1962, 1966, 1970 and 1974. Quite recently, HA has shown by adding the Nimbus 4 BUV data that this oscillation is not restricted to the northern hemisphere but extends over the globe. We call this the four-year oscillation (FYO) for the sake of simplicity. In this spectral region, the period from three to five years is known as a nearly periodic variation of the sea surface temperature in the equatorial eastern Pacific (e.g., Bjerknes, 1966). A comparative study by Angell (1981) indicates that total ozone over North America tends to be a maximum two seasons after the warmest sea surface temperature in that region. In the present article, the evidence of the FYO is put forward following HA with the addition of new results obtained up to the present.

The total ozone fluctuations with period of about four years, obtained by applying the numerical filter (Fig. 1(FYO)), are shown with statistical significance in Fig. 5. This is the same as Fig. 2 except for the FYO instead of the QBO and with sea surface temperature in place of the equatorial zonal wind. Letters C and W respectively indicate the cold and warm anomalies of sea surface temperatures in the region 0-10°S, 180-80°W taken from Angell (1981).

Almost symmetric behavior with respect to the equator is the main characteristic of the FYO. The phase in the equatorial region is opposite to those in mid- and high latitudes with the nodes around 20°N and S. The hemispheric and global mean FYO are seen to be nearly in phase with those of mid- and high latitudes. Though the lag of two seasons between the sea surface temperature and the northern midlatitude total ozone (Angell, 1981) is not apparent, there is possibly a tendency for warm sea surface temperatures to be followed by tropical low and extratropical high ozone in both hemispheres. Similar correspondence of the northern midlatitude FYO to the sea surface temperature could be noticed

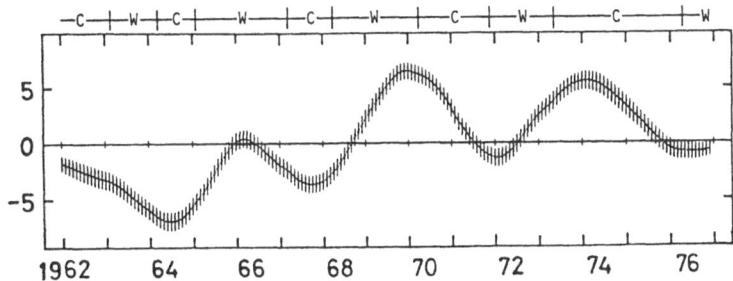

Fig. 6. The long-term variations of zonal mean total ozone for 55°N (after Hasebe, 1980) together with the equatorial sea surface temperature anomaly given by Angell (1981).

in much longer time series derived from the ground-based observations (Fig. 6); here the comparison is made only for the northern midlatitude considering the distribution of observing stations (Hasebe, 1980).

Both the out-of-phase distribution between tropics and extratropics and the phase relation with the equatorial sea surface temperatures suggest that the FYO might be thermally driven by sea surface temperature changes. That is, the warmer sea surface temperature would activate the tropical Hadley circulation; the enhanced upward motion in the tropical stratosphere would lead to tropical low ozone and the stronger Brewer-Dobson circulation lead to extratropical high ozone. Actually some evidence of tropospheric and stratospheric responses to the equatorial sea surface temperature changes have been obtained (e.g., Horel and Wallace, 1981; van Loon et al., 1982).

It would be helpful for our understanding of the FYO to see its spatial distribution. Fig. 7 illustrates typical pictures of such distributions; (a) October 1971 and (b) October 1972. In October 1971, the tropical zonal belt between 20°N and S is covered with positive deviations, while the extratropics show negative ones except for small areas in high latitudes. One year later, the tropical positive belt is cut off by negative deviations around 140°W. This overturning process is remarkably different from the zonally simultaneous phase changes of the QBO. See HA for a series of global patterns related to the FYO and the QBO.

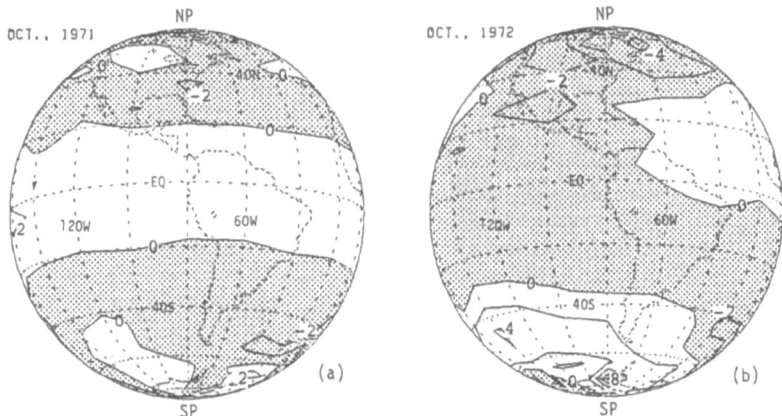

Fig. 7. Hemispheric patterns of the total ozone FYO for (a) October 1971 and (b) October 1972. The interval of the isopleths is 2 matm-cm with the shade on the negative deviations.

The structure of the zonal wave component, similar to Fig. 4 for the QBO, is shown in Fig. 8. In the equatorial region and the southern hemisphere, wavenumbers greater than 2 are very small so that they are not shown. The location of the quasi-standing ozone ridge and trough of wavenumber 1 is near 140°W at 55°S; there is a tendency that the growth of zonal mean value follows the intensification of the ozone ridge

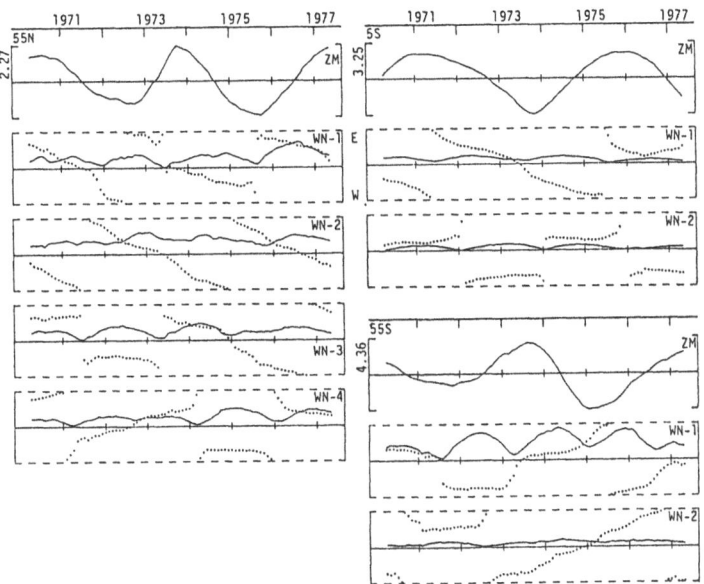

Fig. 8. The same as Fig. 4 but for the FYO.

around 140°W. These features agree well with those of the QBO. In the equatorial region, the amplitude of wavenumber 1 is a maximum near the time of zero deviation of zonal mean values as expected from Fig. 7. In the northern hemisphere, wavenumbers 3 and 4 have large amplitude comparable to those of wavenumbers 1 and 2. The standing oscillation is not so evident except for wavenumber 3. This may imply that the extratropical FYO is associated with relatively lower altitude circulation changes. The driving processes of the FYO is discussed in relation to the lower stratospheric temperature changes in the next section.

4. TOWARD UNDERSTANDING OF THE MECHANISMS

Among the various features of the QBO and the FYO described above, we will focus our discussion on the zonal mean characteristics of the low latitude QBO and the extratropical FYO.

4-1. Low latitude QBO

The dynamic processes of the tropical quasi-biennial oscillation in the stratospheric zonal wind are among those best understood (Lindzen and Holton, 1968; Holton and Lindzen, 1972; Plumb, 1977; see also the excellent review by Plumb in this issue). In this subsection, the total ozone fluctuations in low latitudes originating from the zonal wind QBO are discussed.

454

The vertical wind shear associated with the downward propagating westerly and easterly wind regimes is accompanied by a horizontal temperature gradient satisfying the thermal wind relation. Fig. 9 shows the variations in the meridional gradient of lower stratospheric (approximately 20-100 mb) zonal mean temperature observed by Nimbus 5 SCR(Ch-B4); triangles with thin lines are for 0° minus 12°N, crosses with heavy lines are for 0° minus 12°S. See Barnett et al. (1975) for the weighting function of the SCR. Since only the data for two years are available, these values are derived by subtracting the two-year-mean value for each month. The smoothed curves are the sinusoidal fits of biennial Fourier components. E and W at the top of the diagram indicate equatorial easterly and westerly winds at 50 mb as in Fig. 2. Remembering the downward phase propagation, the warm (cold) anomaly for westerly (easterly) shear would be confirmed.

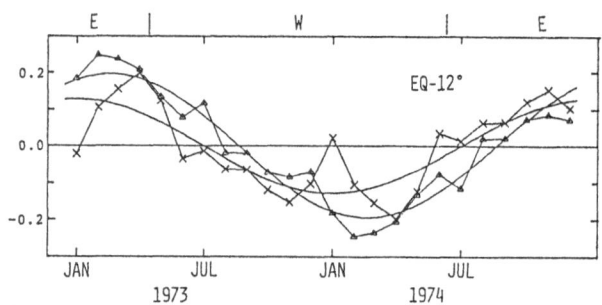

Fig. 9. The biennial component in the meridional gradient (equator minus 12°N and S) of lower stratospheric temperature observed by Nimbus 5 SCR(Ch-B4). See the text for the details.

On the time scale of the quasi-biennial oscillation, these temperature anomalies suffer from radiative damping by infra-red cooling (Holton and Lindzen, 1972). This process can be modeled in the form of Newtonian heating/cooling as follows:

$$\frac{\overline{J_B}}{C_p} = -h \cdot \overline{T_B}, \tag{1}$$

where J_B and T_B are the biennial component of diabatic heating rate per unit mass and temperature, respectively, C_p is the specific heat at constant pressure, h is the Newtonian cooling coefficient, and the over bar represents (Eulerian) zonal mean. The diabatic heating (cooling) in the cold (warm) region drives upward (downward) motion as originally described in Reed (1964). Then, from the thermodynamic equation after Plumb and Bell (1982), we have

$$N^2 \overline{w}^L = \frac{R}{C_p H} \overline{J}, \tag{2}$$

where N^2 is the buoyancy frequency squared, $\overline{w}^L$ is the Lagrangian-mean vertical velocity, R is the gas constant, and H is the scale height.

Such a vertical current is actually obtained in model calculations (e.g., Takahashi and Uryu, 1981; Plumb and Bell, 1982). On the other hand, the time change of a vertically well stratified ozone mass mixing ratio χ is expressed by

$$\frac{\partial \overline{\chi}}{\partial t} = -\overline{w}^L \frac{d\chi_0}{dz} + \frac{\partial}{\partial y}\left(K_{yy}\frac{\partial \overline{\chi}}{\partial y}\right), \tag{3}$$

where y is the northward distance, z is a vertical coordinate, $\chi_0(z)$ is a basic state ozone mass mixing ratio, and K_{yy} is a diffusion coefficient (Holton, 1980). Then the variation of total ozone X is obtained by the vertical integration of eq. (3):

$$\frac{\partial \overline{X}}{\partial t} = \int_{p_1}^{p_2} \frac{\partial \overline{\chi}}{\partial t}\, dp. \tag{4}$$

Combining eqs. (1)-(4), we can estimate the total ozone fluctuations associated with the temperature anomaly depicted in Fig. 9. In this calculation, the second term on the right hand side of eq. (3) is neglected for the first approximation, and χ_0 is assumed to be a linear function of z. As for the $\overline{T_B}$ of each latitude, the biennial Fourier component of the deviations from the mean temperature of 20°N and S is used. This is because (1) the raw temperature field is so strongly influenced by the stratospheric sudden warming of January-February 1973 that it is necessary to eliminate that effect, and (2) the meridional extension of the tropical zonal wind QBO is limited within 20°N and S (Wallace, 1973). The constants used are h = $3 \times 10^{-7} s^{-1}$, H = 7 km, N^2 = $4 \times 10^{-4} s^{-2}$, $d\chi_0/dz = 2$ ($\mu g/g$)km^{-1}, p_1= 20 mb, and p_2= 100 mb. The ozone changes obtained are shown in Fig. 10 together with the observed ones. For comparison, the biennial component of the observed total ozone is re-calculated from the non-filtered values by Fourier expansion. In the equatorial region, very good agreement between the observed and expected total ozone changes is found. However, the meridional phase propagation

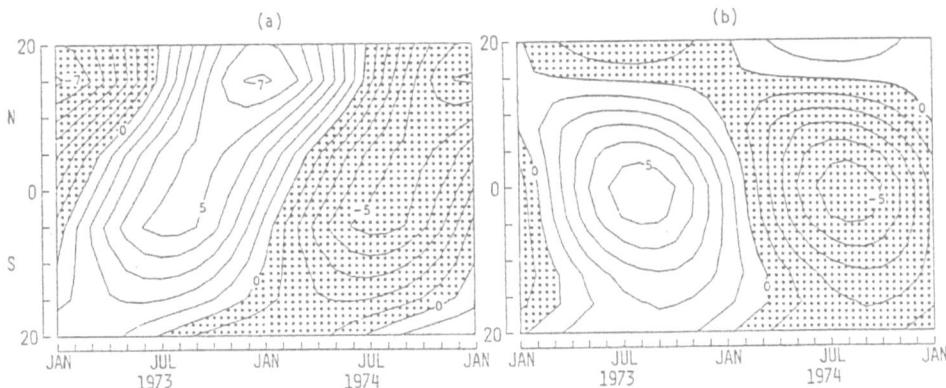

Fig. 10. The biennial oscillations in low latitude total ozone (matm-cm); (a) observed and (b) expected from eqs. (1)-(4). Negative deviations are shaded.

starting from 15°S and crossing over the equator in the observed total ozone is not depicted in Fig. 10(b). Note that the location of the phase reversal in Fig. 10(b) strongly depends on the definition of $\overline{T_B}$.

The cross equatorial phase propagation may be attributable to the transiency of planetary waves which penetrate from the midlatitudes in the winter hemisphere. Fig. 11 shows the global patterns of (a) deviations from the zonal mean of QBO filtered total ozone and (b) non-filtered lower stratospheric temperature of wavenumber 1 observed from Nimbus 5 SCR(Ch-B4) both in January 1973. In both diagrams, a well organized wavenumber 1 distribution is shown crossing the equator from the northern midlatitudes. Cross equatorial waves similar to Fig. 11 are found also for the FYO (see HA for the details) and stratospheric temperature in spring (Barnett, 1975; Hirota, 1976).

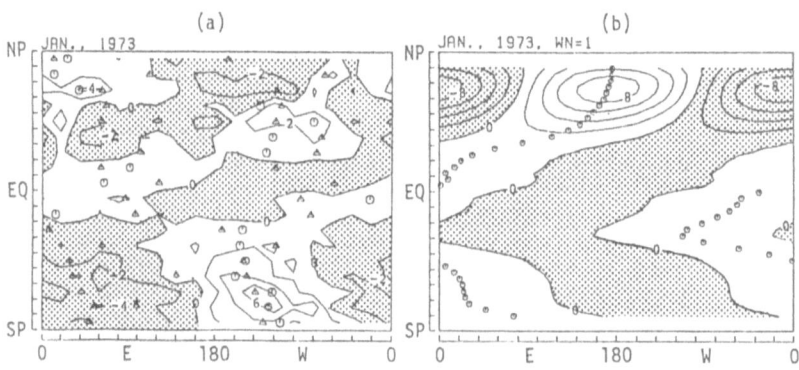

Fig. 11. The global patterns of the zonal wave component of (a) total ozone QBO and (b) monthly mean lower stratospheric temperature (non-filtered) both in January 1973.

The influence of the equatorial quasi-biennial oscillation in the stratospheric zonal wind on the extratropical circulations has been studied extensively. The evidence so far found includes the seesaw pattern in the lower stratospheric geopotential height field, high latitude wave activity (Holton and Tan, 1980, 1982) and the occurrence of stratospheric sudden warmings (Labitzke, 1982). The basic idea for such couplings is that the vertical and meridional propagation of stationary planetary waves in the winter hemisphere should depend on the zero wind critical surface associated with the easterly and westerly phase of the equatorial zonal wind (Holton and Tan, 1982). The scenario including the ozone changes would be as follows:

During the easterly phase of the equatorial zonal wind, tropospheric forcing is concentrated at latitudes higher than the zero wind line in midlatitudes. The stronger forcing brings about earlier and larger amplification of wavenumber 1 in the winter that follows; this drives the midwinter sudden warming. The amplification of the planetary wave, which is so strong as to extend the wave over the equator, is accompanied by poleward transport of ozone (and also potential temperature) in excess of the equatorward transport by the mean meridional circulation. Occasionally an easterly maximum at 50 mb level, together with a negative deviation in total ozone, occur in the northern winter, so that the phase of the QBO shows cross-equatorial northward propagation as far as northern midlatitudes. Owing to the excessive transport by the waves, total ozone and temperature in high latitudes exhibit positive deviations; the out-of-phase relation between middle and high latitudes (the seesaw pattern) of total ozone and temperature is thus realized.

From the features of the ozone variations described in section 2, we infer that the circumstances in the southern hemisphere may be more or less different from those of the northern hemisphere. The node at 15°S might be attributable to the counter current of the equatorial vertical motion. In middle and high latitudes, wave amplification associated with the equatorial zonal wind variation takes place in the winter a half year later than the northern winter. Owing to the weaker tropospheric forcing and/or the miss-matching between the equatorial zonal wind and the extratropical seasonal march, the distinct out-of-phase relation in ozone (and probably in temperature) does not take place, although a similar tendency is observed (Fig. 3).

4-2. Extratropical FYO

For extratropical tracer transport, a near cancellation between the Eulerian-mean meridional circulation and the eddy transport is well demonstrated (e.g., Hunt and Manabe, 1968; Mahlman and Moxim, 1978; Miller et al., 1977). The difference between the Eulerian-mean circulation and the classical Brewer-Dobson circulation comes from the fact that the latter view does not include the idea of waves which is the central concept of the former. Recently it has been noted that the Brewer-Dobson circulation is equivalent to the Lagrangian-mean circulation (e.g., Dunkerton, 1978; Matsuno, 1980), and that we can get a much better view by this formalism (e.g., Kida, 1983). These ideas were initiated by Andrews and McIntyre (1976) who defined a residual mean circulation by eliminating the indirect circulation driven by the eddy heat transport from the Eulerian-mean meridional circulation. The residual circulation, which approximately agrees with the Lagrangian-mean circulation even in the presence of waves, is very convenient for treating tracer transport problems.

In section 3, we show that the meridional structure of the FYO is characterized by an out-of-phase relation between the tropics and extratropics. Such a simple structure, as compared with that of the QBO suggesting strongly the important role of planetary waves, makes us feel that the FYO is a reflection of the modulation of the Brewer-Dobson circulation by sea surface temperature changes. In this case, the transiency of the waves would play only a minor role and the $\overline{w}^L$ in eq. (3)

may be replaced by $\overline{w}^*$, the residual vertical velocity, for the first approximation. Thus, we have

$$\frac{\partial \overline{X}}{\partial t} = -\overline{w}^* \frac{dX_0}{dz} .$$ (5)

In this subsection, the total ozone fluctuations as expected from eq. (5) are discussed and compared with the observed features.

From the thermodynamic equation, Palmer (1981) obtained the approximate expression of $\overline{w}^*$:

$$\frac{\partial \overline{\theta}}{\partial t} = -\frac{\partial \overline{\theta}}{\partial z} \overline{w}^* + \frac{\overline{J}}{C_p},$$ (6)

where θ is the potential temperature. The diabatic heating would be again expressed by the Newtonian heating/cooling as

$$\frac{\overline{J}}{C_p} = -h(\overline{\theta} - \theta^*),$$ (7)

where θ^* is the zonal mean equilibrium potential temperature. With the aid of eqs. (6) and (7) together with

$$\frac{\partial \overline{\theta}}{\partial z} = \frac{N^2}{g} \overline{\theta},$$ (8)

$\overline{w}^*$ is expressed by

$$\overline{w}^* = -\frac{g}{N^2} \left\{ h(1 - \frac{\theta^*}{\overline{\theta}}) + \frac{\partial \ell n \overline{\theta}}{\partial t} \right\},$$ (9)

where g is the acceleration due to gravity. This calculation is made on a daily basis where h is a function of height alone while θ^* is a function of latitude, height and time (day of the year); for both h and θ^* the expressions of Trenberth (1973) are adopted. For the estimation of total ozone fluctuations, the anomaly of monthly mean $\overline{w}^*$, that is, the deviation from the 16 year mean for each calender month, is used in eq. (5). For dX_0/dz, the latitude dependence is introduced as is given in Table 1. Finally the vertical integration in eq. (4) is done using the 7 height levels 300, 200, 150, 100, 50, 30 and 10 mb.

Table 1. The vertical gradient of the basic state ozone mass mixing ratio dX_0/dz ($\mu g/g \cdot km^{-1}$). These are estimated referring to Hering and Borden (1964).

height (mb)	latitude (N)												
	90-80	75	70	65	60	55	50	45	40	35	30	25	20
10	0	0	0	0	0	0	0	0	0	0	0	0.1	0.2
30	0	0	0	0	0	0.1	0.2	0.4	0.6	1.0	1.3	1.5	1.7
50	0.2	0.4	0.5	0.6	0.7	0.8	0.8	0.9	1.0	1.2	1.4	1.4	1.4
100	0.5	0.6	0.6	0.7	0.7	0.6	0.6	0.5	0.5	0.4	0.3	0.2	0.2
150	0.5	0.6	0.6	0.5	0.5	0.4	0.4	0.3	0.2	0.1	0.1	0	0
200	0.2	0.3	0.3	0.4	0.4	0.3	0.3	0.2	0.1	0	0	0	0
300	0.3	0.3	0.3	0.2	0.1	0	0	0	0	0	0	0	0

[matm-cm]

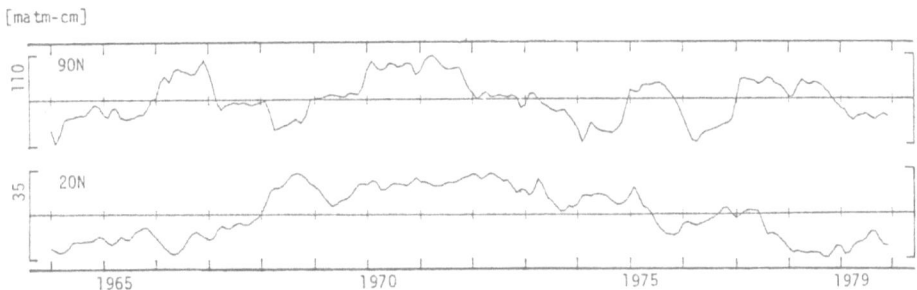

Fig. 12. Fluctuations of total ozone anomalies (matm-cm) for 90°N and 20°N as expected from the computed residual vertical velocity changes (non-filtered).

The data used here are the NMC daily temperature data of 16 years (1964-1979). Unfortunately, the spatial coverage of the data is limited to the region north of 20°N. The values obtained for monthly mean $\overline{w}^*$ are on the order of -0.1 to -1 mm·s^{-1} at 100 mb at mid- and high latitudes in January. These values are in a reasonable range compared with the $\overline{w}^L$ derived from the hemispheric GCM of Kida (1983). The examples of total ozone fluctuations are shown as the time series of monthly mean values for 90°N and 20°N in Fig. 12. It is interesting to see that the periodicity near four years is resolved in both of the time series with a tendency for an out-of-phase relation between them. For the sake of comparison, a time-latitude cross section filtered for the FYO, equivalent to Fig. 5(ZM), is shown in Fig. 13. Though the calculation is limited to north of 20°N, a phase progression similar to Fig. 5(ZM) can be recognized. It is readily seen, however, that the amplitude obtained is very large and that most of the large amplitude for 90°N is contributed by an

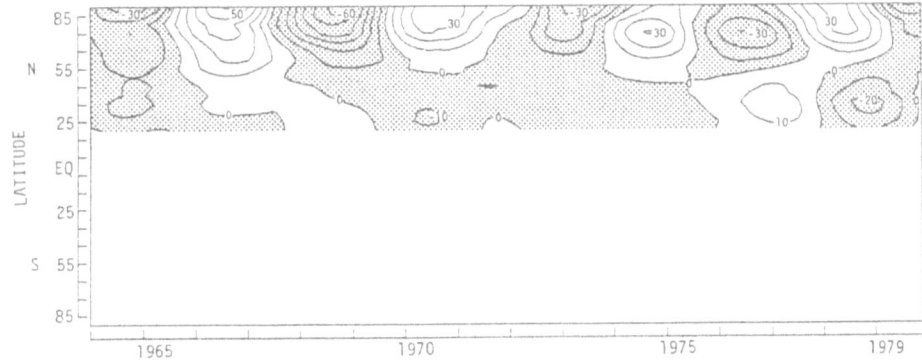

Fig. 13. The FYO in total ozone for latitudes north of 20°N as expected from the computed residual vertical velocity changes (matm-cm).

abrupt change in winter like a step function. Possibilities for such large amplitude are (1) the model does not include any ozone loss process, (2) $\bar{w}^*$ may be considerably different from $\bar{w}^L$ in winter when the wave transiency is not neglected, and (3) the basic state ozone profile may significantly suffer from the enhanced vertical motion in winter. It is apparent that a precise expression of the winter time circulation is required in order to obtain quantitatively reliable ozone fluctuations.

5. CONCLUDING REMARKS

The quasi-biennial and four-year oscillations in total ozone (QBO and FYO), separated with the aid of numerical filters, are investigated by using the Nimbus 4 BUV and ground-based network data. Owing to the improved observing density especially in the tropics and the southern hemisphere, the global structure of the oscillations are revealed in much more detail than before.

Total ozone fluctuations thus obtained are discussed in relation to the lower stratospheric temperature field. The equatorial total ozone QBO is theoretically consistent with the observed temperature QBO. There is a possibility that the cross-equatorial phase propagation and the extratropical structure of the total ozone QBO could be explained by the modulation of tropospheric forcing associated with the location of zero wind critical surface. The FYO could be qualitatively reproduced from the lower stratospheric temperature changes in the northern mid- and high latitudes. However, much more precise treatment is required in order to reproduce quantitatively acceptable ozone changes. In addition, temperature observations for several years in the equatorial region and the southern hemisphere of the lower stratosphere are needed for the further understanding of the FYO.

ACKNOWLEDGEMENTS

The author would like to thank Profs. J. R. Holton, University of Washington, T. Matsuno, University of Tokyo, and I. Hirota, Kyoto University, for giving him an opportunity to participate in such a stimulating meeting. He is also grateful for their helpful comments and discussions. He also wishes to thank Dr. J. C. Gille, NCAR, for discussions at the meeting. Thanks are also due to Prof. R. Yamamoto, Kyoto University, for his critical reading of the manuscript. The author is also indebted to Mr. M. Shiotani for the treatment of Nimbus 5 SCR data and Mr. H. Nasuda for his skillful assistance at the presentations in spherical geometry.

REFERENCES

Andrews, D. G., and M. E. McIntyre, 1976: Planetary waves in horizontal and vertical shear: The generalized Eliassen-Palm relation and the mean zonal acceleration. J. Atmos. Sci., 33, 2031-2048.

Angell, J. K., 1981: Comparison of variations in atmospheric quantities with sea surface temperature variations in the equatorial eastern Pacific. Mon. Wea. Rev., 109, 230-243.

---- and J. Korshover, 1964: Quasi-biennial variations in temperature, total ozone, and tropopause height. J. Atmos. Sci., 21, 479-492.

---- and ----, 1973: Quasi-biennial and long-term fluctuations in total ozone. Mon. Wea. Rev., 101, 426-443.

Barnett, J. J., 1975: Hemispheric coupling - evidence of a cross-equatorial planetary wave-guide in the stratosphere. Quart. J. Roy. Meteor. Soc., 101, 835-845.

----, R. S. Harwood, J. T. Houghton, C. G. Morgan, C. D. Rodgers and E. J. Williamson, 1975: Comparison between radiosonde, rocketsonde, and satellite observations of atmospheric temperatures. Quart. J. Roy. Meteor. Soc., 101, 423-436.

Bjerknes, J., 1966: A possible response of the atmospheric Hadley circulation to equatorial anomalies of ocean temperature. Tellus, 18, 820-829.

Chapman, S., 1930: A theory of upper-atmospheric ozone. Mem. Roy. Meteor. Soc., 3, 103-125.

Cicerone, R. J., R. S. Stolarski, and S. Walters, 1974: Stratospheric ozone destruction by man-made chlorofluoromethanes. Science, 185, 1165-1167.

Coy, L., 1979: An unusually large westerly amplitude of the quasi-biennial oscillation. J. Atmos. Sci., 36, 174-176.

Crutzen, P. J., 1970: The influence of nitrogen oxides on the atmospheric ozone content. Quart. J. Roy. Meteor. Soc., 96, 320-325.

----, 1976: Upper limits on atmospheric ozone reductions following increased application of fixed nitrogen to the soil. Geophys. Res. Lett., 3, 169-172.

Dunkerton, T., 1978: On the mean meridional mass motions of the stratosphere and mesosphere. J. Atmos. Sci., 35, 2325-2333.

Fleig, A. J.,   V. G. Kaveeshwar,   K. F. Klenk,   M. R. Hinman,
P. K. Bhartia, and P. M. Smith, 1981: Characteristics of space and
ground   based   total   ozone   observing   systems   investigated   by
intercomparison of Nimbus 4 backscattered ultraviolet (BUV) data with
Dobson and M-83 results. Proc. Quadrennial Int. Ozone Symp., Boulder,
Colo., August 1980, 9-16.

Funk, J. P., and G. L. Garnham, 1962: Australian ozone observations and
a suggested 24 month cycle. Tellus, 14, 378-382.

Hasebe, F., 1980: A global analysis of the fluctuation of total ozone,
II. Non-stationary annual oscillation, quasi-biennial oscillation,
and long-term variations in total ozone. J. Meteor. Soc. Japan, 58,
104-117.

----, 1983: Interannual variations of global total ozone revealed from
Nimbus 4 BUV and ground-based observations. J. Geophys. Res., 88, in
press.

Heath, D. F.,   C. L. Mateer,   and A. J. Krueger, 1973: The Nimbus-4
backscatter ultraviolet (BUV) atmospheric ozone experiment - two
years' operation. Pure Appl. Geophys., 106-108, 1238-1253.

Hering, W. S., and T. R. Borden, 1964: Ozonesonde observations over
North   America, vol. 2, Environmental Research Papers   No. 38,
AFCRL-64-30(II), Air Force Cambridge Research Laboratories.

Hilsenrath, E., and B. M. Schlesinger, 1981: Total ozone seasonal and
interannual variations derived from the 7 year Nimbus-4 BUV data set.
J. Geophys. Res., 86, 12087-12096.

Hirota, I.,   1976: Seasonal variation of planetary waves in the
stratosphere   observed   by   the   Nimbus   5   SCR.
Quart. J. Roy. Meteor. Soc., 102, 757-770.

Holton, J. R., 1980: Wave propagation and transport in the middle
atmosphere. Phil. Trans. Roy. Soc. London, A296, 73-85.

---- and R. S. Lindzen, 1972: An updated theory for the quasi-biennial
cycle of the tropical stratosphere. J. Atmos. Sci., 29, 1076-1080.

---- and H-C. Tan, 1980: The influence of the equatorial quasi-biennial
oscillation on the global circulation at 50 mb. J. Atmos. Sci., 37,
2200-2208.

---- and ----, 1982: The quasi-biennial oscillation in the northern
hemisphere lower stratosphere. J. Meteor. Soc. Japan, 60, 140-148.

Horel, J. D., and J. M. Wallace, 1981: Planetary-scale atmospheric
phenomena associated with the southern oscillation. Mon. Wea. Rev.,
109, 813-829.

Hunt, B. G., 1966: The need for a modified photochemical theory of the ozonosphere. J. Atmos. Sci., 23, 88-95.

---- and S. Manabe, 1968: Experiments with a stratospheric general circulation model, II. Large-scale diffusion of tracers in the stratosphere. Mon. Wea. Rev., 96, 503-539.

Johnston, H. S., 1971: Reduction of stratospheric ozone by nitrogen oxide catalysts from supersonic transport exhaust. Science, 173, 517-522.

Kida, H., 1983: General circulation of air parcels and transport characteristics derived from a hemispheric GCM, Part 1. A determination of advective mass flow in the lower stratosphere. J. Meteor. Soc. Japan, 61, 171-187.

Labitzke, K., 1982: On the interannual variability of the middle stratosphere during the northern winters. J. Meteor. Soc. Japan, 60, 124-139.

Lindzen, R. S., and J. R. Holton, 1968: A theory of the quasi-biennial oscillation. J. Atmos. Sci., 25, 1095-1107.

Mahlman, J. D., and W. J. Moxim, 1978: Tracer simulation using a global general circulation model: Results from a midlatitude instantaneous source experiment. J. Atmos. Sci., 35, 1340-1374.

Matsuno, T., 1980: Lagrangian motion of air parcels in the stratosphere in the presence of planetary waves. Pure Appl. Geophys., 118, 189-216.

Miller, A. J., R. M. Nagatani, K. B. Labitzke, E. Klinker, K. Rose, and D. F. Heath, 1977: Stratospheric ozone transport during the mid-winter warming of December 1970-January 1971. Proc. Joint Symp. Atmos. Ozone, ed. K. H. Grasnick, Academy of Sciences of German Democratic Republic, Berlin.

----, R. M. Nagatani, J. D. Laver, and B. Korty, 1979: Utilization of 100 mb midlatitude height fields as an indicator of sampling effects on total ozone variations. Mon. Wea. Rev., 107, 782-787.

Molina, M. J., and F. S. Rowland, 1974: Stratospheric sink for chlorofluoromethanes; chlorine atom-catalyzed destruction of ozone. Nature, 249, 810-812.

Oltmans, S. J., and J. London, 1982: The quasi-biennial oscillation in atmospheric ozone. J. Geophys. Res., 87, 8981-8989.

Palmer, T. N., 1981: Diagnostic study of a wavenumber-2 stratospheric sudden warming in a transformed Eulerian-mean formalism. J. Atmos. Sci., 38, 844-855.

Plumb, R. A., 1977: The interaction of two internal waves with the mean flow: implications for the theory of the quasi-biennial oscillation. J. Atmos. Sci., 34, 1847-1858.

---- and R. C. Bell, 1982: A model of the quasi-biennial oscillation on an equatorial beta-plane. Quart. J. Roy. Meteor. Soc., 108, 335-352.

Reed, R. J., 1964: A tentative model of the 26-month oscillation in tropical latitudes. Quart. J. Roy. Meteor. Soc., 90, 441-466.

Takahashi, M., and M. Uryu, 1981: The Lagrangian-mean motions forced by steady, dissipating equatorial waves: Part 1. J. Meteor. Soc. Japan, 59, 781-800.

Tolson, R. H., 1981: Spatial and temporal variations of monthly mean to-tal columnar ozone derived from 7 years of BUV data. J. Geophys. Res., 86, 7312-7330.

Trenberth, K. E., 1973: Global model of the general circulation of the atmosphere below 75 kilometers with an annual heating cycle. Mon. Wea. Rev., 101, 287-305.

van Loon, H., C. S. Zerefos, and C. C. Repapis, 1982: The southern oscillation in the stratosphere. Mon. Wea. Rev., 110, 225-229.

Wallace, J. M., 1973: General circulation of the tropical lower stratosphere. Rev. Geophys. Space Phys., 11, 191-222.

Wilcox, R. W., G. D. Nastrom, and A. D. Belmont, 1977: Periodic varia-tions of total ozone and of its vertical distribution. J. Appl. Meteor., 16, 290-298.

MODELING

J. R. Holton and T. Matsuno, Dynamics of the Middle Atmosphere, 467-500.
Copyright © 1984 by Terra Scientific Publishing Company.

MODELING THE MIDDLE ATMOSPHERE CIRCULATION*

Marvin A. Geller

NASA/Goddard Space Flight Center
Laboratory for Planetary Atmospheres

ABSTRACT

Some of the motivations for constructing models of the middle
atmosphere circulation are given. These are as follows: (1) to provide
a better understanding of middle atmosphere dynamics; (2) to study the
coupling of middle atmosphere dynamics with radiation and chemistry; (3)
to study the sensitivity of tropospheric climate modeling and/or weather
forecasting to changes in the middle atmosphere; (4) to better understand
the limitations of more simplified models; (5) to supply a proxy for
atmospheric data for diagnostic analysis; and finally, (6) for forecast-
analysis of data. Different types of models are discussed in relation
to their anticipated use. Various model simplifications, such as using
the quasi-geostrophic set of equations and simplified radiative transfer,
are discussed as are some of the consequences of these simplifications.
Some of the accomplishments of middle atmosphere circulation modeling
are presented as are some of the difficulties in existing models.
Finally, some of the problems in constructing and verifying middle
atmosphere circulation models are discussed.

1. INTRODUCTION

There are several groups in the world today that are involved in
constructing models of the middle atmosphere circulation. A variety
of model structures are being used by these groups, and this diversity
in models reflects, in many cases, the different types of studies for
which the models are intended. In this introduction, we will briefly
review several reasons for constructing models of the middle atmosphere
circulation.

One reason may be stated in rather idealistic basic research terms.
It is to provide a better understanding of middle atmosphere dynamics.
One way of providing this understanding is to construct models to test
physical hypotheses in a quantitative manner. Matsuno's (1971) classic

*Contribution Number 6 of the Stratospheric General Circulation with
 Chemistry Project at NASA/Goddard Space Flight Center

paper used a model to illustrate the role of transient forcing of stationary planetary waves in giving rise to sudden stratospheric warmings. Another application of using models to advance our understanding of middle atmospheric dynamics is to use model results as "perfect data" for diagnostic studies. For instance, Dunkerton et al. (1981) performed very revealing diagnostic studies on an application of Holton's stratospheric warming model that was reported by Hsu (1980) for planetary wavenumber two forcing.

Another application of middle atmosphere circulation modeling is to study the coupling of middle atmosphere dynamics with radiative and chemical processes. Recently, for example, both Wehrbein and Leovy (1982) and Apruzese et al. (1982) have examined the consequences of including explicit calculations of infrared transfer in zonally symmetric models of the middle atmosphere circulation instead of using the Newtonian cooling approximation as was done in the earlier models of Holton and Wehrbein (1980) and Schoeberl and Strobel (1978), respectively. Examples of model studies of the coupling of middle atmospheric dynamics and chemistry are the works of Cunnold et al. (1980) and Mahlman et al. (1980) in which model dynamics were used together with various levels of simplification of ozone production and loss processes to study the processes that maintain stratospheric ozone distributions.

Middle atmosphere models have also been merged with tropospheric models to study the sensitivity of tropospheric climate modeling and/or weather forecasting to inclusion of the middle atmosphere. For instance, Geller and Alpert (1980) used a stationary planetary wave model to look into the possible changes in tropospheric planetary waves to changes in the stratospheric circulation. Simmons and Strüfing (1983) have looked at the sensitivity of tropospheric forecasts to both extending the top of the European Center for Medium-Range Weather Forecasting (ECMWF) model from 50 mb to 10 mb and using a hybrid vertical coordinate system instead of sigma coordinates. Mechoso et al. (1982) used the UCLA general circulation model to study the sensitivity of numerical forecasts to moving the top level of the model from the lower stratosphere up to the stratopause.

Models of the middle atmosphere can also be used to understand the limitations of more simplified models. For instance, Mahlman (1975) and Tuck (1979) have used tracer transport experiments with general circulation models to study some limitations of the simplified transport formulations in one- and two-dimensional photochemical models. Also, Fels et al. (1980) used a general circulation model to assess the validity of one- and two-dimensional radiative-convective and fixed-dynamical-heating models.

While most scenarios of anthropogenic chemical perturbations have been addressed with one- and two-dimensional photochemical or radiative-convective models, some models of middle atmosphere dynamics have also been used for this purpose. For instance, Pyle (1980) has used his two-dimensional model, with self-consistently calculated mean meridional motion, to calculate ozone depletion by chlorofluorocarbons, and Fels et al. (1980) have calculated the effects on middle atmosphere structure for uniform doubling of CO_2 amounts and halving of O_3 amounts.

The results of general circulation models can be used as a proxy for atmospheric data to examine the representativeness of present observational networks as well as to look at the impact of future systems. For instance, Mahlman (1980) has used the results of general circulation transport studies to look at the representativeness of global averaged ozone amounts that are inferred by the ground-based ozone network.

Finally, one area in which middle atmosphere dynamics models are beginning to be used is in analyzing data. This forecast-assimilation method is already used extensively for tropospheric studies (e.g., McPherson et al.,1979). In this system, data is analyzed, a forecast is run, and the results of the forecast are used to help analyze the data at a later time. The cycle is then repeated. This type of analysis yields a complete dynamically and energetically consistent data set that is constrained by observations. Stratospheric forecast-analysis methods are just now being applied to stratospheric satellite data by the British Meteorological Office and groups in the United States.

Depending on the anticipated use of middle atmospheric models, different model structures are required. This is discussed in more detail in the next section.

2. TYPES OF MODELS

The various types of middle atmosphere dynamics models have been reviewed in WMO (1981). The most comprehensive of these models reviewed has been constructed at the NOAA Geophysical Fluid Dynamics Laboratory in Princeton, New Jersey. This is a gridpoint general circulation model with forty vertical levels extending up to 80 km. It includes full radiative transfer, topography, a hydrological cycle, surface boundary layer processes, and a Richardson number dependent formulation for vertical diffusion of heat and momentum. Thus, the tropospheric forcing is modeled with a full state-of-the-art tropospheric general circulation formulation. This comprehensive model structure is motivated by a desire to develop a three-dimensional fully interactive model for investigating middle atmosphere processes as well as to use the results from a comprehensive model as a proxy for atmospheric dynamics for transport studies.

Another class of middle atmosphere dynamics model has been referred to as mechanistic circulation models. Examples of this type of model are those at NASA's Langley Research Center and the model that is being developed cooperatively at the Massachusetts Institute of Technology and the Georgia Institute of Technology. Both of these models include topography, have relatively simple formulations of radiative transfer, and do not include any hydrological cycle. Thus, instead of explicitly modeling the tropospheric forcing by means of a full general circulation model in the troposphere, an attempt is made to specify an empirical tropospheric forcing to give what looks like a realistic troposphere (see Cunnold et al., 1975). This is considered an acceptable strategy since the goal of these types of models is to investigate the interaction among chemistry, dynamics

and radiation <u>within</u> the stratosphere. Thus, if this idealized troposphere formulation forces a proper stratosphere this goal can be realized.

A third type of model for middle atmosphere dynamics is the so-called
mechanistic model. In this type of model, the stratospheric forcing is
specified at a lower boundary (100 mb for example). The models of Holton
and Wehrbein (1980) and Schoeberl and Strobel (1978) are of this type. In
some cases, these models may use radiative heating parameterizations and
gravity wave turbulence parameterizations at least as realistic as those
used in the more comprehensive general circulation models e.g., Holton
(1983). This type of model is usually used to investigate middle atmosphere dynamics and transport characteristics.

We see then that it has been traditional to consider various simplifications in middle atmosphere dynamics depending on the class of problems to
be explored with the model.

3. MODEL SIMPLICATIONS AND THEIR EFFECTS

In this next section, we will examine some of the more common model
simplifications made in middle atmosphere models along with some of the
consequences of these simplifications.

Quasi-Geostrophic

One simplification that has been used in several middle atmosphere
models is to integrate the quasi-geostrophic set of equations rather than
the primitive equations. By making this simplification, gravity waves are
filtered out of the system, thus enabling a longer time step to be used in
the integration. Examples of middle atmosphere models that make this
simplification are those of Cunnold et al. (1975) and Schoeberl and
Strobel (1978). The obvious problem in using the quasi-geostrophic set
of equations is that they do a poor job of representing the motions near
the equator.

Another problem in using most of these quasi-geostrophic set of
equations arises through their use of a prescribed global static stability.
This could be a serious shortcoming if one is modeling future scenarios in
which there is a large change in the global distribution of the radiative
gases which play a large role in determining the globally averaged temperature, and hence stability, profile. Another shortcoming can be seen
by looking at these authors' thermodynamic equations (Cunnold et al.'s
1975, eqn. (4) and Schoeberl and Strobel's, 1978, eqn. (5)). In these
quasi-geostrophic models, the local heating rate due to vertical motions
results from the local vertical motion interacting with the global static
stability profile. It is known, however, that the static stability varies
on the order of 50-100% in the middle and high latitude mesosphere from
winter to summer, at high latitudes especially (e.g., see Dickinson, 1966,
pp.33 and 34). This is a result of the warmer (cooler) stratopause and
the cooler (warmer) mesopause in summer (winter). Since, by continuity
considerations, the net vertical motion in the summer hemisphere must

be equal and opposite to that in the winter hemisphere, the use of a constant static stability with latitude and season may lead to a considerable distortion in the distribution of heating and cooling by vertical motions in the mesosphere with lesser distortions in the stratosphere.

Radiative Transfer

All middle atmosphere circulation models use some level of approximation to full line-by-line radiative transfer for reasons of computational efficiency and simplicity. The simplest approximation that is commonly used is that of Newtonian cooling in which the local heating/cooling rate is taken to be proportional to the negative of the temperature deviations from a reference equilibrium state. Examples of middle atmosphere circulation models that have used this Newtonian cooling approximation are the British Meteorological Office model discussed by O'Neill et al. (1982) and the models of Schoeberl and Strobel (1978) and Holton and Wehrbein (1980). There have been several comparisons of middle atmosphere circulation models with Newtonian cooling versus full radiative transfer. We will briefly look at some of these to see the effects of the Newtonian cooling approximation to radiative transfer.

Ramanathan and Grose (1978) did such a comparison with a nine-level quasi-geostrophic circulation model that extended from the ground up to about 70 km. They found that in the lower stratosphere the latitudinal variations in the long wave heating due to exchange of radiation between atmospheric layers was very significant. This effect is ignored in the Newtonian cooling approximation. They also found that in the upper stratosphere the effective Newtonian cooling coefficient undergoes significant latitudinal and seasonal variations. They concluded that while the Newtonian cooling approximation may be valid for mechanistic studies, it is a poor approximation for models attempting to simulate the climatology of the lower stratosphere.

Another type of comparison between Newtonian cooling and full radiative transfer can be seen by comparing the results of Holton and Wehrbein (1980) and Schoeberl and Strobel (1978) with those of Wehrbein and Leovy (1982) and Apruzese et al. (1982), respectively.

Both the Holton and Wehrbein (1980) and Wehrbein and Leovy (1982) papers used the same primitive equation model formulation to compare zonally symmetric middle atmosphere circulation model results where the earlier works used Newtonian cooling and the latter works used a long-wave radiative transfer model. The comparisons between these two calculations are shown in Figure 1 for their net radiative heating and their mean zonal winds. It is obvious that the calculations with infrared radiative transfer gave a much larger differential net radiative heating, and this leads to much stronger mean zonal winds given the same zonal momentum drag. One of Wehrbein and Leovy's (1982) conclusions is that the model with infrared radiative transfer requires more momentum drag to bring the modeled mean zonal winds into agreement with observations,

472

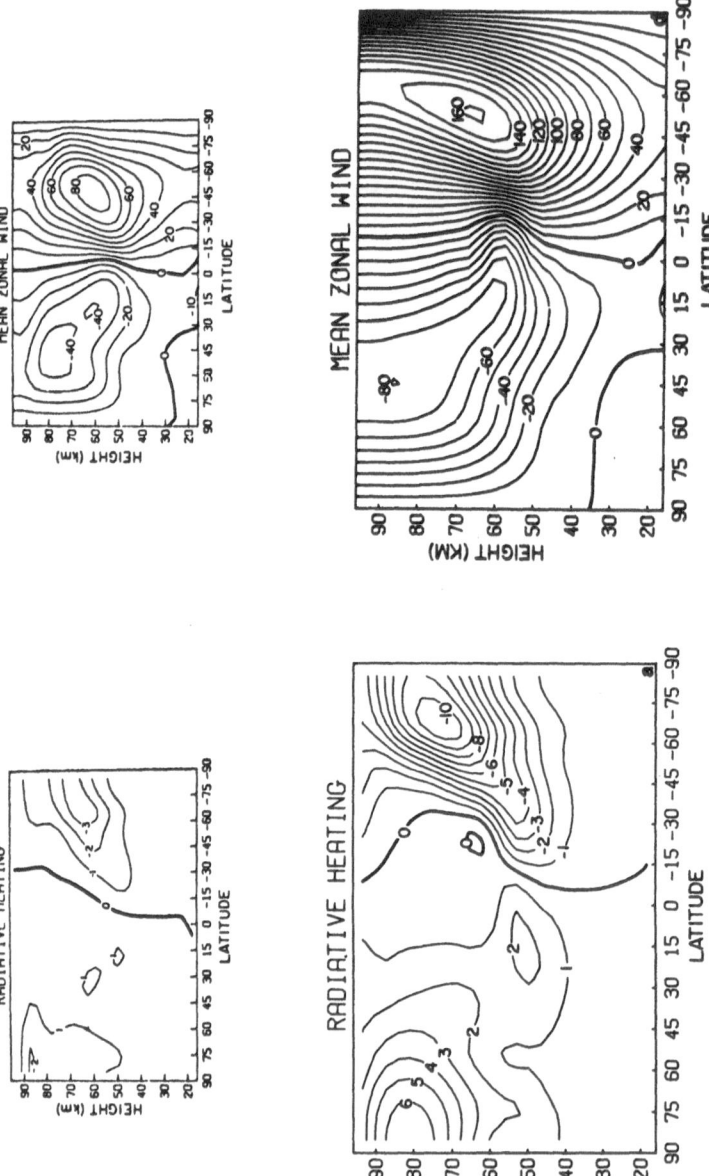

Figure 1: Top: Latitude-height section of the zonal mean net radiative heating rate in K day^{-1} (left) and mean zonal winds in ms^{-1} (right) for the summer solstice. Model results from Holton and Wehrbein (1980).

Bottom: Net radiative heating rate in K day^{-1} (left) and mean zonal wind in ms^{-1} (right) at solstice. Model results from Wehrbein and Leovy (1982).

and that this will lead to a much stronger accompanying meridional circulation.

The comparison between the calculated mean zonal winds of Schoeberl and Strobel (1978) and those of Apruzese et al. (1982) also gave modeled mean zonal winds that were stronger when infrared radiative transfer rather than Newtonian cooling was used but not by nearly as much as was the case for the Wehrbein and Leovy (1982) calculation. Apruzese et al. (1982) found a strengthening of the winter westerlies by about 10% when infrared radiative transfer was used. Thus, while the conclusion that using full infrared radiative transfer instead of Newtonian cooling requires stronger momentum drag to bring the modeled zonal winds into agreement with observations is common to both sets of investigations, there is a substantial conflict between the magnitude of this effect in the two studies cited above.

Another somewhat controversial point about radiation treatments in middle atmosphere circulation modeling is the subject of a paper by Ramanathan et al. (1983). In this paper, it is pointed out that the problems of the too cold winter polar lower stratosphere with the accompanying lack of tropopause jet closure and the excessively strong polar night jet was largely solved in the NCAR spectral general circulation model given certain improvements in the radiation calculations. The comparison of the zonally averaged temperatures and the mean zonal winds with improved versus unimproved radiation packages are shown in Figure 2. The improvements cited by Ramanathan et al. (1983) were as follows:

(1) More careful treatment of the upper boundary condition for O_3 solar heating;

(2) Inclusion of the correct temperature dependence of long wave cooling by the CO_2 15 µm bands;

(3) Use of a latitudinal distribution of lower stratosphere H_2O mixing ratios that falls off toward the winter pole; and

(4) Inclusion of the dependence of cirrus emissivity on cloud liquid water content.

Drs. J. Mahlman and S. Fels of the NOAA Geophysical Fluid Dynamics Laboratory in Princeton, New Jersey (personal communication) feel that this result can be a misleading one since they feel that models with tops higher than 30 km (the top of Ramanathan et al.'s model) show more serious cold winter pole problems that cannot be done away with by any reasonable changes in radiation treatment. Mahlman and Fels feel that these problems at the winter stratopause are of the same type as the problem in the lower stratosphere, and the proper explanation for these deficiencies lies in improved dynamics rather than radiation (see Mahlman and Umscheid, 1983).

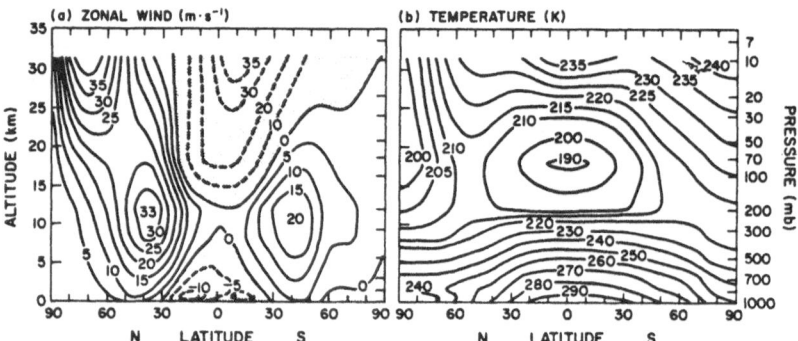

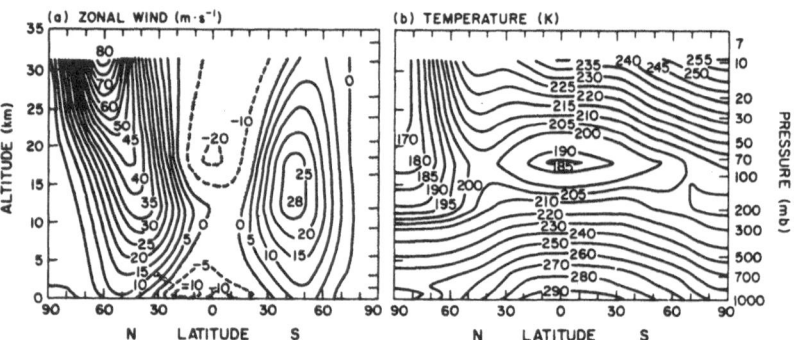

Figure 2: Mean zonal winds in ms^{-1} and zonally averaged temperatures as modeled by Ramanathan et al. (1983) with improved (top) and degraded (bottom) treatments of radiation.

Momentum Drag and Diffusion

It is by now well appreciated that in addition to differential
diabatic heating, mechanisms for zonal momentum drag are required to
explain the mean zonal circulation of the middle atmosphere. This has
been shown by the modeling work of Leovy (1964), Schoeberl and Strobel
(1978), and Holton and Wehrbein (1980), among others. In these mechanistic
models of the middle atmosphere circulation, momentum dissipation was put
into their models by means of Rayleigh drag, that is to say, a deceleration toward a state of rest that is proportional to the magnitude of the
mean zonal wind. The physical basis of this procedure lies in the picture
of gravity waves that grow with altitude such that they break at a level
where they become unstable by either shear or static instability. At
these levels, momentum flux divergences or convergences will occur that
will tend to decelerate the mean flow toward the wave's phase velocity.
Since most middle atmosphere gravity waves are assumed to have their
source in the troposphere, they will have small zonal phase speeds
characteristic of tropospheric zonal winds. Lindzen (1981) has given a
derivation of these effects using internal gravity wave theory, and
Geller (1983) has given a simple heuristic discussion of these processes.
The principal reason why it is felt that such gravity wave decelerations
are necessary is that the other candidate mechanism for such decelerations, planetary waves, cannot provide the necessary deceleration
mechanism in summer where planetary waves are virtually absent whereas
such zonal flow decelerations are required in both seasons (see Geller,
1983, for example).

Parameterizations of gravity waves are now being used in mechanistic
models of the middle atmosphere (e.g., Holton, 1982 and 1983; Dunkerton,
1982; Matsuno, 1982; Miyahara, 1983). General circulation models of the
middle atmosphere have not included these type of parameterizations yet.
One reason for this is that, in most cases, the wave drag formulations
that have been used in mechanistic middle atmosphere models have been
tuned to give reasonable looking circulations. While such tuning is
easily done in simple models, it is a very time consuming and expensive
proposition in general circulation models.

Middle atmosphere general circulation models that extend to the
mesosphere sometimes use fixed diffusion profiles of the type that appear
in one-dimensional photochemistry models, but with smaller magnitudes
(see Cunnold et al., 1980, Figure 5, and Hunt, 1981a, Figure 1). Fels
et al. (1980) use a Richardson number dependent parameterization for the
vertical mixing of heat and momentum that is dependent on vertical grid
resolution.

One expects a good deal of research activity in the area of determining gravity wave and diffusion parameterizations for middle atmosphere
models in the near future given the level of theoretical interest and the
measurement capabilities of new techniques such as VHF radar (e.g.,
Vincent and Reid, 1983) and lidar (e.g., Chanin and Hauchecorne, 1982).

Hydrological Cycle

Several middle atmosphere models that include a relatively full set of physical parameterizations in their formulation choose not to include the tropospheric hydrological cycle. This includes the models of NASA/ Langley Research Center, Hunt (1981a) and Cunnold et al. (1980). In this section, we inquire into what some of the consequences might be when hydrological processes are neglected. Very little work has been done on this subject since the work of Smagorinsky et al. (1965) and Manabe et al. (1965). Figure (3) shows some annual mean simulations from these works with and without the inclusion of a hydrological cycle. Note that the tropopause jet intensity decreases with altitude more markedly in the case with a hydrological cycle.

Another result from these works has been shown in Manabe et al. (1970) where they compared the mean meridional circulations from the observed atmosphere to those of the moist and dry models. They show that the tropical Hadley circulation is about four times stronger in the moist model (in better agreement with observations) than for the dry model. From this, we conclude that middle atmosphere circulation models whose goal includes studying constituent transport from the troposphere to the stratosphere might be underestimating these effects significantly if they neglect the hydrological cycle.

Effects of Mountains

At least one middle atmosphere model (Hunt, 1981a) does not include orographic forcing. The influences of including or not including oro- graphic forcing in models of the middle atmosphere circulation have been studied by Kasahara et al. (1973) and Manabe and Terpstra (1974). It is interesting that the results of these two investigations differed in some important respects, and that this type of investigation has not been carried out since. Some of the results of these two investigations can be seen in Figure (4) that show the modeled mean zonal winds with and without mountains. Note that both calculations show that the inclusion of mountains leads to a decrease in the polar night jet intensity, but that Kasahara et al. (1973) found that the inclusion of mountains decreased their polar night jet intensity by more than a factor of two whereas Manabe and Terpstra found that the inclusion of mountains decreased their polar night jet intensity by only about 15%. This result may be due to the fact that while Manabe and Terpstra (1974) spun their model up to a steady state which had a very excessive polar night jet intensity in relation to that which is observed, Kasahara et al. (1973) showed results when their model had not yet reached a steady state but had more reasonable looking polar night jet intensities. This difference in their wind structures will affect the planetary wave propagation appreciably (see Schoeberl and Geller, 1976, for example). Another interesting point is that Kasahara et al. (1973) found that the inclusion of mountains increased the modeled planetary wave amplitudes to be more in line with observations but that the inclusion of mountains made the planetary wave phases worse relative to the observed phases than did the no mountain case.

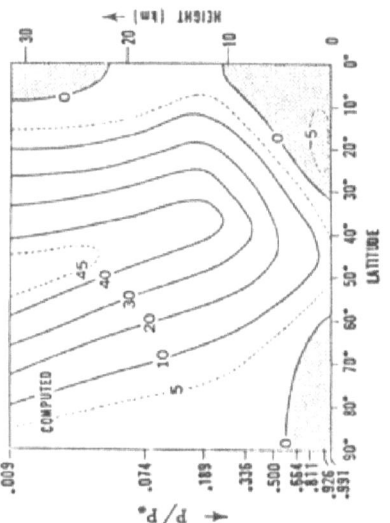

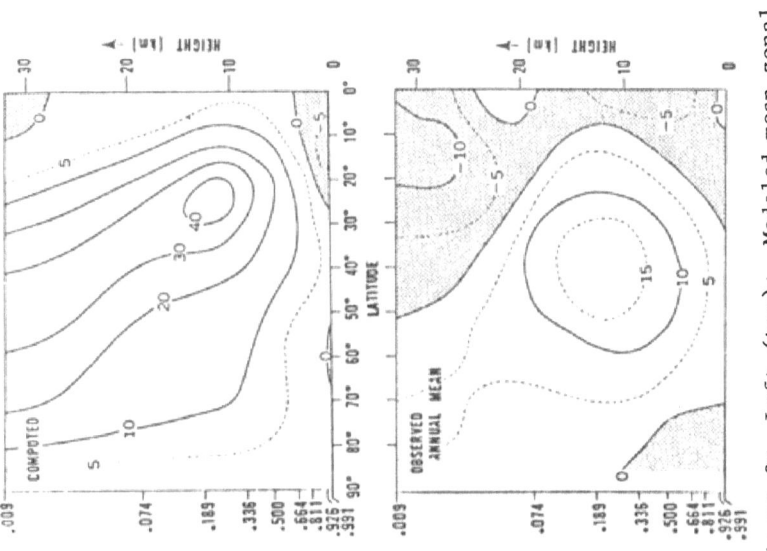

Figure 3: Left (top): Modeled mean zonal winds (ms^{-1}) with no land-sea thermal contrast, no orography, with a hydrological cycle parameterization for annual mean insolation.
Left (bottom): Observed annually averaged mean zonal winds (ms^{-1}). From Manabe et al. (1965).
Right: Modeled mean zonal winds (ms^{-1}) with no land-sea thermal contrast, no orography, and no hydrological cycle for annual mean insolation. From Smagorinsky et al. (1965).

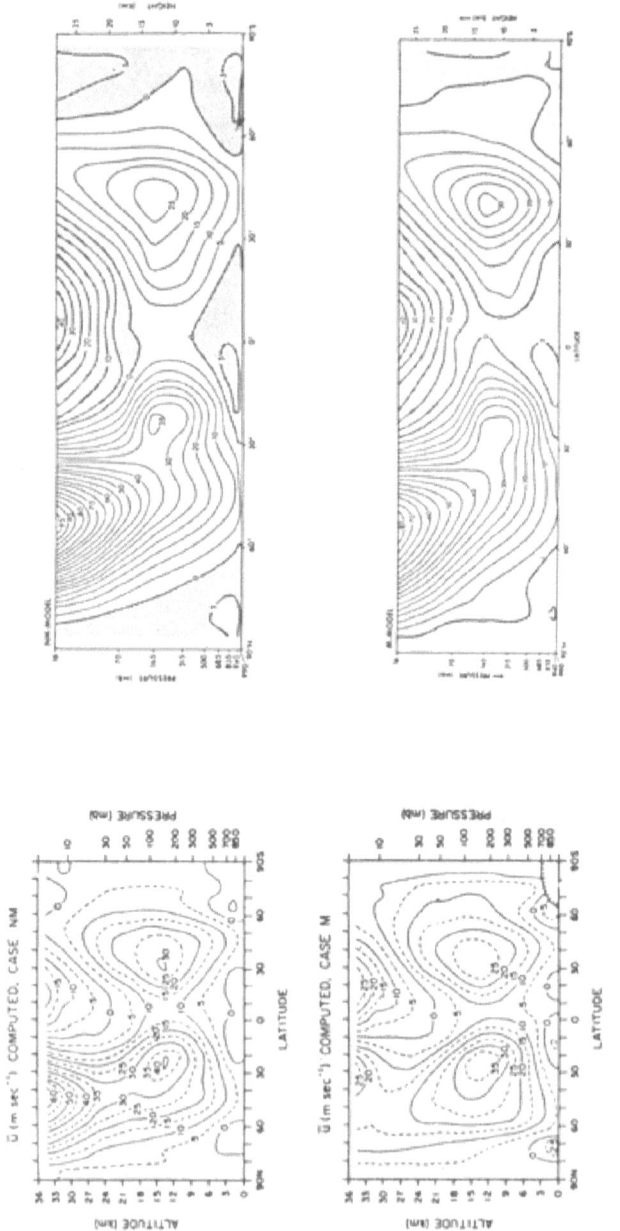

Figure 4: Modeled mean zonal winds for perpetual January insolation for the case of no mountains (top) and with mountains (bottom). Results from Kasahara et al. (1973) are shown on the left while the results from Manabe and Terpstra (1974) are shown on the right.

There have been several recent papers using steady state models to investigate the roles of topographic and diabatic heating in forcing planetary waves (e.g., Huang and Gambo, 1982; Lin, 1982; Lindzen et al. 1982; and Alpert et al. 1983). All of these papers conclude that topographic forcing plays a dominant role in forcing middle atmosphere planetary waves. This being the case, it is clear the surface topography must be included in any model of the middle atmosphere that includes the troposphere. The manner of including mountains in relatively coarse grid resolution middle atmosphere models such as now exist is not perfectly clear, however, (Dickinson, 1980).

4. ACCOMPLISHMENTS AND DIFFICULTIES

General circulation models of the middle atmosphere have modeled several aspects of observed middle atmospheric structure. In the following, we will look into some of the contributions of middle atmosphere circulation models to our understanding of middle atmosphere behavior. We will also look at some model shortcomings.

Climatology

There have been several general circulation models of the middle atmosphere that have appeared recently. These include the model results presented by Schlesinger and Mintz (1979), Hunt (1981a), and O'Neill et al. (1982), for example. Figure (5) shows the computed mean zonal winds for perpetual January integrations in the cases of Schlesinger and Mintz (1979) and Hunt (1981a) and for the February 2 day of an annual cycle integration for the case of O'Neill et al. (1982). Also shown in this figure is the "observed" mean zonal wind for northern hemisphere winter from Newell (1968) as was reproduced in Hunt's (1981a) paper. Note that all three simulations reproduce the winter westerlies and the summer easterlies, but the middle atmosphere winter westerlies are excessive in Schlesinger and Mintz's (1979) and Hunt's (1981a) simulations but not in O'Neill et al.'s (1982) results. The problem of excessive modeled westerlies is common to all present middle atmosphere general circulation models that calculate radiative heating explicitly and extend at least up to 50 km. The model reported in O'Neill et al. (1982) did not compute radiative heating directly from radiative transfer but instead used an empirical relation between radiative heating and cooling rates and zonally averaged temperatures. Schlesinger and Mintz (1978) used full radiative heating and cooling calculations below 30 km but used Dickinson's (1973) Newtonian cooling approximation to radiative cooling above this level. Hunt (1981a) used full radiative heating and cooling calculations below 67.5 km and used a Newtonian cooling relaxation to observed values above.

The present state-of-the-art is that all middle atmosphere general circulation models with relatively complete internal physics calculations (including radiative transfer) give excessive winter westerlies.* Furthermore, there is an indication that these excessive winter jet intensities

---

*This subject is covered in more detail by the Mahlman and Umscheid (1983) paper in this volume.

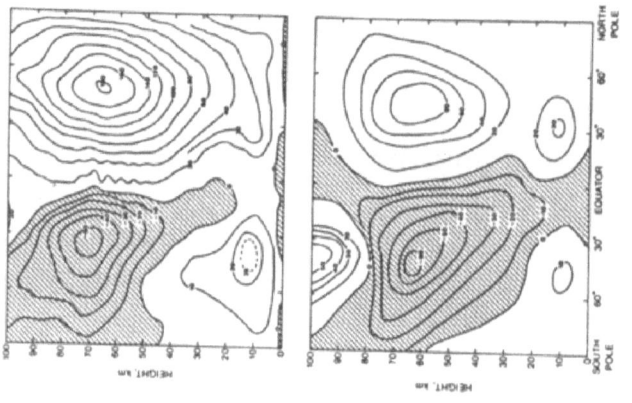

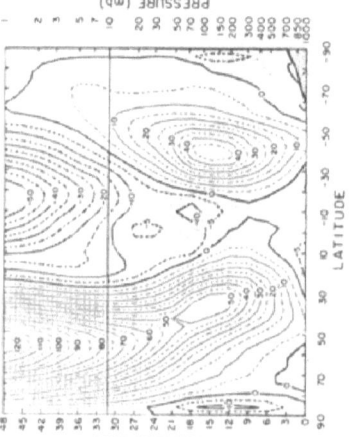

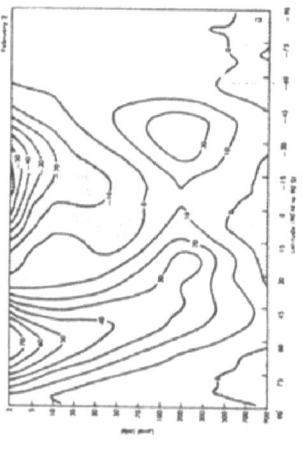

Figure 5. Mean zonal wind in ms^{-1} as modeled with perpetual January insolation by Schlesinger and Mintz (1979) – upper left; mean zonal wind for February 2 of an annual cycle integration from O'Neill et al. (1982)– lower left; mean zonal wind as modeled with perpetual January insolation by Hunt (1981a) – upper right; and "observed" mean zonal wind for northern hemisphere winter conditions from Newell (1968) – lower right.

prevent spontaneous stratospheric warmings from taking place. For instance, the model reported by O'Neill et al. (1982) produced such a warming but the Schlesinger and Mintz (1979) and Hunt (1981a) models did not. This is probably due to two reasons. One is that the excessively strong westerly winds tend to trap the planetary waves too low down in the middle atmosphere, and second, the very strong westerlies cannot be reversed easily.

Perturbation Studies

Middle atmospheric circulation models have also been used to test for their response to perturbation scenarios. Fels et al. (1980) tested what the effects of CO_2 doubling and O_3 halving gave to their modeled results. This was done using a low horizonal resolution (9° latitude x 10° longitude) model with 40 levels extending from the Earth's surface to about 80 km for annual mean insolation conditions. Figure (6) shows their modeled zonally averaged temperatures for the control, or unperturbed, case as well as the differences in the zonally averaged temperatures for the $2xCO_2$ and $1/2xO_3$ cases.

Fels et al. (1980) found for the $2xCO_2$ case that the resulting middle atmospheric cooling was quite independent of latitude. This does not imply that much of a change will take place in the mean zonal wind field. The $1/2xO_3$ case gives much more latitude structure in the middle atmosphere cooling than is the case for $2xCO_2$ which, in turn, gives a much greater effect on the mean zonal flow. These mean zonal flow changes are found to take place mainly above 30 km, however. Geller and Alpert (1980) have shown that changes in the mean zonal flow above 30 km will not give rise to any appreciable change in tropospheric planetary waves through altering transmission-reflection properties. No changes in tropospheric planetary wave structure were found by Fels et al. (1980) in the $2xCO_2$ and $1/2xO_3$ cases from the control case. One very interesting feature of the $1/2xO_3$ experiment was the strong cooling at the tropical tropopause which should lead to much less stratospheric water vapor by increased the freezing out of water in the rising branch of the Hadley circulation.

Another middle atmosphere perturbation study has been carried out by Hunt (1981b) who tried to look at the sensitivity of his model results to polar cap ozone depletions of the type that occur during strong polar cap absorption events. He found a large change ($\sim$25%) in mean zonal wind speed in the mid-latitude upper troposphere occurring about 20-24 days after the event in his model. The reasons for this are not at all clear.

Comparison with Simpler Models

Another application of general circulation models of the middle atmosphere is in comparing the results of full general circulation models with more simplified models for the purpose of seeing the limitations of these simple models. A very nice example of this type of work is found in Fels et al. (1980). They used their general circulation model to explore the

482

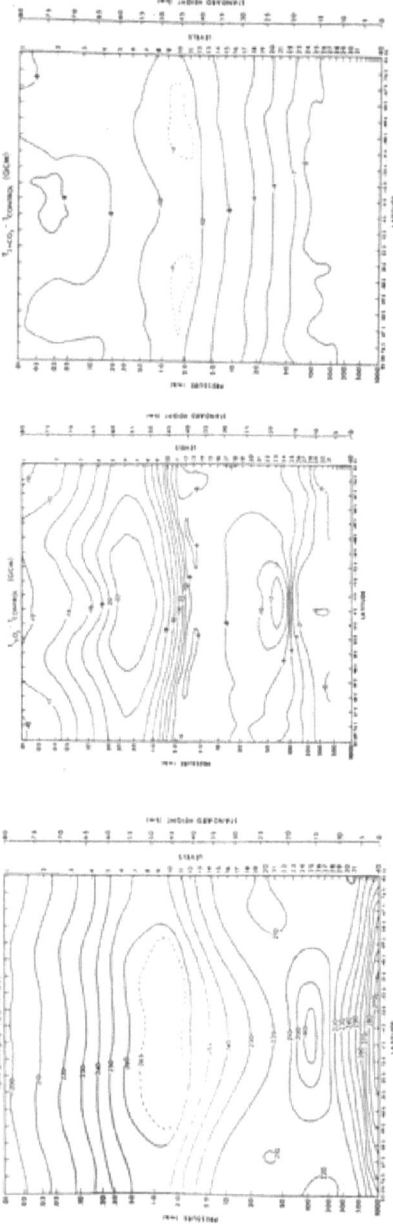

Figure 6. (Left) Modeled zonally averaged temperature ($^{\circ}$K); (middle) difference between modeled zonally averaged temperatures and those with uniformly halved O_3 concentrations; (right) difference between modeled zonally averaged temperatures and those with uniformly doubled CO_2 concentrations. All are general circulation model results from Fels et al. (1980).

applicability of the simpler Radiative-Convective-Equilibrium (RCE) and Fixed-Dynamical-Heating (FDH) models to the $2xCO_2$ and $1/2xO_3$ experiments discussed in the previous section.

The RCE model is one in which the solar heating is taken to balance the long wave cooling at each point in the model except where the temperature lapse rate exceeds some critical value in which case the lapse rate is fixed at this value. Usually this critical lapse rate is taken to be 6.5 K km^{-1}. In this type of model, the surface temperature is calculated self-consistently. Fels et al. (1980), in applying the RCE model to the $2xCO_2$ and $1/2xO_3$ sensitivity experiments, fixed the surface temperatures, however, so that the response to the imposed perturbation is purely radiative above the convective zone and purely dynamical below.

The FDH model is one in which the local equilibrium is taken to be the situation where the sum of the solar heating, long wave cooling and dynamical heating rates is everywhere zero. Thus, given Fels et al. (1980) general circulation results, the local radiative heating can be calculated thus determining the distribution of the dynamical heating rate. In performing the perturbation experiments, this dynamical heating rate is assumed to remain unchanged, and the new FDH temperature distribution is one for which the new local radiative heating rate plus the unchanged dynamical heating rate is everywhere zero.

Figure (7) shows the RCE and FDH modeled temperature differences for the $1/2xO_3$ perturbation experiment. This is to be compared with full general circulation results for this case that was shown in Figure (6). Looking at Figures (6) and (7), we see that above 35 km there is qualitatitative similarity among all three model results with maximum coolings of 22-25°K near the tropical stratopause; however, larger horizonal gradients in the amount of cooling are seen in the FDH results. Below about 25 km there are substantial differences in the three results that can be attributed, in part, to the fixed lapse rate constraint in the troposphere and to the different "control" temperature distribution of the RCE model (see Fels et al. 1980). There is a qualitative difference between the FDH and GCM results in the tropics between about 55 to 75 km where the dynamical cooling has apparently changed in the GCM case. There are also significant differences in the cooling results just above the tropical tropopause among the GCM, FDH, and RCE results. Fels et al. (1980) point out that this is a particularly sensitive region to very small changes in heating or cooling rates. Quite similar results are obtained in all of the three models (GCM, RCE, and FDH) for the $2xCO_2$ case (not shown here). The only very significant difference is in the shape of the perturbation response in the lower stratosphere in the RCE model with those of the GCM and FDH models. This is due to the fixed lapse rate constraint in the RCE case. The fact that the FDH model does so well in the $2xCO_2$ case is consistent with the rather flat response in the GCM results with latitude. This flat response should not produce much altered dynamics.

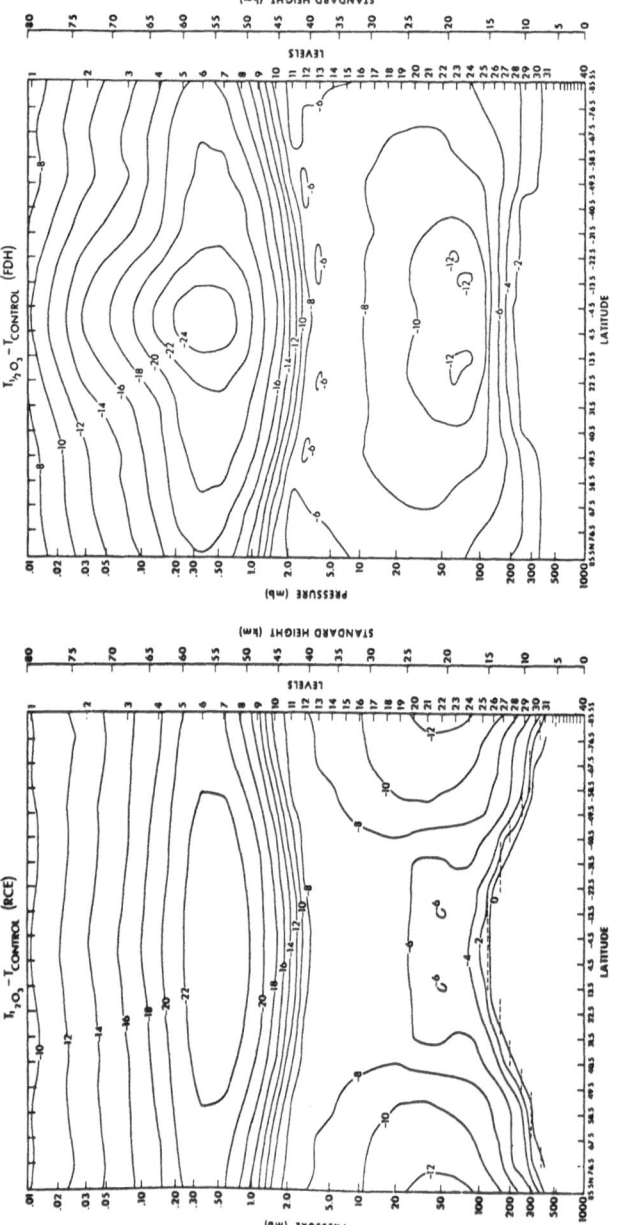

Figure 7. Differences in zonally average temperatures (^{O}K) between $1/2 \times O_3$ case and control case using a radiative-convective-equilibrium (RCE) model (left) and fixed-dynamical-heating (FDH) model (right). From Fels et al. (1980).

Thus, the Fels et al. (1980) investigation showed that for perturbation experiments involving changes in concentrations of long wave radiatively active constituents such as CO_2 which give a flat temperature change with latitude, the FDH, and, in fact, the RCE models give pretty good simulations of the temperature change. In cases where the altered constituent is a strong solar radiation absorber such as is the case for O_3, the GCM gives a temperature response with more latitudinal structure and more altered dynamics. Even in this case, however, the FDH model does well except in the tropical lower stratosphere and mesosphere.

Investigations of this type show the utility of middle atmosphere GCM's in seeing the limitations of simpler models.

Transport and Photochemistry Studies

Middle atmosphere circulation models have also been very useful for studies of transport and photochemistry. Two very different types of studies in this regard have been accomplished by the use of marked parcels and by tracer transport formulations some of which have involved photochemistry while others did not. Examples of middle atmosphere transport studies using marked parcels are those of Kida (1977), Hsu (1980), and Kida (1983). Examples of transport studies without photochemistry include Mahlman and Moxim (1978) while those with some degree of photochemistry include Cunnold et al. (1975, 1980), Mahlman et al. (1980), and Levy et al. (1979).

As examples of these two types of investigations, we will briefly discuss the work of Kida (1983) and Mahlman et al. (1980). Kida (1983) used a 12-level hemispheric general circulation model extending from the ground to 1 mb. Its horizontal grid is 3° (longitude) by 2.5^o (latitude). It has no topography but does include a parameterization of thermal forcing of planetary waves. Kida (1983) examined the very long term motions of air parcels into and out of the stratosphere by performing trajectory analyses on a large number of marked air parcels. In this work, he defined the age of an air parcel as the length of time elapsed since it first enters the stratosphere. Figure (8) shows parcel age spectra for 5° (latitude) x 1 km (altitude) domains for three separate latitude bands: $5-10^o$ (tropics), $45-50^o$ (mid-latitudes), and $75-80^o$ (polar latitudes). These are given for five altitude ranges. At the start of this experiment, the parcels are all located just beneath the tropical tropopause. In the tropics, the tropopause altitude is about 17 km, so we may interpret the age spectra as showing strongly peaked distributions above the tropopause with the peak in the age spectra occurring at longer times with increasing altitude consistent with the speed of the rising motion in the Hadley circulation ∿5 km/year. Below the tropopause, the distribution is flat indicating that after about a year the marked parcels start reentering the troposphere and build up to a steady state number density. At middle latitudes, one sees less sharp stratospheric distribution peaks than was the case for the tropics. The broader peak in the age spectra occurs at longer times with increasing altitude showing that

486

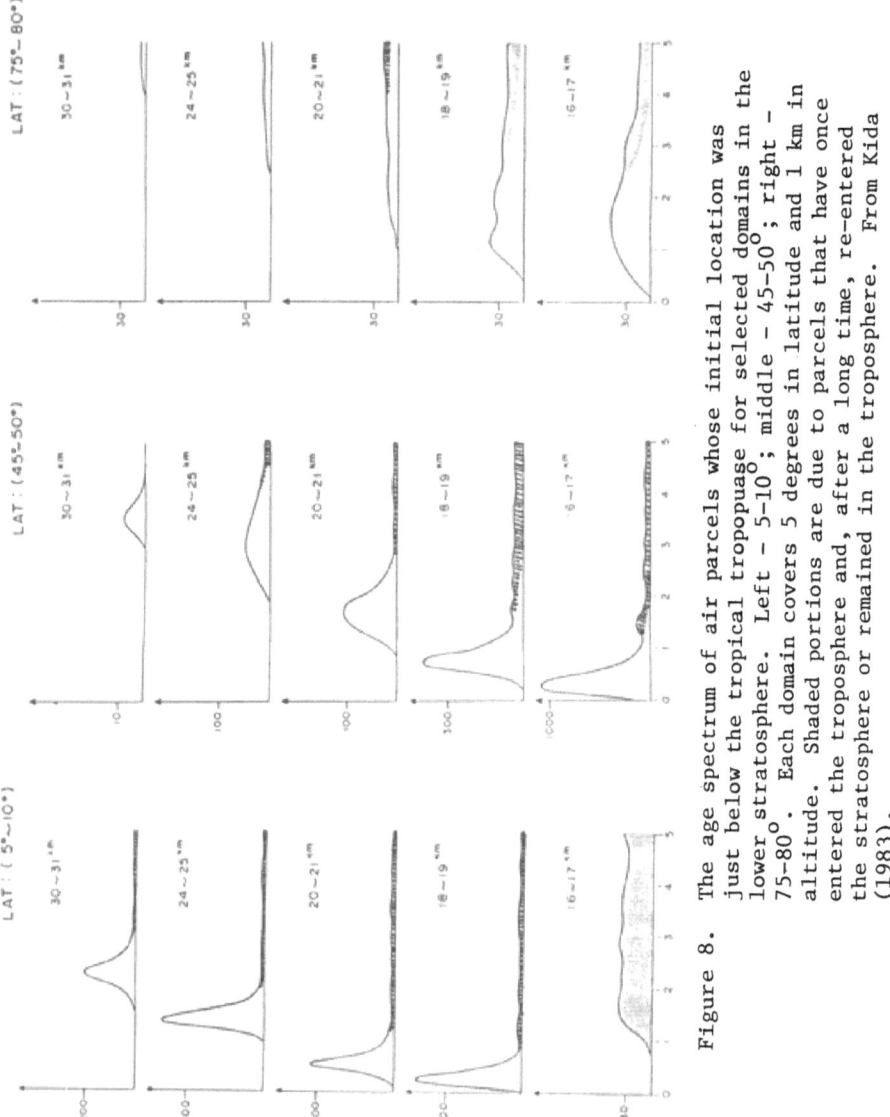

Figure 8. The age spectrum of air parcels whose initial location was just below the tropical tropopuage for selected domains in the lower stratosphere. Left – 5-10°; middle – 45-50°; right – 75-80°. Each domain covers 5 degrees in latitude and 1 km in altitude. Shaded portions are due to parcels that have once entered the troposphere and, after a long time, re-entered the stratosphere or remained in the troposphere. From Kida (1983).

it has taken longer to reach these higher altitudes, and that the trajectories reaching the higher altitudes are more diverse. Note that all
five altitudes are in the stratosphere at middle latitudes. Finally, the
high latitude distributions have very flat distributions reflecting the
situation that it takes a long time for "new" stratospheric parcels to
reach the polar stratosphere, and that the trajectories followed were
very diverse.

Another very different use of a middle atmosphere circulation model
is that of Mahlman et al. (1980) who used the GFDL 11-level general cir-
culation/ tracer model (see Manabe and Mahlman, 1976 and Mahlman and
Moxim, 1978) for two idealized ozone experiments. The first of these
experiments, the Stratified Tracer Experiment, specified instant relaxation of the ozone concentration at the top model level (10 mb) to
7.5 ppmv. Ozone is then treated as an inert tracer at all levels below
this top level until it is removed in the lower troposphere. In the
second experiment, the Simple Ozone Experiment, a simplified ozone photochemistry is used at the top model level, and again ozone is taken to be
inert at lower levels and is removed in the lower troposphere. Mahlman
et al. (1980) found that both experiments gave remarkably similar results.
For example, Figure (9) shows the zonal-mean ozone mixing ratio (ppmv)
from the fourth year of both experiments. These results suggest that the
details of middle stratosphere ozone chemistry exert very little influence
on the distribution of ozone in the lower stratosphere compared with the
influence of transport processes.

Modeling of Stratospheric Warmings

There has been little work undertaken on actual prediction of middle
atmospheric motions; however, there have been a few recent attempts at
using middle atmosphere circulation models to model the evolution of
observed stratospheric warmings since the pioneering effort of Miyakoda
et al. (1970). Butchart et al. (1982) used a version of the British
Meteorological Office GCM with 5° (latitude) x 10° (longitude) horizontal
resolution with 17 levels in the vertical from z = 16 km (100 mb) to 80
km. Observational data was used to drive the model at the lower boundary
of 100 mb. The model was started with observed conditions on February 16,
1979. Both the modeled and observed mean zonal winds are shown in Figure
(10). One sees a reasonable simulation of the onset of the polar easterlies during the stratospheric warming event. Butchart et al. (1982) also
did idealized modeling to see what factors were responsible for their
successful simulation compared with simpler stratospheric warming models.
They found that the initial mean zonal wind distribution and the inclusion of the traveling wave forcing at 100 mb were the most critical
factors in this regard.

Another forecasting experiment for stratospheric warmings was
carried out by Simmons and Strüfing (1982). They used an ECMWF model with
1.875° x 1.875° resolution with three different vertical structures – 18-
levels up to 10 mb, 16-levels up to 25 mb, and 14-levels up to 50 mb.

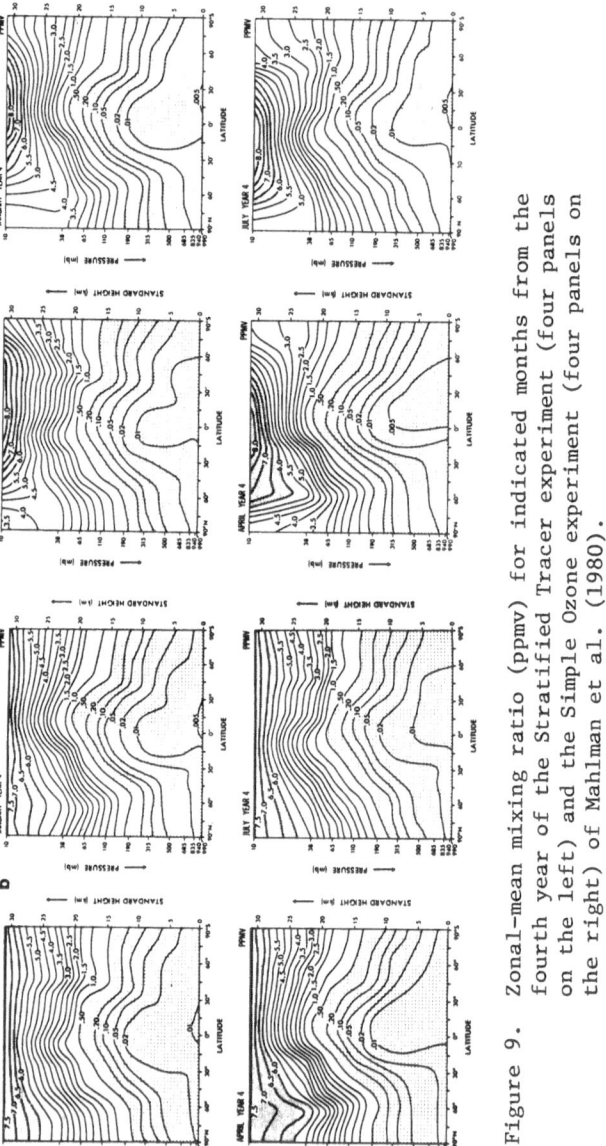

Figure 9. Zonal-mean mixing ratio (ppmv) for indicated months from the
fourth year of the Stratified Tracer experiment (four panels
on the left) and the Simple Ozone experiment (four panels on
the right) of Mahlman et al. (1980).

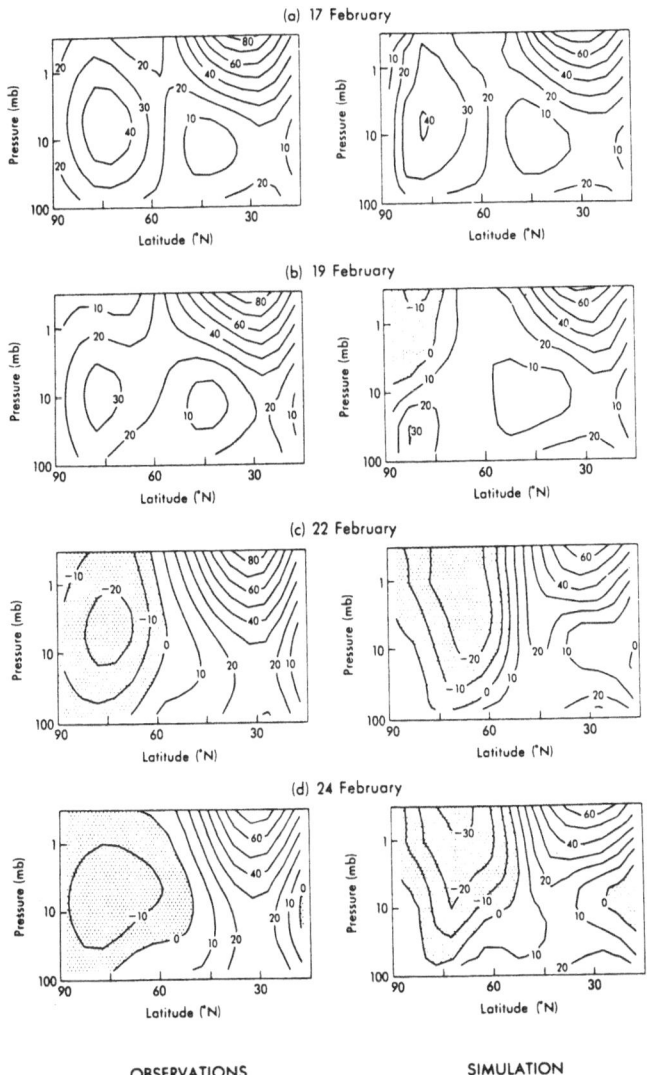

OBSERVATIONS   SIMULATION

Figure 10. Mean zonal wind (ms^{-1}) for observed (left hand column) and model simulation (right hand column) for; (a) February 17, (b) February 19, (c) February 22, and (d) February 24, 1979. From Butchart et al. (1982).

Figure (11) shows some of their results from 100-10 mb using the ECMWF
18-level model for 8 day forecasts that verify on February 21 and 25,
and March 5, 1979. As one can see, the forecast of the occurrence of
polar easterlies was quite successful as was the subsequent restoration
to westerlies. Simmons and Strüfing (1982) also showed considerable
success in forecasting the geopotential patterns at 50 and 10 mb.

5. PROBLEMS IN CONSTRUCTING AND VERIFYING MIDDLE ATMOSPHERE CIRCULATION
   MODELS

There are certain features of middle atmosphere behavior that bring
about some problems in constructing or verifying middle atmosphere circulation models that tend to be more severe than for the lower atmosphere.
In this section, we will briefly discuss some of these difficulties.

Vertical Coordinate

Most lower atmosphere models use $\sigma(= \dfrac{p - p_T}{p_s - p_T})$ as the vertical
coordinate (Phillips, 1957), where p is pressure, p_s is surface pressure,
and p_T is the pressure at the top of the model. Use of this vertical
coordinate allows a simple formulation of a terrain-following lower
boundary condition. For middle atmosphere models, there are some dis-
advantages of using this σ-coordinate system. For instance, many physical
processes require the time consuming transformations between σ and p to
be done continually. Numerical inaccuracies are introduced as a result
of these transformations (Mahlman and Moxim, 1976). This approach gives
rise to significant undesirable distortions between p and σ surfaces in
the stratosphere and above. Approaches used in middle atmosphere models
include using log σ as a vertical coordinate and using either a hybrid
σ-p (e.g., Arakana and Lamb, 1977) or a stretched vertical coordinate
(Simmons and Strüfing, 1983).

Upper Boundary Condition

General circulation models generally use a $\dfrac{d\sigma}{dt} = 0$ or a $\dfrac{dp}{dt} = 0$ upper
boundary circulation. This causes spurious reflections of wave energy
(Lindzen et al. 1968). This is particularly serious in the middle atmosphere where vertically propagating waves achieve high amplitudes.
Approaches used to avoid model contamination by these reflections include
extending the model domain upward to where physical processes will
dissipate the wave energy sufficiently so that they do not appear in the
domain of interest (Fels et al. 1980). Another approach is to use sponge
layer(s) at the top of the model. A difficulty with using an extended
sponge region at the top of the model is that the dissipation must be
introduced slowly in altitude with respect to the vertical wavelengths of
the propagating waves to avoid reflections, or in the case of a single
sponge layer (Schlesinger and Mintz, 1979) not all wave modes are absorbed.

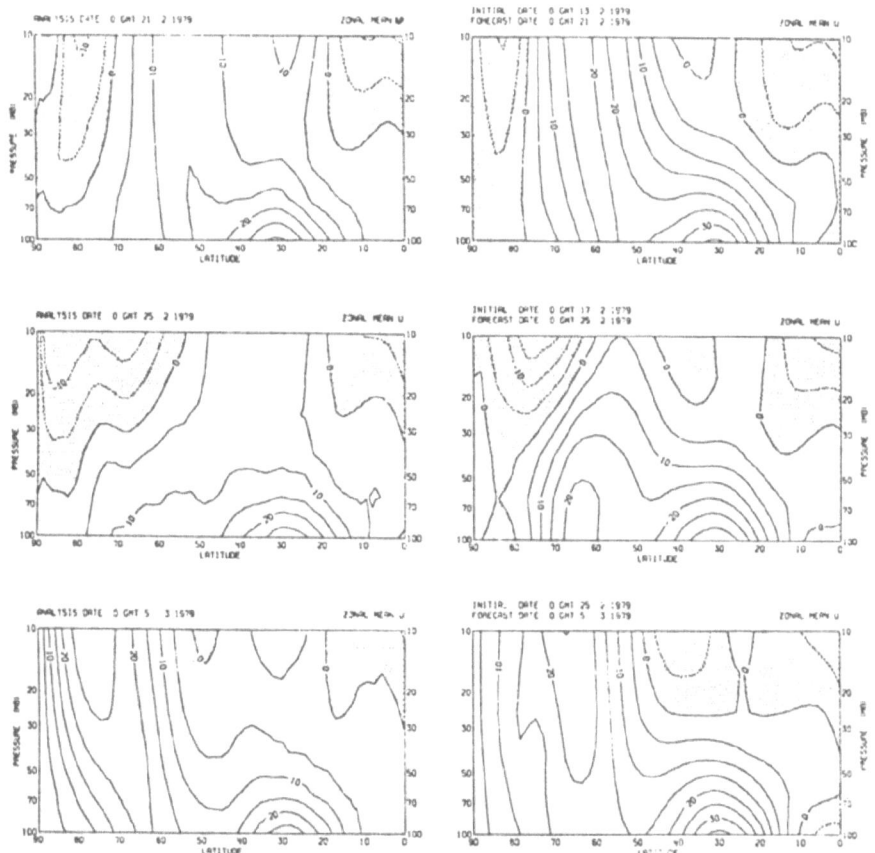

Figure 11.   Observed meridional cross sections for the mean zonal wind
for February 21 (upper left), February 25 (middle left),
and March 5, 1979 (lower left).   Eight-day forecasts
verifying on these dates are shown in the right-hand
sections.   From Simmons and Strüfing (1983).

Vertical and Horizontal Resolution

The choice of horizontal and vertical resolutions in a model always represents a compromise between computational accuracy and expense. In lower atmosphere models, the emphasis has tradionally been more on horizontal resolution than on vertical resolution. This has been dictated, in part, by the belief that one must model baroclinic instability properly, including its interaction with moisture processes, and nonlinear cascade in lower atmosphere models. In middle atmosphere models, one needs to pay relatively more attention to vertical resolution since it is believed that this is a region where vertically propagating waves play a dominant role. For middle atmosphere circulation models, which include explicit modeling of the troposphere, the minimum horizontal scale is probably set by the need to resolve motions on the scale of the Rossby radius of deformation which is about 1000 km in middle latitudes. This would dictate a horizonal grid resolution of about 250 km or about 3° at middle latitudes (or alternatively in spectral terms, zonal wavenumber 28). The minimum vertical resolution required would be set by the need to accurately resolve a vertical scale height, which dictates a vertical grid distance no more than 3 km. These resolution requirements imply that very great computational resources are required to model the troposphere/middle atmosphere system. If only the middle atmosphere is being modeled with the tropospheric forcing being specified, the horizontal resolution requirements are not so great. Depending on the specific problem being attacked, the vertical resolution requirements may be even more severe, however. For instance, modeling the middle atmosphere tropical waves may require a vertical grid resolutions of 1 km or less.

Topography

Airflow over topography is thought to be the dominant forcing mechanism for extratropical planetary waves in the middle atmosphere (Lindzen et al., 1981; Alpert et al., 1983). Because of this, one expects a smoothed surface topography to underestimate this forcing and thus to provide too little forcing for the extratropical planetary waves. When comparing the modeled planetary waves in different middle atmosphere circulation models, one must keep in mind the effective topographic forcing used which may be a function of the model horizontal resolution.

Polar Difficulties

Due to the convergence of meridians, special computational techniques are required near the poles in gridpoint models. This is because the CFL condition (see Haltiner and Williams, 1980, page 119) places an upper limit on the time increment that can be used for numerical stability, and this maximum time increment is proportional to the horizontal grid spacing. If one is using a latitude-longitude grid, and if the time increment is set according to the middle latitude grid spacing, the numerical scheme will be unstable near the poles. If, on the other hand, the time increment is set according to the polar grid spacing, this puts an unreasonable demand on computational resources. There are at least three ways to deal with this problem. One can use a horizontal gridding that attempts to use approximately equal grid spacings as a function of latitude (Kurihara, 1965). One may zonally Fourier filter either the predicted variables in the polar regions (Kálnay-Rivas, et al. 1977), or, alternatively, one may apply such a filter to the terms in the governing equations that are responsible for these linear instabilities (Arakawa and Lamb, 1977). These difficulties are present in lower atmosphere gridpoint models, but they are particularly troublesome in middle atmosphere models since high latitude disturbances (e.g., stratospheric warmings) are of particular interest. One advantage of spectral models is that they avoid these pole problems.

Wave Growth with Altitude

Middle atmosphere models typically span from eight to fifteen scale heights in vertical extent. Thus, wave disturbances are expected to grow by factors of about 50-1500 in the absence of dissipation and wave ducting. Even with dissipation and ducting effects present, waves will show very great amplitude growth with height. Primitive equation models of the middle atmosphere will look quite noisy at high levels for this reason. Damping these waves out artifically may lead to undesired momentum and heat depositions. This is one reason why deep middle atmosphere models impose more stringent computational demands than do the more shallow lower atmosphere models.

Verification of Middle Atmosphere Models

Middle atmosphere general circulation data is just begining to be presented for multiyear data sets. For example, zonally averaged temperatures, mean zonal winds, planetary wave structures, and eddy flux quantities from the ground to 55 km have been presented by Geller et al. (1983a) for four Northern Hemisphere winters. These values are not representative for individual years, however, as can be seen from the large interannual variability in the individual monthly mean zonal winds that went into this four year average (Figure 12 from Geller et al., 1983b). Such interannual variability is also seen in the planetary waves and in the standing and transient eddy fluxes. For instance, Smith (1983) has looked at four different winters than did Geller et al. (1983a,b), and Smith (1983) found the January wavenumber one to be less than in December and February while Geller et al. (1983a) find wavenumber one to be largest in January. There are at least two implications here. One is the requirement that relatively complete troposphere/middle atmosphere models should be able to model the observed interannual variability, and the second is that comparison of monthly mean model statistics with observed statistics by itself is not a fruitful strategy for model verification. Geller et al. (1983b) suggest certain features of the observations that persist from year to year that should be compared with model behavior. Of course, model statistics should also be in a range that is spanned by the observations.

494

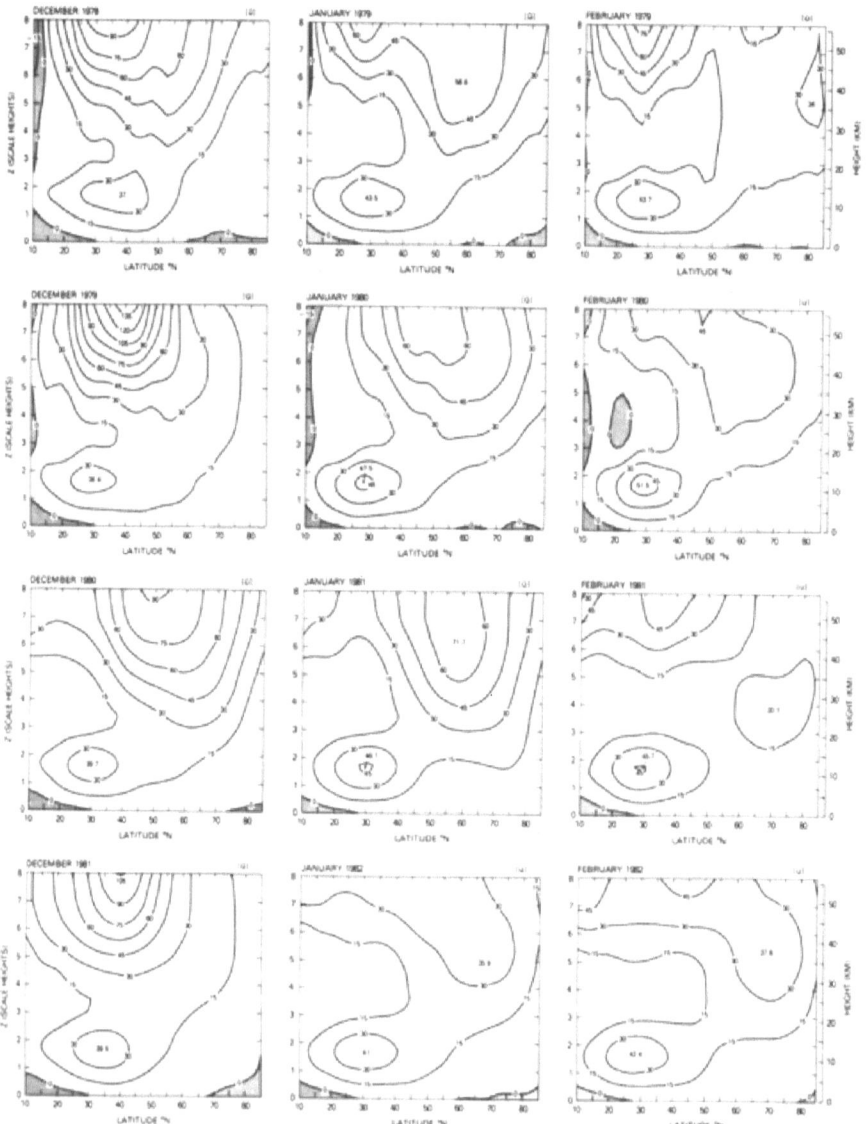

Figure 12. Northern Hemisphere average December (left column), January
(middle column) and February (right column) mean zonal winds,
[ū] in ms⁻¹, for the winters of 1978-79 (top row), 1979-80
(second row from top), 1980-81 (third row from top), and
1981-82 (bottom row). From Geller et al. (1983b).

Acknowledgement

I thank Drs. T. Matsuno, T. Tokioka, and M. Schoeberl for very valuable
comments on an earlier version of this paper.

REFERENCES

Alpert, J. C., M. A. Geller, and S. K. Avery, 1983: The response of
    stationary planetary waves to tropospheric forcing. (to appear in
    J. Atmos. Sci.)

Apruzese, J. P., M. R. Schoeberl, and D. F. Strobel, 1982: Parameteri-
    zation of IR cooling in a middle atmosphere dynamics model 1. Effects
    on the zonally averaged circulation. J. Geophys. Res., 87, 8951-8966.

Arakawa, A., and V. R. Lamb, 1977: Computational design of the basic
    dynamical processes of the UCLA general circulation model. Methods
    in Computational Physics, 17, J. Chang (ed.), Academic Press, New
    York, 173-265.

Butchart, N., S. A. Clough, T. N. Palmer, and P. J. Trevelyan, 1982:
    Simulations of an observed stratospheric warming with quasi-
    geostrophic refractive index as a model diagnostic. Quart. J. R.
    Met. Soc., 108, 475-502.

Chanin, M. L., and A. Hauchecorne, 1982: Lidar observation of gravity
    and tidal waves in the middle atmosphere. J. Geophys. Res., 86,
    9715-9721.

Cunnold, D., F. Alyea, N. Phillips, and R. Prinn, 1975: A three-
    dimensional dynamical chemical model of atmospheric ozone. J. Atmos.
    Sci., 32, 170-194.

Cunnold, D. M., F. N. Alyea, and R. G. Prinn, 1980: Preliminary calcu-
    lations concerning the maintenance of the zonal mean ozone distri-
    bution in the northern hemisphere. Pure. Appl. Geophys., 118,
    329-354.

Dickinson, R. E., 1966: Propagators of Atmospheric Motions, Rept. No. 18,
    Planetary Circulations Project, Massachusetts Institute of Technology,
    Dept. of Meteorology, Cambridge, Massachusettes, 243 pp.

Dickinson, R. E., 1973: Method of parameterization for infrared cooling
    between altitudes of 30 and 70 kilometers. J. Geophys. Res., 78,
    4451-4457.

Dickinson, R. E., 1980: Planetary waves: Theory and observation.
    Orographic Effects in Planetary Flows, GARP Publ. Ser. No. 23, WMO,
    ICSU, 56-84.

Dunkerton, T. J., 1982: Stochastic parameterization of gravity wave stresses. J. Atmos. Sci., 39, 1711-1725.

Dunkerton, T. J., C.-P.F. Hsu, and M. E. McIntyre, 1981: Some Eulerian and Lagrangian diagnostics for a model stratospheric warming. J. Atmos. Sci., 38, 819-843.

Fels, S. B., J. D. Mahlman, M. D. Schwarzkopf, and R. W. Sinclair, 1980: Stratospheric sensitivity to perturbations in ozone and carbon dioxide: Radiative and dynamical response. J. Atmos. Sci., 37, 2265-2297.

Geller, M. A., 1983: Dynamics of the Middle Atmosphere. Space Sci. Rev., 34, 359-375.

Geller, M. A., and J. C. Alpert, 1980: Planetary wave coupling between the troposphere and the middle atmosphere as a possible sun-weather mechanism. J. Atmos. Sci., 37, 1197-1215.

Geller, M. A., M.-F. Wu, and M. E. Gelman, 1983a: Troposphere-strato-sphere (surface 55km) monthly winter general circulation statistics for the northern hemisphere-four year averages. J. Atmos. Sci., 40, 1334-1352.

Geller, M. A., M.F. Wu, and M. E. Gelman, 1983b: Troposphere-strato-sphere (surface 55km) monthly winter general circulation statistics for the northern hemisphere-interannual variability. (submitted to J. Atmos. Sci.).

Haltiner, G. J. and R. T. Williams, 1980: Numerical Prediction and Dynamic Meteorology (second edition), John Wiley & Sons, New York, 477 pp.

Holton, J. R., 1982: The role of gravity wave induced drag and diffusion in the momentum budget of the mesosphere. J. Atmos. Sci., 39, 791-799.

Holton, J. R., 1983: The influence of gravity wave breaking on the general circulation of the middle atmosphere. J. Atmos. Sci., 40, (in press).

Holton, J. R., and W. M. Wehrbein, 1980: A numerical model of the zonal mean circulation of the middle atmosphere. Pure Appl. Geophys., 118, 284-306.

Hsu, C.-P., 1980: Air parcel motions during a numerically simulated sudden stratospheric warming. J. Atmos. Sci., 37, 2768-2792.

Huang, R.-H., and K. Gambo, 1982: The response of a hemispheric multi-level model atmosphere to forcing by topography and stationary heat sources (1) Forcing by topography. J. Meteor. Soc. Japan, 60, 78-107.

Hunt, B. G., 1981b: The maintenance of the zonal mean state of the upper atmosphere as represented in a three-dimensional general circulation model extending to 100 km,. J. Atmos. Sci., 38, 2172-2186.

Hunt, B. G., 1981b: An evaluation of a sun-weather mechanism using a general circulation model of the atmosphere. J. Geophys. Res., 86, 1233-1245.

Kálnay-Rivas, E., A. Bayliss, and J. Storch, 1977: The 4th order GISS model of the global atmosphere. Beitr. Atmos., 50, 299-311.

Kasahara, A., T. Sasamori, and W. M. Washington, 1973: Simulation experiments with a 12-layer stratospheric global circulation model. I. Dynamical effect of the earth's orography and thermal influence of continentality. J. Atmos. Sci., 30, 1229-1251.

Kida, H., 1977: A numerical investigation of the atmospheric general circulation and stratospheric-tropospheric mass exchange: I. Long-term integration of a simplified general circulation model. II. Lagrangian motion of the atmosphere. J. Meteor. Soc. Japan, 55, 52-88.

Kida, H., 1983: General circulation of air parcel and transport in stratosphere and troposphere derived from GCM. I. Mean mass flow in the lower Stratosphere. II. Very long-term motions of air parcels. (to appear in J. Meteor. Soc. Japan).

Kurihara, Y., 1965: Numerical integration of the primitive equations on a spherical grid. Mon. Wea. Rev., 93, 399-415.

Leovy, C. B., 1964: Simple models of thermally driven mesospheric circulations. J. Atmos. Sci., 21, 327-341.

Levy, H., II, J. D. Mahlman, and W. J. Moxim, 1979: A preliminary report on the numerical simulation of the three-dimensional structure and variability of atmospheric N_2O. Geophys. Res. Lett., 6, 155-158.

Lin, B.-D., 1982: The behavior of winter stationary planetary waves forced by topography and diabatic heating. J. Atmos. Sci., 39, 1206-1226.

Lindzen, R. S., 1981: Turbulence and stress owing to gravity wave and tidal breakdown. J. Geophys. Res., 86, 9707-9714.

Lindzen, R. S., T. Aso, and D. Jacqmin, 1982: Linearized calculations of stationary waves in the atmosphere. J. Meteor. Soc. Japan, 60, 66-77.

Lindzen, R. S., E. S. Batten, and J. W. Kim, 1968: Oscillations in atmospheres with tops. Mon. Wea. Rev., 96, 133-140.

Mahlman, J. D., 1975: Some fundamental limitations of simplified transport models as implied by results from a three-dimensional general circulational tracer model. Proc. Fourth Conference Climatic Impact Assessment Program, T. M. Hard and A. J. Broderick, editors, DOT-TSC-OST-75-38, U.S. Department of Transportation, Washington, D.C., 132-146.

Mahlman, J. D., 1980: Coupling of atmospheric observation with comprehensive numerical models. Proc. of ICMUA Sessions and IUGG Symposium 18, XVII IUGG General Assembly, Canberra, Australia, 253-259.

Mahlman, J. D., H. Levy II, and W. J. Moxim, 1980: Three-dimensional tracer structure and behavior as simulated in two ozone precursor experiments. J. Atmos. Sci., 37, 655-685.

Mahlman, J. D., and W. J. Moxim, 1976: A method for calculating more accurate budget analyses of "sigma" coordinate model results. Mon. Wea. Rev., 104, 1102-1106.

Mahlman, J. D., and W. J. Moxim, 1978: Tracer simulation using global general circulation model: Results from a midlatitude instantaneous source experiment. J. Atmos. Sci., 35, 1340-1374.

Mahlman, J. D., and L. J. Umscheid, 1983: Circulation of the middle atmosphere: successes and problems of three-dimensional modeling. (in this volume).

Manabe, S., J. L. Holloway, Jr., and H. M. Stone, 1970: Tropical circulation in a time-integration of a global model of the atmosphere. J. Atmos. Sci., 27, 580-613.

Manabe, S., and J. D. Mahlman, 1976: Simulation of a seasonal and inter-hemispheric variations in the stratospheric circulation. J. Atmos. Sci., 33, 2185-2217.

Manabe, S., J. Smagorinsky, and R. F. Strickler, 1965: Simulated climatology of a general circulation model with a hydrologic cycle. Mon. Wea. Rev., 93, 769-798.

Manabe, S., and T. B. Terpstra, 1974: The effects of mountains on the general circulation of the atmosphere as identified by numerical experiments. J. Atmos. Sci., 31, 3-42.

Matsuno, T., 1971: A dynamical model of the stratospheric sudden warming. J. Atmos. Sci., 28, 1479-1494.

Matsuno, T., 1982: A quasi one-dimensional model of the middle atmosphere circulation interacting with internal gravity waves. J. Meteor. Soc. Japan, 60, 215-226.

McPherson, R. D., K. H. Bergman, R. E. Kistler, G. E. Rasch, and D. S. Gordon, 1979: The NMC operational global data assimilation system. Mon. Wea. Rev., 107, 1445-1461.

Mechoso, C. R., M. J. Suarez, K. Yamazaki, J. Spahr, and A. Arakawa, 1982: A study of the sensitivity of numerical forecasts to an upper boundary in the lower stratosphere. Mon. Wea. Rev., 110, 1984-1993.

Miyahara, S., 1983: A numerical simulation of the zonal mean circulation of the middle atmosphere including effects of solar diurnal tidal waves and internal gravity waves: solstice condition. (in this volume).

Miyakoda, K., R. F. Strickler, and G. D. Hembree, 1970: Numerical simulation of the breakdown of a polar-night vortex in the stratosphere. J. Atmos. Sci., 27, 139-154.

Newell, R. E., 1968: The general circulation of the atmosphere above 60 km. Meteor. Monogr., No. 31, Amer. Meteor. Soc., 98-113.

O'Neill, A., R. L. Newson, R. J. Murgatroyd, 1982: An analysis of the large-scale features of the upper troposphere and the stratosphere in a global, three-dimensional, general circulation model. Quart. J. R. Met. Soc., 108, 25-53.

Phillips, N. A., 1957: A coordinate system having some special advantages for numerical forecasting. J. Meteor., 14, 184-185.

Pyle, J. A., 1980: A calculation of the possible depletion of ozone by chlorofluorocarbons using a two-dimensional model. Pure Appl. Geophys., 118, 355-377.

Ramanathan, V., and W. L. Grose, 1978: A numerical simulation of seasonal stratospheric climate: Part I. Zonal temperatures and winds. J. Atmos. Sci., 35, 600-614.

Ramanathan, V., E. J. Pitcher, R. C. Malone, and M. L. Blackmon, 1983: The response of a spectral general circulation model to refinements in radiative processes. J. Atmos. Sci., 40, 605-630.

Schlesinger, M. E. and Y. Mintz, 1979: Numerical simulation of ozone production, transport and distribution with a global atmospheric general circulation model. J. Atmos. Sci., 36, 1325-1361.

Schoeberl, M. R., and M. A. Geller, 1976: The structure of stationary planetary waves in winter in relation to the polar night jet intensity. Geophys. Res. Lett., 3, 177-180.

Schoeberl, M. R., and D. F. Strobel, 1978: The zonally averaged circulation of the middle atmosphere. J. Atmos. Sci., 35, 577-591.

Simmons, A. J., and R. Strüfing, 1983: Numerical forecasts of stratospheric warming events using a model with a hybrid vertical coordinate.
Quart. J. R. Met. Soc., 109, 81-111.

Smagorinsky, J., S. Manabe, and J. L. Holloway, 1965: Numerical results
from a nine-level general circulation model of the atmosphere. Mon.
Wea. Rev., 3, 727-768.

Smith, A. K., 1983: Stationary waves in the winter stratosphere: seasonal
and interannual variability. J. Atmos. Sci., 40, 245-261.

Tuck, A. F., 1979: A comparison of one- two- and three-dimensional model
representations of stratospheric gases. Phil. Trans. Roy. Soc. Lond.,
A290, 477-494.

Vincent, R. A., and J. M. Reid, 1983: Radar measurements of gravity wave
breaking and stress in the mesosphere. J. Atmos. Sci., 40, 1321-1333.

Wehrbein, W. M., and C. B. Leovy, 1982: An accurate radiative heating and
cooling algorithm for use in a dynamical model of the middle atmosphere.
J. Atmos. Sci., 39, 1532-1544.

WMO, 1981 The Stratosphere 1981: Theory and Measurements. WMO Global
Ozone Research and Monitoring Project Report No. 11, World Meteorological Organization, Geneva, Switzerland.

J. R. Holton and T. Matsuno, Dynamics of the Middle Atmosphere, 501-525.

DYNAMICS OF THE MIDDLE ATMOSPHERE: SUCCESSES AND PROBLEMS OF THE
GFDL "SKYHI" GENERAL CIRCULATION MODEL

J. D. Mahlman and L. J. Umscheid

Geophysical Fluid Dynamics Laboratory
Princeton University

ABSTRACT

The seasonal variation of the circulation of the middle atmosphere
is investigated using a medium resolution (5° latitude by 6° longitude) version of the GFDL "SKYHI" general circulation model. The
model includes rather complete tropospheric physical and dynamical
processes as well as a state-of-the-art radiative transfer algorithm.

Simulation successes in the model middle atmosphere include: the
cold equatorial tropopause; the midlatitude warm belt in the winter
lower stratosphere; clear separation of the subtropical and polar
night jet streams; stratospheric summer easterlies of the proper speed
and extent; a strong sudden warming type event just above the stratopause; closing off of the polar night jet stream in the mesosphere;
and a pronounced equatorial semi-annual oscillation.

Model results which must be regarded as failures include: excessively cold polar night vortices with associated overly intense zonal
winds; an underestimate of the amount of wave activity escaping the
troposphere; a mesospheric summertime easterly jet which does not
close off; and no evidence of a quasi-biennial oscillation in the
equatorial lower stratosphere.

1. INTRODUCTION

Meteorologists have been working on the problem of explaining the
dynamics of the winter extratropical stratosphere for more than 25
years. Much of the motivation for this interest has been inspired by
observations of the sudden stratospheric warming phenomenon.

Perhaps surprisingly, nearly 20 years ago efforts were already
underway to construct self-consistent numerical models of the stratosphere (Smagorinsky et al., 1965; Manabe et al., 1965; Peng, 1965).
Of all the models currently employed, we will emphasize here our own

general circulation model (GCM). In the present context, a GCM is a model which attempts to include all major physical processes self-consistently. For practical purposes here, the important differences between the GCM's and the "mechanistic" models relate to the inclusion of realistic radiative transfer, a self-determined troposphere and, perhaps, a "sufficient" model spatial resolution.

The first GCM to resolve the stratosphere adequately was that of Manabe and Hunt (1968). Although that model used annual mean insolation, and thus was inappropriate for winter conditions, it nevertheless was the first model to identify an important simulation short-coming; the structure of this annual mean model was actually closer to winter than to yearly mean conditions. Most notably, the simulated higher latitude temperatures were more typical of the winter months.

Later, GCM's with seasonal insolation, orography and "realistic" stratospheric radiative transfer were constructed (Manabe and Mahlman, 1976; Schlesinger and Mintz, 1979). In these simulations the winter polar temperatures are considerably colder ($\sim25°C$) than corresponding observations. Associated with these excessively cold temperatures are zonal westerlies which are about a factor of two stronger than observed. Manabe and Mahlman (1976) could not identify the cause of this phenomenon, but they speculated that it might have been related to the model's relatively coarse vertical resolution and use of a "lid" boundary condition above the top model level in the middle stratosphere.

In view of these modeling difficulties, it is interesting to note the presence of a parallel theoretical development that implicitly assumed a radically different view of stratospheric dynamics than implied by the GCM's. This was the application of so-called "mechanistic" models of the stratosphere to investigate its dynamical behavior (e.g., Matsuno, 1971; Geisler, 1974; Holton, 1976; Hsu, 1980; Schoeberl and Strobel, 1980). Such models are generally simplified by "switching" on disturbances through a lower boundary condition at the tropopause. Traditionally, they have included only one or two harmonics of resolution in the zonal direction. Perhaps most significantly in the present context, such models have assumed all stratospheric zonal-mean diabatic processes to either be zero, or to be governed by a simple linear damping to an observed temperature.

Both of the above treatments of diabatic processes invoke an implicit assumption that the winter (pre-sudden warming) climatological temperature distribution (and its concomitant zonal winds) are very near a state of radiative equilibrium. This is because such model's initial states are usually set close to observed values, but with an assumption that such a "basic state" is unforced. That is, in the absence of any imposed forcing, no systematic changes to the initial state will occur.

Such a view requires that, in the "pre-sudden warming strato-sphere," the integrated effect of upward propagating disturbances on the mean flow be <u>zero</u>. It is our view that this perspective was sustained by a serious misapplication of the non-acceleration theorem; that is, it was assumed that the large degree of mean-cell, eddy can-cellation in stratospheric zonal mean balances leads to no systematic effect on the zonal wind. Thus, in this view, planetary waves and other disturbances contribute significantly to zonal balances only during sudden warming events.

In this paper, we will investigate these two widely divergent viewpoints through use of the GFDL "SKYHI" GCM. In our model we have made every effort to incorporate state-of-the-art radiative transfer in the calculation of model heating and cooling rates (Fels and Schwarzkopf, 1981). Furthermore, the modeling of tropospheric pro-cesses has been constructed to the level of our current capability and resources. For more detailed description of model processes, see Fels et al. (1980), Levy, Mahlman, and Moxim (1982) and Andrews, Mahlman, and Sinclair (1983) (hereafter AMS).

This version of the "SKYHI" model has 40 levels in the vertical (see Fig. 1 tic marks for level spacing) with the highest level near 80 km, and 5° latitude by 6° longitude horizontal resolution. Thus, the historically stated reasons for the "cold pole" problem probably should not be a factor here (i.e., faulty radiative transfer, coarse vertical resolution or "lid" boundary conditions in the middle strato-sphere).

2. EXPERIMENTAL DESIGN

The physical processes in the "SKYHI" GCM for this experiment are the same as described in AMS. In the current experiment, however, an annual cycle of insolation is introduced. Also, the prescribed sea surface temperature is set at observed, seasonally varying values.

The model integration begins on astronomical date 22 September 1981 using instantaneous model data from day 1229 of the annual mean insolation experiment described by AMS. Because conditions at the equinox resemble those of the annual mean, the integration becomes rather close to its statistical (seasonal) equilibrium state within a few months. This integration has been completed out to model month May "1983."

3. MODEL RESULTS

In this paper we will present only those model results that relate to some current issues in simulating stratospheric dynamical processes. Accordingly, we will concentrate here on the zonal-mean behavior and defer topics dealing specifically with disturbance structure to a later paper.

We present in Fig. 1 the zonal mean temperature ($\overline{T}^\lambda$) averaged over the first January after beginning the experiment. This figure shows the following features in the lower stratosphere ($\sim$100 mb): a cold equatorial tropopause with the lowest temperatures near 188°K (a few degrees below observed); a well developed midlatitude warm belt near 55°N; a relatively cold north polar region (about 15° below observed); and a monotonic increasing $\overline{T}^\lambda$ from the equator to the south pole. The cold north polar region becomes even colder relative to observations (e.g., Murgatroyd, 1969) in the middle stratosphere ($\sim$30° too cold) and near the stratopause ($\sim$50° too cold). At the stratopause $\overline{T}^\lambda$ decreases monotonically from the summer pole ($\sim$280°K) to the winter pole ($\sim$180°K). In the mesosphere, Fig. 1 shows $\overline{T}^\lambda$ increasing from the equator toward both poles; a $\overline{T}^\lambda$ maximum is seen near 70°N, with decreasing $\overline{T}^\lambda$ between 70°N and the north pole.

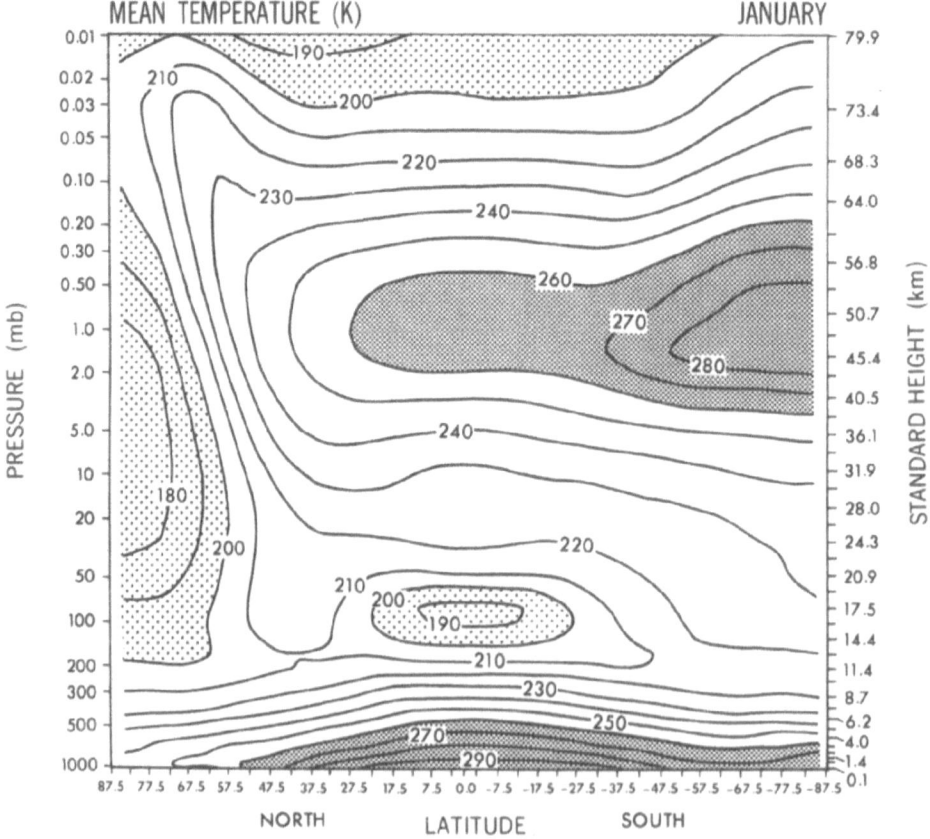

Fig. 1. Zonally averaged temperature ($\overline{T}^\lambda$) in degrees K for the astronomical) month of January "1982". Samples taken once per model day.

Fig. 2 shows the mean zonal wind $(\overline{u}^{\lambda})$ averaged over simulated January "1982". This $\overline{u}^{\lambda}$ field is very close to gradient wind balance with the $\overline{T}^{\lambda}$ field shown in Fig. 1. In the troposphere, the maximum zonal winds are somewhat too strong and appear somewhat equatorward of their observed positions; this leads to upper tropospheric equatorial westerlies which are stronger than observed. Previous modeling experience (e.g., Manabe and Mahlman, 1976) indicates that these simulation deficiencies are largely rectified in simulations with somewhat higher horizontal resolution.

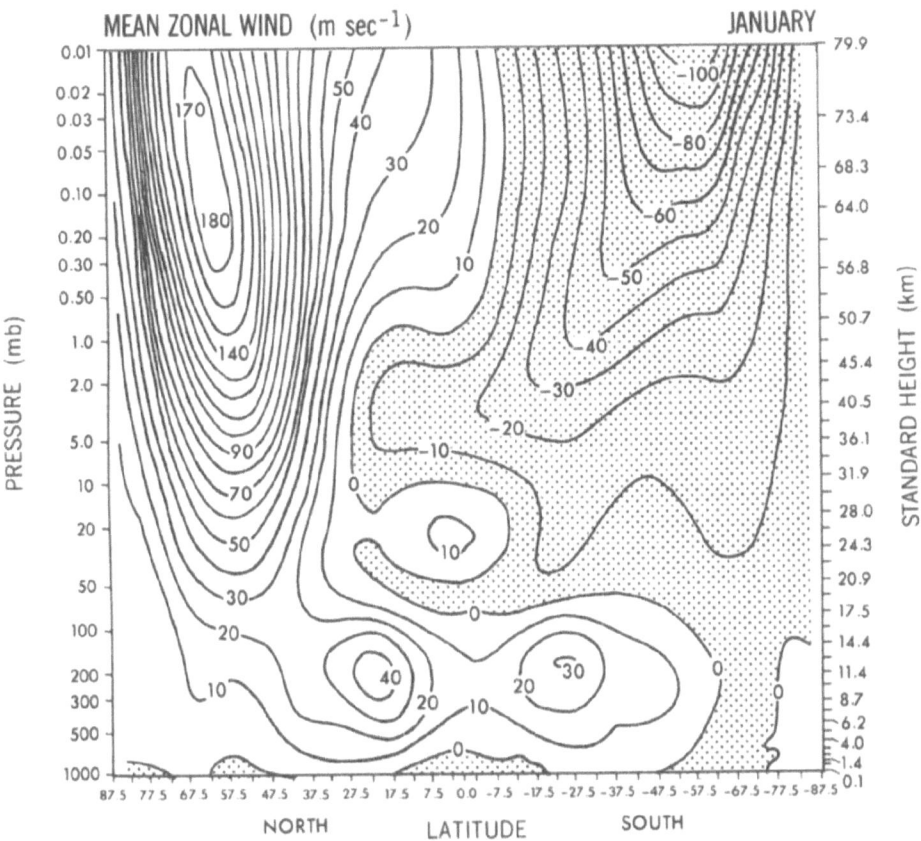

Fig. 2. Zonally averaged zonal wind $(\overline{u}^{\lambda})$ in m sec^{-1} for January "1982."

In the stratosphere and mesosphere, Fig. 2 shows Southern Hemisphere easterlies everywhere above the approximately correct zero line near 50 mb. The speed of these easterlies is rather realistic up to about .1 mb(∿65 km). However, above this level, the easterlies continue to increase with altitude to the top of the model. This is in sharp contrast to observations which show a closed off easterly jet near 65 km, accompanied by very cold polar temperatures near 80 km. In this simulation, the **warmest** mesopause temperatures are found near the south pole. This has to be regarded as a significant simulation failure.

Note in Fig. 2 that the Southern Hemisphere easterlies penetrate all the way across the equator and virtually surround the equatorial westerly jet near 20 mb (∿25 km). At other times of the year, the equatorial jet is considerably stronger. The result shown in Fig. 2 is related to the easterly phase of the model's simulated semi-annual oscillation (for more details, see Section 5). Also, this equatorial jet is a pronounced feature in the companion annual mean results of AMS. Its structure and dynamics are described there and in Hayashi, Golder, and Mahlman (1984).

A dominant feature of Fig. 2 is the very strong polar night jet stream with a $\bar{u}^{\lambda}$ maximum near 65° and .1 mb. Note that this jet is clearly separated from the simulated tropospheric subtropical jet. This is a marked improvement over previous January simulations (e.g., Manabe and Mahlman, 1976; Schlesinger and Mintz, 1979). However, as expected from Fig. 1, a serious discrepancy remains; the zonal wind speeds near the core of the jet are too large by more than a factor of two. An encouraging result, though, is that the jet is "closed off" such that the maximum speeds are near 65 km, rather than at the highest level as is typical of previous GCM's. To the best of our knowledge, this is the first GCM simulation to accomplish this with use of a proper radiation scheme or without addition of a large **external** retarding force on the zonal wind.

In spite of some of the above simulation successes, we nevertheless are forced by these results to address the obvious deficiency of the excessively cold polar night vortex and its concomitant excessively strong westerly winds in higher latitudes. We begin by posing the question as to what temperature distibution the January middle atmosphere would have in the absence of any in-situ dynamical processes.

To investigate the implication of the radiative effects, we appeal to the time-dependent, "radiative-convective" model results of Fels and Schwarzkopf (1984). The radiative transfer algorithm is exactly the same as used in this GCM study. In their calculation they set the surface temperature at the observed zonal-mean, seasonally varying values and prescribe the ozone (below 35 km) and cloudiness distribution at the annual mean values used in Fels et al. (1980).

The ozone above 35 km in both the radiative-convective model and this
GCM is allowed to "float" (in response to temperature variations)
toward a crude ozone photochemical equilibrium. This is accompanied
by the method outlined in Fels et al. (1980, Appendix A), but with
improved chemical rate coefficients and a correction for density
variations.

The temperature is time-marched through the annual cycle so that,
strictly speaking, radiative equilibrium is only attained in the
annual average. The calculations are performed independently at each
5° latitude. Thus no meridional coupling is present except that
implied in the prescribed quantities.

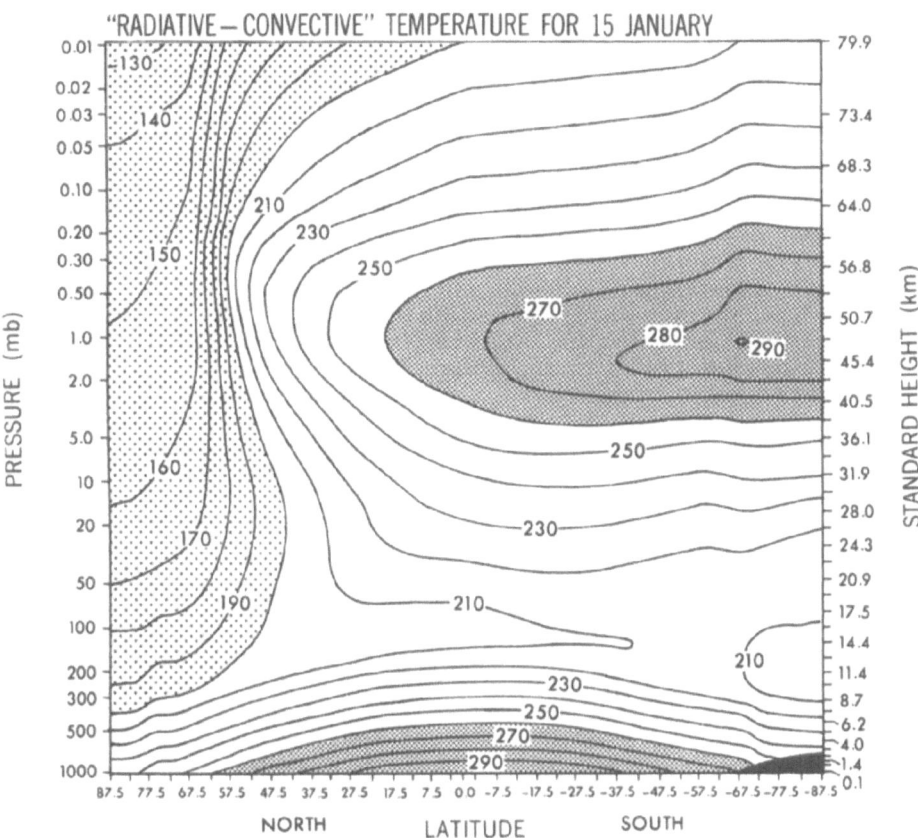

Fig. 3. Time-dependent "radiative-convective" temperature ($T_{R.C.}$) for
15 January "1982" from the calculation of Fels and Schwarzkopf (1984).
The surface temperatures are prescribed at their seasonally varying
observed values. Ozone and clouds are the same as used in Fels et al.
(1980). Details of the water vapor prescription are relatively standard and are described in Fels and Schwarzkopf (1984).

The results of the Fels and Schwarzkopf (1984) radiative-convective temperature for astronomical date 15 January are given in Fig. 3. The most noteworthy middle atmosphere feature of Fig. 3 is the very strong pole-to-pole temperature gradient, with extremely low values near the north pole. Of special interest is the contrast between these radiative-convective temperatures and the GCM temperatures in Fig. 1. The GCM north polar temperatures are, at all altitudes, warmer than the radiative-convective temperatures. In the lower to middle stratosphere the GCM temperatures are about 15° warmer. Near the stratopause and middle mesosphere, the GCM temperatures are about 30° to 60° warmer, respectively.

A further comparison of Figs. 1 and 3 indicates that the lower stratosphere and the Northern Hemisphere middle mesosphere thermal structures must be under strong dynamical control. It is in these two regions of the model that the meridional temperature gradients are opposite to those present in the radiative-convective atmosphere. Thus, the "dynamical heating" effect (Fels et al., 1980) is of major importance in the model heat balance of these regions.

In the northern higher latitudes, radiative control is implied to a first approximation. However, the above comparisons have shown that the GCM temperatures there are considerably warmer than the radiative-convective temperatures. Further, the observed winter polar temperatures (e.g., Murgatroyd, 1969) are warmer than the GCM temperatures by a more or less equal amount. This implies that the real winter polar middle atmosphere is under significantly stronger dynamical control than is the GCM. In other words, the GCM needs considerably more "dynamical heating" in the polar night to allow the model temperatures to exist at the relatively large departures from the radiative temperatures that appear to be typical of the real atmosphere.

The inference here is that the fundamental cause of the GCM's excessively cold winter polar temperatures is dynamical in origin. This contention implies that it is not a problem in our understanding of radiative transfer that is the dominant contributor to the model's deficiency. According to Fels[1] the absolute accuracy of "properly modeled" stratospheric infra-red cooling rates for a given distribution of absorbers should be less than $\pm$ 20% of the actual cooling rates. (Uncertainties in short wave heating rates are much less a factor in this question because the major problem arises in the polar night.) For the GCM to obtain the proper polar night temperatures, the infra-red cooling rates would have to be perhaps half the values typical of the current model. Thus, we attribute the essential cause of the discrepancy to insufficient "dynamical heating" of the model polar night middle atmosphere. Accordingly, we seek here to identify the causes of this dynamical deficiency.

---

[1]Fels, S. B., 1982: personal communication

A large body of recent literature (e.g., Andrews and McIntyre, 1976; Edmon, Hoskins, and McIntyre, 1980; Dunkerton, Hsu, and McIntyre 1981; AMS) has shown the quantity called the Eliassen-Palm (E-P) flux and its divergence plays a fundamentally important role in the dynamics of the interplay of zonal-mean flows and "eddies." More importantly, the so-called transformed Eulerian mean equations appear to provide a meaningful interpretation of the influence the eddies are exerting on the zonal mean flow (see AMS). In the transformed zonal momentum equation, the Eliassen-Palm flux divergence (EPFD) appears explicitly as a force per unit mass exerted on the zonal flow. Thus, in this study its magnitude and structure may be interpreted in terms of the adequacy of the model forcing of the mean flow by the eddies. Specifically, we anticipate a decelerative EPFD because the "radiative-convective" atmosphere of Fig. 3 indicates that the radiative forcing is attempting to relax to an atmosphere with considerably stronger wintertime meridional temperature gradients, and thus, a considerably stronger polar night jet stream.

Fig. 4 shows a cross section of the direction (not magnitude) of the E-P flux and its divergence (EPFD). In the Northern Hemisphere mesosphere, the EPFD shows negative (or westerly reducing) magnitudes ranging from 8 to 40 x 10^{-5}m sec^{-2}. The Northern Hemisphere stratosphere shows typical values ranging from -.5 to -4 x 10^{-5}, but a small positive "spot" appears near 3 mb between 50 and 60°N. Such a "spot" has been seen in the observational results of Palmer (1981), Hamilton (1982a) and Geller, Wu, and Gelman (1983). To the best of our knowledge, an interpretation of this phenomenon has not yet been published.[2] Note that these values of negative EPFD are typically a factor of four larger in magnitude than those reported by AMS for the annual mean insolation version of this "SKYHI" GCM.

The tropospheric EPFD structure is rather similar to that in AMS. As a lengthy analysis of that structure is offered in their paper, no further discussion will be presented here. Also, note in Fig. 4 the

---

[2]R. A. Plumb (1983, personal communication) offers the following hypothesis: Since the "spot" of positive EPFD occurs near the base of the polar night jet (see Figs. 2 and 4), it is probably associated with a reversed meridional gradient of quasi-geostrophic potential vorticity. If the waves are acting to produce a down gradient flux of potential vorticity (to maintain wave amplitudes against dissipation), the meridional flux of potential vorticity must be locally poleward. Because of the equivalence between quasi-geostrophic EPFD and the flux of potential vorticity (Dickinson, 1969), the EPFD must then be locally positive.

downward directed E-P arrows in the (summer) Southern Hemisphere lower
latitudes. At first glance, this appears to indicate downward propagation of wave activity because the vertical and meridional direction
of the E-P flux is reversed there. This probably relates to the
upward-propagating waves there having dominantly <u>eastward</u> phase speeds
relative to the basic (easterly) zonal winds. These phenomena will be
investigated in a later study.

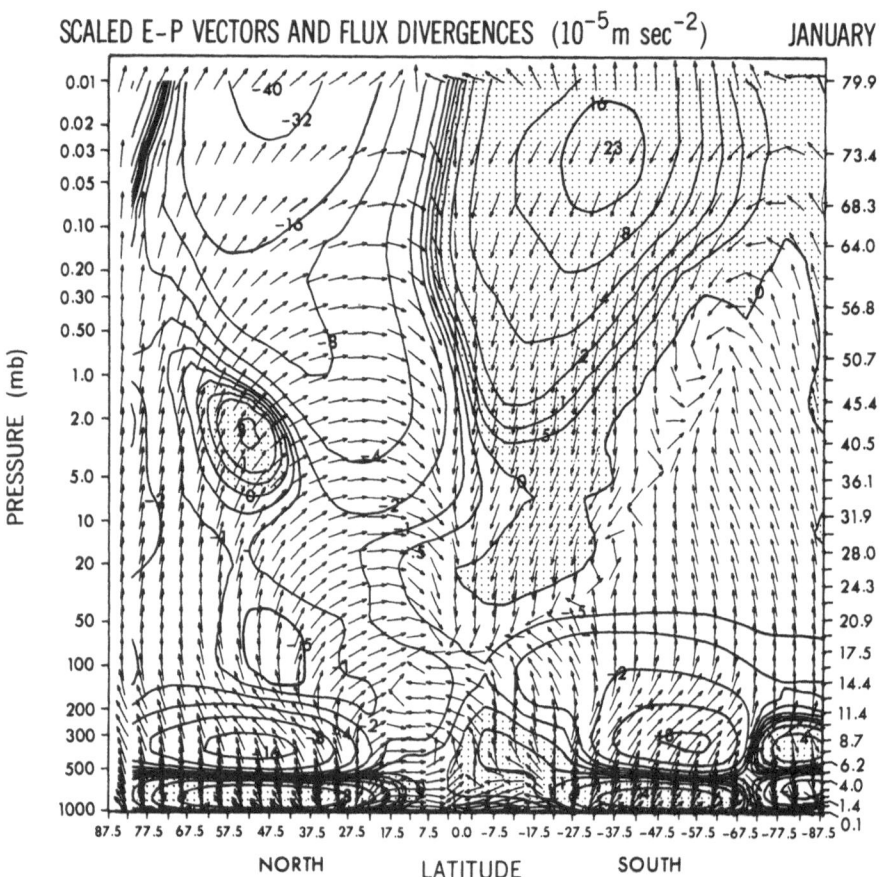

Fig. 4. Meridional cross section of Eliassen-Palm flux vector directions (arrows) and contours of Eliassen-Palm flux divergence (EPFD)
normalized as zonal force per unit mass (10^{-5} m sec^{-2}) for January
1982. Regions of positive EPFD (representing an eastward force per
unit mass) are shaded. To ease interpretation, the EPFD field has
been subjected to a 1-2-1 smoothing operator in the meridional and
vertical.

Overall, the Northern Hemisphere structure of the EPFD in Fig. 4 forces us to conclude that the reason for the model's excessive zonal winds (Fig. 2) is attributable to the insufficient decelerative force per unit mass exerted by the eddies on the zonal flow. Thus, the magnitudes of negative EPFD should be larger in the model at virtually all altitudes above 10 km in the winter hemisphere; otherwise the zonal wind speeds will remain too intense. What remains unclear, however, is the amount and character of additional forcing required to bring the winter middle atmosphere closer to the observed structure. This uncertainty arises because of the effect of mean flow changes in the receptivity of the middle atmosphere to further tropospheric forcing on the mean flow. A spectacular example of such a positive feedback is the sudden stratospheric warming phenomenon. An example of a SKYHI sudden warming event is described in the following section because of its obvious relevance to the "cold bias" deficiency under discussion here.

4. A SUDDEN WARMING-TYPE EVENT

As pointed out in the previous section, the winter high-latitude SKYHI temperatures are considerably colder than observed. However, in late January the high-latitude temperatures of the lower mesosphere begin to warm appreciably.

Fig. 5 shows a time series of 5-day average $\overline{T}^\lambda$ at 85.5°N and 0.13 mb ($\sim$64 km standard altitude), at 0.76 mb ($\sim$50 km) and 12.2 mb ($\sim$30 km). Note that the temperature at 0.13 mb rises by about 43° over a 15-day period (even more in the daily values). At 0.76 mb the temperature rise is nearly as large, while at 12.2 mb the warming is very much smaller. These increases are large enough to be classified as "major" in any reasonable sense, but they do not penetrate very far into the stratosphere. Thus, we will call this model event a "minor warming," in agreement with traditional nomenclature.

The numbers plotted along the curves of Fig. 5 represent the difference between the local zonal-mean temperature and the radiative-convective temperature of Fels and Schwarzkopf (1984) corresponding to the indicated date ($\overline{T}^\lambda-T_{R.C.}$). These numbers show also a remarkable increase in $\overline{T}^\lambda-T_{R.C.}$ at the peak of the warming to values greater than 100°. This produces a large increase in the radiative cooling rates at the pole up to values more than 15×10^{-5} deg sec^{-1} at 0.13 mb. (During January before the warming event, the cooling rate at the same place was about 2.5×10^{-5} deg sec^{-1}.)

512

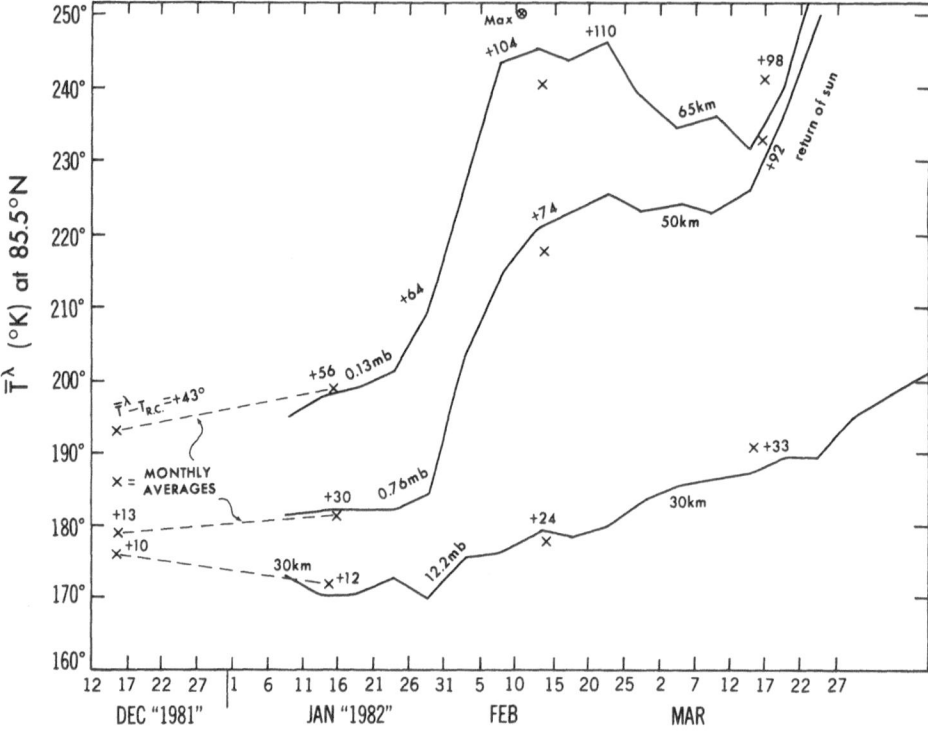

Fig. 5. Time series of 5-day averages (solid lines) of $\overline{T}^\lambda$ at 85.5°N and 0.13 mb (∿64 km <u>standard</u> altitude), 0.76 mb (∿50 km) and 12.2 mb (∿30 km) for indicated dates. Monthly mean $\overline{T}^\lambda$'s are given by x's. Values of $\overline{T}^\lambda$ - $T_{R.C.}$ for the date are given above each solid line.

As a diagnostic of the model behavior that led to this "minor warming" event, we present in Fig. 6 the same type of E-P cross section given in Fig. 4, but for the period 1-5 February. This figure shows the E-P arrow directions to be very similar to that for the January pre-warming state in Fig. 4. However, in the Northern Hemisphere, the magnitude of negative EPFD is much larger above 50 mb than for January. There are now values more negative than -20×10^{-5} m sec^{-2} near the polar night jet at 1 mb (nearly -40×10^{-5} for the largest value before smoothing) and a very large region near 5 mb that are more negative than -8×10^{-5} m sec^{-2}. These are well over an order of magnitude greater than found by AMS in the same region for annual mean conditions. This indicates a very strong wave-induced decelerative force on the zonal wind and a concomitant polar warming produced mainly by the required increased strength of the residual circulation, with descent over the polar region.

SCALED E-P VECTORS AND FLUX DIVERGENCES $(10^{-5}\,\mathrm{m\ sec^{-2}})$ 1-5 FEBRUARY

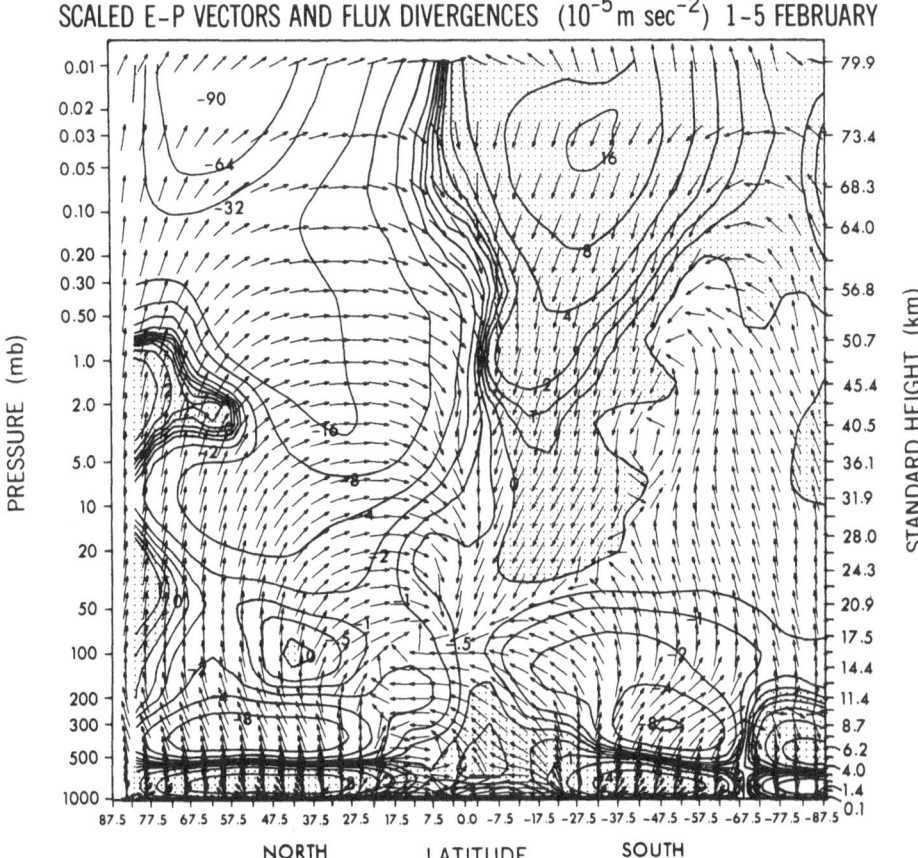

Fig. 6. Eliassen-Palm cross section as described in Fig. 4, but for peak of "minor warming" period 1-5 February "1982".

Perhaps more significantly, the largest negative value of p-component of E-P flux at 25 mb (located at 57.5°N) for the 1-5 February average is -0.35×10^{-2} mb m sec^{-2}. This is associated with a poleward eddy heat flux of about 75 deg m sec^{-1}. For the January mean the p-component of E-P flux is $-.12 \times 10^{-2}$ mb m sec^{-2} and the poleward eddy heat flux is about 25 deg m sec^{-1}. The corresponding observed January mean value of eddy heat flux from Newell et al. (1974) at the same location is about 40 deg m sec^{-1}. This suggests that the model is somewhat deficient in the average magnitude of disturbances propagating out of the troposphere. It is also probably relevant that the observations show the largest meridional eddy heat fluxes to be nearly 10° latitude poleward of the largest model values. If the model were to have its largest vertical E-P flux 10° latitude further poleward, this could allow a more efficient propagation of wave activity into the polar regions.

Thus during warming periods, the model appears capable of generating disturbances propagating into the stratosphere which are considerably more intense than the climatological values. On the average, however, the disturbance intensities appear to be somewhat weak in the model -- say two-thirds as intense as in the real lower stratosphere.

During the peak of the model "minor warming" on 1-5 February, the dominant longitudinal scale of the disturbance is planetary wave number 1. Associated with this disturbance are very large values of "traditional" diagnostic quantities in the lower and middle polar mesosphere. For example the entire polar cap above 3 mb and below 0.02 mb shows meridional eddy heat flux convergences greater than $+50 \times 10^{-5}$ deg sec^{-1} with a number of values exceeding 70×10^{-5} deg sec^{-1}. Just south of the polar cap the meridional eddy momentum flux convergences are greater than $+150 \times 10^{-5}$ m sec^{-2}, with peak values exceeding $+180 \times 10^{-5}$ m sec^{-2}. By comparison, the January averages for this region shows eddy heat flux convergences of $\sim +12 \times 10^{-5}$ deg sec^{-1} and eddy momentum flux convergences of $\sim +30 \times 10^{-5}$ m sec^{-2}.

Accompanying these large eddy flux convergences are appreciable changes in the zonal mean structure. The entire polar cap above 2 mb warms by more than 25° over the 5-day period, with peak $\overline{T}^{\wedge}$ changes exceeding 40°C near 0.35 mb. Over the same 5-day period, the zonal wind decelerates by about 25 m sec^{-1} in the middle and upper mesosphere. These results illustrate the completely inappropriate use of eddy momentum flux convergence as a predictor of deceleration. As suggested by others (e.g., Dunkerton, Hsu, and McIntyre, 1981), the structure of EPFD provides a better indicator of zonal wind forcing and strong deceleration. Contrary to their results, the analysis of AMS suggests that it is incremental EPFD above its natural background value which should usually give a better indicator of actual zonal wind decelerations.[3] On the other hand, the incremental meridional eddy heat flux convergence does correlate rather well with the actual temperature increases (but with much larger magnitudes), just as observed by Mahlman (1969).

The zonal mean temperature for 11 February (just following the peak of the warming) is given in Fig. 7. Note the temperatures are quite similar to those in Fig. 1 except in the north polar 3.0-0.02 mb region. Here, the effect of the warming is clearly evident with temperatures 30-55° warmer than the mean of the previous month.

---

[3]One must be cautious with such statements, however, because the realized zonal wind deceleration is generally non-local in the EPFD (see AMS, App. B, Eq. B.8).

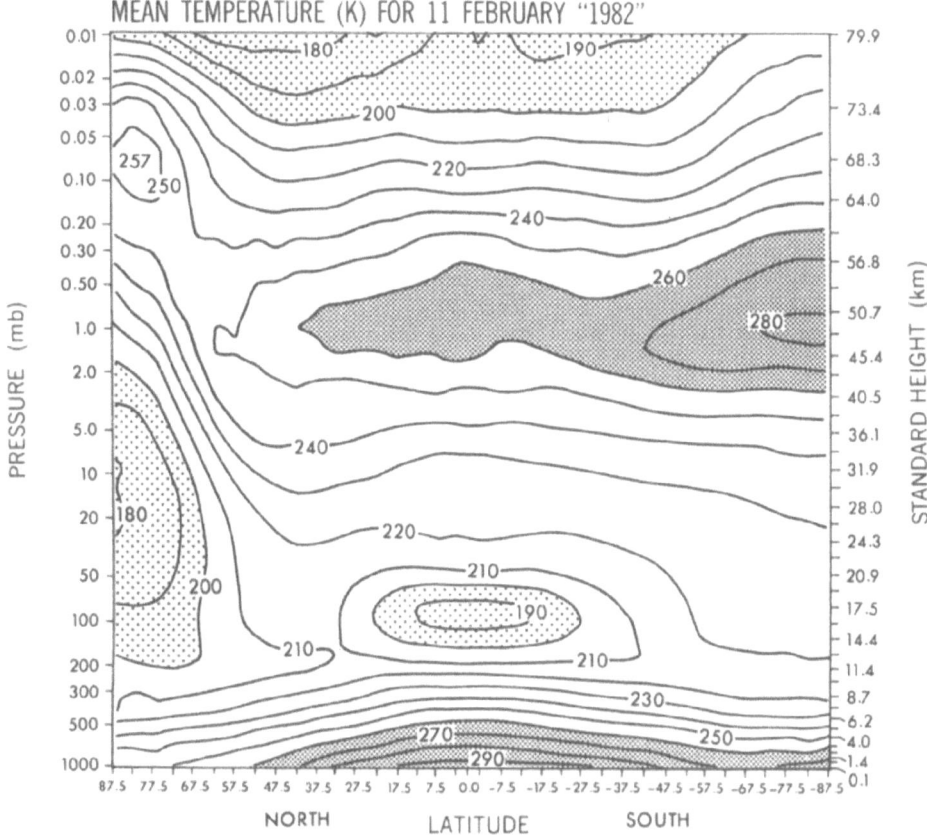

Fig. 7. Zonally averaged temperature $(\overline{T}^{\lambda})$ in degrees K for model date 11 February "1982."

The zonal wind cross section corresponding to the temperature field of Fig. 7 is given in Fig. 8. Comparison of this figure with Fig. 2 shows $\overline{u}^{\lambda}$ above the polar night jet has decelerated by over 80 m sec^{-1}. The jet core has dropped from 0.1 mb down to 0.5 mb and its core speed has diminished by over 30 m sec^{-1}. In the north polar middle stratosphere, the zonal winds have diminished only slightly, indicating how this warming does not significantly improve the middle stratospheric "cold bias" problem.

In lower Northern Hemisphere latitudes, Fig. 8 shows an isolated peak of easterlies near 25–30°N and 5–1 mb. This peak appears to be only partially explained by the advection of easterly angular momentum across the equator from the summer hemisphere giving the easterly phase of the semi-annual oscillation (Holton and Wehrbein, 1980; Mahlman and Sinclair, 1980). Model analysis shows it to be mainly caused by an easterly force per unit mass exerted by planetary waves propagating from midlatitudes. This inference is compatible with the E-P flux directions and large negative values of EPFD seen in Fig. 6 at this region of growing easterlies. The onset of easterlies is also compatible with the sudden warming mechanism given by Dunkerton, Hsu, and McIntyre (1981) who suggest the importance of low latitude easterly winds in helping to "channel" the E-P flux to higher latitudes. In our "minor warming," there is no obvious shift in E-P flux directions between Figs. 4 and 6 as the subtropical easterlies intensify. However, the magnitude of E-P flux into stratopause-level high latitudes does seem to be enhanced by this effect.

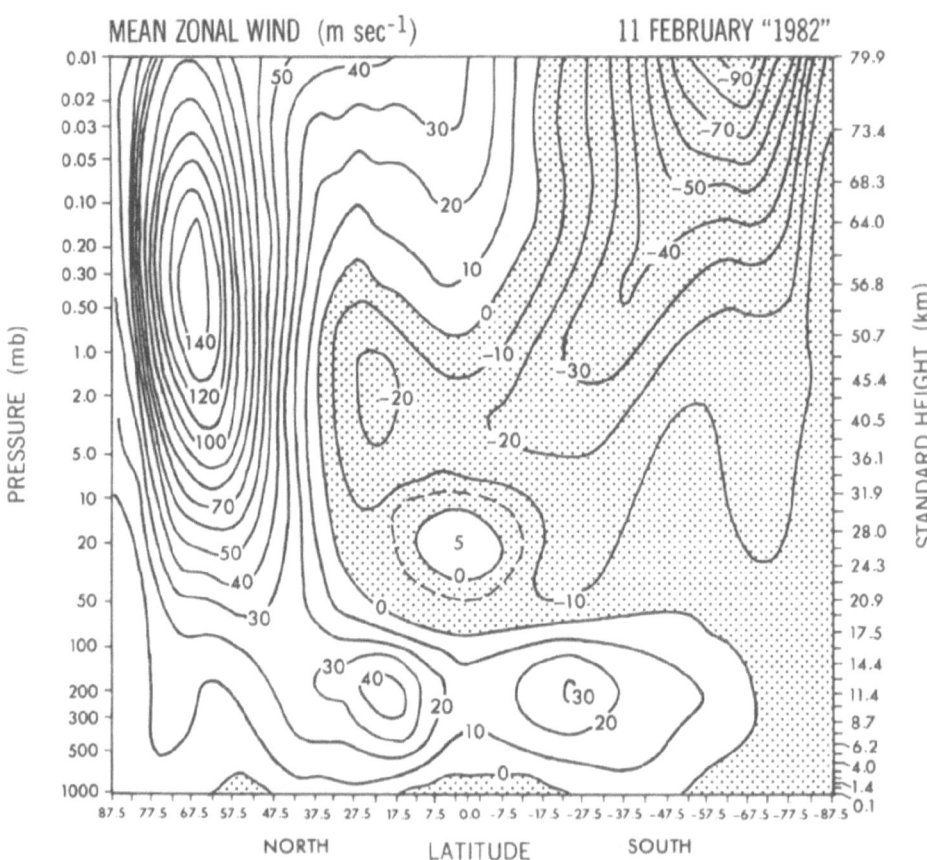

Fig. 8. Zonally averaged zonal wind $(\bar{u}^\lambda)$ in m sec^{-1} for model date 11 February "1982".

Also, the equatorial latitudes in Fig. 8 show the lower stratosphere westerly jet (Fig. 2 and AMS) is now nearly eliminated by the
easterlies. Note also, the appearance of isolated equatorial westerly
winds above 1 mb. Both of these features are related to the model's
simulated semi-annual oscillation (SAO) which will be discussed in
more detail in the next section.

5. SIMULATION OF THE SEMI-ANNUAL OSCILLATION

Another topic which has attracted the widespread attention of
theorists is the equatorial zonal wind and its temporal variations.
Since the pioneering theory of Holton and Lindzen (1972) on the quasi-
biennial oscillation (QBO), there have been many attempts to explain
these fluctuations. Later, the semi-annual oscillation (SAO) became
the center of such efforts (e.g., Hirota, 1976; Dunkerton, 1979).

More recently, some success has been obtained in modeling the SAO
by Holton and Wehrbein (1980) and by Mahlman and Sinclair (1980).
Holton and Wehrbein, using an axially symmetric model, show how advection of easterly angular momentum across the equator from the summer
hemisphere can give significant easterly accelerations. The same
mechanism is shown to be dominant in the lower resolution "SKYHI" GCM
study of Mahlman and Sinclair (1980). In addition, their study gave a
strong westerly phase to the SAO mainly due to preferential absorption
of equatorial Kelvin waves. The major weaknesses in that simulation
is that the westerly phase was too strong and the easterly phase was
too weak. Also, their SAO was essentially coherent in phase from the
middle stratosphere to the upper mesosphere. This disagrees with some
observational results (Hirota, 1980; Hamilton, 1982b) which suggest
the mesospheric SAO to be out of phase with the stratospheric SAO.
For explanation of the differences (besides resolution) between the
Mahlman and Sinclair (1980) model and the "SKYHI" GCM presented here,
see AMS.

Fig. 9 shows the time series of monthly average $\overline{u}^\lambda$ at 2.5°N for
the first 20 months of this model integration. This figure shows a
pronounced SAO between about 30 and 55 km (standard altitude). The
model exhibits distinct easterly and westerly phases with a peak
amplitude of 20 m sec^{-1} between 40 and 50 km. This agrees well with
observations, although the westerly winds are slightly larger
(5 m sec^{-1}) than observed, while the easterlies are slightly smaller
by the same amount. There is some indication of a phase reversal into
the mesosphere, but the model semi-annual component is quite weak compared to that of the upper stratosphere.

In the lower stratosphere, this model still shows little tendency
to produce a QBO, although the prerequisite capability to produce
equatorial westerlies and easterlies is clearly present. Mechanistic

518

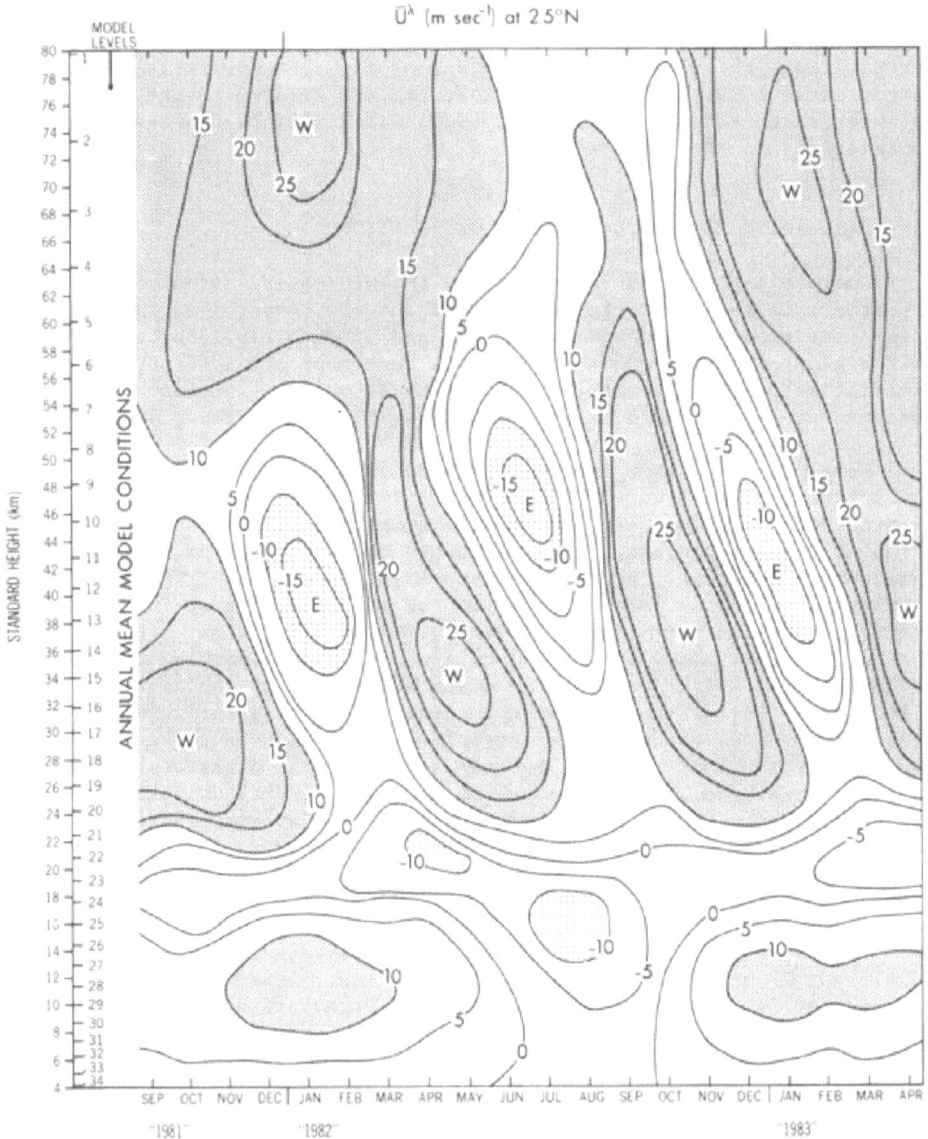

Fig. 9. Time-height cross section of monthly averaged $\overline{u}^\lambda$ at 2.5°N. Abscissa is standard altitude for the model isobaric levels. Grid point positions are as indicated. The earliest date shown (September "1981") gives the $\overline{u}^\lambda$ for the annual mean insolation model described by AMS.

model studies (e.g., Plumb, 1977) suggest that perhaps greater vertical grid resolution than used here will be required to make further
progress on the QBO problem.

The westerly acceleration appears very similar to the mechanism
simulated earlier by Mahlman and Sinclair (1980). Equatorial Kelvin
waves[4] are strongly deposited at the base of the westerly shear layer
(showing positive EPFD). This produces downward propagation of the
westerly maximum (for example, in March). As the easterlies interrupt
the cycle, the Kelvin waves penetrate deeper into the stratosphere and
reinitiate the process from above. The "fast" Kelvin waves (found
observationally by Hirota, 1979) appear to be important contributors
to this simulated westerly acceleration.

The model easterly acceleration appears to be more complex than in
the previous low resolution "SKYHI" simulation of Mahlman and Sinclair
(1980) and in the symmetric model of Holton and Wehrbein (1980). The
cross-equatorial advection mechanism appears to be limited in the
sense that it is difficult to get much easterly acceleration on the
winter hemisphere side of the equator. For the (now) poleward moving
symmetric circulation to accelerate easterlies, it is necessary that
the flow be inertially unstable (or negative absolute vorticities in
the Northern Hemisphere). It does not appear possible to sustain an
inertially unstable region over the width of the observed easterlies.
An inertially unstable configuration was present in the Mahlman and
Sinclair (1980) simulation, but it was confined to a rather narrow
zone near the equator.

Perhaps more plausibly, on the winter hemisphere side of the equator the effect of planetary waves propagating into low latitudes and
inducing easterly accelerations appears as an important effect in this
simulation. See Section 4 and Figs. 6 and 8 for a discussion of this
effect in the context of the "minor warming" event. This mechanism
has been proposed earlier by Hopkins (1975) and Hirota (1980) and more
recently by Dunkerton (1982) and Takahashi (1983). The obvious advantage of the combined mechanism is the easterly acceleration can then
be spread out over a wider equatorial region with much less asymmetry
than can be readily provided by either mechanism working in isolation.
Moreover, the model shows a stronger easterly phase north of the
equator during Northern Hemisphere winter than during the Southern
Hemisphere counterpart. This is compatible with Hopkins' (1975) hypothesis that the Rossby wave propagation effect being stronger there
explains the similar asymmetry found in the observations.

Overall, this simulation of the stratospheric SAO must be regarded
as reasonably successful. The mesospheric SAO simulation is uncertain
but promising, while the QBO remains virtually absent in the current
model.

---

[4]For a detailed analysis of the "SKYHI" Kelvin wave structure, see
Hayashi, Golder and Mahlman (1984).

6. SUMMARY AND DISCUSSION

A series of simulation results from an annual cycle version of the 5° latitude by 6° longitude GFDL "SKYHI" GCM has been presented. State-of-the-art tropospheric dynamical and physical processes have been included in the model, as has the best available radiative transfer algorithm (Fels and Schwarzkopf, 1981).

Simulation successes include: the cold equatorial tropopause; mid-latitude warm belt of the winter lower stratosphere; clear separation between the subtropical and polar night jet streams; reversed meridional temperature gradient in the winter mesosphere (and closed off polar night jet); stratospheric summer easterlies of the proper magnitude and depth; a sudden warming ("minor warming") in the model lower mesosphere; and a pronounced equatorial semi-annual oscillation.

Simulation failures in the middle atmosphere include: a very cold polar night vortex with corresponding excessively strong zonal winds; a modest but significant underestimate of the magnitude of the vertical component of the Eliassen-Palm flux escaping the troposphere; an easterly jet stream in the summertime mesosphere which does not close off; and no evidence of a quasi-biennial oscillation in the tropical lower stratosphere.

The "minor warming" event exhibits many features of observed warmings which do not penetrate into the middle stratosphere. In the lower mesosphere, the polar cap warms by over 43° in 15 days. The higher temperatures increase the polar cap diabatic cooling rates to more than 15×10^{-5} deg sec^{-1}. Accompanying the warming is a deceleration of the mesospheric zonal wind by more than 80 m sec^{-1}. This "minor warming" was initiated by a large increase in the vertical component of Eliassen-Palm flux emanating from the troposphere. That flux leads to flux divergences exceeding -40×10^{-5} m sec^{-2} just above the stratopause. Such a level of model forcing is sufficient to induce the large deceleration (and its associated warming). The process appears to have been facilitated by an onset of easterly winds in the Northern Hemisphere subtropics of the upper stratosphere as suggested by Dunkerton, Hsu, and McIntyre (1981).

Simulation of the semi-annual oscillation has been successful in a number of respects. The amplitude maximum of about 20 m sec^{-1} between 40 and 50 km and the phase of the oscillation agrees rather well with observations. The westerly phase is driven by absorption of dissipating Kelvin waves, just as in the previous simulation by Mahlman and Sinclair (1980). For the easterly phase, however, the model mechanism is more complex. On the summer hemisphere side of the equator, the model easterly acceleration is provided by meridional advection of easterly angular momentum in the cross equatorial flow from the summer to the winter hemisphere. On the winter hemisphere side, however, the dominant model mechanism for the easterly acceleration is provided by absorption of planetary waves propagating in from higher latitudes as hypothesized by Hopkins (1975). Thus, according to this simulation, the easterly phase of the semi-annual oscillation depends upon the cooperative contributions of the advection and wave-propagation mechanisms.

The most serious simulation difficulty appears to be the tendency for the winter polar vortex to be significantly too cold on the average. This cold polar bias exists in the model virtually from the tropopause to the mesopause. We contend here that the basic cause for this problem is not radiative in origin. Rather, we argue the problem must arise from a dynamical deficiency. Specifically, additional dynamical mechanisms must be simulated (or parameterized) which are capable of exerting additional decelerative force per unit mass on the westerly zonal flow.

The most obvious candidate for momentum deposition in the present model context is a more vigorous planetary wave activity. The model appears to be somewhat deficient in the amount of wave activity propagating out of the troposphere. Moreover, the largest model values of vertical component of Eliassen-Palm flux in the winter lower stratosphere are found nearly 10° latitude equatorward of the observed maximums. This assumed need for increased planetary wave activity is plausible for two reasons. First, the "SKYHI" GCM shows a dramatic response to bursts of wave activity from the troposphere (Section 4). Second, the observed stratosphere shows large variations in zonal wind associated with fluctuations in planetary wave amplitudes (e.g., Geller, Wu and Gelman, 1983). Currently, we are investigating various possible causes for this model deficiency through a new series of "SKYHI" GCM test experiments.

Another possibility for increased momentum deposition in the GCM is that arising from gravity waves currently unresolved by this model. This effect is almost certainly important in the mesosphere (e.g., Lindzen, 1981; Holton, 1982; Matsuno, 1982). In fact this GCM can, in a sense, be thought of as an indirect test of the need for such a mechanism, as we attempt to include "everything else" in the model explicitly. What remains unclear, however, is the relative importance of this effect throughout the stratosphere. At these levels, in contrast to the mesosphere, considerably smaller magnitudes of gravity wave produced Eliassen-Palm flux divergence might be significant. The potential importance of smaller-scale gravity waves in the stratosphere momentum balance has been suggested by Weinstock (1982).

The possibility of important stratospheric gravity wave effects may be indirectly suggested by a vexing, yet intriguing, result of this model. In the transition season between September and October, the mean upper zonal winds at 60°N increase by more than 100 m sec^{-1}.

So, essentially at the onset of winter, the zonal winds are already stronger than winter observations. This is at a time when the tropospheric planetary waves are rather weak. Thus, the stratosphere "sees" the onset of winter radiative conditions long before the thermally buffered lower troposphere has time to react through increasing planetary wave amplitudes. In the real stratosphere, the zonal wind response to onset of winter insolation is quite significant, but apparently much weaker than found in our model. Thus, we hypothesize that the required "missing decelerator" may be provided by smaller-scale gravity waves.

Finally, there is a distinct possibility that substantial progress may result from a combination of relatively modest effects due to enhanced planetary wave activity and a proper parameterization of stratospheric gravity wave effects. In both cases, we are going to need an improved understanding of the nature of mechanical dissipation of disturbances of all scales, and also of the scale interactive character of such dissipation. Obviously, very careful research on a number of modeling, observational and theoretical fronts will be required to provide a definitive answer to these problems.

ACKNOWLEDGMENTS

We are especially grateful to S. B. Fels for his efforts to create the best possible radiative transfer model for the "SKYHI" GCM, and also for his active interest and involvement in the model's polar "cold bias" problem. Thanks are due to M. D. Schwarzkopf, R. W. Sinclair and J. L. Holloway, Jr. for their contributions to programming this model. We are indebted to D. G. Andrews, Y. Hayashi, C.-P. F. Hsu, S. Manabe, M. E. McIntyre, R. A. Plumb, and M. Salby for their support, input and advice on various aspects of the work reported here. Finally, we appreciate the manuscript assistance of J. Kennedy.

REFERENCES

Andrews, D. G., J. D. Mahlman, and R. W. Sinclair, 1983: Eliassen-Palm diagnostics of wave, mean-flow interaction in the GFDL "SKYHI" general circulation model. J. Atmos. Sci. 40, (in press).

_____, and M. E. McIntyre, 1976: Planetary waves in horizontal and vertical shear: The generalized Eliassen-Palm relation and the mean zonal acceleration. J. Atmos. Sci., 33, 2031-2048.

Dickinson, R. E., 1969: Theory of planetary wave, zonal-flow interaction. J. Atmos. Sci., 26, 73-81.

Dunkerton, T., 1979: On the role of the Kelvin wave in the westerly phase of the semi-annual zonal wind oscillation. J. Atmos. Sci., 36, 32-41.

_____, 1982: Theory of the mesopause semiannual oscillation. J. Atmos. Sci., 39, 2681-2690.

_____, C.-P. F. Hsu, and M. E. McIntyre, 1981: Some Eulerian and Lagrangian diagnostics for a model stratospheric warming. J. Sci., 38, 819-843.

Edmon, H. J., Jr., B. J. Hoskins, and M. E. McIntyre, 1980: Eliassen-Palm cross sections for the troposphere. J. Atmos. Sci., 37, 2600-2616. (See also Corrigendum, J. Atmos. Sci., 38, 1115.)

Fels, S. B., J. D. Mahlman, M. D. Schwarzkopf, and R. W. Sinclair, 1980: Stratospheric sensitivity to perturbations in ozone and carbon dioxide: Radiative and dynamical response. J. Atmos. Sci., 37, 2265-2297.

_____, and M. D. Schwarzkopf, 1981: An efficient, accurate algorithm for calculating CO_2 15μm cooling rates. J. Geophys. Res., 86C, 1205-1232.

_____, and M. D. Schwarzkopf, 1984: (Manuscript in preparation).

Geisler, J. E., 1974: A numerical model of the sudden stratospheric warming mechanism. J. Geophys. Res., 79, 4989-4999.

Geller, M. A., M.-F. Wu, M. E. Gelman, 1983: Troposphere-stratosphere (surface-55 km) monthly winter general circulation statistics for the Northern Hemisphere - Four year averages. J. Atmos. Sci., 40, 1334-1352.

Hamilton, K., 1982a: Some features of the climatology of the Northern Hemisphere stratosphere revealed by NMC upper atmosphere analysis. J. Atmos. Sci., 39, 2737-2749.

_____, 1982b: Rocketsonde observations of the mesospheric semiannual oscillation at Kwajelein. Atmos.-Ocean, 20, 281-286.

Hayashi, Y., D. G. Golder and J. D. Mahlman, 1984: Stratospheric and mesospheric Kelvin waves simulated by the GFDL "SKYHI" general circulation model. J. Atmos. Sci. (submitted).

Hirota, I., 1976: Seasonal variation of the planetary waves in the stratosphere observed by the Nimbus 5 SCR. Quart. J. Roy. Meteor. Soc., 102, 757-770.

_____, 1979: Kelvin waves in the equatorial middle atmosphere observed by the Nimbus SCR. J. Atmos. Sci., 36, 217-222.

_____, 1980: Observational evidence of the semiannual oscillation in the tropical middle atmosphere - A review. Pure Appl. Geophys., 118, 217-238.

Holton, J. R., 1976: A semi-spectral numerical model for wave, mean flow interactions in the stratosphere: application to sudden stratospheric warmings. J. Atmos. Sci., 33, 1639-1649.

_____, 1982: The role of gravity wave induced drag and diffusion in the momentum budget of the mesosphere. J. Atmos. Sci., 39, 791-799.

_____, and R. S. Lindzen, 1972: An updated theory for the quasibiennial oscillation of the tropical stratosphere. J. Atmos. Sci., 29, 1076-1080.

_____, and W. M. Wehrbein, 1980: A numerical model of the zonal mean circulation of the middle atmosphere. Pure. Appl. Geophys., 118, 284-306.

Hopkins, R. H., 1975: Evidence of polar-tropical coupling in upper stratospheric zonal wind anomalies. J. Atmos. Sci., 32, 712-719.

Hsu, C.-P. F., 1980: Air parcel motions during a numerically simulated sudden stratospheric warming. J. Atmos. Sci., 37, 2768-2792.

Levy, H., II, J. D. Mahlman, and W. J. Moxim, 1982: Tropospheric N_2O variability. J. Geophys. Res., 87, 3061-3080.

Lindzen, R. S., 1981: Turbulence and stress owing to gravity wave and tidal breakdown. J. Geophys. Res., 86, 9707-9714.

Mahlman, J. D., 1969: Heat balance and mean meridional circulations in the polar stratosphere during the sudden warming of January 1958. Mon. Wea. Rev., 97, 534-540.

_____, and R. W. Sinclair, 1980: Recent results from the GFDL troposphere-stratosphere-mesosphere general circulation model. Collection of Extended Abstracts Presented at ICMUA Sessions and IUGG Symposium 18, XVII IUGG General Assembly, Canberra, Australia, December 1979, 11-18. (Available from S. Ruttenberg, Secretary General, IAMAP, NCAR, Boulder, Colo. 80307, price $3.00.)

Manabe, S., and B. G. Hunt, 1968: Experiments with a stratospheric general circulation model. I. Radiative and dynamic aspects. Mon. Wea. Rev., 96, 477-539.

_____, and J. D. Mahlman, 1976: Simulation of seasonal and interhemispheric variations in the stratospheric circulation. J. Atmos. Sci., 33, 2185-2217

_____, J. Smagorinsky, and R. F. Strickler, 1965: Simulated climatology of a general circulation model with a hydrologic cycle. Mon. Wea. Rev., 93, 769-798.

Matsuno, T., 1971: A dynamical model of the stratospheric sudden warming. J. Atmos. Sci., 28, 1479-1494.

_____, 1982: A quasi-one-dimensional model of the middle atmosphere circulation interacting with internal gravity waves. J. Meteor. Soc. Japan, 60, 215-226.

Murgatroyd, R. A., 1969: The structure and dynamics of the stratosphere. The Global Circulation of the Atmosphere, G. A. Corby, Ed., Royal Meteorological Society, London, 159-195.

Newell, R. E., G. F. Herman, J. W. Fullmer, W. R. Tahnk and M. Tanaka, 1974: Diagnostic studies of the general circulation of the stratosphere. Proc. Intern. Conf. Structure, Composition and General Circulation of the Upper and Lower Atmosphere and Possible Anthropogenic Perturbations. Vol. I., Melbourne, Australia, 17-82. [Available from Office of the Secretary, IAMAP.]

Palmer, T. N., 1981: Diagnostic study of a wavenumber-2 stratospheric sudden warming in the transformed Eulerian-mean formalism. J. Atmos. Sci., 38, 844-855.

Peng, L., 1965: A simple numerical experiment concerning the general circulation in the lower stratosphere. Pure. Appl. Geophys., 61, 191-218.

Plumb, R. A., 1977: The interaction of two internal waves with the mean flow: Implications for the theory of the quasi-biennial oscillations. J. Atmos. Sci., 34, 1847-1858.

Schlesinger, M.E., and Y. Mintz, 1979: Numerical simulation of ozone production, transport and distribution with a global atmospheric general circulation model. J. Atmos. Sci., 36, 1325-1361.

Schoeberl, M., and D. F. Strobel, 1980: Numerical simulation of sudden stratospheric warmings. J. Atmos. Sci., 37, 214-236.

Smagorinsky, J., S. Manabe and J. L. Holloway, Jr., 1965: Numerical results from a nine-level general circulation model of the atmosphere. Mon. Wea. Rev., 93, 727-768.

Takahashi, M., 1983: A numerical model of the semi-annual zonal wind oscillation. In this volume.

Weinstock, J., 1982: Nonlinear theory of gravity waves: Momentum deposition, generalized Rayleigh friction, and diffusion. J. Atmos. Sci. 39, 1698-1710.

J. R. Holton and T. Matsuno, Dynamics of the Middle Atmosphere, 527-537.
Copyright © 1984 by Terra Scientific Publishing Company.

ON THE JANUARY SIMULATION OF STRATOSPHERIC CIRCULATIONS WITH
THE MRI GENERAL CIRCULATION MODEL: PRELIMINARY RESULTS

Tatsushi Tokioka and Isamu Yagai

Meteorological Research Institute

1. INTRODUCTION

A twelve-layer global general circulation model which has seven
layers in the stratosphere (100mb-1mb) has been developed at the Meteorological Research Institute. The purpose of this study is to investigate the performance of the model in simulating global stratospheric
circulations. A long-term integration was performed for the period from
October to January. After the end of January, the position of the sun
was artificially moved back to that of Jan. 1, and the integration was
continued for one more month under January conditions. Then analyses
have been made for the two January months. Some preliminary results,
including the behaviour of tidal modes, are presented below.

2. MODEL

The MRI general circulation model is basically identical to the
UCLA model described by Arakawa and Mintz (1974) and Schlesinger and
Mints (1979) with minor changes in both dynamical and physical processes.
In the model, the following variables are predicted: two components of
the horizontal wind, temperature, surface pressure, mixing ratios of
water vapor and ozone, ground temperature, ground wetness, snow depth,
depth of the atmospheric boundary layer, and the gaps in physical quantities at the top of the boundary layer. The surface albedo is determined
by the model as a simple function of surface conditions. The cloud amount
is also determined diagnostically by the model and is used in the calculation of radiation. The model includes the earth's topography and land-
sea distribution. The geographical distributions of sea surface temperature and sea ice are prescribed, based on the climatological data as
a function of time. The diurnal and seasonal variations of the solar
insolation are considered. Thus, the model atmosphere undergoes a diurnal cycle as well as an annual cycle.

Heating rates due to the solar and terrestrial radiations are computed based on the formulation by Katayama (1972) as modified by Schlesinger (see Arakawa and Mintz, 1974). The absorption of the solar radiation by O_3 and H_2O and absorption and emission of terrestrial radiation

528

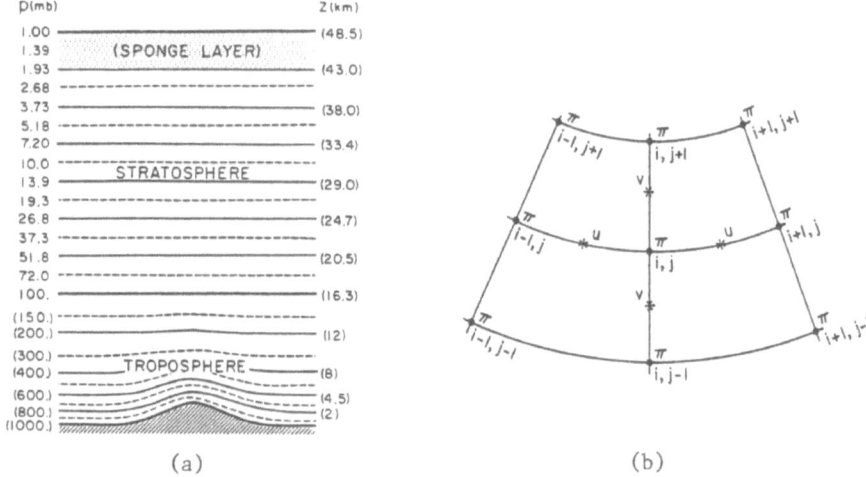

(a)   (b)

Fig. 1.   (a) Vertical structure of the model.   Prognostic variables are
defined on the σ-surfaces shown by the dashed lines.
(b) Distribution of variables on the spherical surface.   "i"
and "j", and "u" and "v" indicate indices and velocity components in the longitudinal and latitudinal directions, respectively.   Thermodynamic variables are defined at π-points.

by H_2O, O_3 and CO_2 are included in the radiation calculation at levels
below 26.8mb.   Above that level, the cooling due to terrestrial radiation is parameterized following Dickinson's (1973) approximation.

In the vertical direction, a modified σ-coordinate is adopted; both
the lower boundary and a prescribed pressure level near the tropopause
(100mb in the present simulation) are coordinate surfaces.   The model
atmosphere between 100mb level and the top (1mb) is divided into seven
layers with an equal interval in ln p scale (see Fig.1a) for better simulation of internal waves (Tokioka, 1978).

In the uppermost layer of the model, a special sponge term is included in the thermodynamic equation for sppressing artificial reflection of waves due to the upper boundary condition ω = 0, where ω is the
vertical p-velocity.   The sponge term is based on the formulation by
Tokioka and Arakawa (see Arakawa and Mintz, 1974) and is designed in
such a way as to change both the amplitudes and the phases of waves.

The model is a grid model on the spherical coordinate with resolution of 5° and 4° in longitudinal and latitudinal directions, respectively.   Variables are distributed in the horizontal plane as shown in Fig.
1b.   Time interval of integration is 7.5 min and all the diabatic forcing terms are updated at every one hour.

3. ZONAL MEAN STATE

Figs. 2 and 3 show the zonally averaged monthly mean zonal wind
velocity and temperature, respectively. There exist easterlies in the
equatorial area throughout the whole depth of the model atmosphere. In
the stratosphere, the easterly extends to the southern (summer) hemisphere. However, the tropospheric westerly in the southern hemisphere
penetrates far into the stratosphere at latitudes higher than 50°S. This
corresponds to relatively cold south pole in the lower stratosphere.
Although the tropical tropopause is well simulated in its height and temperature, the temperature in the north polar region in the stratosphere
is much colder than that observed, as most of the simulations so far
reported (e.g. Manabe and Mahlman, 1976; Schlesinger and Mintz, 1979;
Hung, 1981; see Geller in this volume). The separation between the tropospheric subtropical jet and the polar night jet is not clear, and the
latter has the maximum speed exceeding 140m/s at the highest level around
55°N in accord with the colder north pole.

Fig.4 shows the zonally averaged monthly mean total diabatic heating rate. The heating rate in the troposphere is reasonable with deep
heating in the low latitudes due to cumulus activity, and with shallow
heating in the subtropical and mid-latitudes due to the turbulent energy
flux from the surface and the condensation heating associated with cyclone activity. There is a pronounced contrast of cooling and heating
in the stratosphere between the winter and summer hemispheres. The

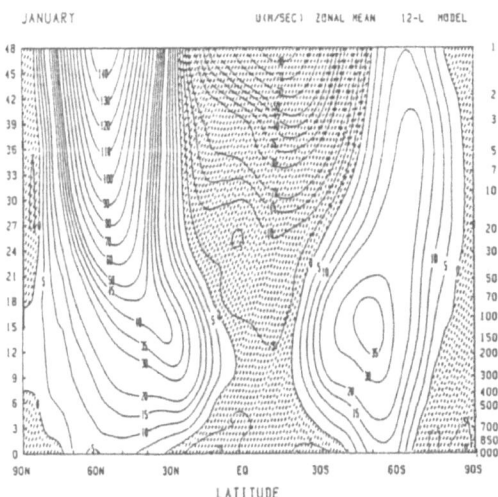

Fig. 2.   The zonally averaged monthly mean zonal wind velocity.   Regions
of easterly are hatched.   (unit: m/s)

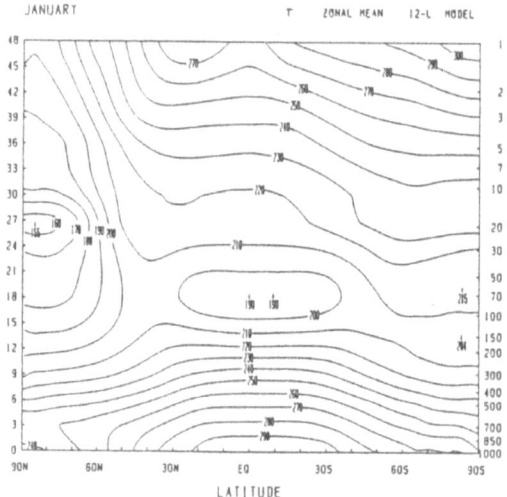

Fig. 3. The zonally averaged monthly mean temperature. (unit: K)

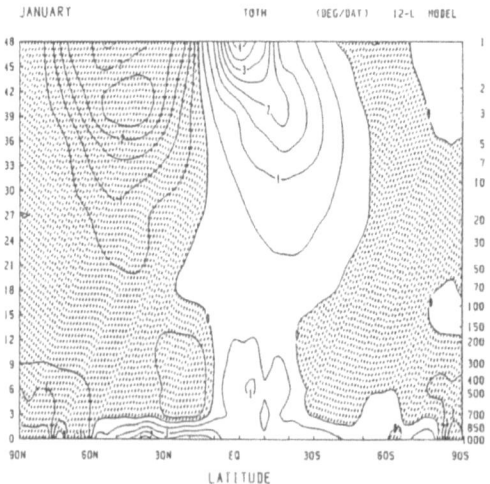

Fig. 4. The zonally averaged monthly mean total diabatic heating. Regions of cooling are hatched. (unit: K/d)

heating region is, however, confined to a region between 10°N and 40°S.
There exists a weak cooling region poleward of 40°S. This is caused by
the strong cooling calculated by Dickinson's parameterization. This
cooling seems to be unrealistic as there exists no major contribution of
the heat flux convergence there. In the present model, the cooling is
balanced with the adiabatic heating due to subsidence. Some unrealistic
features mentioned previously, i.e., the relatively colder stratosphere
at high latitudes in the southern hemisphere and the penetration of westerly winds through the high latitudes of the southern stratosphere, seem
to be connected with the excessive radiative cooling given by Dickinson's
parameterization.

4. PLANETARY WAVES

Fig.5 shows the amplitude and the phase of geopotential height of
the stationary wave with wavenumber one and corresponding observations.
The amplitude in the winter stratosphere is much larger than that in the

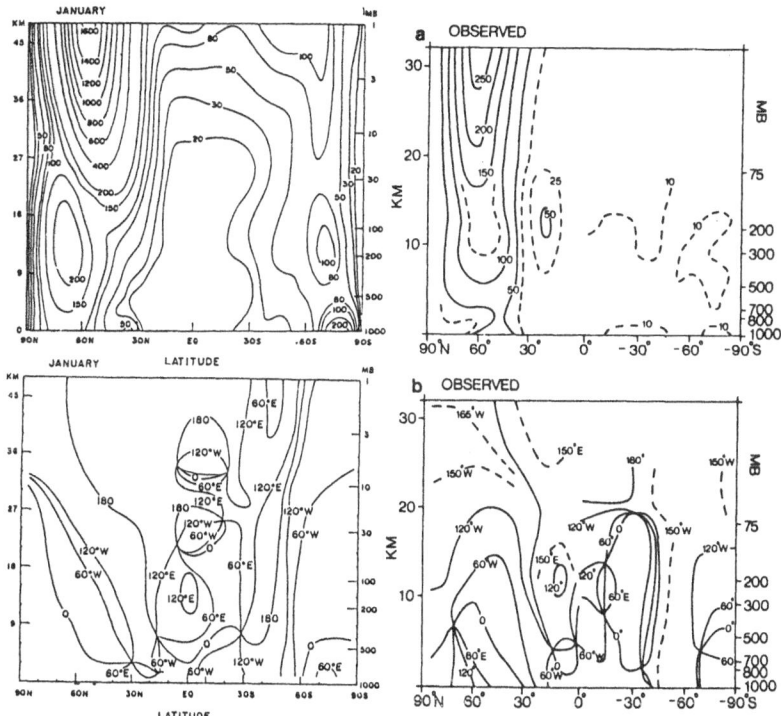

Fig. 5.   The amplitude and the phase of geopotential height for the
          stationary wave of wavenumber 1, together with those observed
          (van Loon and Jenne (1972) and van Loon et al. (1973) as com-
          piled by McAvaney et al. (1978)) (unit of the amplitude: g.p.m.)

532

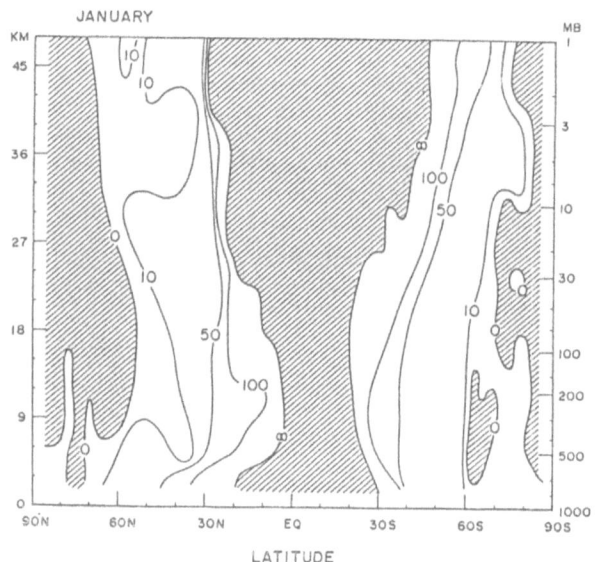

Fig. 6. The index of refraction squared introduced by Matsuno (1970)
for the stationary wave of wavenumber 1.

summer hemisphere, and the amplitude peak coincides fairly well with the
axis of the polar night jet. The index of refraction squared introduced
by Matsuno (1970) is shown in Fig.6 for the stationary wave of wavenumber one. The zero line is located in the slightly poleward position of
the polar night jet axis. Most of the region poleward of 60°N is of the
negative sign. There is another peak in amplitude inside the negative
index region with its peak located around 60°N and 200mb. This peak is
considered to be caused by the trapping of the wave energy within the
model atmosphere where the westerly is still intense in the polar region.
This is different from the climatological structure of the wave shown in
Fig.5. On the other hand the vertical structure of the wave between 30°N
and 60°N where propagations are allowed is compared well with the climatological one.

Fig.7 shows the time changes of the amplitude of wavenumber one at
50°N at various levels. As the phase of the wavenumber 1 changes only
slightly, it is not shown here. There is a tendency of gradual increase
of the amplitude within the stratosphere. The cause of this increase
has not been identified yet. This might be related somehow to the repe-
tition of January condition in the present simulation, suggesting that
a longer time is necessary for establishing a quasi-equilibrium state
for the perpetual January conditions.

Fluctuations with time scales of 10 to 20 days are superposed upon
the gradual increase trend. The amplitude of the fluctuations increases

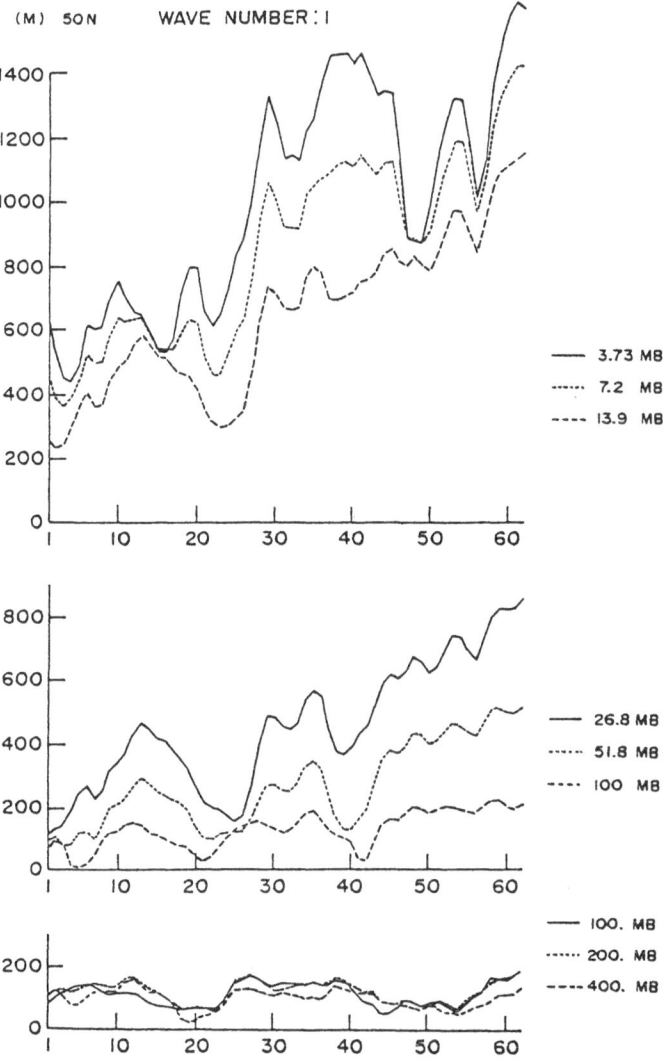

Fig. 7. Time variation of the amplitude of wavenumber 1 at 50°N at various vertical levels. Units of the ordinate and the abscissa are g.p.m. and day.

with height. It is noted that most maxima and minima found in the stratosphere occur almost simultaneously at different levels, like barotropic disturbances. In cases of sudden warmings (including minor ones) the peak in geopotential amplitude can be traced from the lower to the upper stratosphere with a lag from several days to a few weeks. Such phenomena have not occured in the present simulation.

5. TIDAL WAVES

As the present model allows the diurnal variation of the solar insolation, many tidal modes are excited in the model. A brief sketch of
them will be given in this section.

Well known tidal waves are ($m=1$, $\sigma=-1$ d^{-1}) and ($m=2$, $\sigma=-2$ d^{-1})
where m is the zonal wavenumber and σ is the frequency defined as positive for eastward propagating waves. Such modes can be identified in
the model results. Besides them, we can identify the modes, ($m=5$, $\sigma=$
-1 d^{-1}), ($m=3$, $\sigma=1$ d^{-1}), ($m=5$, $\sigma=1$ d^{-1}) etc.. Fig. 8a shows the power
spectral analysis of the zonal wind at 30°S and 19.3mb ($\approx$27km). In this
diagram, we can recognize those spectral peaks mentioned above, clearly
separated from the power corresponding to slowly varying large-scale
phenomena. The power associated with the tidal waves increases with
altitude, while that of the slowly varying large-scale phenomena damps
in the summer hemisphere, consistent with the propagation theory of
planetary waves.

Fig. 8b shows the power spectrum of the total diabatic heating rate
at the 700mb level at 30°S. It is evident that the strong spectral
peaks at the diurnal frequency in Fig. 8a have their counterparts in
Fig. 8b. It is interesting that large heating rates are found not only
for the westward moving wave with zonal wavenumber one but for many
higher wavenumbers and also for eastward moving waves. In order to
search for the cause of this spectral distribution, a hypothetical heating distribution is considered and its spectrum is shown in Fig. 8c.
This was calculated by assuming that heating occurs only over sunlit
parts of the continents existing at 20° $\sim$ 30°S, and the magnitude of
heating is proportional to cosine of the solar zenith angle. There is
a close resemblance between Fig. 8b and Fig. 8c so far as the diurnal
cycle is concerned. Therefore we confirm that the tidal peaks in Fig.
8a really have corresponding peaks in the diabatic heating in the lower
troposphere, mainly due to the geographical distribution of land and
sea.

Fig. 9a and b show the horizontal structures of ($m=1$, $\sigma=-1$ d^{-1})
and ($m=5$, $\sigma=-1$ d^{-1}) at t=06Z and at the level p=5.18mb respectively.
Arrows show wind vectors and contours show geopotential field. Areas
of negative geopotential deviation are shaded. Fig. 9a is compared
fairly well with the m=1 diurnal ground mode with a negative equivalent depth found by Lindzen (1966) and Kato (1966). The amplitude is
pronounced in the summer hemisphere. Both the amplitude and the phase,
also, roughly agree with those observed and theoretically calculated
(see Reed et al., 1969 for example).

The ($m=5$, $\sigma=-1$ d^{-1}) mode has a typical character of gravity waves
with its convergence zone located ahead of the ridge. This mode,

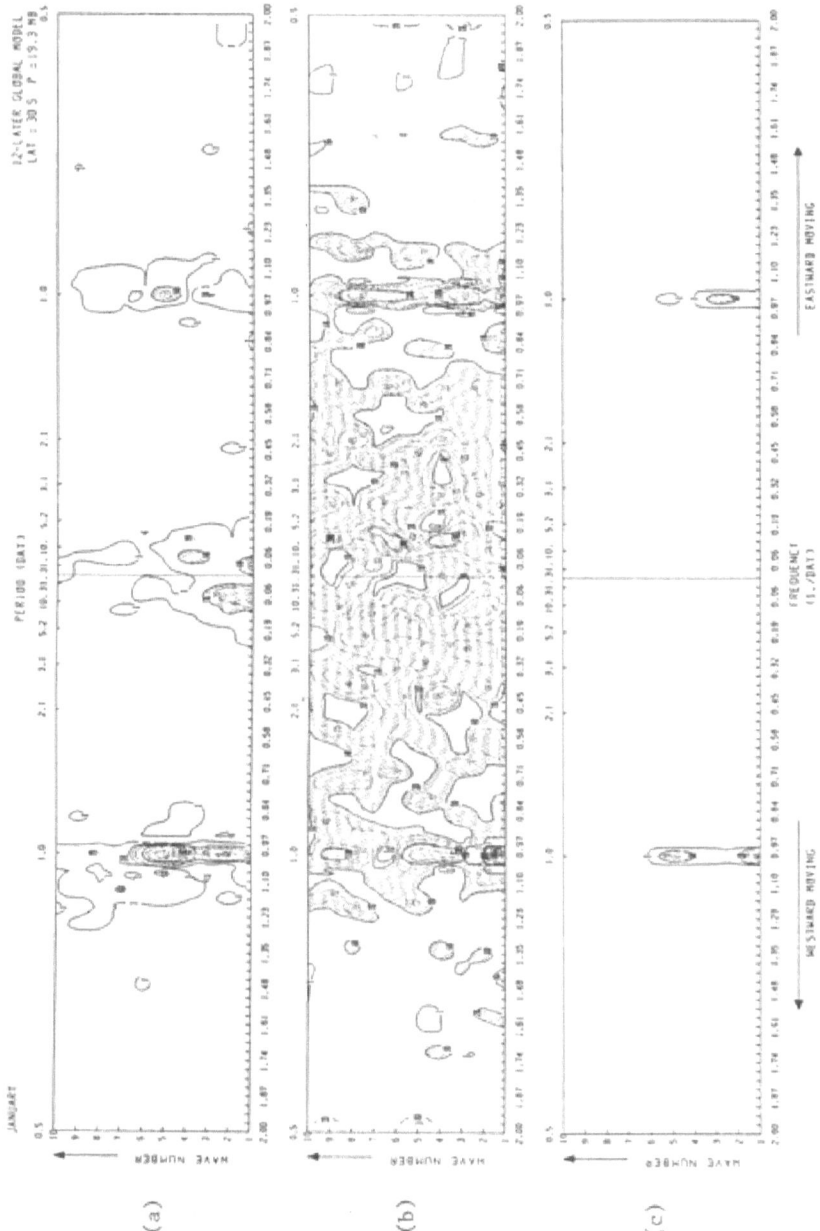

Fig. 8. Power spectral analysis of (a) the zonal wind at 30°S and 19.3mb, (b) the total diabatic heating rate at 30°S and 700mb, and (c) the hypothetical heating where heating proportional to the cosine of the solar zenith angle is assumed over the continent along 20°-30°S latitudes only when it is positive.

536

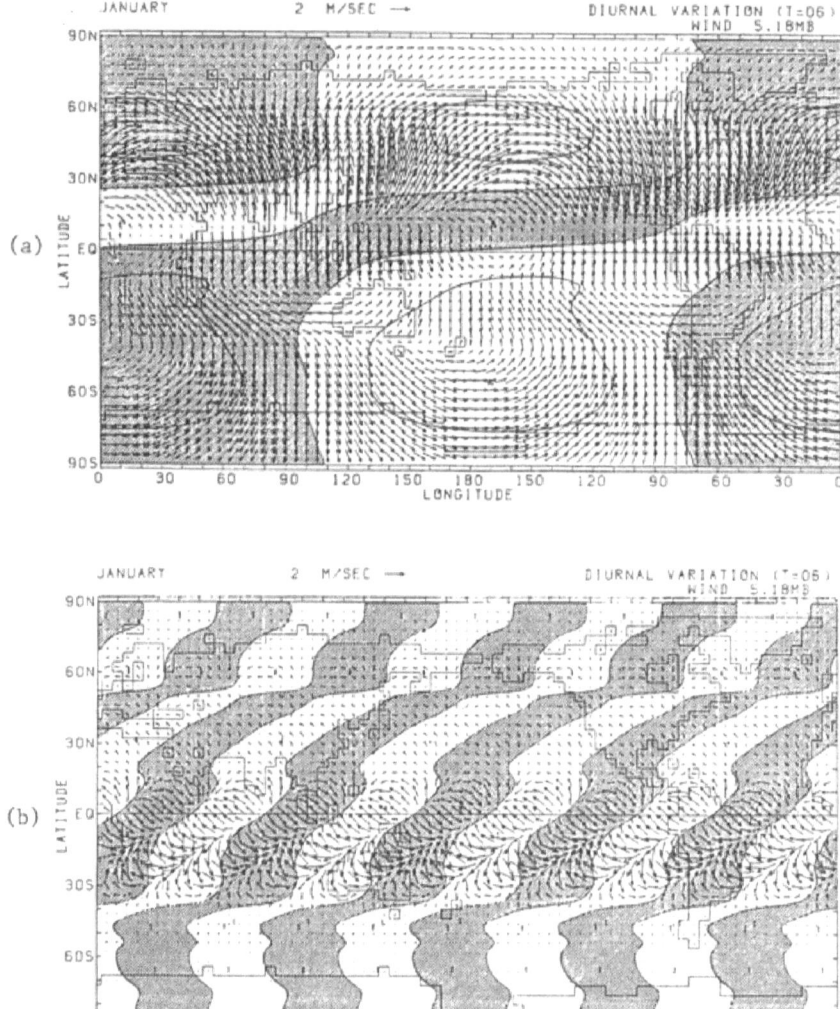

Fig. 9. Horizontal structure of (a) (m=1, σ =-1 d^{-1}) mode at p=5.18mb
and (b) (m=5, σ =-1 d^{-1}) mode at p=5.18mb at t=06z, where m is
the zonal wavenumber and σ is the frequency defined as positive
for eastward propagating mode.

originated from the tropical and subtropical latitudes in the southern hemisphere, tends to have its peak in the northern hemisphere near the top of the model, and has non-negligible southward transport of zonal momentum there.

REFERENCES

Arakawa, A. and Y. Mintz, 1974: Workshop notes on the UCLA atmospheric general circulation model (24 March - 4 April 1974). Dept. of Meteorology, Univ. of California, Los Angeles, 404pp.

Dickinson, R.E., 1973: Method of parameterization for infrared cooling between altitudes of 30 and 70 kilometers. J. Geophys. Res., 78, 4451-4457.

Hunt, B.G., 1981: The maintenance of the zonal mean state of the upper atmosphere as represented in a three-dimensional general circulation model extending to 100km. J. Atmos. Sci., 38, 2172-2186.

Kato, S., 1966: Diurnal atmospheric oscillation, I. eigenvalues and Hough functions. J. Geophys. Res., 71, 32 1-3209.

Katayama, A., 1972: A simplified scheme for computing radiative transfer in the troposphere. Tech. Rep. No. 6, Dept. of Meteorology, Univ. of California, Los Angeles, 77pp.

Lindzen, R.S., 1966: On the theory of the diurnal tide. Mon. Wea. Rev., 94, 295-301.

Manabe, S. and J.D. Mahlman, 1976: Simulation of a seasonal and inter-hemispheric variations in the stratospheric circulation. J. Atmos. Sci., 33, 2185-2217.

Matsuno, T., 1970: Vertical propagation of stationary planetary waves in the winter northern hemisphere. J. Atmos. Sci., 27, 871-883.

Reed, R.J., M.J. Oard and M. Sieminski, 1969: A comparison of observed and theoretical diurnal tidal motions between 30 and 60 km. Mon. Wea. Rev., 97, 456-459.

Schlesinger, M.E. and Y. Mintz, 1979: Numerical simulation of ozone production, transport and distribution with a global atmospheric general circulation model. J. Atmos. Sci., 36, 1325-1361.

Tokioka, T., 1978: Some considerations on vertical differencing. J. Meteor. Soc. Japan, 56, 98-111.

Author Index

Subject Index